Battery Technology

Battery Technology

Fundamentals of Battery Electrochemistry, Systems and Applications

Alexander Börger and Heinz Wenzl

WILEY-VCH

Authors

Dr. Alexander Börger
An der Schule 32
38165 Lehre
Germany

Prof. Dr. Heinz Wenzl
Technical University of Clausthal
Institute of Electrical Power Engineering
and Energy Systems
Leibnizstr. 28
38678 Clausthal-Zellerfeld
Germany

Cover Images: © Kashtanowww/ Shutterstock (background), Courtesy of the Authors (circular element)

Typesetting: Straive, Chennai, India

Printing and Binding: CPI Group (UK) Ltd, Croydon, CR0 4YY

Library of Congress Card No.: applied for

British Library Cataloguing-in-Publication Data A catalogue record for this book is available from the British Library.

Bibliographic information published by the Deutsche Nationalbibliothek
The Deutsche Nationalbibliothek lists this publication in the Deutsche Nationalbibliografie; detailed bibliographic data are available on the Internet at <http://dnb.d-nb.de>.

Print ISBN: 978-3-527-35412-2
ePDF ISBN: 978-3-527-84805-8
ePub ISBN: 978-3-527-84804-1
oBook ISBN: 978-3-527-84806-5

C9783527354122_100925

Contents

Preamble

Global transformation processes have led to a significant increase in the number and energy content of batteries as the most adaptable and scalable energy storage system. The driving factors include electric road vehicles, the growing number of conventional road vehicles, including trucks, and their need for increased energy efficiency; the growth of digital infrastructure with the requirement for data centers and their large uninterruptible power supply systems; the flood of portable electronic devices; and the impact of photovoltaics and wind turbines on electricity systems. These drivers will lead to considerable technical development and increased economic importance for batteries in the future. Industry expectations are that the dominant battery technologies, lead–acid and lithium-ion, will both benefit from this growth. Lead–acid batteries may not be the focus of scientific and public interest but will continue to be an important technology for many years to come.

The use of the term "battery" follows the established scientific convention. Differentiations are made, whenever appropriate, between individual cells, modules consisting of a number of cells connected in series and/or parallel, and battery systems that include all safety-relevant auxiliary components and a monitoring system.

The electrochemical principles of all batteries are identical, regardless of whether they are rechargeable (accumulators, secondary galvanic systems) or non-rechargeable (primary galvanic systems). For this reason, the basics of battery technology for lead–acid and lithium-ion batteries are discussed together. Examples of other battery technologies are used where they provide further insight. Fuel cells, on the other hand, are energy conversion devices and not energy storage systems and are therefore not included.

Battery technology requires the combination of many scientific and technical disciplines. Our readers will therefore have a wide range of backgrounds: electrochemistry, chemistry, physics, electrical engineering, material sciences, and software engineering, to name just a few. In addition, some may still be students, while others are experienced scientists and engineers who already have some knowledge of batteries.

The aim of this book is to explain how batteries work and how they should be used for safe and reliable operation. We bridge the gap between a black-box approach to batteries for non-electrochemists and enable electrochemists to understand the many technical and practical problems that arise from connecting many cells in series and parallel as part of complex energy systems.

- Why do lithium-ion batteries require extensive monitoring, for example, single-cell voltage measurements and complex, real-time data evaluation, but lead–acid batteries do not?
- Why is the internal resistance of a battery a function of the current amplitude, current direction, and state of charge?
- Why is the heat conductivity of cells anisotropic and heat generation inhomogeneously distributed, depending on current amplitude and direction?
- Why do cells cool down during discharging or charging?
- Why is the calculation of state of charge so difficult?

This book deals with batteries but does not attempt to add to the already very good selection of books on electrochemistry. We hope that our descriptions will also provide some new insights into battery technology for electrochemists, for example, our use of the Butler–Volmer equation to explain the voltage curve of a battery during discharging and charging and our use of equivalent circuit diagrams that show that an ideal voltage source and a resistor, under certain conditions, are a correct representation of a battery and not a simplification. We are also sure that our discussion of a safe zone that will prevent the occurrence of thermal runaway will be of interest.

For many applications, lead–acid and lithium-ion batteries offer the best combination of properties and thus dominate the market. In the second part of the book, they are therefore described in detail. Other battery systems that are still used because of very special properties and/or established market structures are described briefly. Upcoming battery systems that may have the potential to compete with either lead–acid or lithium-ion batteries in the future are dealt with in a separate chapter. The path from a laboratory cell with highly attractive properties to a complete battery system which meets all requirements is long and uncertain.

The main aspect that we would like to stress in the third part of the book is the wide range of operating conditions which batteries are subjected to. In many applications, the battery current is highly dynamic with fast changes from discharging to charging. Phases with constant or only slowly varying currents are likely to be the exception. Particularly when troubleshooting, knowledge of the past operating conditions is crucial and no effort should be spared to understand how a particular battery has been used. We refrain from comparisons between different energy storage systems – flywheels do not compete with pumped hydro storages and capacitors do not compete with batteries. When selecting a battery system for an application, a decision can only be made when all technical and economic constraints are known. In some applications, lead–acid batteries and lithium-ion batteries are alternatives or may be used in parallel. It is therefore necessary to have an understanding of both battery technologies. In addition, we strongly believe that knowledge of both technologies leads to a much deeper understanding of the advantages and disadvantages of each technology.

The appendix contains terms used in battery technology, an overview of safety issues and precautions, and an overview of norms and standards.

Each chapter contains a list of relevant references and tasks for further study which have been used by us as part of our teaching experience at Clausthal University of Technology and the Technical University of Brunswick, both in Germany.

To write this book, we had invaluable discussions with colleagues who kindly provided us with diagrams and measurements to illustrate important aspects of battery technology. Special thanks go to research assistants and the staff of the Institute for Electrical Power Engineering and Energy Systems of Clausthal University of Technology who helped us in different ways to complete this book.

Writing this book has been very time-consuming and demanded patience from our families. We are very grateful for their support.

Abbreviations

BMS	Battery management system
BV	Butler-Volmer
CT	Charge transfer

About the Companion Website

Battery Technology: Fundamentals of Battery Electrochemistry, Systems and Applications, is accompanied by a companion website.

www.wiley.com/go/Börger/Battery_Technology

The website includes:

- Notes on tasks answers and solutions
- Tasks
- Chapter tasks

1

Introduction

Batteries and other energy storage systems are options for the technical and economic optimization of an energy supply system and, in many cases, indispensable for ensuring the required functions. Very often, however, batteries are in competition with other technologies, which impact the development and market opportunities of batteries.

A comparison of batteries with other energy storage technologies is of little value without precise knowledge of the application and limitations of competing technologies.

All batteries are based on the same physical and chemical principles. Different electrochemically active materials and designs lead to major differences in properties, including the necessity of additional components required for safe and long-lasting operation.

Batteries are usually categorized according to their bridging time and application areas: portable, mobile, and stationary.

1.1 Energy Supply in General

Energy storage systems are an option for the technical and economic optimization of an energy supply system because they allow energy generation[1] to be quickly and efficiently adapted to energy consumption. Without energy storage systems that can both store and release energy, generation and consumption units would always have to adapt to each other with very high dynamics. Fast response times are often not possible or only possible at great expense. Energy storage systems also serve as an energy source for technical systems that do not have their own energy supply from primary energy sources, as well as for starting up systems that in most cases cannot be started without the provision of electrical energy from an energy storage system or the electrical grid.

1 The terms "energy production" and "energy consumption" correspond to common usage and are therefore used here. In the literal sense of the terms, of course, energy cannot be generated and consumed, but only converted.

Battery Technology: Fundamentals of Battery Electrochemistry, Systems and Applications,
First Edition. Alexander Börger and Heinz Wenzl.

Companion website: www.wiley.com/go/Börger/Battery_Technology

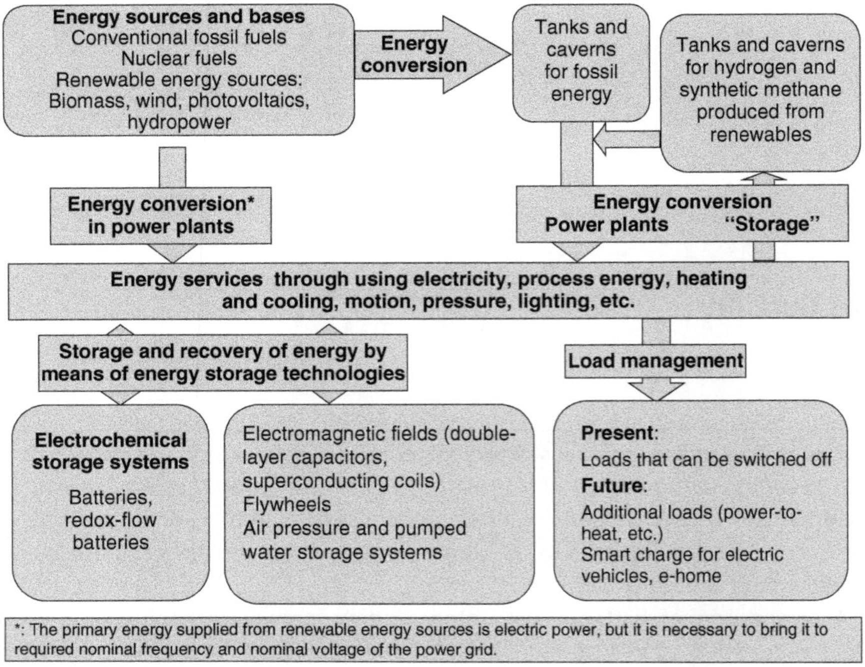

Figure 1.1 Electrochemical energy storage as part of the power supply system.

The use of batteries is indispensable in many cases to ensure the required functions. The following examples of electromobility and the electricity supply system will show that the overall technical and economic context must always be carefully considered when estimating and forecasting the future importance of batteries.

Figure 1.1 shows the integration of energy storage systems into the overall electrical energy supply system and illustrates in particular that energy storage[2] competes with many technical alternatives to ensure the required functions. In addition to highly dynamic generation units, which, unlike conventional thermal power plants, can adapt their power output very quickly to demand, alternatives for quickly balancing power generation and consumption are primarily load management systems and switchable loads, in particular heat generators (power-to-heat).

Electricity from a photovoltaic system that is not consumed immediately at the site of installation can be used locally or, for example, it is to be

- stored in a battery,
- used as thermal energy for space heating or hot water supply via an electrically operated heating cartridge,

2 In Figure 1.1, it should be noted that storage facilities for materials (tanks for processed energy sources) are only referred to as energy storage facilities if they are produced using electricity that is not used directly for energy services. Without this conceptual differentiation, crude oil, heating oil, or gasoline tanks and underground salt caverns for fossil fuels would otherwise also have to be referred to as energy storage facilities.

- used by switching on household loads such as washing machines or refrigerators depending on the supply, or
- made available to the electrical grid for loads elsewhere.

From a system perspective, these alternatives are equivalent and therefore are often referred to as storage-equivalent systems or functional storage systems.

Electrochemical energy storage systems are also in technical and economic competition with other energy storage technologies, see Ref. [1].

1.2 Electrochemical and Non-electrochemical Energy Technologies

In principle, energy can be stored in very different ways, that is, in different forms of energy, namely

a) mechanically, for example, in the form of potential energy in pumped storage power plants or in the form of rotational energy in flywheels,
b) magnetically, for example, in the magnetic field of a superconducting coil,
c) electrically, for example, in the electric field of double-layer capacitors,
d) chemically, for example, by conversion to hydrogen,
e) thermally, for example, in the form of hot water storage tanks or in steam boilers, and
f) electrochemically, that is, by converting electrical energy into chemical energy.

Table 1.1 provides a short summary of these technologies and the basic physical formulas.

Table 1.1 Comparison of different energy storage technologies.

Energy type	Principle	Examples
Potential energy	$E = mg\Delta h$	Pumped storage power plants
Gas pressure	$E = p\Delta V$	Compressed air storage
Rotational energy	$E = 0.5J\omega^2$	Flywheels
Magnetic energy	$E = 0.5LI^2$	Loss-free direct current flowing in a superconducting coil (so-called SMES)
Electrical energy	$E = 0.5CU^2$	Double-layer capacitors (ultracapacitors, electrolytic capacitors, etc.)
Chemical energy	$E = n\Delta_r G$	Hydrogen storage
Thermal energy	$E = C_i\Delta T$	Hot water tank
Electrochemical energy	$E = \int UI\mathrm{d}t$	Batteries

Explanation of symbols: m: mass, g: gravity constant, h: height, p: pressure, V: volume, J: moment of inertia, ω: rotational speed, L: inductance, I: current, C: capacitance (in Farad), U: voltage, n: amount of substance, $\Delta_r G$: free enthalpy of reaction, C_i: heat capacity of substance i, ΔT: temperature difference, t: discharge time.

With some energy storage technologies, particularly thermal storage, the stored energy cannot be made available to the overall system as electrical energy or only at great cost. Despite various limitations, different energy storage systems compete with each other in certain applications. Before discussing electrochemical energy storage systems in detail in the following chapters, here are some comments on non-electrochemical energy storage systems.

1.2.1 Capacitors and Ultracapacitors

The energy content of capacitors is very low, even for the group of so-called ultracapacitors or supercapacitors[3] (ultracaps) with very high capacitances (unit Farad: 1 F = 1 As/V). At a nominal voltage of 2.5 V and a capacity of 3000 F, for example, the energy content is only approx. 2.6 Wh, of which normally only 75% can be technically extracted, compared to approx. 9 Wh for a small 2.5 Ah lithium-ion cell, which is significantly more compact, lighter, and cheaper. In terms of specific power (W/kg), however, ultracapacitors can deliver significantly higher electrical power and are therefore used in special applications.

Ultracapacitors have a high self-discharge rate (they are often completely discharged within 24 hours) and therefore a high energy loss in standby mode. They require a similarly complex charge control as lithium-ion batteries.

Other types of capacitors, such as classic electrolytic capacitors, only have a capacitance in the micro- or millifarad range and less and are therefore not able to store large amounts of energy, even if their rated voltage is very high.

1.2.2 Superconducting Coils

The magnetic field present in a current-carrying coil stores usable amounts of energy at high currents. The losses are only sufficiently low if the resistance of the coil is minimized by superconductivity. However, energy storage systems based on superconductivity require complex cooling and therefore have high standby losses. In the 1990s, superconducting magnetic energy storage (SMES) systems with an output of 1 MW for 10 seconds and an energy content of several kilowatt-hours were built to stabilize the power grid.

1.2.3 Flywheels

The energy content stored in flywheels depends on the square of the rotational speed and is proportional to the moment of inertia. Slowly rotating flywheels (with up to approx. 4000 rpm) are commercial products for uninterruptible power supply (UPS) systems with an output of 1.6 MW for 15 seconds (Powerbridge, Piller GmbH), corresponding to 6.7 kWh energy content, of which only 75% can be technically

3 In this textbook, the term "ultracapacitor" or "supercapacitor" is always used for this class of energy storage. To avoid confusion, the term "double-layer capacitor" always refers to the electrode/electrolyte interface.

extracted. These flywheels are a technical and economical alternative to batteries for bridging times of a few seconds until the starting of diesel power generation units.

Very fast rotating flywheels (up to 100,000 rpm) are light and very powerful, whereby the power is a function of the generator coupled to it. Such flywheels have been used in motorsport.

All flywheels are characterized by very high self-discharge rates, even when using vacuum and magnetic bearings to minimize friction. The stored energy is generally lost after approx. 24 hours. Use in vehicles as a replacement for starter batteries is therefore not possible.

1.2.4 Compressed Air and Pumped Water Storage

Small compressed air energy storage, excluding compressed air storage for optimizing air supply in factories, is sometimes used for starting diesel engines for emergency power generation. A few large-scale compressed air energy storage systems for utility-scale energy storage (100 MWh range) have been built worldwide in the past but have a low energy efficiency and do not seem to be an option in the future.

Pumped water storage systems have energy capacities of around 100–10,000 MWh and are relevant energy storage systems in the electricity industry worldwide for covering peak loads and storing energy during times of low electricity demand. They are characterized by very high initial investment and environmental impact but low operating costs.

Very large battery storage systems with energy contents of several hundred MWh, which are currently being built in some regions, could become technical alternatives in the future.

1.3 Basic Properties of Batteries: Similarities and Differences

Batteries are electrochemical energy converters. The fundamental principles of all battery technologies are based on the same physical and chemical laws. The basic requirements for their design and operation are also very similar.

Batteries are DC voltage sources whose voltage is a function of the materials used and their local concentrations as well as the current flowing through them. In most cases, it is necessary to connect individual cells in series and integrate them into a power supply system using converters (DC/DC or DC/AC converters) and/or use loads with a wide range of input voltages. The maximum charging voltage is often approx. 1.5 times the minimum discharge voltage and even higher for some battery technologies and applications.

There are many different battery technologies, which differ in many other characteristics apart from their voltage level. Figure 1.2 shows an overview of the functional principle and different battery systems.

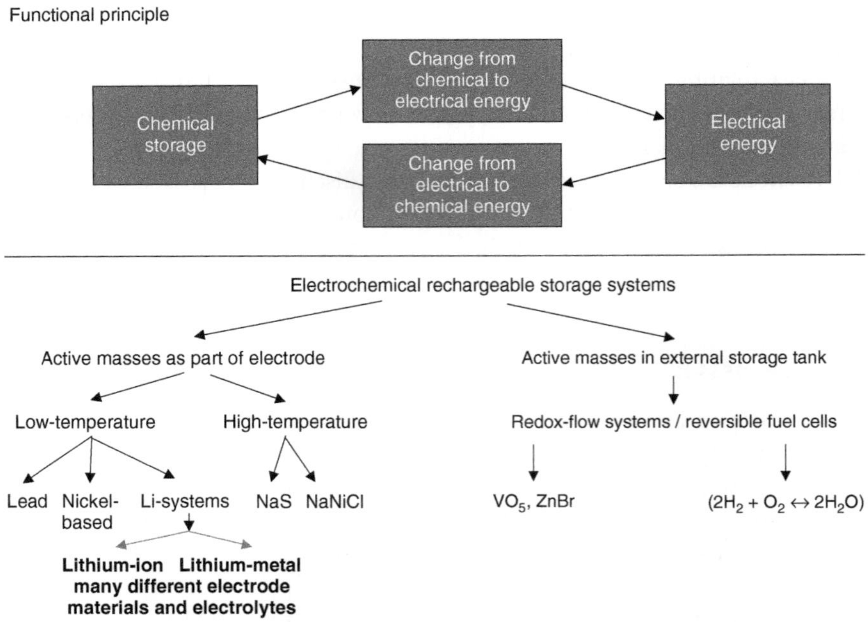

Figure 1.2 Overview of the functional principle of batteries and various materials. For details, see chapters 12–14.

In all cases, chemical energy is converted directly into electrical energy. Only in the case of accumulators, that is, rechargeable galvanic secondary elements, it is possible to convert electrical energy back into chemical energy (and vice versa) easily and frequently. All such systems are referred to below as batteries.

An important differentiation between batteries concerns systems in which the active material is an integral part of the electrodes, and therefore energy content (Wh) and power (W) are always coupled, and battery systems in which the active materials are stored externally in tanks and pumped into the space between the electrodes – so-called flow batteries or redox flow batteries. In flow batteries, the energy content can be increased as required using large tanks without increasing the power provided by the reaction chamber. Flow batteries are related to fuel cells in which one of the active materials (hydrogen) is supplied from a tank and the other active material (oxygen) is taken from the air. Metal–air systems are also related to flow systems and fuel cells, except that one of the active materials (the metal) is in solid form (and can be renewed by mechanical exchange, if necessary), while the oxygen is taken from the air – possibly requiring a purification stage.

Battery technologies differ in terms of internal resistance, energy content, specific energy (Wh/kg), energy density (Wh/l), specific power (W/kg), power density (W/l), permissible temperature range, options for monitoring the battery system, calendar and cyclical service life, and the system components required for safe and economical operation and their operating conditions. Modifications of design and materials also lead to considerable differences in properties within a battery technology.

Most of the materials used in batteries are toxic and/or corrosive. If environmental impacts during production, use, and disposal as well as investment costs and the total costs of operation are taken into account, the range of properties to be considered when choosing a battery system expands considerably. The dominance of lead-acid and lithium-ion batteries is due to their favorable overall properties for many different applications. Current market research shows that at the time of writing this book (2024), the value of lithium-ion and lead-acid batteries constitutes over 80% of all battery sales, with the share of lead-acid batteries slightly below that of lithium-ion batteries since 2023. Growth rates are expected to be slightly higher for lithium-ion batteries than for lead-acid batteries.[4] In terms of energy content, lead-acid batteries dominate lithium-ion batteries by a huge margin.

1.4 Bridging Time

A common classification of batteries is based on their bridging time, the ratio of energy content to the power that can be constantly drawn from a battery until the end-of-discharge voltage is reached.

For all batteries, the amount of energy that can be drawn decreases with increasing power. The diagram of the characteristic curve of the specific energy content (kWh/kg) at different specific outputs (kW/kg) is known as the Ragone diagram. Figure 1.3 is a Ragone diagram based on data sheet information for various battery technologies, see Ref. [2] for details of the calculation.

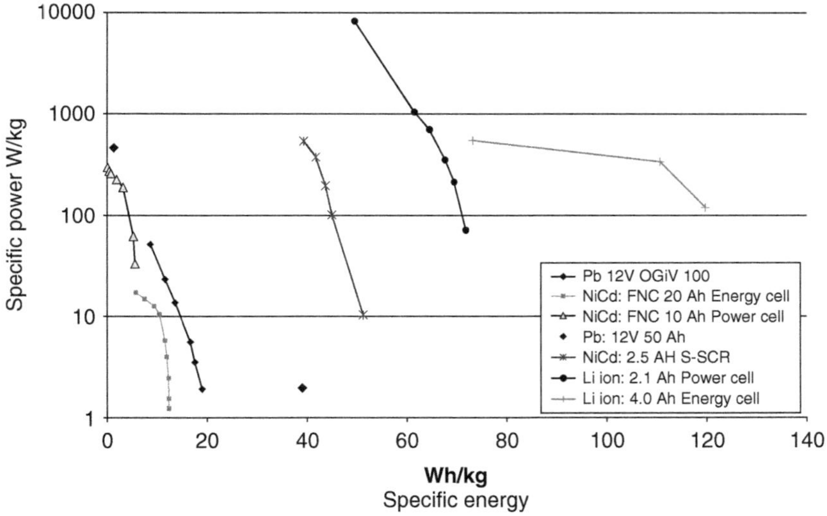

Figure 1.3 Ragone diagram of various battery technologies based on data sheet information (see Ref. [2] for further details).

4 www.grandviewresearch.com.

For more precise sizing of a battery system and to optimize the technical and economic use of batteries, a method was proposed in Ref. [3] to create a Ragone diagram based on measurements of discharge curves at different power levels. The method described in Ref. [3] results in deviations from the diagram shown in Figure 1.3. These can be significant for the design of battery systems.

Figure 1.4 shows a different representation of power and energy content. This makes the frequent division of storage systems into short-term and long-term storage systems particularly clear. The straight lines in the figure represent equal bridging times regardless of the energy content of the energy storage system under consideration. Bridging times of a few seconds are relevant for power quality applications to stabilize the voltage in the electricity grid during short interruptions of power. Bridging times of 15 minutes to a few hours are required in applications for ensuring the power supply in the event of a power failure (emergency lighting systems, UPS systems) for computers and telecommunications equipment, etc. In the field of renewable energies, bridging times of approx. one day to several days are required, depending on the system design, and a bridging time of several months would be required for seasonal balancing between generation and consumption (energy systems based primarily on hydropower, wind power, or photovoltaics). Task 1.1 shows how large electricity storage systems would have to be designed for the Federal Republic of Germany if the electricity generation system were to be based exclusively on wind power and photovoltaic systems.

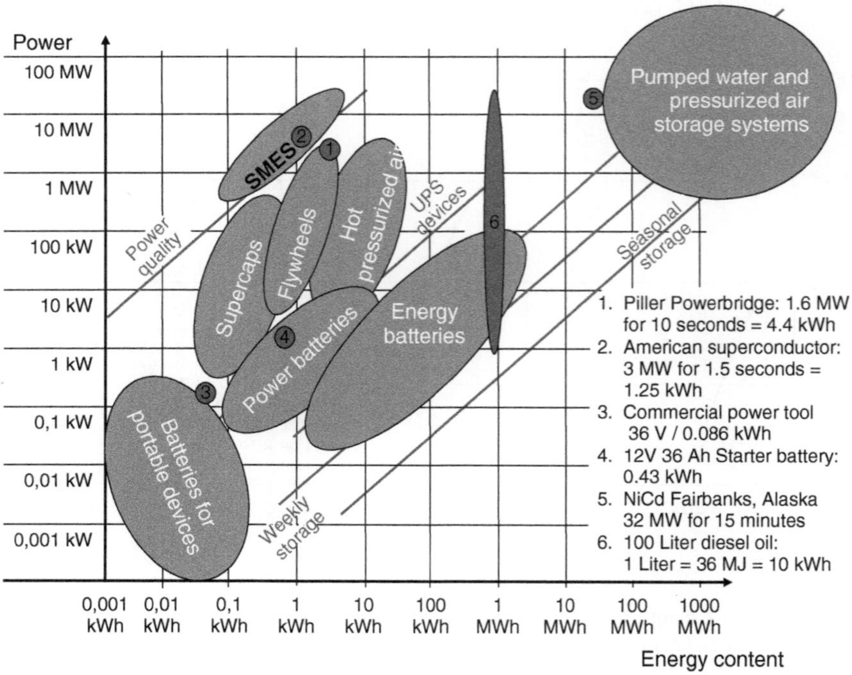

Figure 1.4 Comparison of energy storage systems based on the respective bridging time.

With batteries and many other energy storage technologies, the relationship between energy content and power is determined by technical and economic factors. In fuel-based energy systems, on the other hand, power and energy content are independent of each other. For a better understanding, Figure 1.4 shows the energy content of 100 l of diesel fuel. Since there is no technical link between energy content (proportional to the size of the tanks) and power (dimensioning of the engines), the associated power range extends over several orders of magnitude, and the term "bridging time" is no longer applicable without further information.

The points shown represent various energy storage systems – the marked areas are only intended as a guideline.

Figure 1.4 shows that few energy storage systems are technical and economic alternatives for any specific application (bridging time and power).

1.5 Comparison of Battery Technologies

When comparing different battery technologies, it is necessary to know the specific application so that the data presented in a comparison can also be used to select the most suitable storage technology. The following list provides an overview of the most important points to consider:

1) Definition of the system environment:
   Energy storage systems require various additional components to ensure safe, reliable, and economical operation. When comparing large lead-acid and lithium-ion batteries, the comparison of cells is often only of minor importance because, for safety reasons, lithium-ion batteries must be equipped with various monitoring and protective devices in an overall system, which are not required for lead-acid batteries. All safety and monitoring devices that are necessary for safe and long-lasting operation must be included. The comparison of systems with different safety standards and expected service lives is only meaningful if the differences are carefully evaluated from a technical and economic point of view.
2) Costs:
   The costs of the overall system must be considered with comparable system delimitation.
3) Energy content:
   Only the energy content that can actually be used during operation and with which the desired service life can be achieved should be considered. In many battery systems, for operational and/or service life reasons, the usable energy content must be significantly limited compared to the energy content of the fully charged battery.
4) Volume and weight:
   The overall system must be considered.
5) Other technically relevant properties:
   - response time to performance requirements;
   - temperature range within which the system can be operated;

- additional heating or cooling in the system must be taken into account because they affect weight, volume, and costs; and
- maintenance and monitoring options.

6) Other economically relevant properties:
   - maintenance costs over the service life,
   - energy consumption when stationary and in use, and
   - efficiency.

General comparisons of energy storage systems, as shown in Figure 1.4 or as shown in tabular form in many studies, can only serve as a guideline. Detailed comparisons without a description of the specific application often do not allow any useful conclusions to be drawn. When describing the applications and additional conditions to be considered, the number of suitable energy storage technologies is often considerably reduced. For technical and economic reasons, there is often only one energy storage technology that can be considered for the application in question. Pumped storage plants do not compete with flywheel mass storage systems!

1.6 Applications and Integration of Batteries into Overall Systems

Batteries are usually coupled in parallel with generation systems and consumers. In many cases, this arrangement leads to a highly dynamic current load on the battery with frequent changes between charging and discharging currents. This behavior is immediately clear in hybrid vehicles, whose battery constantly alternates between absorbing braking energy and releasing energy to support acceleration. However, batteries for autonomous power supply systems or in standby parallel operation are also often subject to a rapid change from charging to discharging currents due to the residual current ripple and the control characteristics of the chargers and consumers. Measurements often show a superposition of the battery current with an alternating current, the amplitude of which can significantly exceed the value of the direct current. This is discussed in more detail in the overview of applications for electrochemical energy storage systems from chapter 15 onward.

It is often useful to differentiate between portable, mobile, and stationary applications, whereby a distinction between grid-connected and autonomous energy supply systems is helpful for stationary applications.

1) Mobile applications:
   In some traction applications, the battery (e.g. industrial trucks and electric vehicles) is discharged during use, interrupted only by occasional regenerative braking, etc. and then recharged in a separate step.

   However, hybrid vehicles have their own power generation unit (combustion engine, etc.) as part of the overall mobile system, which can also charge the battery during use. The battery current and state of charge fluctuate greatly depending on the operating strategies.

   In mobile applications, apart from traction, the battery is often only used for self-starting (so-called black start) of the overall system (power supply for the engine and onboard electronics, starting the combustion engine), independent of other energy sources.

The requirements for energy content, weight, and many other properties differ greatly between these areas of application.

2) Stationary applications:

In autonomous, non-grid-connected energy systems, the battery serves as the sole source of energy when primary energy generation is not available (no wind, no solar radiation, generator motor not in operation) and to absorb electrical energy when generation exceeds consumption. A constant alternation between charging and discharging processes must be assumed.

In grid-connected overall systems, the battery is usually used to secure the power supply in the event of grid failures and voltage problems (emergency power systems, UPS systems) and to provide and absorb power fluctuations to stabilize the power grid. With increasing electricity generation from wind power and photovoltaic systems, energy storage systems are also used to temporarily store electricity from renewable energies.

3) Portable applications:

The batteries in a device are designed to only be discharged during use and then recharged on a charger. Use during charging is the exception rather than the rule. The obvious criteria are low volume and weight.

References

1 Energy Research Center of Lower Saxony (2016). *Potentials of Electrochemical Storage in Electrical Grids in Competition with Alternative Technologies and System Solutions (ESPEN (Original Germna title: Potenziale elektrochemischer Speicher in elektrischen Netzen in Konkurrenz zu alternativen Technologien und Systemlösungen (ESPEN), Abschlussbericht), Final Report).* Goslar: EFZN.

2 Wenzl, H., Benger, R., and Hauer, I. (2024). Power. In: *Encyclopedia of Electrochemical Power Sources*, 2e, vol. I (ed. J. Garche). Amsterdam: Elsevier.

3 Sarpal, I., Bensmann, A., Mähliß, J. et al. (2018). Characterisation of batteries with E-P curves: quantifying the impact of operating conditions on battery performance. *Electrical Power and Energy Systems* 99: 722–732.

Tasks

Task 1.1

Design of electricity storage systems with electricity exclusively generated from photovoltaic and wind power plants

Germany consumes around 500 TWh of electricity per year. An electricity system based exclusively on-wind and photovoltaics must be able to cover periods of

(Continued)

Task 1.1 (Continued)

no wind and no sun, sometimes referred to as "dunkelflaute." The maximum duration of no wind and no sun is usually given as 21 days.

Assume that the daily generation capacity is 10 GW even during such phases (generation by hydropower plants, combined heat and power plants that have to be operated due to heat demand, hazardous waste incineration plants, etc.).

- What quantities of energy need to be stored in electricity storage systems?
- Can you assume that the electricity storage units are fully charged at the beginning of a no wind, no sun period? What quantities of energy should still be available in the electricity storage systems at the end of the no wind, no sun period so that the supply remains secure in the event of an immediate shortfall in electricity demand from the generation units?
- What safety margins for the duration of a no wind, no sun period do you assume (maximum expected duration in the coming decades)? What percentage of the installed electricity storage systems will not be used at all or only once over the period under consideration?
- Which industries or groups of households will be the first to be shut down if the storage facilities are not large enough?

Task 1.2

System comparison of drivetrains

The drivetrain of an electrically powered vehicle consists of a vehicle battery, including all necessary components such as battery management system (BMS) and fuses; an electric motor, possibly with a single two-stage gearbox and differential; cables; and inverter. The drivetrain of a vehicle powered by an internal combustion engine consists of a tank, fuel lines and feed pump, starter, alternator, internal combustion engine, radiator with fan and water pump, multistage gearbox, clutch, differential, and exhaust system with exhaust gas purification.

a) Estimate the total weight of the two powertrains for comparable vehicles (e.g. Tesla Model S and BMW 5 Series) with the same range and engine power. To do this, use 75% of the specified range of the Tesla and weight data from the Internet.
b) How do the two drivetrains differ in terms of their specific energy?
c) Estimate the energy consumption of both cars per kilometer, including losses when the cars are parked and charging efficiency.
d) Is this comparison admissible? What conclusions can and cannot be drawn from it?

2

Electrochemical Basics

In this chapter, the basic processes that take place in batteries are discussed using examples from the field of lead-acid and lithium-ion batteries. The electrochemical principles are only explained to the extent that they are necessary for an understanding of batteries. For further explanations, please refer to the relevant textbooks, for example, Refs. [1–3].

Batteries are electrochemical systems based on the following electrochemical principles:

1) As soon as an electron-conducting material comes into contact with an electrolyte (i.e. an ion-conducting material), one or more reactions take place at the interface, which lead to a charge carrier separation. At the anode, electrons remain in the electron-conducting material without countercharges, and ions remain in the electrolyte without the corresponding counterions. At the cathode, electrons are missing in the electrode material, and ions are missing in the electrolyte. The interface forms a capacitor (double-layer capacitor). A voltage is created immediately. As a result, there is always a voltage difference between the terminals of a battery, with all its possible safety hazards (short-circuit currents, arcing, etc.).
2) The voltage generated by the reaction at the interface depends on the materials involved and the properties of the electrolyte. The equilibrium voltage can be calculated using the free enthalpy (Gibbs enthalpy) of the reaction and the Nernst equation which describes the influence of the concentration-dependent activities of the reactants on the voltage (Nernst voltage). The voltage generated by the reaction can be modeled as a voltage source in an equivalent circuit diagram.
3) If different electrochemical reactions can take place simultaneously at the electrode interface, then each of the reactions has its own specific Nernst voltage. The voltage levels of the electrodes or the cell then change to a slightly different value called a mixed potential. This leads to self-discharge effects and may cause corrosion.

(Continued)

Battery Technology: Fundamentals of Battery Electrochemistry, Systems and Applications, First Edition. Alexander Börger and Heinz Wenzl.

Companion website: www.wiley.com/go/Börger/Battery_Technology

(Continued)

4) When current flows through the battery, electrochemical reactions take place, whose reaction rates are directly proportional to the current (Faraday's law). A charge quantity of 96,480 As corresponds to the conversion of 1 mol of the electrochemically active materials of an electrode if the reaction only transfers one electron.
5) The voltage of the battery is changed by current flow. The voltage difference between the equilibrium potential and the electrode potential caused by the reaction is referred to as the charge transfer voltage η_T and can be understood as an electrical resistance with a very large current dependence. The voltage drop can be calculated using the Butler-Volmer (BV) equation. At low currents, the charge transfer voltage is the dominant voltage deviation from the equilibrium voltage of the electrodes; at very high currents, this voltage drop is of lesser importance compared to the voltage drop in the electrically conductive components of the cell (ohmic voltage drop η_Ω) and restrictions on mass transport (diffusion overvoltage η_D).
6) The charge transfer voltage or the associated resistance can be understood electrically as the internal resistance of the voltage source formed by the respective reaction.
7) Current can only flow if the positive (cathode) and negative (anode) electrodes are connected to each other. Apart from this, the two electrodes are independent of each other.

2.1 Basic Electrochemical Terms

2.1.1 Some Definitions

Strictly speaking, an *electrode* is the point of contact between an electron conductor, respectively semiconductor, and an ion conductor. In practice, the term "electrode"[1] usually refers to the metallic or at least semiconducting solids that conduct the current.

An *electrolyte*, on the other hand, is strictly speaking a chemical compound that is dissociated into ions in a solid or liquid. In practice, the entire ion-containing solid or liquid is usually referred to as an electrolyte.[2]

A *half-cell* is the combination of one electrode and electrolyte.

---

1 There are very different types of electrodes. They can be classified according to the phases involved (solid/liquid, solid/solid), according to the number of particles passing through (single: only one particle, multiple: more than one particle), or according to the type of particle passing through (redox electrode: electron passes through; electrode of the first type: the voltage-forming ion passes through; and second-type electrode: an ion that is in an upstream equilibrium passes through the phase boundary).

2 There are also very different types of electrolytes, which can be divided into solutions (aqueous or nonaqueous), molten salts, ionic liquids, solid electrolytes, gel electrolytes, or polymer electrolytes according to the phases involved, or into strong and weak electrolytes according to the degree of dissociation.

A(n) *(electrochemical) cell* is created by an ion-conducting connection between two half-cells, usually using a separator.

The *anode* is the place of oxidation. In the case of rechargeable batteries, the discharge reaction is used to define which electrode is the anode. Electrically, the anode is the negative terminal of a cell.

The *cathode is* the place of reduction. In the case of rechargeable batteries, the direction of the discharge reaction is used to define which electrode is the cathode. Electrically, the cathode is the positive terminal.

A *separator* separates the electrolyte space from the anode (anolyte) and cathode (catholyte), but allows ionic conduction between the electrodes.

A *battery* is the commonly used term for a single cell or a group of cells connected electrically in parallel and/or in series.

A *primary cell* is a cell that can only release electrical energy once during discharge.

An *accumulator* (also known as a secondary cell) is a battery that can be discharged and then recharged. Typical accumulators allow many hundreds of such charging and discharging cycles.

A *flow cell*, often also called a *redox flow cell*, is a cell in which the electrochemically active reactants circulate as part of the electrolyte in the anode compartment (anolyte) and cathode compartment (catholyte) and are pumped into a storage tank after or during discharge. The electrolytes are usually also recharged in the anode or cathode chamber.

In a *fuel cell*, gases (usually hydrogen on the one hand, and oxygen or air on the other) are continuously supplied, and the reaction products (usually water or water vapor) are removed from the reaction chamber. Fuel cells are conceptually related to flow batteries, but cannot be charged and are therefore not energy storage devices.

In battery terms, the capacity (usually Ah for ampere-hour) of a battery is the amount of charge that it releases during discharge under precisely defined conditions.

Annex A of this book contains a more comprehensive list of terms used in battery science.

2.1.2 Voltage and Charge Carrier Distribution

A voltage is created by an inhomogeneous distribution of charge carriers. Using Maxwell's basic equations of electrodynamics, the electric field $\vec{E}$ caused by a charge carrier distribution $\rho(\vec{x})$ can be calculated:

$$\vec{E} = k \int \rho(\vec{x}) \mathrm{d}\vec{x}, \tag{2.1}$$

where $\rho(\vec{x})$ is the spatial charge carrier distribution of positive and negative charges, the integral is performed over the entire space, and k is a proportionality constant.

The potential difference between two points is formed as an integral over the electric field along any path between the two points:

$$\Delta U = k \int \vec{E} \mathrm{d}\vec{s}. \tag{2.2}$$

The presence of a voltage at the poles of a battery therefore means that an uneven distribution of charge carriers must have been formed in the cell. The voltage is independent of the integration path, and it therefore does not matter how large the electrode surface is. Electrochemical cells therefore always have the same voltage at rest and under otherwise identical conditions regardless of their size.

2.1.3 The Main Reactions – The Voltage Defining Reactions

In contrast to chemical reactions, in which the reactants and products of the reaction are uncharged, in electrochemical reactions, ions and electrons are part of the reaction and are explicitly mentioned in reaction equations.

In general terms (ions are referred to as Me^+ and electrons as e^-), the reactions for the electrodes of a cell are as follows:

$$\textbf{Anode:}\quad a\mathrm{A} + b\mathrm{B} \rightleftarrows c\mathrm{C} + d\mathrm{D} + n\mathrm{Me}^{l+} + m\mathrm{e}^-, \tag{2.3}$$

$$\textbf{Cathode:}\quad e\mathrm{E} + f\mathrm{F} + p\mathrm{Me}^{l+} + q\mathrm{e}^- \rightleftarrows g\mathrm{G} + h\mathrm{H}, \tag{2.4}$$

where A, B, E, and F stand for the reactants and C, D, G, and H for the products of the reactions. The lowercase letters denote the respective number of molecules or atoms either as individual particles or as a quantity of substance (unit: mol). The number l denotes the charge number of the ion involved. To ensure the electroneutrality, it must be $n \cdot l = m$ and $p \cdot l = q$. Coupling of the two reactions in a battery is only possible if the ion is the same in the anodic reaction (where it is made available during discharge) and in the cathodic reaction (where it is consumed during discharge). m and n or p and q are only the same if the ion formed has one charge only ($l = 1$). Since the number of charge carriers and the ion type must always be the same on both electrodes during discharge and charging, the anodic and cathodic reactions must take place in the ratio m/q.

In thermodynamic equilibrium, the reaction proceeds at the same rate in both directions because the total energy on both sides of the reaction equation, including the energy stored in the electric field, is the same.

Table 2.1 shows the main anodic and cathodic reactions for various electrochemical systems in this general form.[3] In addition to the electrode reactions, the sum of both reactions, the cell reaction, is also shown.

Figure 2.1 shows a schematic representation of the charge carrier distribution at the interfaces of an electrode, the resulting electric field, and the voltage. At the anode, the negative electrode, positively charged ions are released into the electrolyte, and electrons associated with the ions remain in the metallic conductive electrode. At the cathode, the corresponding charge carriers are missing. The resulting charge carrier distribution is shown in the middle. Missing electrons and positive excess charges lead to the same charge carrier distribution as in the anode. The voltage curve resulting from this charge carrier distribution is shown in Figure 2.1c. The values shown correspond approximately to those of a fully charged

3 The notation chosen here for reactants is particularly easy to understand, but must be modified for more detailed considerations.

Table 2.1 Overview of the main anodic and cathodic reactions (simplified representation) and the cell reactions of important electrochemical systems.

	Anode (negative electrode)	**Cathode (positive electrode)**	**Sum of reactions of the cell**
Lead-acid batteries	$Pb + H_2SO_4 \rightleftarrows PbSO_4 + 2H^+ + 2e^-$	$PbO_2 + H_2SO_4 + 2H^+ + 2e^- \rightleftarrows PbSO_4 + 2H_2O$	$Pb + PbO_2 + 2H_2SO_4 \rightleftarrows 2PbSO_4 + 2H_2O$
Lithium-ion batteries[a)]	$LiC_6 \rightleftarrows C_6 + Li^+ + e^-$	$MeO_x + Li^+ + e^- \rightleftarrows LiMeO_x$	$LiC_6 + MeO_x \rightleftarrows C_6 + LiMeO_x$
H_2-fuel cells	$H_2 \rightleftarrows 2H^+ + 2e^-$	$\frac{1}{2}O_2 + 2H^+ + 2e^- \rightleftarrows H_2O$	$H_2 + \frac{1}{2}O_2 \rightleftarrows H_2O$

a) MeO_x refers to the different metal oxides that make up many cathodes in lithium-ion batteries. Graphite is the most commonly used anode material in lithium-ion batteries.

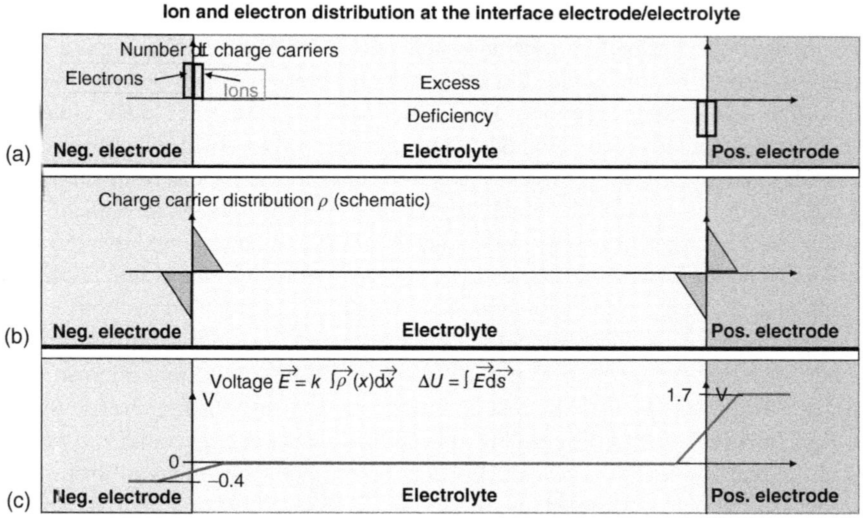

Figure 2.1 (a) Schematic representation of the number of charge carriers, (b) charge carrier distribution, and (c) resulting voltage curve at the positive and negative electrodes of a cell. The voltage values (c) shown correspond approximately to the voltage level of a lead-acid battery.

lead-acid battery. The voltage level of the electrolyte is often assumed to be zero in such considerations. Similar representations are used to explain the voltage level of a pn junction in a semiconductor. At the interface between n- and p-conducting materials, a comparable charge carrier distribution results, whereby *p* (defect electron or hole) can be regarded as a positive charge, analogous to an ion.[4]

The further reaction of the ions and electrons to form an uncharged particle does not take place due to the potential barrier and the spatial separation (electrons in the metallic conductive electrode, ions in the electrolyte) between them. Electrochemical systems can therefore have a very long storage time without any

---

4 Further information on the physics of semiconductors can be found in Ref. [4].

noticeable loss of the stored charge, for example, 10 years of storage time or more. A shorter storage time and a technically relevant self-discharge of batteries are always due to side reactions (see the individual battery systems chapters 12–14).

2.1.4 Double-Layer Capacitor and Exchange Current Density

The missing or excess electrons in the electrode and the missing or excess ions in the electrolyte form two layers of charge with opposite polarity and constitute an electrical capacitor. Together, they are referred to as a double-layer capacitor. The double-layer capacitor model can be used to understand the voltage behavior of cells with rapid current changes. For more in-depth electrochemical questions, it is necessary to have a more precise picture of the structure of the charge carrier distribution on the surface. The first model for the double layer was presented by Helmholtz (Helmholtz layer). Refinements are known, for example, under the name Gouy-Chapman. In simplified form, these models state that there are two types of layers: a rigid one, formed by ions adsorbed on the electrode surface, and a diffuse layer, formed by ions under the influence of the electric field but moving in the solution due to particle motion. For further details, please refer to the general textbooks on electrochemistry already cited and to Ref. [5] for further details.

There is a strong electrostatic force between the electrons in the electrode material and the ions in the electrolyte, which are released or consumed as a result of the reactions taking place. At an electrode voltage of 1 V and an assumed distance of 10 nm between the two charge carrier layers of the double-layer capacitor, the field strength is in the range of 10^8 V/m, which is a very high electric field strength.

The strong electric field prevents further reactions at the interface or limits the number of charge carrier pairs that can form per unit area (effective microscopic area), as further charge carriers must be formed against the existing electric field. The high electric field and the equilibrium conditions between reactants and products of the reaction lead to a reverse reaction. Macroscopically, the equilibrium of the forward and backward reactions results in a stable voltage and charge carrier density, which do not change over time without external influences.

2.1.5 Faraday Number

The number of electrons that are made available in an electrochemical reaction, that is, the amount of charge Q (unit: coulomb, 1 C = 1 As) transported by them, and the number of particles that are converted per mole are proportional to each other; the proportionality factor is referred to as Faraday's number F. Its value is 96,500 As/mol.

Faraday's law states that the amount of charge generated and the number z of electrons exchanged during the conversion of a certain amount of substance n of an electrode is proportional to the amount of charge generated and the number z of electrons exchanged. z is the number of electrons exchanged, that is,

$$Q = nzF. \tag{2.5}$$

zF is a frequently used conversion factor between chemistry and electrical engineering.

Table 2.2 Theoretical specific capacity of a lead-acid battery and a lithium-ion battery with NMC 111 cathode and graphite anode.

	Cell reaction	Molecular weight [g/mol]	z	Theoretical specific capacity [Ah/g]
Lead-acid batteries	$Pb + PbO_2 + 2H_2SO_4 \rightleftarrows 2PbSO_4 + 2H_2O$	642	2	83.5
Lithium-ion batteries	$LiC_6 + Ni_{1/3}Mn_{1/3}Co_{1/3}O_2 \rightleftarrows C_6 + LiNi_{1/3}Mn_{1/3}Co_{1/3}O_2$	168	1	159.4

2.1.6 Theoretical Specific Capacity of Electrodes or Cells

Assuming that the entire electrochemically active material of an electrode or a cell (as an interconnection of anode and cathode) is converted, the theoretical specific capacity (capacity per unit weight) can be calculated. To do this, zF is divided by the molecular weight of the material required for the sum of the reactions of the cell. Only either the charged or discharged material may be taken into account. Table 2.2 shows the calculation for lead-acid batteries and lithium-ion batteries with a NMC 111 cathode (NMC: nickel, manganese, cobalt; 111 refers to the mixture of these metals, that is, the same number of atoms or moles of the three metals in the cathode) and graphite anode. In Task 2.1, the theoretical specific capacity of various lithium-ion batteries is calculated.

The specific capacity of technically designed battery systems is much smaller because a complete conversion of the active material is not possible for technical reasons or for reasons of service life and safety, and numerous additional materials of a complete cell must also be taken into account [6].

2.2 Electrochemical Thermodynamics

The following discussion is restricted to the presence of only one reaction at the electrode/electrolyte interface. In most electrochemical systems, additional, parallel side reactions occur at many technically relevant operating voltages, which are of great importance during operation. Their influence on the equilibrium voltage and the properties of cells is explained in section 2.5 and chapter 3.

2.2.1 Energy Balance and Equilibrium Voltage

The energy balance of chemical reactions is expressed by the change in free or Gibbs enthalpy ΔG during the reaction:

$$\Delta G = \Delta H - T \cdot \Delta S. \tag{2.6}$$

ΔH is the change in enthalpy, T is the temperature, and ΔS is the change in entropy of the overall system, whereby the entropy S is a measure of the disorder or randomness of the system.

The calculation of energy released by chemical reactions is based on the thermodynamic values of its components, which are tabulated for so-called standard conditions – room temperature, normal pressure, and one-molar aqueous solution (so-called standard free enthalpy of formation). The enthalpy of reaction is determined by the difference enthalpies of formation of the products and the reactants.[5] The enthalpy of reaction therefore does not depend on the reaction paths and any intermediary products, but only on the end products.

The following sign convention applies: a negative value of ΔG means that the system loses energy and the reaction occurs spontaneously. The values are therefore negative for the discharging reaction and positive for charging.

The free enthalpy ΔG_{00} of the reaction under standard conditions is directly proportional to the equilibrium voltage E_{00} of the electrode in electrochemical systems:

$$E_{00} = -\frac{\Delta G_{00}}{zF}. \tag{2.7}$$

For aqueous systems, the equilibrium voltage E_{00} of an electrochemical reaction can therefore be calculated with high precision under standard conditions.

The reaction conditions in a technically usable cell do not correspond to the standard conditions, so the actual conditions, in particular the concentrations of the reactants, must be taken into account. This is done using the Nernst law.

2.2.2 Concentration Dependence of the Equilibrium Voltage (Nernst Voltage)

In general, the reactants are not present in a one-molar solution and, the higher the concentration, the greater the interactions between the ions and the solvation shell surrounding the ions. It is therefore clear that the reactivity of the ions, their activity, is a function of their concentration in the electrolyte. To calculate the dependence of the equilibrium voltage on nonstandard conditions, the activity a_i of the ions and not their concentration c_i must be used.

$$a_i = f_i \cdot c_i. \tag{2.8}$$

The generally temperature and concentration-dependent proportionality factor f_i is called the activity coefficient.[6]

The activity of homogeneous solids is constant and is set equal to 1 by definition.

The relationship between free enthalpy and the activities of the substances involved is as follows:

$$\Delta G_0 = \Delta G_{00} + RT \cdot \ln \left(\prod a_i^{m_i} \right), \tag{2.9}$$

where a_i is the activity of reactant type i and m_i is the number of particles involved in the reaction. m_i is positive for products of the reaction and negative for reactants.

5 The calculation of such values in nonaqueous systems, such as for all lithium-ion systems, is complicated because a thermodynamically correct link to aqueous systems must be found. For details of electrochemistry in non-aqueous solvents, see Ref. [7].

6 Activity coefficients must normally be determined experimentally. For dilute aqueous electrolyte solutions, the Debye-Hückel theory exists for calculating activity coefficients. However, electrolyte concentrations in real batteries are normally much higher.

Converted into voltages, this results in the so-called Nernst equation for the concentration-dependent cell voltage:

$$E_0 = E_{00} - \frac{RT}{zF} \cdot \ln\left(\prod a_i^{m_i}\right). \tag{2.10}$$

E_0 is referred to as the Nernst voltage and is the equilibrium voltage of the reaction under the given concentration conditions of the electrode. In the equilibrium state of a cell (no current flow), the Nernst voltage is the same at all locations of the electrode as long as there are no differences in concentration. This requirement is not always met for lead-acid batteries in particular (see the topic of acid stratification in section 12.9 and appendix E).

For lead-acid batteries, Nernst's law can be applied directly and the equilibrium voltage can be calculated for any concentration of aqueous sulfuric acid. For lithium-ion batteries, however, the application of Nernst's law to calculate the equilibrium voltage is more difficult due to the nonaqueous solvent normally used.

2.2.3 Temperature Dependence of the Equilibrium Voltage

The temperature dependence of the equilibrium voltage around the operating point is calculated using the energy balance[7]:

$$\frac{\mathrm{d}\Delta G_{00}}{\mathrm{d}T} = -\Delta S_{00} \tag{2.11}$$

and therefore

$$\frac{\mathrm{d}\Delta E_{00}}{\mathrm{d}T} = \frac{\Delta S_{00}}{zF}. \tag{2.12}$$

The remarkable result is, therefore, that changes in entropy (which are otherwise more difficult to measure) can be determined from the temperature dependence of the cell voltage.

These explanations show that the thermodynamic data of electrochemical systems can be measured directly and are of immediate technical relevance. In the discussion of the state of charge (chapter 7), an approach is shown as to how the state of charge can be estimated by knowledge of the thermodynamic data.

2.2.4 Entropy Term and Temperature – Reversible Heat

The energy balance of the reaction can also be used to calculate the heat dissipation. Depending on the sign, the term "$T\Delta S$" leads to a heat release or a heat demand during charging (the cell cools down when current flows). In the case of discharging, the heat balance is reversed, hence the term "reversible heat" (see Task 2.2). This term, which is proportional to the current, must be used in detailed studies of battery temperatures. Figure 2.2 shows the temperature curve of a lithium-ion cell during cycling. It can be seen very clearly that the temperature increases at the end of discharge and falls below the ambient temperature at the beginning of charging. Whether the heating or cooling of the cell must be taken into account must

7 In the following, the usual assumption is made that there is only a slight temperature dependence, which can be neglected in most cases.

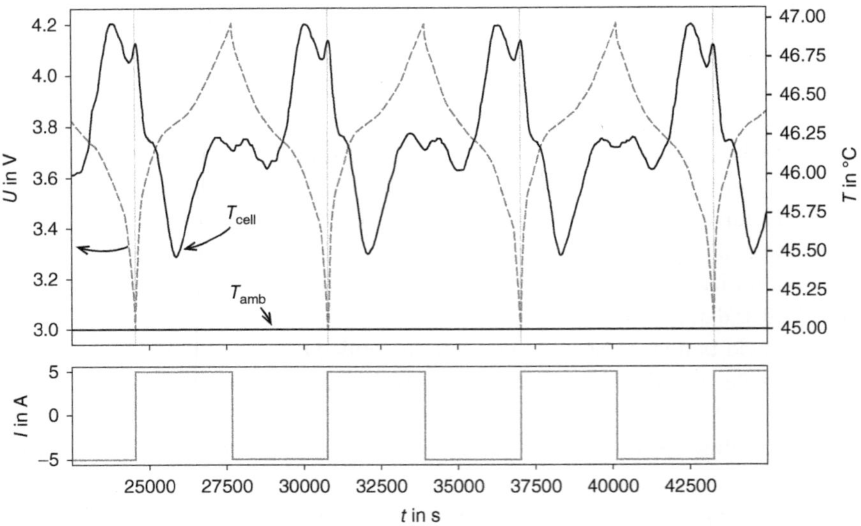

Figure 2.2 Measurement of the reversible heat of a lithium-ion cell during cycling. Source: Courtesy of Jacob Klink, TU Clausthal.

be examined in detail; in some alkaline nickel-based batteries, the effect is implicitly used to control important charging characteristics.

2.2.5 Electrochemical Series

When tabulating values for the equilibrium voltage of different materials, a value related to a half-cell potential must be determined.

By international agreement, the standard hydrogen electrode was chosen as the reference point, and the equilibrium voltage value of 0 V was assigned to it.

This makes it possible to tabulate electrochemical systems in the so-called electrochemical series, which can be found in many electrochemistry textbooks. From these tables, indications can be derived as to which material combinations could be particularly suitable for the construction of batteries. However, technically usable materials for battery electrodes must fulfill numerous other requirements, so the electrochemical voltage series is not explained further here.

2.2.6 Limits of Thermodynamic Considerations

For the use of thermodynamic data for calculating the equilibrium voltage, the following limits must be observed in addition to the restriction for use with aqueous solutions only:

1) Are the reactants and products clearly defined?
   This condition is usually fulfilled if the chemical substances involved have a solid composition. However, this is often not the case for solids with poorly defined phases (intercalation compounds, daltonide compounds). For details, see Refs. [8, 9].

2) Is there thermodynamic data available for the materials that exist in the electrodes?
   Similar to the first question, this uncertainty applies in particular to non-stoichiometric compounds, such as those found in many alkaline nickel batteries.
3) Is the crystal structure of the active material known?

However, the changes in the equilibrium voltage and the reversible heat of reaction caused by such uncertainties are technically negligible in many cases.

In principle, these thermodynamic considerations also apply to lithium-ion batteries. However, the impossibility of working with aqueous solvents leads to considerable difficulties, so the considerations presented here are significantly more complex, and comparable values, such as those that exist for lead-acid batteries for calculating the voltage level, are not available in many cases.

In all battery systems, the equilibrium voltage or Nernst voltage is higher in the fully charged state than in the partially discharged or fully discharged state, as can be seen directly from the Nernst equation.

2.2.7 Theoretical Specific Energy

Using the theoretical specific capacity of a battery and the equilibrium voltage, the theoretical specific energy of a cell can be calculated. It is based on the assumption that the entire active material can be completely converted to generate electrical energy, that the conversion process takes place at the voltage of thermodynamic equilibrium (i.e. without any loss of energy), and that there are no other materials that are necessary for the construction of a cell. Despite these unrealistic assumptions, the calculation can serve as a guide, for example, to compare different battery systems in principle. In Task 2.3, values are calculated for some battery systems, and it is shown that there are considerable differences in the theoretical specific energy between the various electrochemical systems.

2.2.8 Reference Electrode

The voltage difference that is measured between the terminals of a cell is the sum of both electrode potentials caused by the spatial separation of negative and positive charge carriers on the electrode/electrolyte interface. To measure individual electrode potentials, a reference potential must be defined in the electrolyte that is independent of the operating conditions of the cell (state of charge, current, etc.). It is only possible to measure the voltage difference between an electrode and the electrolyte directly with a metallic measuring tip if the electrochemical reactions of the measuring tip with the electrolyte are well known.

To precisely measure the potential of a single electrode in relation to the electrolyte, reference electrodes must have the following properties:

a) The potential of the reference electrode in relation to the electrolyte must be independent of all changes (state of charge, current, concentration changes in the electrolyte, etc.) that are to be investigated.
b) Operation for voltage measurement only, that is, without current flow.

For battery systems with aqueous electrolytes, the potential of a hydrogenelectrode ($H_2 \rightleftarrows 2H^+ + 2e^-$) in the electrolyte against the cathode or anode is used. For lithium-ion batteries, the potential of a lithium electrode ($Li \rightleftarrows Li^+ + e^-$) in the electrolyte is used. The hydrogen electrode consists of a platinum wire inserted into the electrolyte acid and is surrounded by hydrogen from a hydrogen source, while the lithium electrode consists of a lithium wire that is inserted into the electrolyte. The reference electrode is then connected to one electrode of the cell to be measured via a high resistance, so that the influence of current through the reference electrode does not have to be taken into account in the further analysis.

If the voltage between the terminals of a cell is not given, but, for more in-depth analyses, only the voltage of the individual electrodes in relation to the electrolyte is given, then the type of reference electrode used in the analysis must always be specified as well.

2.3 Electrochemical Kinetics

2.3.1 Types of Overvoltage

The term "overvoltage" is generally used to describe deviations of the measured voltage from the equilibrium or open-circuit voltage (OCV) of a cell.

In the following, E denotes the cell voltage and E_o the equilibrium voltage under the existing concentration and temperature conditions. If necessary, the indices + for the cathode, – for the anode, and c for the cell are used when distinctions are made between an individual electrode or the entire cell. U indicates the terminal voltage of the cell. Without current through the cell and after all equalizing currents and diffusion effects have ended, the values for $E_{o,Z}$ and U are identical if there are no side reactions.

When current flows through two positive and negative electrodes (= cell) connected in series, a voltage deviation from the equilibrium voltage occurs. The voltage U that can be measured at the terminals is essentially[8] determined by the following factors:

- η_T: charge transfer voltage of the positive $\eta_{T,+}$ or negative electrode $\eta_{T,-}$ (topic of this section),
- η_Ω: ohmic voltage drop (see chapter 4 on the structure of active materials and the structure of cells and batteries),
- η_D: diffusion effects or mass transport phenomena (see section 2.3.5), and
- η_K: crystallization overvoltage[9] (see chapter 12 on lead-acid batteries).

8 In addition to the types of overvoltage mentioned here, electrochemistry also recognizes other types, such as reaction overvoltage in the case of inhibited upstream or downstream chemical reactions (in lead-acid batteries at the end of charging), and adsorption overvoltage when adsorption processes are involved (voltage hysteresis in nickel-metal hydride batteries).

9 The crystallization overvoltage results from the energy difference between particles in a crystallite and particles outside the crystallite or in other crystal structures.

The terminal voltage of a cell U is therefore related to the electrode voltage E_0 in the following relationship:

$$U = E_{0,Z} - \eta_{T,+} - \eta_{T,-} - \eta_{\Omega} - \eta_{D} - \eta_{K}. \tag{2.13}$$

Voltage deviations due to diffusion effects cannot be completely assigned to one or the other electrode and are summarized here, as is common in the literature. The crystallization overvoltage can be neglected except in certain operating phases of lead-acid batteries (see chapter 12); in lithium-ion batteries, it does not exist for the main reactions, as there are no crystallization effects.

It is usual to combine the various voltage terms and deviations from the equilibrium voltage as follows:

$$U = E_{0,Z} - \eta_{T} - \eta_{\Omega} - \eta_{D} - \eta_{K} \tag{2.14}$$

During current flow, mass transport phenomena (see section 2.3.5) influence the concentration of the reactants at the interface, and the Nernst voltage can then no longer be calculated precisely.

2.3.2 Charge Transfer Voltage

For a discharge to begin, the two electrodes must be connected to each other using an electron conductor. Figure 2.3 shows the process schematically. The freely available electrons of the negative electrode can compensate for the lack of electrons at the positive electrode. At the same time, ions are transported from the negative electrode to the positive electrode. The number of charge carriers at the electron-electrolyte interface is thus reduced, but immediately replenished by the formation of new charge carriers as a result of electrochemical reactions until the availability of reactants decreases. The capacitance of the double-layer capacitor reaches a current-dependent equilibrium. Therefore, batteries with a high reaction rate of the electrochemical reaction, that is, all lead-acid and lithium-ion batteries, can have very high short-circuit currents for a sustained period of time, leading to various safety issues.

2.3.3 Butler-Volmer Equation

The BV equation describes the relationship between the current and electrode potential during charging or discharging. The difference between the thermodynamic equilibrium voltage and the electrode potential is also referred to as charge transfer overvoltage. The BV equation is based on thermodynamic considerations and therefore equilibrium states. Nevertheless, the equation can also be used to describe the relationship between current and voltage outside the equilibrium state. There is no explicit dependence on time.

The BV equation is central to the understanding of batteries because it describes many properties, see also chapter 3. The difference between the equilibrium voltage and the electrode voltage that occurs during current flow can be understood

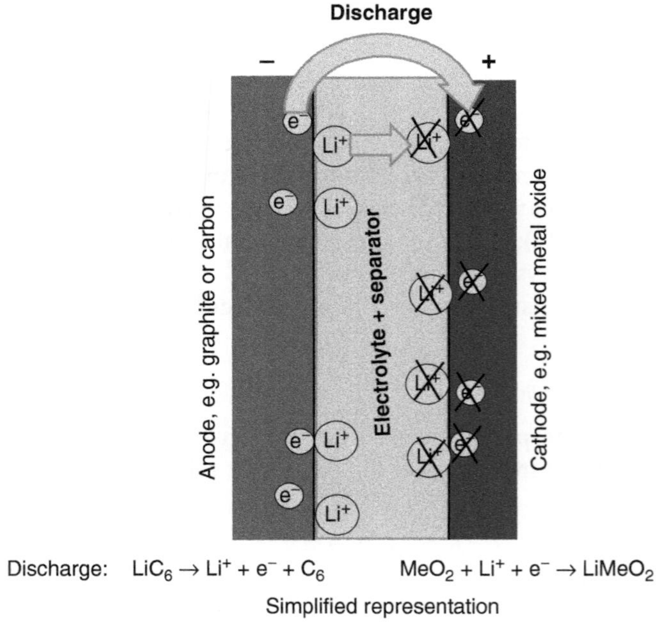

Figure 2.3 Schematic representation of the discharging process of a lithium-ion battery. The charge carriers consumed by the electrochemical reactions at the cathode are marked with a cross as "no longer present." The charge carrier transport from the negative to the positive electrode is shown by arrows in the upper part of the diagram. The voltage-generating reaction causes new charge carriers to be formed immediately when current flows.

as the energy that is necessary for the removal of charge carriers from the double layer or for the passage of charge carriers through the potential barrier at the electrode-electrolyte interface. The term "charge transfer voltage" can be understood in this way.

The behavior of a metal electrode in an electrolyte (acid) can be used as a model for the processes involved in discharging and charging an electrode. The reaction at the interface is:

$$Me \rightleftarrows Me^+ + e^-. \tag{2.15}$$

Dissolution of the metal and deposition of metal ions on the electrode always occur simultaneously.

If electrons can flow from the metal electrode to a counter electrode via a load, electrons are conducted from the metal electrode to the counter electrode, and the metal ions are transported from the surface of the metal electrode to the counter electrode, where they react and can be deposited. The dissolution reaction then continues – the lower the resistance of the load, the faster. When a voltage is applied, the metal is deposited on the electrode (galvanic process). The transfer of ions through the electrolyte/electrode interface is an electric current. The reaction speed corresponds to the amount of metal that is deposited, and the total amount deposited is proportional to the amount of charge that has passed through the electrodes.[10]

In a state of equilibrium, the rate of the dissolution reaction (forward reaction) is equal to the rate of the deposition (reverse or backward reaction). The current associated with the dissolution of the metal electrode and the current associated with the deposition of metal ions are called the exchange current I_o.

The reaction rate for the forward and reverse reaction can be converted into its associated current I_{hin} and $I_{rück}$ values by multiplying the reaction constant k [s^{-1} m^{-2}] under standard conditions (1 M concentration, per m^2) with the concentrations of the reactants [mol] with the area at which the reaction takes place, and the Faraday constant of the reaction zF [As/mol].

$$I_{hin} = zF \cdot A \cdot C \cdot k_{hin}, \tag{2.16}$$

with A: reaction area, C: product of the concentration of reactants A, B, C, etc. (c_A, c_B, c_C, etc.) on the surface of the electrode material, k_{hin}: area-related reaction constant.

In electrochemical processes, the reaction rate depends on the voltage level:

$$k_{hin} = k_{o,hin} \cdot \exp(\alpha zF(E - E_o)/RT), \tag{2.17}$$

with R as the general gas constant and α as the transfer coefficient, which describes the effect of the voltage difference on the forward and reverse reaction. α can also be regarded as a fit parameter for experimental results and, like the exchange current density, cannot be calculated from thermodynamic data.[11]

α can assume values between 0 (no effect of an applied voltage on the reaction) and 1. A value above 0.5 means that the reaction takes place more easily, that is, with less deviation of the electrode voltage from the equilibrium voltage (see Figure 2.4). The assignment of the transfer coefficient α and $-(1 - \alpha)$ is selected so that the total current I is positive when charging $E > E_o$

The same applies to $I_{rück}$ with

$$k_{rück} = k_{o,rück} \cdot \exp(-(1 - \alpha)zF(E - E_o)/RT). \tag{2.18}$$

To be able to specify data independently of the electrode size, the exchange current or the reaction rate is based on one unit of surface area and called the exchange current density i_o. Its value for lead-acid batteries is approx. 0.5×10^{-2} A/m^2 [10]. It should be noted that in electrochemistry, the exchange current density is often related to standard conditions, whereas the exchange current density used here is already related to the reaction conditions present in a battery.

The current flowing through an electrode is the difference between I_{fw} and $I_{rev}I_{Batt}$:

$$I_{fw} - I_{rev} = I_{Batt}, \tag{2.19}$$

$$I_{Batt} = i_o \cdot A \cdot \exp(\alpha zF(E - E_o)/RT) - i_o \cdot A \cdot \exp(-(1 - \alpha)zF(E - E_o)/RT). \tag{2.20}$$

---

10 In electrochemistry, the term "Faraday current" refers to the current associated with electrochemical reactions, as opposed to the current through a capacitor.

11 These considerations are based on the assumption that the activation energy, which has to be overcome in any chemical reaction, is affected differently for the electrochemical forward and reverse reactions when an external voltage is applied.

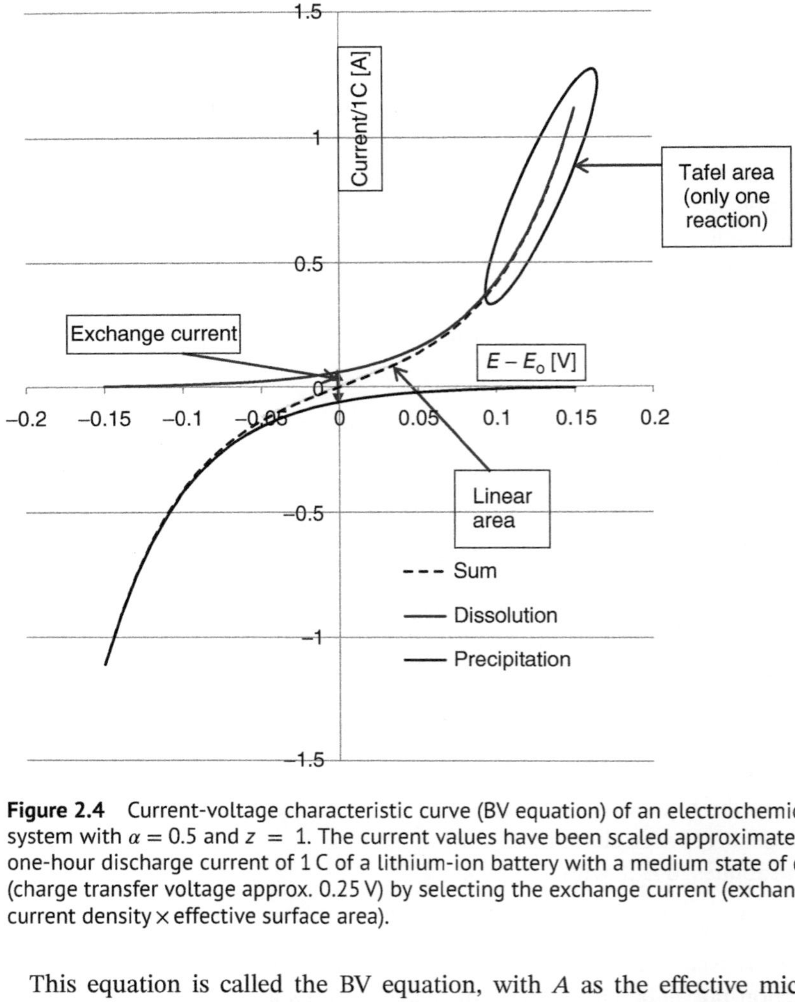

Figure 2.4 Current-voltage characteristic curve (BV equation) of an electrochemical system with $\alpha = 0.5$ and $z = 1$. The current values have been scaled approximately to a one-hour discharge current of 1 C of a lithium-ion battery with a medium state of charge (charge transfer voltage approx. 0.25 V) by selecting the exchange current (exchange current density × effective surface area).

This equation is called the BV equation, with A as the effective microscopic electrode area. It enables the calculation of the charge transfer voltage, the difference between the equilibrium potential and the electrode potential for an external current. It applies to conditions close to equilibrium.[12]

If $\alpha = 0.5$, which can often be approximately assumed for the main reaction of accumulators, then the BV equation can be simplified using the hyperbolic sine sinh:

$$I_{\text{Batt}} = i_0 \cdot A \cdot 2 \cdot \sinh((0.5\, zF/RT)(E - E_0)). \tag{2.21}$$

Two approximations of the BV equation are relevant in practice: the linear approximation at low current and low voltage deviation from E_0 and the range of large currents and voltage deviations from the equilibrium voltage, where only

12 The Nernst voltage is derived on the basis of thermodynamic equilibrium considerations; when deriving the BV equation, a disturbance of the thermodynamic equilibrium state is assumed. When charging and discharging batteries, however, there is no longer an equilibrium. Nevertheless, the BV equation is also used for high currents and has proven itself for practical work.

the reaction in one direction needs to be considered (Tafel approximation). When using batteries, the Tafel approximation can almost always be used; the linear approximation must only be used for very small deviations of the terminal voltage from the equilibrium voltage (approx. <0.08 V at $z = 1$ and 0.04 V at $z = 2$) or very small currents (Figure 2.4).

In the linear approximation, all terms of higher order of a series expansion are neglected, and the result is

$$I_{\text{Batt}} = i_o \cdot A \cdot (\alpha zF/RT)(E - E_o) \tag{2.22}$$

The battery current is therefore proportional to the exchange current or the effective surface area of the electrodes, which is proportional to the capacity of a battery to a first approximation (see chapter 4).

With the Tafel approximation for high currents and large deviations of the electrode voltage from the equilibrium voltage, one of the exponential function terms in the BV equation can be neglected:

$$I_{\text{Batt}} = i_o \cdot A \cdot \exp(\alpha zF/RT)(E - E_o). \tag{2.23}$$

Drawing the logarithm gives the following result:

$$E - E_o = RT/\alpha zF \cdot \ln(I/(i_o \cdot A)) = a + b \ln I, \tag{2.24}$$

with $a = -RT/\alpha zF \cdot \ln(i_o \cdot A)$ and $b = RT/\alpha zF$. The parameter b can be considered constant for a battery system if no factors need to be taken into account that may change the value of α due to catalytic effects or side reactions.

Figure 2.4 shows the BV equation with the two approximation ranges, linear range, and Tafel range.

If the voltage drop due to ohmic effects is neglected (resistance of the electrolyte, electrodes, etc.), $E - E_o$ of both electrodes corresponds to the deviation of the terminal voltage from the equilibrium voltage. In the case of high currents and a detailed analysis of current-voltage curves of batteries, this simplification must not be made, but the electrode voltage must be written as

$$E = U - IR \tag{2.25}$$

with R as the ohmic resistance of all components of the cell. It is therefore useful to distinguish between the ohmic resistance of the battery and the internal resistance of the battery, which includes the charge transfer voltage and other electrochemical effects (diffusion effects).

The following aspects are relevant when applying the BV equation:

1) There is no explicit time dependence.[13] The change in voltage over time with changes in current cannot be represented by the BV equation and is dealt with in section 2.4 and chapter 8.
2) Current and charge transfer voltage are clearly linked. An external influence on the current/voltage dependence of the BV equation can only be achieved by varying the area of the electrode and the concentration of the reactants.

---

13 The equation provides no information about the current or voltage of real battery systems changing over time, i.e. the equation is implicitly based on the assumption of a stationary state.

2.3.4 Dependence of the BV Equation on Important System Parameters

The above equations show that the general curve shape of the battery current versus the charge transfer voltage essentially depends on the number z of electrons transferred, the transfer coefficient of the exchange current, and thus the effective surface area of the electrodes. The general shape is therefore independent of the electrochemical system under consideration. The behavior of batteries is determined to a large extent by the charge transfer voltage, so that the BV equation is of great importance for understanding battery systems. In addition, a BV equation may even be able to describe the processes mathematically correctly if the conditions for its derivation are not actually met [11].

2.3.4.1 Transfer Coefficient

For batteries that can be discharged as well as charged, the transfer coefficient is $\alpha \approx 0.5$. A transfer coefficient different from c. 0.5 means that the system is inhibited for either the charging or discharging reaction. The transfer coefficient shown in Figure 2.5 of 0.2 corresponds approximately to the conditions of the water

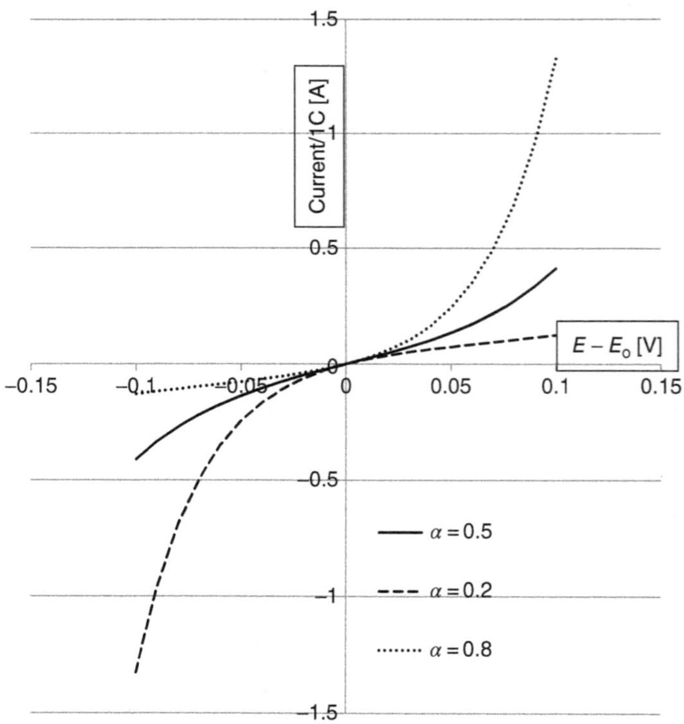

Figure 2.5 BV equation for different values of the transfer coefficients. The current values have been scaled to a one-hour discharge current of a lithium-ion battery in an average state of charge by appropriately selecting the exchange current (exchange current density multiplied by effective surface area) as in Figure 2.4.

decomposition reaction in lead-acid batteries (see section 2.5, chapters 3 and 12). Changes in the transfer coefficient α can be caused by catalytic effects, for example.

2.3.4.2 Number of Electrons Transferred *z*

The number of electrons z transferred differs between lithium-ion batteries ($z = 1$) and lead-acid batteries ($z = 2$). Under otherwise identical conditions, in particular the product of exchange current density and effective surface area, the current in lead-acid batteries will change twice as much as with lithium-ion batteries in the range of the Tafel approximation with the same change in electrode potential. In Figure 2.6, the current-voltage characteristics and the Tafel curves for both cases are shown in relation to a current I_1 or 1 C with the same exchange current. The curve shown in Figure 2.6 is also shown for such low charge transfer voltages that the nonlinear range becomes visible, that is, the Tafel approximation is no longer valid.

2.3.4.3 Exchange Current

Figure 2.7 shows the current-voltage characteristics and Tafel curves for charging of cells with different exchange currents. The lower the exchange current, the higher the charge transfer voltage at the same current. The fact that the Tafel curves run in parallel is due to the fact that with a constant number of electrons z in the reaction equation and a constant transfer coefficient α, the constant b in Eq. (2.24) remains constant as the slope of the curve. The Tafel curves shown in Figure 2.7 were calculated with a constant exchange current.

2.3.5 Charge Transfer Voltage as Resistance

When using equivalent circuit diagrams (see section 2.5), it is useful to express the current-voltage ratio described by the BV equation as a resistance $(E - E_0)/I$, referred to below as the BV resistance R_{BV}.

The result for the linear approximation is

$$R_{BV,linear} = (E - E_0)/I = RT/((i_0 \cdot A) \cdot (\alpha z F)) \tag{2.26}$$

and for the Tafel approximation:

$$R_{BV,Tafel} = (E - E_0)/I = (RT/\alpha z F) \cdot \ln(I/(i_0 \cdot A))/I = a/I + b \cdot (\log I)/I. \tag{2.27}$$

For $I = 0$, the resistance cannot be calculated using the Tafel approximation.

The resistance that can be attributed to the transfer voltage in the linear range is therefore constant, while the resistance in the range of the Tafel approximation depends on the current. As the logarithm of a number increases more slowly than the number, this resistance decreases with increasing current, although the charge transfer voltage continues to increase. For the same product of exchange current density and surface area, the BV resistance for lithium-ion batteries is twice as high as for lead-acid batteries. The lower internal resistance of lithium-ion batteries compared to lead-acid batteries with the same capacity is due to other effects, see chapters 12 and 13.

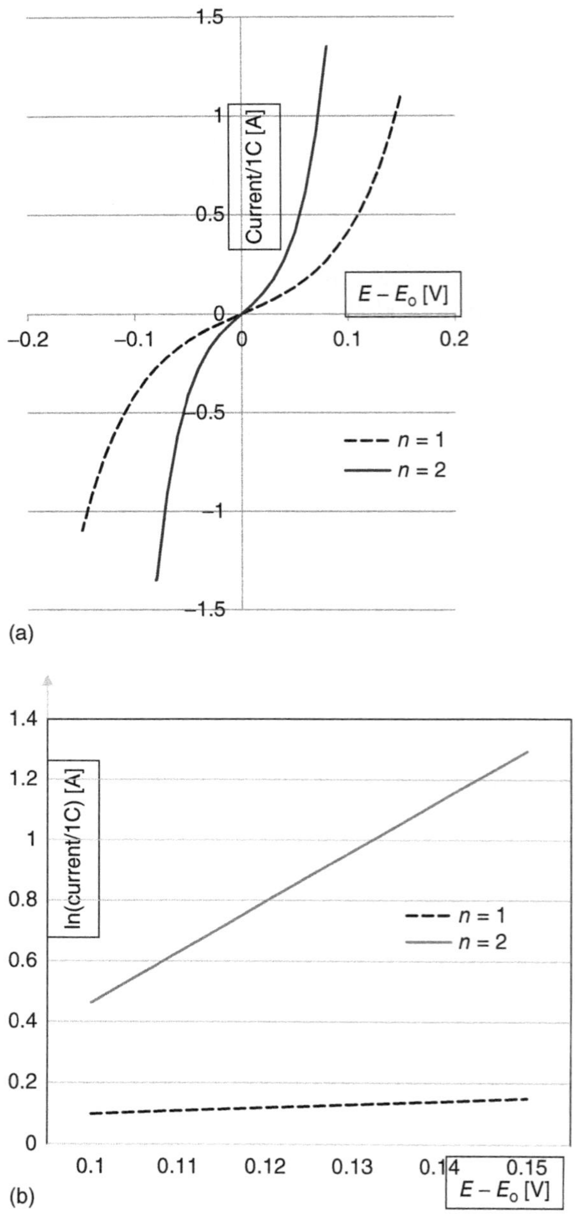

Figure 2.6 Illustration of the (a) BV equations and (b) the straight line of the Tafel approximation for charging for $z = 1$ and $z = 2$ with $\alpha = 0.5$ for both cases. The current values have been scaled approximately to a one-hour discharge current of a lithium-ion battery with a medium state of charge by appropriately selecting the exchange current. A negative value on the Tafel curve means that the battery current is lower than the exchange current.

Figure 2.8 shows the distinctly nonlinear course of the resistance of batteries with two electrons per reaction (lead-acid battery). The scaling of the resistance in the linear approximation also results in relative units for the current. In relation to the charge transfer voltage, the curve is always the same. The range of 0.15 V per electrode selected in Figure 2.8 corresponds to a charge transfer voltage of the battery of 0.3 V, which corresponds to comparatively high charging or discharging currents.

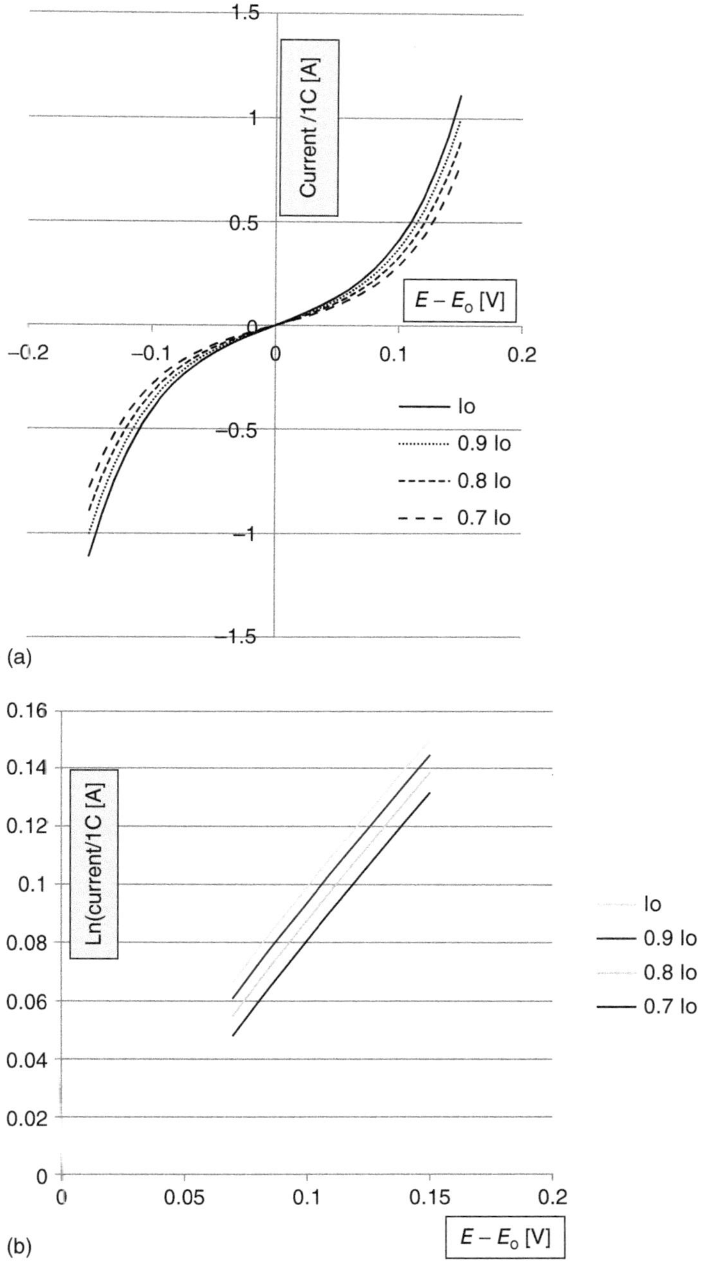

Figure 2.7 (a) Current-voltage characteristics and (b) Tafel curves for the charging range for cells with different exchange currents. The current values are determined by the choice of exchange current (exchange current density multiplied by effective surface area) as in Figure 2.4, scaled to approximately a one-hour discharge current of a lithium-ion battery in a medium state of charge.

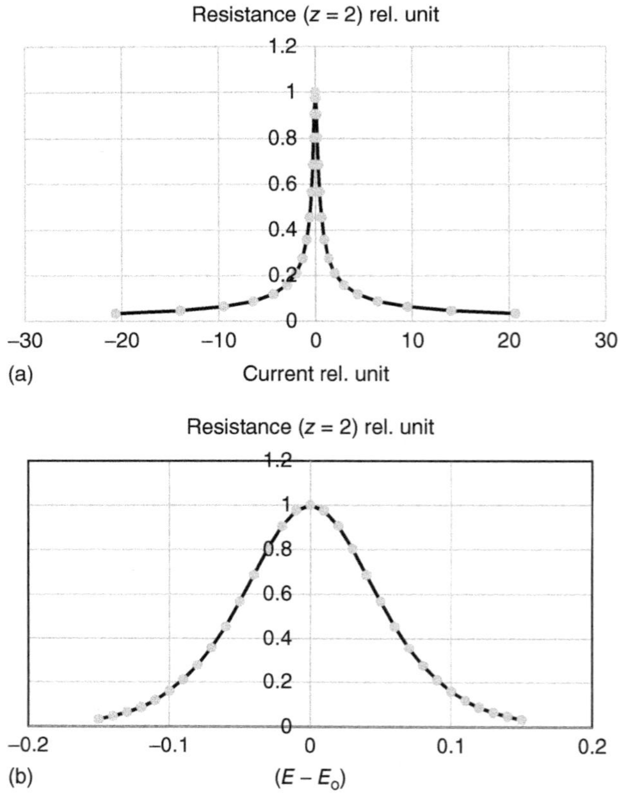

Figure 2.8 Current dependence (a) and voltage dependence (b) of the BV resistance of electrodes with two electrons per reaction. To calculate the resistance at equilibrium voltage, $(E - E_0) = 0$, the value for the linear approximation was used and scaled to the value of 1. For values of 0.01 V or greater, the full BV equation was used. The maximum current in relative units corresponds to a transfer voltage of 0.15 V.

2.3.6 Effects of Temperature

For all chemical reactions, including electrochemical reactions, there is a temperature dependence according to the Arrhenius law. The Arrhenius law predicts an exponential dependence of the reaction rate k on temperature:

$$k = A \cdot \exp(-E_A/RT). \tag{2.28}$$

The energy E_A in this equation is called activation energy. For most reactions, the value of the activation energy is such that the reaction rate doubles for every 10 K increase in temperature. The temperature dependence of the reaction rate can therefore be written as follows:

$$k = A \cdot 2^{(T - T_0/10\,\mathrm{K})}, \tag{2.29}$$

where T is the temperature of the reactants and T_0 is the reference temperature, for example, room temperature 293 K.

However, because T appears in the denominator of the exponent of the BV equation and implicitly in the equilibrium voltage E_0, the change in current with temperature under otherwise identical conditions is not trivial; in particular, it is complex at very high temperatures and in the case of coupled electrochemical reactions, that is, the existence of two parallel electrochemical reactions at the same electrode (section 2.5). Figure 2.9 shows the current-voltage characteristic and Tafel curve for charging currents for $\alpha = 0.5$ and $z = 1$ without taking into account a temperature dependence of the equilibrium voltage.

2.3.7 *U-I* Characteristic Curve of Electrochemical Systems

U-I characteristic curves, terminal voltage as a function of the current, are used to characterize electrochemical systems. In the case of batteries, however, it should be noted that the state of charge of the system changes during the current flow, that is, during the measurement. The voltage depends on the initial state of charge and the current direction, as well as on mass transport processes and the change in the state of charge during the measurement. *U-I* characteristic curves are therefore shown less frequently for batteries than for fuel cells, for example. There is no characteristic curve, but a set of curves must be used for characterization. Figure 2.10a shows schematic characteristic curves of a lead-acid battery for two different states of charge, which are used to determine the state of charge during operation.

The shape of the *U-I* curve is characterized by the BV equation and is usually divided into three areas:

a) Range at low currents, often referred to as the polarization range.
   The voltage drops very quickly as the current increases.
b) Table range for normal operating currents, often referred to as the linear range because the total internal resistance of designed systems is characterized not only by the BV equation but also increasingly by the ohmic resistances of the battery's passive components as the current amplitude increases.
c) Mass transport-limited range in which the availability of the reactants is no longer sufficient and the charge transfer resistance increases. This range is only reached in batteries at very high current amplitudes and very low temperatures.

Figure 2.10b shows measurements on a lithium-ion battery at an initial state of charge of 50%. The measurement points represent the voltage at the end of 5 second discharge pulses followed by a 15 minute pause, charging until the amount of charge removed is recharged with a maximum of 15 A and a 15 minute pause, and then discharging again with the next current amplitude. The discharge pulses were interrupted at a cell voltage of 2 V, well below the usual voltage limits. The measurement only shows the Tafel range, as the initial discharge pulse already had too high a current amplitude, and the mass transport-limited range was not yet reached even at a discharge current of 10 C and a voltage of 2 V. Diffusion effects are not pronounced in this cell, at least in normal operation. The dashed curve shows the expected values at higher currents.

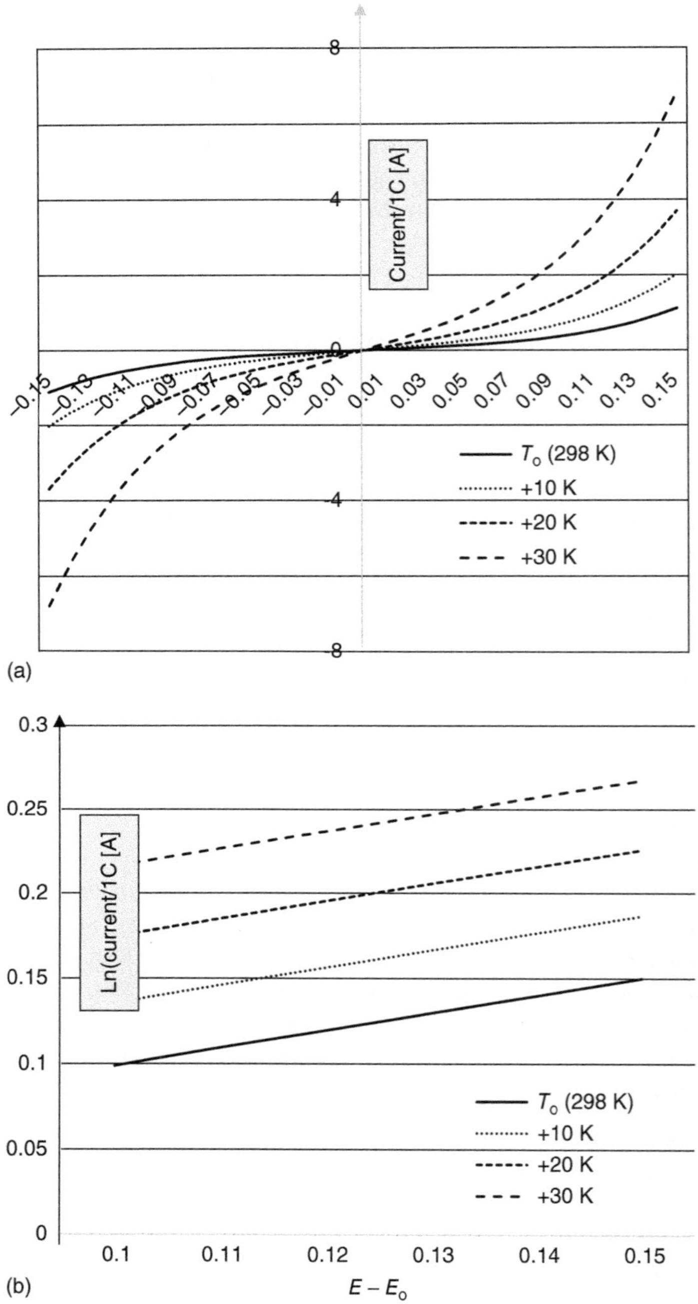

Figure 2.9 (a) Current-voltage characteristics and (b) Tafel curve for the charging direction at different temperatures with $T_o = 298$ K, $\alpha = 0.5$, and $z = 1$, the temperature dependence of the equilibrium voltage was neglected. The current values have been scaled to a one-hour discharge current of a lithium-ion battery in a medium state of charge by selecting the exchange current (exchange current density multiplied by effective surface area) as in Figure 2.4.

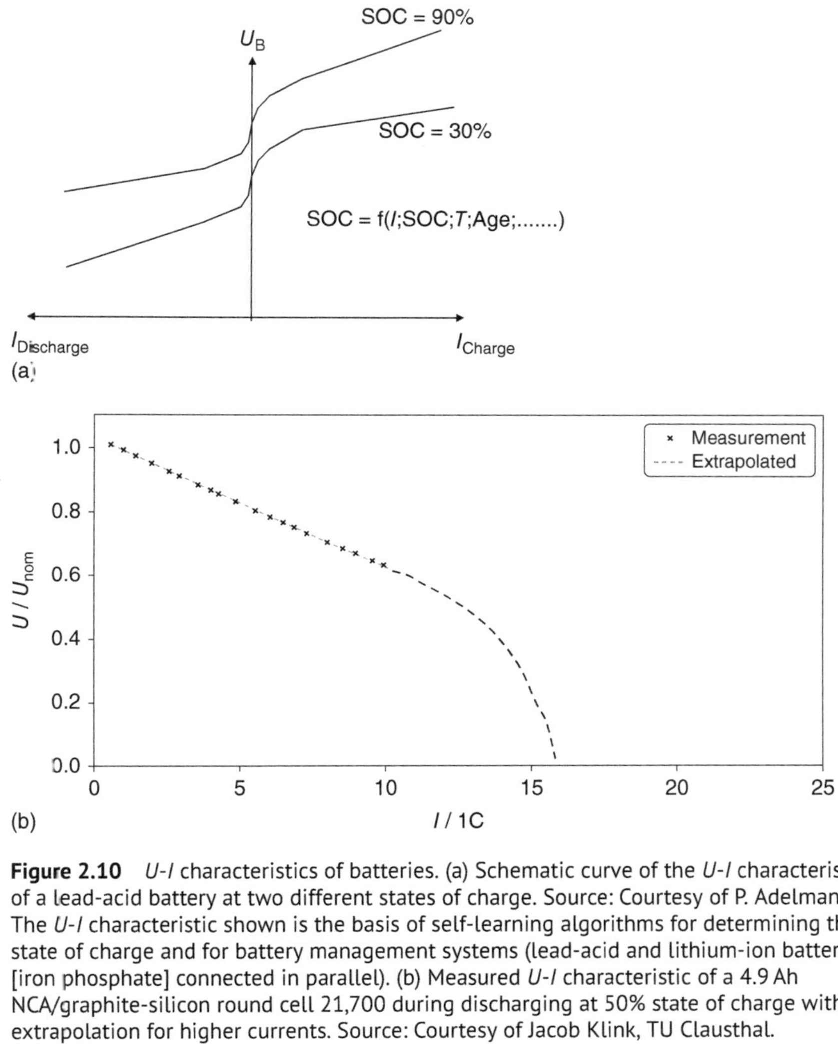

Figure 2.10 *U-I* characteristics of batteries. (a) Schematic curve of the *U-I* characteristics of a lead-acid battery at two different states of charge. Source: Courtesy of P. Adelmann. The *U-I* characteristic shown is the basis of self-learning algorithms for determining the state of charge and for battery management systems (lead-acid and lithium-ion batteries [iron phosphate] connected in parallel). (b) Measured *U-I* characteristic of a 4.9 Ah NCA/graphite-silicon round cell 21,700 during discharging at 50% state of charge with an extrapolation for higher currents. Source: Courtesy of Jacob Klink, TU Clausthal.

2.4 Equivalent Circuit Diagrams

2.4.1 Basics of Equivalent Circuit Diagrams for Electrochemical Systems

Equivalent circuit diagrams are well suited to understanding and modeling the electrical behavior of batteries and are frequently used in this textbook to illustrate properties. In the following, only equivalent circuit diagrams whose components can be directly linked to physical effects are used. The equivalent circuit diagrams used for modeling and the characterization of cells using impedance spectroscopy (see chapter 10 and appendix D) will be introduced later. The symbols used are based on the standard IEC 60617 [12].

The following components are required:

1) Capacitor: In the description of the voltage-forming reaction, it was explained that charge carrier separation takes place at the interface between the electrode and electrolyte. On the anode, this leads to an excess of ions in the electrolyte and free electrons in the electrode, and on the cathode, to a lack of ions in the electrolyte and a lack of electrons in the electrode. This double layer of charge carriers of different polarity is electrically a capacitor and is usually referred to as a double-layer capacitor. The capacitance of such double-layer capacitors [As/V] is large compared to conventional capacitors even for cells with low capacity.
2) Voltage source: The voltage-forming reactions correspond electrically to a voltage source, which always has the thermodynamic equilibrium voltage without current flow in the equilibrium state.
3) BV resistance: The charge transfer overvoltage according to the BV equation, which describes the deviation of the equilibrium voltage from the voltage under current flow, is represented as a resistance (BV resistance). The charge transfer resistance can be understood analogously to the internal resistance of electrical voltage sources.
4) Resistance of passive components: The current-carrying components of a cell (electrolyte, active masses, and current collectors) represent a resistance to the current, which can be calculated using Ohm's law and is independent of the current amplitude.
5) Inductance: All electrical circuits and systems exhibit inductive behavior at very high rates of current change and at very high frequencies. Inductance is neglected in this chapter due to its very low value and is only considered in the discussion of impedance spectroscopy, applications with very high power changes, and loads with high currents at high frequency.

In the following, effects due to diffusion and different electrochemical conditions in different areas of the electrode are not represented in the equivalent circuit diagram. More comprehensive equivalent circuit diagrams, adapted for the discussion of specific issues, are introduced in the following chapters and sections. In this chapter, only the Nernst voltage, the charge transfer overvoltage or the BV resistance, the double-layer capacitor, and the ohmic resistance of the various components are included in the equivalent circuit diagrams.

2.4.2 Basic Equivalent Circuit Diagrams of an Electrode and a Cell

Figure 2.11 shows equivalent circuit diagrams of an electrode (a) and (b) as well as an entire cell (c). The equivalent circuit diagrams (a) and (b) are two different but electrically equivalent representations of the electrode and are electrically connected to the electrolyte and the electrode. Equivalent circuit diagram (c) is a complete equivalent circuit diagram of a cell, with the exception of the inductance, in which the two resistors R contain all electron-conducting passive components, including the electron-conducting material of the electrodes, and the electrolyte resistor $R_{Electrolyte}$ is shown separately.

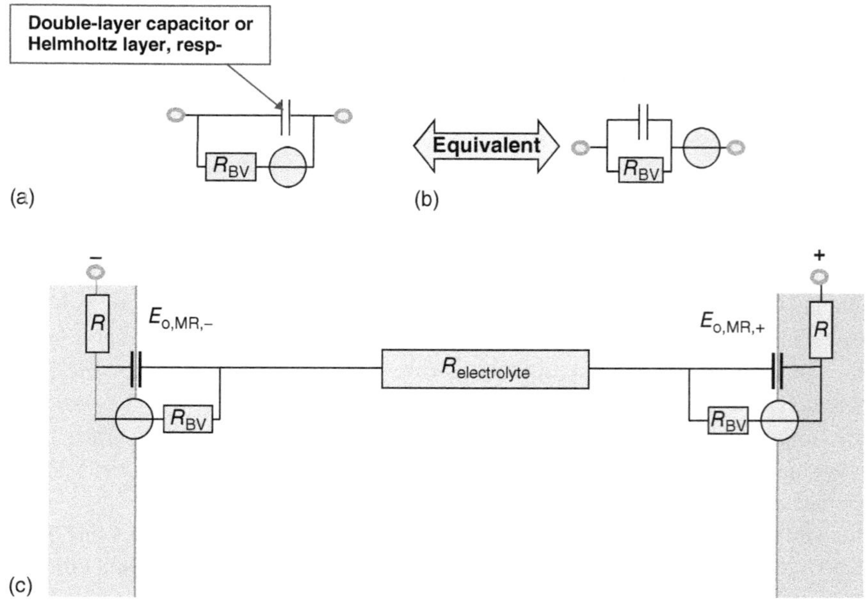

Figure 2.11 Equivalent circuit diagrams showing the voltage-forming reaction of an electrode in two different ways (a) and (b) and of both electrodes of a cell (c). The shaded areas indicate the position of the electrodes. The double-layer capacitor is located at the electrode/electrolyte interface. $E_{o,\,MR+}$ denotes the equilibrium potential of the voltage-forming main reaction of the positive electrode and, $E_{o,\,MR,\,-}$ of the negative electrode. R_{BV} is the BV resistance, and R the sum of the resistances of all passive, electrically conductive components of a cell on the positive and negative electrodes.

The equivalent circuit diagram (b) is usually used to represent the electrode. For the following reasons, the electrically equivalent circuit diagram (a) is always preferred in this book, except for the representation of impedance spectroscopy (see chapter 10 and appendix D).

1) When there is no current flow, the capacitor in the diagram (b) discharges via the resistor connected in parallel. However, the voltage-forming reaction leads to the formation of a voltage in the double-layer capacitor even without current flow. This behavior is shown in (a); the voltage source always keeps the double-layer capacitor at the equilibrium voltage of the reaction when there is no current.
2) When considering multiple reactions occurring in parallel at the electrode/electrolyte interface, for example, side reactions (see section 2.5), the equivalent circuit diagram (a) is easier to use and, in particular, allows the calculation of mixed potentials more easily (see section 2.5).

Equivalent circuit diagram (c) is used for a complete view of the cell, which can be simplified with regard to the ohmic resistors because the three ohmic resistors shown can be combined in series to form a resistor element. Due to the different dependencies on the state of charge and temperature, the three resistors are often shown separately in this book. The two electrode reactions are shown correctly provided there are no side reactions. Combining the RC (resistor R and capacitor C connected in parallel) elements of both electrodes into one RC element is a frequently made simplification, but it is no longer electrically equivalent to the representation (c).

2.4.3 Equivalent Circuit Diagram with Constant Current

Under constant current conditions, the capacitors representing the electrochemical double layers of the two electrodes do not contribute to the battery current as long as slow changes in the Nernst voltage and the charge transfer voltage, which result from changes in the availability of reactants, do not have to be taken into account. In the quasi-stationary range, the equivalent circuit diagram can thus be considerably simplified because all resistors can be combined into a single resistance and both voltage sources can be combined into a single voltage source (Figure 2.12). This simplified equivalent circuit diagram is fully equivalent.

This simplification can also be used for cells with side reactions, however, it is then no longer electrotechnically equivalent, but only a simplified equivalent circuit diagram which may be sufficient for many discussions.

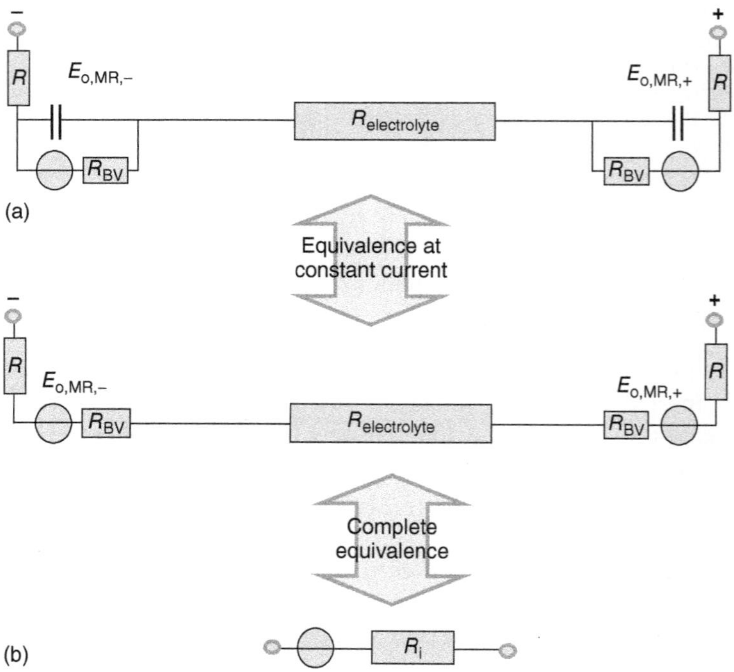

Figure 2.12 Equivalent circuit diagram at constant current. (a) No current flows through the double-layer capacitor at constant current. (b) In the equivalent circuit diagram, both voltage sources are combined into one voltage source and all resistors are combined into one internal resistance R_i.

2.5 Side Reactions

In the discussion so far of the theoretical principles and the equivalent circuit diagram, it was assumed that only one reaction can take place between the ions of the electrolyte and the materials of the electrode, the main voltage-forming reaction, which leads to a discharge when a load is connected to the poles and to a recharge when a charger is connected to the discharged electrodes.

In addition to the main reaction, additional reactions, usually referred to as side reactions, may exist and are introduced below. These occur in most battery systems and play an important role in operation. The previous explanations of the equilibrium voltage and the charge transfer voltage apply to all electrochemical reactions and therefore also to all side reactions. Depending on the potential of the electrode, side reactions take place at a more or less high reaction rate, which can be calculated using the BV equation. Side reactions are characterized by very low exchange currents or a very high BV resistance compared to the main reaction at the voltage level determined by the main reaction. The term "kinetically inhibited reaction" is also used to describe these conditions.

In lithium-ion batteries, side reactions usually represent a safety and service life problem and are often quite complex. With lead-acid batteries, on the other hand, they are very well understood, so the following explanations are based on lead-acid batteries. Side reactions in lead-acid batteries facilitate and even simplify operation under certain circumstances.

The reactions at the negative Pb electrode are:

$$Pb + HSO_4^- \rightleftarrows PbSO_4 + H^+ + 2e^- \quad \text{(anodic main reaction)}, \tag{2.30}$$

with a potential of approx. −0.4 V with respect to the electrolyte, depending on the activities of the reactants involved. The most important side reaction is:

$$2H^+ + 2e^- \rightarrow H_2 \quad \text{(side reaction)}, \tag{2.31}$$

with a potential of 0 V relative to the electrolyte. This reaction always takes place at the negative electrode due to the availability of protons.

The two reactions (2.29) and (2.30) together lead to a self-discharge reaction of the negative electrode.

$$Pb + HSO_4^- + H^+ \rightarrow PbSO_4 + H_2 \quad \text{(self-discharge reaction of anode)}. \tag{2.32}$$

The main anodic reaction supplies the energy, ions, and electrons so that the side reaction can take place. The two reactions are electrically coupled. A lead electrode is therefore thermodynamically unstable in aqueous sulfuric acid; after a long time, it will convert into lead sulfate with the formation of hydrogen.

The discharge current of the main reaction (conversion of lead into lead sulfate) corresponds to the charge current of the side reaction (formation of hydrogen from protons). The voltage between points A and B is the same for both current paths. Due to the close proximity of the reaction sites, no resistance element needs to be taken into account for the transfer of ions and electrons from the main reaction to the side reaction. By using the equivalent circuit diagram, the current amplitude

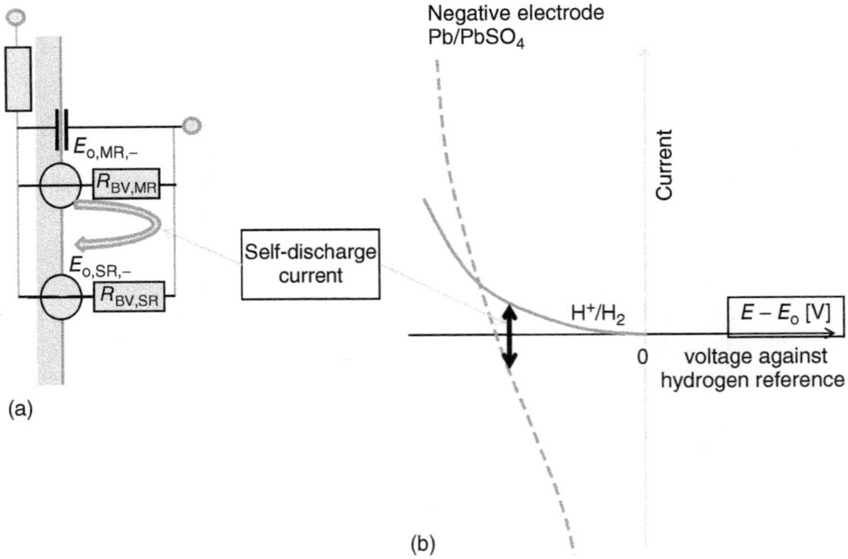

Figure 2.13 Mixed potentials. (a) Equivalent circuit diagram of the negative electrode of a lead-acid battery with the voltage-forming main reaction and the hydrogen-generating side reaction. (b) Current-voltage characteristics of both reactions based on the BV equation. The discharge current for the negative electrode drives the hydrogen evolution at the electrode. No current flows via the battery terminals. The measurement of the electrode voltage against the electrolyte leads to a deviation from the equilibrium voltage.

can be calculated with the help of Kirchhoff's rules if the thermodynamic and kinetic data of both reactions are known. In Task 2.5, it is shown how to calculate the electrode potential as a result of the two coupled reactions. There is a slight shift compared to the equilibrium voltage of the main reaction, which can be measured using a reference electrode or at the terminals of the cell. The new potential is called the mixed potential. The voltage of the cell then no longer corresponds to the equilibrium voltage of the main reaction and is referred to as the OCV.

The formation of the mixed potential can also be shown using the current-voltage characteristics of the two reactions, as shown in Figure 2.13b. The mixed potential is at the point where the discharging current of the main reaction corresponds to the charging current of the side reaction. A quantitative calculation of the mixed potential is therefore also possible.

It should be noted that the difference between the thermodynamic equilibrium voltage and the equilibrium voltage and the mixed potential is negligible for most batteries and most operating and aging conditions. Nevertheless, a distinction should always be made between the terms equilibrium voltage and open-circuit voltage (OCV).

References

1 Atkins, P. and de Paula, J. (2013). *Physical Chemistry*, 5e. Weinheim: Wiley-VCH.
2 Hamann, C. and Vielstich, W. (2005). *Electrochemistry*. Weinheim: Wiley-VCH.

3 Kortüm, G. (1957). *Textbook of Electrochemistry*, 2e. Weinheim: Verlag Chemie.
4 Memming, R. (2015). *Semiconductor Electrochemistry*, 2e. Weinheim: Wiley-VCH.
5 Sabo, L. and Jacob, T. (2014). Fundamentals of electrochemical double layers. *Bunsen Magazine* 16: 260–266.
6 Wenzl, H., Benger, R., and Hauer, I. (2024). *Capacity: Encyclopedia of electrochemical Power Sources*, 2e. Elsevier.
7 Izutsu, K. (2009). *Electrochemistry in Nonaqueous Solutions*. Weinheim: Wiley-VCH.
8 Fueki, K. and Kudo, T. (1990). *Solid State Ionics*. Weinheim: VCH.
9 Maier, J. (2000). *Solids – Defects and Function (Original German title: Festkröperfehler und Funktion)*. Stuttgart: Teubner.
10 Sauer, D.-U. (2000). *Optimization of the Use of Lead-Acid Accumulators in Photovoltaic Hybrid Systems with Special Consideration of Battery Ageing (Original German title: Optimierung des Einsatzes von Blei-Säure-Akkumulatoren in Photovoltaik-Hybrid-Systemen unter spezieller Berücksichtigung der Batteriealterung)*, Dissertation. University of Ulm.
11 Forker, W. (1994). *Electrochemical Kinetics (Original German title: Elektrochemische Kinetik)*. Berlin: Akademie-Verlag.
12 IEC 60617-2/-11:1996 Graphical symbols for diagrams – Part 2: Symbol elements, qualifying symbols and other symbols having general application.
13 Berndt, D. (1993). *Maintenance Free Batteries*. New York: Wiley.

Tasks

Task 2.1

Calculate the theoretical specific capacity of different lithium-ion batteries using the following simplified cell reaction equations (molecular weights can be taken from a periodic table).

a) $LiC_6 + CoO_2 \rightleftarrows C_6 + LiCoO_2$
b) $LiC_6 + NiO_2 \rightleftarrows C_6 + LiNiO_2$
c) $LiC_6 + MnO_2 \rightleftarrows C_6 + LiMnO_2$
d) $LiC_6 + FePO_4 \rightleftarrows C_6 + LiFePO_4$
e) $LiTiO_2 + CoO_2 \rightleftarrows TiO_2 + LiCoO_2$
f) $LiTiO_2 + FePO_4 \rightleftarrows TiO_2 + LiFePO_4$

Task 2.2

Calculate the equilibrium voltage and heating effect of the reactions for the following cell reactions and the single-electrode reactions.

a) Calculation based on the simplified notation of the reactions

$$\text{Negative electrode: } Pb + H_2SO_4 \rightleftarrows PbSO_4 + 2H^+ + 2e^-$$
$$\text{Positive electrode: } PbO_2 + H_2SO_4 + 2H^+ + 2e^- \rightleftarrows PbSO_4 + 2H_2O$$
$$\text{Sum of both reactions: } Pb + PbO_2 + 2H_2SO_4 \rightleftarrows 2PbSO_4 + 2H_2O$$

(Continued)

Task 2.2 (Continued)

b) In aqueous sulfuric acid, the sulfuric acid molecule H_2 SO_4 is almost completely dissociated to form HSO_4^- and H^+, whereby H^+ is present as H_3O^+; the second protolysis stage (i.e. the release of the second proton), on the other hand, is only incomplete. A simplified, chemically more correct notation of the reaction is therefore:

$$\text{Negative electrode: } Pb + HSO_4^- \rightleftarrows PbSO_4 + H^+ + 2e^-$$

$$\text{Positive electrode: } PbO_2 + HSO_4^- + 3H^+ + 2e^- \rightleftarrows PbSO_4 + 2H_2O$$

$$\text{Sum of both reactions: } Pb + PbO_2 + 2HSO_4^- + 2H^+ \rightleftarrows 2PbSO_4 + 2H_2O$$

c) How can the differences be explained?

Required thermodynamic data according to Ref. [13]:

	$\Delta_f H$ [kJ/mol]	$\Delta_f G$ [kJ/mol]
Pb	0	0
PbO_2	−277.4	−217.4
$PbSO_4$	−919.9	−813.2
H^+	0	0
HSO_4^-	−887.3	−755.4
H_2O	−285.8	−237.2
H_2SO_4	−814	−690.1

Task 2.3

Using the information from the table above, calculate the equilibrium voltage and reversible heat of reaction for the formation of water!

$$\text{Oxyhydrogen reaction: } 2H_2 + O_2 \rightleftarrows H_2O$$

$$\text{Reaction at the negative electrode: } 2H^+ + 2e^- \rightleftarrows H_2$$

$$\text{Reaction at the positive electrode: } 2H^+ + \frac{1}{2}O_2 + 2e^- \rightleftarrows H_2O$$

Task 2.4

Calculate the theoretical specific energy of different battery technologies by using the theoretical specific capacities from Task 2.1. Since the thermodynamic equilibrium voltages are not given in the literature or cannot be calculated from thermodynamic data, use the nominal voltage of the cells instead.

What significance do these values have for battery systems? What can they be used for?

Task 2.5

Calculate the errors that are made when using the linear approximation and the Tafel approximation instead of the BV equation.

Assume the following values:

Transfer coefficient:	**0.5**
Exchange current:	1

Number of electrons transferred: a) 1 and b) 2.

Up to what charge transfer voltage can you use the linear approximation and when should you start using the Tafel approximation?

Task 2.6

Calculate the mixed potential of the negative electrode of a lead-acid battery.

Data:

	Transfer coefficient	**Exchange current density i_0 [A/cm²]**	**Equilibrium voltage [V]**
Main reaction $Pb/PbSO_4$	0.5	$0.5 \cdot 10^{-2}$ A/cm²	−0.336
Side reaction H^+/H_2	0.2	$5 \cdot 10^{-9}$ A/cm²	0

Assume 1 m² effective surface area. What is the difference between the equilibrium voltage and the mixed potential that can be measured using a reference electrode?

3

Charging and Discharging Cells and Batteries

In the previous chapter, the electrochemical principles for the behavior of single electrodes were explained. In this chapter, some basic electrochemical parameters of cells are first defined. Next, discharge and charge processes are discussed in general terms, followed by an examination of the behavior of a cell as the sum of the two individual electrodes. The Butler-Volmer (BV) equation again plays an important role (see chapter 2). The voltage differences between cells with slightly different capacities and internal resistances, connected in series and/or in parallel to form modules and complete battery systems, are explained in detail. This is one of the most important issues in battery science.

The most important basic parameters of a battery are the amount of charge it can deliver, its capacity (Ah), and its internal resistance. A distinction must be made between different values for the term "capacity" – the nominal capacity is a value assigned to the battery, while the actual capacity indicates the amount of charge that can be removed, depending on the operating conditions and aging.

Batteries can be charged and discharged using different methods. One very important charging method is the so-called IU or CC/CV method, in which the battery is first charged with a constant current until the end-of-charge voltage is reached, and then with a constant voltage until the current falls below a threshold value. When cathodes and anodes are connected in series in a cell, a change in the current/voltage characteristics of one electrode has a direct effect on the other electrode in the cell during current flow.

According to the BV equation, the voltage of electrodes is characterized by the availability of the reactants. At the end-of-discharge and at the end of charging, the availability of the reactants, and therefore the voltage, changes particularly quickly. Voltage curves during the discharging and charging of electrochemical systems can be derived from the BV equation if the availability of reactants, which changes during discharging and charging, is taken into account in the BV equation.

In electrical terms, a series connection means that the total current flowing through both electrodes of a cell and all cells in series is always the same at all

(Continued)

Battery Technology: Fundamentals of Battery Electrochemistry, Systems and Applications, First Edition. Alexander Börger and Heinz Wenzl.

Companion website: www.wiley.com/go/Börger/Battery_Technology

(Continued)

times. Differences between individual electrodes or cells caused by production variations or aging processes lead to significant voltage differences, despite the same current. This can have catastrophic consequences. A distinction must be made between electrochemical systems with side reactions causing little degradation and having high ampacity (e.g. lead-acid batteries) and electrochemical systems without appropriate side reactions causing significant degradation (e.g. lithium-ion batteries).

For systems with side reactions causing little degradation: the capacity of the battery string can be utilized more or less completely. However, the end-of-discharge voltage of the string is reached earlier due to the reduced voltage of a cell that has a lower capacity due to aging or production differences. The larger the number of cells connected in series, the less the capacity of the battery is reduced. Side reactions during charging make it possible for each cell in a string to be fully charged to the maximum of its respective capacity, without the need for a battery management system. In addition, each electrode in each cell can be fully charged.

For systems with side reactions causing significant degradation: batteries can only be discharged as far as the electrode with the lowest capacity in any cell allows. The capacity of all other cells remains unused. Further discharge driven by other cells in the battery leads to destruction and possibly external damage (fire, explosion). The same applies when charging. Cell balancing systems are often used so that all cells can be fully charged, and the charging process is not limited by the weakest cell. Appropriate safety precautions must therefore be taken for all such electrochemical systems, that is, in particular all lithium-ion batteries.

3.1 Definitions of Capacity and Internal Resistance

3.1.1 Capacity

In battery technology, the amount of charge stored in a battery is called capacity, with the unit Ah (ampere hour).

However, there are numerous different terms to be distinguished when using the term "capacity," and there is a sometimes large dependency on the discharge current and the temperature:

a) **Nominal capacity** C_n, assigned to a cell by the battery manufacturer.
For lithium-ion batteries, the value is usually based on a discharge duration of 1 h at room temperature (20 or 25 °C). For lead-acid batteries, the nominal capacity is based on different discharge durations at nominal temperatures, depending on the application, because the capacity depends significantly on the discharge current. The discharge duration n in hours then provides more realistic information on the capacity of the cell to be expected, for example, C_{20} as the nominal

capacity of batteries for road vehicles, which are required to provide lighting for a parked car overnight. The end-of-discharge voltage must also be specified.

The nominal discharge current is related to the nominal capacity. For lithium-ion batteries, the nominal discharge current usually corresponds to "1 C_1/1 h," that is, a fully charged cell with a nominal capacity of 25 Ah will be discharged in one hour with 25 A. This current is the nominal discharge current "1 *C*" often referred to as the 1C-rate. Significantly lower discharge currents are common for lead-acid batteries. The following nomenclature is used: I_n is calculated as C_n/n A, that is the nominal current of a lead-acid battery with a nominal capacity C_{20} of 60 Ah is 60 Ah/20 h = 3 A.

b) **Actual capacity** C_{act}. To determine the actual capacity, that is, the real capacity of a cell, the battery must first be fully charged under the conditions specified by the battery manufacturer. The battery is then discharged under conditions specified by the manufacturer; in particular, the nominal discharge current I_{n} assigned to the battery, the nominal temperature, and, if necessary, other ambient conditions (e.g. climatic chamber, often also a water bath for lead-acid batteries in the laboratory). The final discharge voltage must also be observed. If the actual capacity is to be determined as an average value from several individual measurements, the resting time before the next full charge and the subsequent charging process must also be specified.
c) The **initial capacity** C_0 is the actual capacity of a new battery. It is often required by law or standards that it must exceed a percentage of the nominal capacity or that it must be reached at least once when the actual capacity is determined three times.
d) The **residual capacity** C_{r} is the amount of charge that can still be discharged from a battery under defined conditions at any time during operation. The discharge conditions must be specified in each case. The usual current and temperature conditions are either the nominal conditions or the relevant operating conditions, for example, the current of the connected loads. As the residual capacity changes with the battery condition, it can only be determined once.
e) The **practical capacity** C_{p} or **operating capacity** is the capacity of a battery that can be drawn from the battery under application-specific conditions (full charge, current, temperature, and end-of-discharge voltage). This term is useful for applications where there are, for example, unfavorable voltage limits for chargers and loads, insufficient charging time and charging energy, and discharge conditions that may differ greatly from those required under nominal conditions.

The ratio of residual capacity under nominal discharge conditions to actual capacity is the only way to experimentally determine the state of charge (*SoC*):

$$SoC = \frac{C_{\text{r}}}{C_{\text{act}}}. \tag{3.1}$$

The state of charge can also be based on the nominal or initial capacity. In all cases, states of charge above 100% or below 0% can result from Eq. 3.1. The term "state of charge" is explained in detail in chapter 7 (see also Ref. [1]).

The difference between two states of charge is also known as *DoD* (depth of discharge), particularly during cycling.

The ratio of actual capacity to initial capacity is referred to as state of health (SoH) in relation to the decrease in capacity:

$$SoH = \frac{C_{\text{act}}}{C_0}. \tag{3.2}$$

However, the state of health can also be based on the nominal capacity. There are also definitions (albeit rarely used) where the state of health refers to the development of the internal resistance, in addition to the change in capacity – see chapter 7 for a more precise definition.

In all applications, the amount of energy delivered by a battery is more important than the amount of charge. The dominance of the term "capacity" rather than "energy" has historical reasons. Terms such as "nominal energy," "residual energy," etc. are more meaningful but are not used in battery science.

3.1.2 Internal Resistance

The internal resistance of a battery depends significantly on the state of charge, temperature, aging condition, and the history of the battery. It can be determined using short high-current pulses or AC measurement methods. The internal resistance R_i is calculated using the definition of differential resistance[1] when using high-current pulses:

$$R_i = \frac{U_0 - U_1}{I_0 - I_1}. \tag{3.3}$$

From the discussions of the BV equation and mass transport phenomena in chapter 2, it is immediately obvious that the value of internal resistance depends on the current amplitude and the duration of the high-current pulse. For this reason, a distinction is made in this book between the internal resistance defined as differential resistance and the resistance of the cell components, which can be calculated by applying Ohm's law.

3.2 Terms Used for Charging and Discharging Batteries

Discharging and charging of batteries are carried out using different procedures, which are, however, based on a few general types. The resulting current and voltage curves are due to the properties of the BV equation in section 3.4 and are essentially similar for all battery systems.

3.2.1 Discharging

In practice, batteries are discharged with very different current profiles depending on the loads or application-specific requirements, whereas inverters and loads with inductive or capacitive properties may cause frequent short charging pulses during discharging. Three basic types can be distinguished in the laboratory:

1 Note the sign convention in the equation: discharge currents are given a negative sign.

a) discharge with constant current, corresponding to a constant reaction rate;
b) discharge with constant resistance; and
c) discharge with constant power.

The current and voltage curves for such discharges are shown in Figure 3.1. Discharge curves with constant current and constant power are often provided in data sheets.

As can be seen from Figure 3.1, the voltage drops more quickly during a constant current discharge than in the case of a discharge with constant resistance, because the discharge current decreases when keeping the ratio U/I constant, and thus the voltage drops more slowly. For a constant power discharge, on the other hand, the voltage decrease must be compensated for by a higher discharge current since the product $U \cdot I$ must be kept constant.

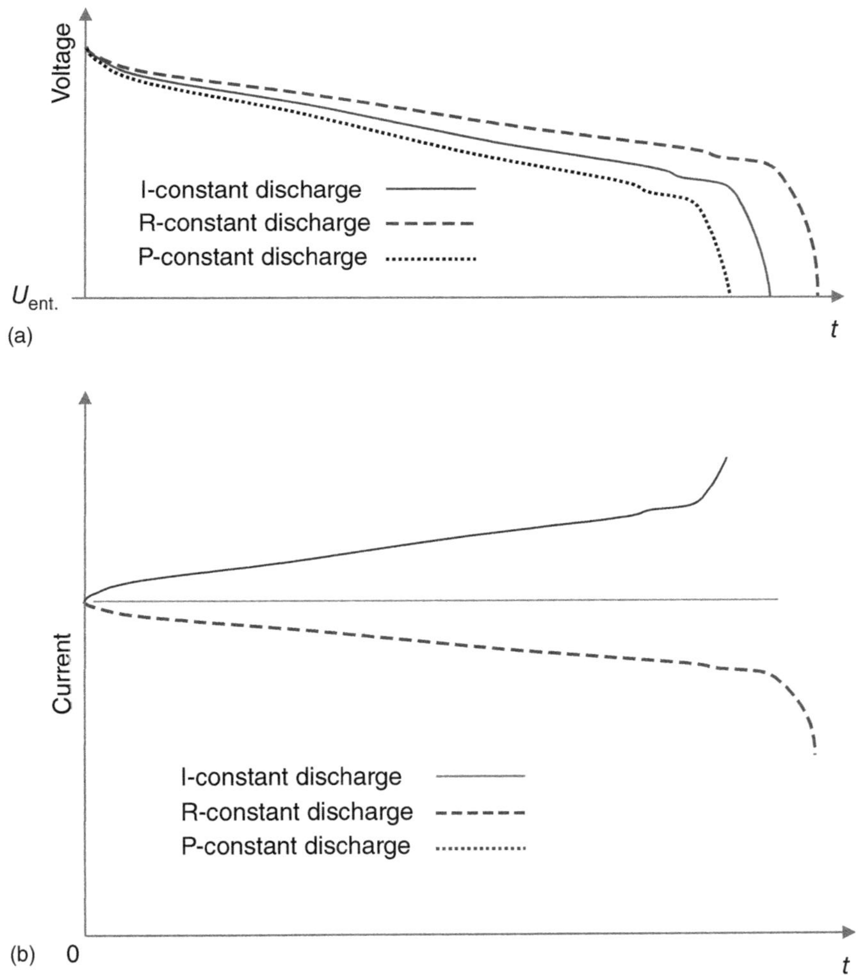

Figure 3.1 (a) Voltage and (b) current curves for different types of discharge.

All discharges end at the so-called end-of-discharge voltage, but the time at which the end-of-discharge voltage is reached varies. The end-of-discharge voltage depends on the battery type and is specified by the manufacturer. In general, however, the end-of-discharge voltage is significantly higher than 0 V, as very low cell voltages (i.e. deep discharges) cause side reactions with high degradation.

3.2.2 Charging

While discharging is impossible to standardize in battery practice (except in the laboratory), charging is, in many applications, a distinct phase and can be controlled precisely. Charging methods are ultimately core know-how in battery development. The faster and more gently a battery can be fully recharged, the more useful the charging method is, and the more widely it can be used.

As with discharging, there are several basic types of charging methods:

d) charge with constant current,
e) charge with constant voltage, and
f) charge with constant power.

An overview of these basic procedures, which are described in various standards, is provided in Table 3.1. Notation 1 is specified in DIN 41772 and explained in greater detail in Charger Assignments for Traction Batteries, 2004-04, leaflet no. 11 www.zvei.org.

In this book, we will use Notation 1 for lead-acid batteries, as this seems to be widely used, whereas for lithium-ion batteries, we will use Notation 2, as the dominant charging characteristic to date is CC/CV, and other methods are not common.

An IUa characteristic means that a battery is initially charged with a constant current until a switchover point is reached, and then continues to be charged with a constant voltage until a switch-off criterion (e.g. a time or charging current) is reached. For lithium-ion batteries, Notation 2 is usually used, so a CC/CV process corresponds to an IU process.

The current-voltage curves for some typical charging methods are shown in Figure 3.2.

In practice, however, in addition to these basic types, there are also charging methods that use comparatively high currents and pulses, known as pulse charging,

Table 3.1 Charging characteristics in different notations.

Notation 1	Notation 2	Description
I (I-charge)	CC (constant current)	Constant current charging
U (U-charge)	CV (constant voltage)	Constant voltage charging
W (W-charge)	tp (taper charge)	Charging with decreasing current when the voltage increases (analogous to transformers or constant power)
0	[no equivalent term]	Switching to another charging parameter within the same charging mode (e.g. IU0U)
a	[no equivalent term]	End of charging after reaching an end-of-charging criterion

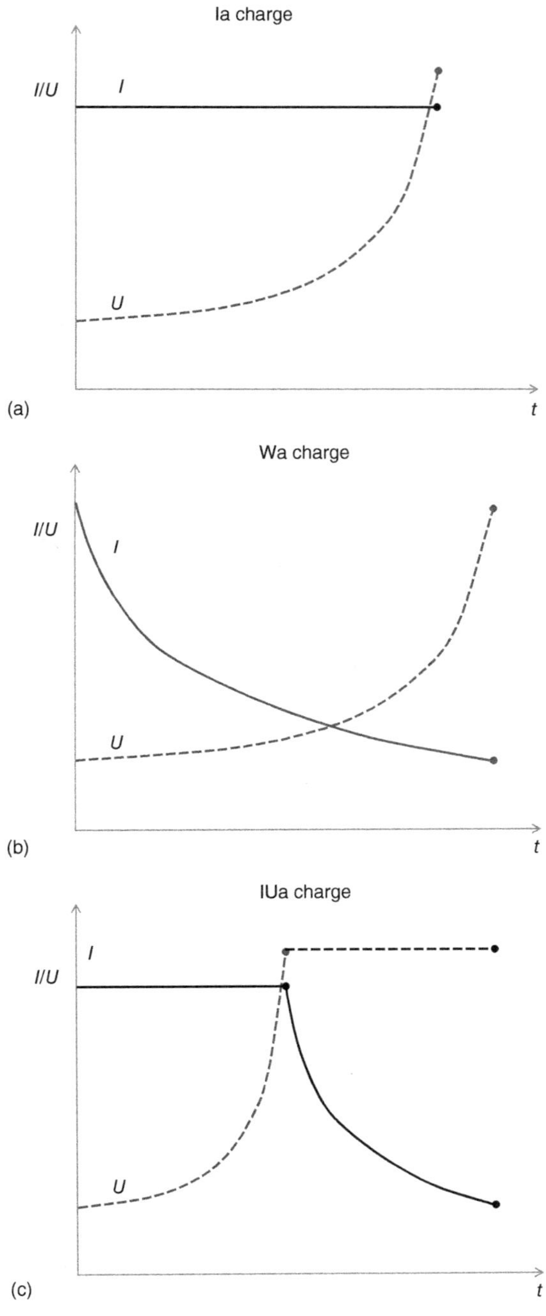

Figure 3.2 Current-voltage curves for important charging methods: (a) Ia charging, (b) Wa charging, (c) IUa charging.

where the individual pulse sequence is battery (manufacturer) specific. An example is shown in Figure 3.3.

The end-of-charge criteria for a charging method can vary greatly. A voltage limit or the reaching of a minimum charging current (e.g. 1/20 of the nominal current for lithium-ion batteries) is often specified, but temperatures, temperature derivatives after time (dT/dt), or the second voltage derivative after time (d^2U/dt^2) are also used,

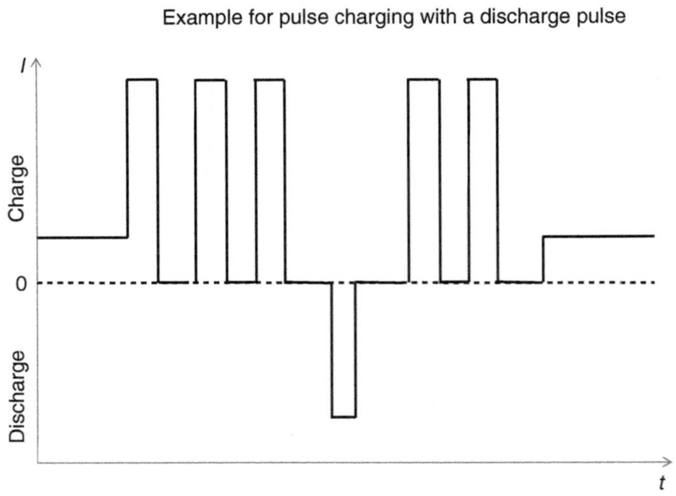

Figure 3.3 Schematic diagram of a pulse charging method in which a discharge pulse is integrated into the actual charging pulse.

especially for alkaline-nickel batteries. Whether a battery is actually fully charged when the charging process is terminated depends on the criterion used, the age and temperature of the battery, and must be checked carefully in case of doubt.

3.2.3 Charge Factor and Efficiency

The ratio of the amount of charge stored during charging (integral over the charging current I_{cha} for a whole cycle) to the amount of charge that can be removed during discharging (integral over the discharging current I_{dis} for a whole cycle) is called the charge factor CF. The charge factor is always greater than 1 (mainly due to side reactions). The reciprocal of the charge factor is the Coulomb efficiency or charging efficiency η_{C} of the charging process, which is always less than 1.

$$CF = \frac{1}{\eta_C} = \frac{\int I_{\mathrm{Cha}} dt}{\int I_{\mathrm{Dis}} dt} \tag{3.4}$$

For lithium-ion batteries, both values are relatively close to 1 (1.005 and 0.995, respectively), showing that only a very small proportion of the battery's charging current flows via side reactions. With lead-acid batteries and other technologies, the values are usually significantly different from 1 due to side reactions, with values for new lead-acid batteries ranging from approx. 1.05 to approx. 1.20 depending on the type, corresponding to approx. 95–83% charging efficiency.

In addition to the charge factor, which is important for electrochemical considerations, the (energy) efficiency is important for cost considerations and designs.

$$\eta_{\mathrm{e}} = \frac{\int U_{\mathrm{ela}} I_{\mathrm{ela}} \mathrm{d}t}{\int U_{\mathrm{Cha}} I_{\mathrm{Cha}} \mathrm{d}t} \tag{3.5}$$

Energy efficiency also takes into account the different voltages of the battery during discharging and charging, which depend on many factors. A distinction must be made between the efficiency of the battery with and without a charger, and with and without system components that are required for the safe and reliable operation of the battery. The ratio of energy efficiency to Coulombic efficiency is called voltage efficiency, a seldom-used term:

$$\eta_V = \frac{\eta_e}{\eta_C} \tag{3.6}$$

3.3 Discharging and Charging the Electrodes of a Cell

In the following, the charging and discharging processes at the level of the individual electrodes and thus at a more fundamental electrochemical level are discussed.

3.3.1 Significance of the BV Equation for Changes in Current and Voltage

When discharging electrodes, charged active material is converted, and discharged material is formed. When charging, the discharged material is converted back.[2] For all electrodes, the availability of the required reactants changes significantly at the end of discharging or charging. At the end of discharging, this leads to a rapid voltage drop, and when charging with a constant current, to a rapidly increasing charging voltage. Mathematically, the behavior can be described using the BV equation.

Quasi-stationary states are always assumed for the following discussion. At the end of charging or discharging and at very high currents, the change in concentration of the reactants in the electrode is rapid and must be taken into account. In particular, there is virtually no discharged material present at the end of charging and no charged material at the end of discharging. By applying the BV equation, it is shown below that this leads to a fast voltage drop at a constant discharge current and to a rapidly increasing charging voltage at a constant charging current.

Due to the dominance of one current direction, the BV equation can be written for the following discussion using the Tafel approximation, that is, only one exponential term of the function is considered, and the other can be neglected. The BV equation is modified by adding a time-dependent term for the rapid changes in concentration at the end of charging or discharging. This corresponds to the changeover time of a sequence of quasi-stationary states.

At constant current, the concentration of the reactants $c(t)$ decreases both during discharging and charging according to the following formula:

$$c(t) = c_0 - \frac{It}{zFV}, \tag{3.7}$$

---

2 In fuel cells, redox flow systems, and air cathodes of metal-air batteries, charged material is constantly added to the reaction area, while discharged material is removed. Consequently, the terms "charging" and "discharging" therefore have a different meaning for these systems with regard to the availability of the electrochemically active material.

with c_0 as the initial concentration of the reactants. This equation can also be used for active materials that are present as solids at the electrode-electrolyte interface.[3] However, it does not take into account that the electrochemically available reactants can also be reduced by mass transport restrictions and that the effective surface area on which reactions take place does not necessarily have to be directly proportional to the available active material. Mass transport restrictions are technically relevant for fast charging processes. At low states of charge, there are no restrictions on the amplitude of charging currents originating from electrochemical processes. At high states of charge, the restrictions are primarily electrochemical in nature: solubility of lead sulfate and surface area of lead sulfate crystals for lead-acid batteries, and diffusion rate of lithium ions during the intercalation processes in active mass particles on both the cathode and anode (see chapters 12 and 13 for details). As mass transport phenomena cannot be accelerated by applying a higher charging voltage, the conversion, respectively intercalation rate, of active material slows down. The strong dependence of mass transport on temperature leads to further restrictions at low temperatures. These issues are not discussed further in the following.

The modified BV equation with constant current and only one reactant during charging or discharging is, for the charging direction:

$$I_{\text{Cha}} = zFk_{\text{Cha}} \cdot \left(c_{0,\text{Disch}} - \frac{I_{\text{Cha}} t}{zFV} \right) \cdot e^{\frac{\alpha RT(E-E_0)}{zF}} \tag{3.8}$$

and, for the discharging direction:

$$I_{\text{Disch}} = zFk_{\text{Disch}} \cdot \left(c_{0,\text{Cha}} - \frac{I_{\text{Disch}} t}{zFV} \right) \cdot e^{-\frac{(1-\alpha)RT(E-E_0)}{zF}}. \tag{3.9}$$

k_{Cha} and k_{Disch} are the reaction constants in the charging and discharging directions, $c_{0,\text{ Disch}}$ and $c_{0,\text{ Cha}}$ are the concentrations of the reactants.

At a constant current, the concentration becomes smaller and smaller in proportion to time and the deviation of the electrode voltage from the equilibrium voltage must then become greater and greater. The electrode voltage therefore drops quickly at the end of discharge and increases steeply at the end of charging. This means that the BV resistance increases during constant current flow and reaches very high values at the end of discharge or charge. When the current direction changes at the end of charge or discharge, the BV resistance is very low because sufficiently high concentrations of reactants have been formed beforehand. The BV resistance therefore has a very different value depending on the current direction, particularly at the end of charging and discharging – very low in the discharging direction and very high in the charging direction when fully charged. This behavior of the modified BV equations (3.2) and (3.3) is not taken into account in the BV equation shown in BV equation introduced in section 2.3.

---

3 Since the concentration of a solid is of course constant, the mass or amount of active material must be used in the equation.

3.3.2 Discharging and Charging with Constant Current

The rate of change of the concentration of the reactants $\mathrm{d}c(t)/\mathrm{d}t$ is constant at constant current, however, the ratio between the amount of the available active material in relation to the amount being consumed during current flow decreases faster and faster. Everything else being equal (all parameters of the BV equation are considered to be constant for the period under consideration during discharging or charging, no changes in the electrical conductivity of other components, no change in the equilibrium voltage E_0), the difference between the electrode voltage and the equilibrium voltage must increase at constant current faster and faster. The voltage difference in relation to the equilibrium voltage of the electrode changes the same way for all electrochemical systems at the end of charging or discharging. Solving Eq. (3.9) to $E - E_0$ leads directly to

$$E - E_0 = 1/(-(1-\alpha)zF/RT) \cdot \ln(I_{\mathrm{Disch}}/(k_{\mathrm{Disch}} \cdot c_{0,\mathrm{Cha}} \cdot (1 - (I_{\mathrm{Disch}} \cdot t)/zFV))). \tag{3.10}$$

The same considerations apply to charging

$$E - E_0 = 1/(\alpha zF/RT) \cdot \ln(I_{\mathrm{Cha}}/(k_{\mathrm{Cha}} \cdot c_{0,\mathrm{Disch}} \cdot (1 - (I_{\mathrm{Cha}} \cdot t)/zFV))). \tag{3.11}$$

If all constants or terms assumed to be time invariant are summarized, the result for the charging direction is

$$E - E_0 = A \cdot \ln(B/(C \cdot (1 - Dt))), \tag{3.12}$$

with $A = RT/(\alpha zF)$, $B = I_{\mathrm{Cha}}$, $C = k_{\mathrm{Cha}} \cdot c_{0,\,\mathrm{Disch}}$, $D = I_{\mathrm{Cha}}/zFV$.

The same applies to Eq. (3.11). The general form of the function $\ln(1/(1-t))$ is shown in Figure 3.4: Figure 3.4a for discharging and Figure 3.4b for charging. At time $Dt = 1$, there is no more charged or discharged material, the BV equation no longer applies respectively Eq. (3.12) then no longer has a solution.

Figure 3.5 shows the measured voltage curves of lead-acid and lithium-ion batteries in which the discharge time was scaled to the same value, and the voltage was normalized to the nominal voltage (a) and to the voltage difference between the float charge voltage and the end-of-discharge voltage (b).

3.3.3 Charging with Constant Voltage

To avoid impermissibly high voltages at the end of charging, the charging current is reduced at a certain voltage threshold so that the charging voltage remains constant. The change in current over time $\mathrm{d}I/\mathrm{d}t$ results from Eq. (3.8) to:

$$\mathrm{d}I_{\mathrm{Cha}}/\mathrm{d}t = \mathrm{d}((k_{\mathrm{Cha}} \cdot c_{0,\mathrm{Disch}} \cdot (1 - (I_{\mathrm{Cha}} \cdot t)/zFV)) \cdot \exp((\alpha zF/RT) \cdot (E - E_0)))/\mathrm{d}t. \tag{3.13}$$

If all constants are summarized as K, including the charge transfer overvoltage $E - E_0$ which is equivalent to assuming a constant electrode voltage when the equilibrium voltage E_0 is constant or changes only slowly over time:

$$\mathrm{d}I_{\mathrm{Cha}}/\mathrm{d}t = -K \cdot I_{\mathrm{Cha}}. \tag{3.14}$$

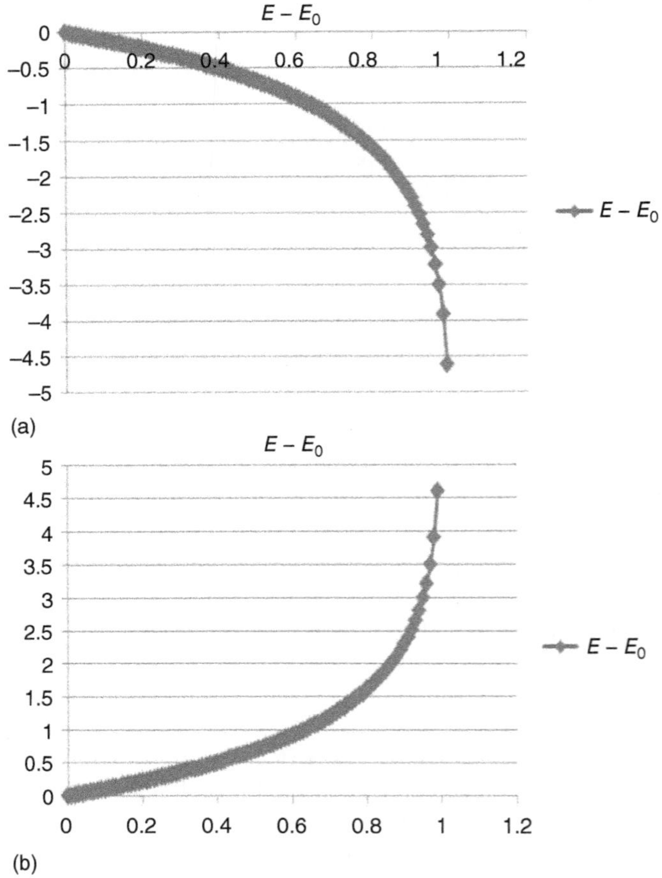

Figure 3.4 Representation of the function $\ln(1/(1-t))$ for the range of values $0 < t < 0.99$: (a) discharging, (b) charging. The two curves are mirrored in the charging and discharging directions due to the different signs of the exponents of the exponential function.

By integrating

$$I_{\mathrm{Cha}}(t) = I_{\mathrm{Cha,o}} \exp(-Kt), \tag{3.15}$$

with $I_{\mathrm{Cha,\,o}}$ as the initial value of the integration corresponding to the constant charging current at the point in time at which the voltage is kept constant.

This exponential decrease of the charging current when the charging voltage is limited is also characteristic of all electrochemical cells if the voltage is kept constant during charging. The exponential drop in current only applies exactly at the beginning of the current reduction, as long as all other effects can be regarded as time invariant, but represents a good approximation of the observed current curves.

The explanations in this section show that for a CC/CV charging characteristics, the values for $\mathrm{d}U/\mathrm{d}t$ at the end of the constant current charging phase and $\mathrm{d}I/\mathrm{d}t$ at the beginning of the constant voltage charging phase are a function of the concentration of discharged active material and thus the state of charge. The analysis of the

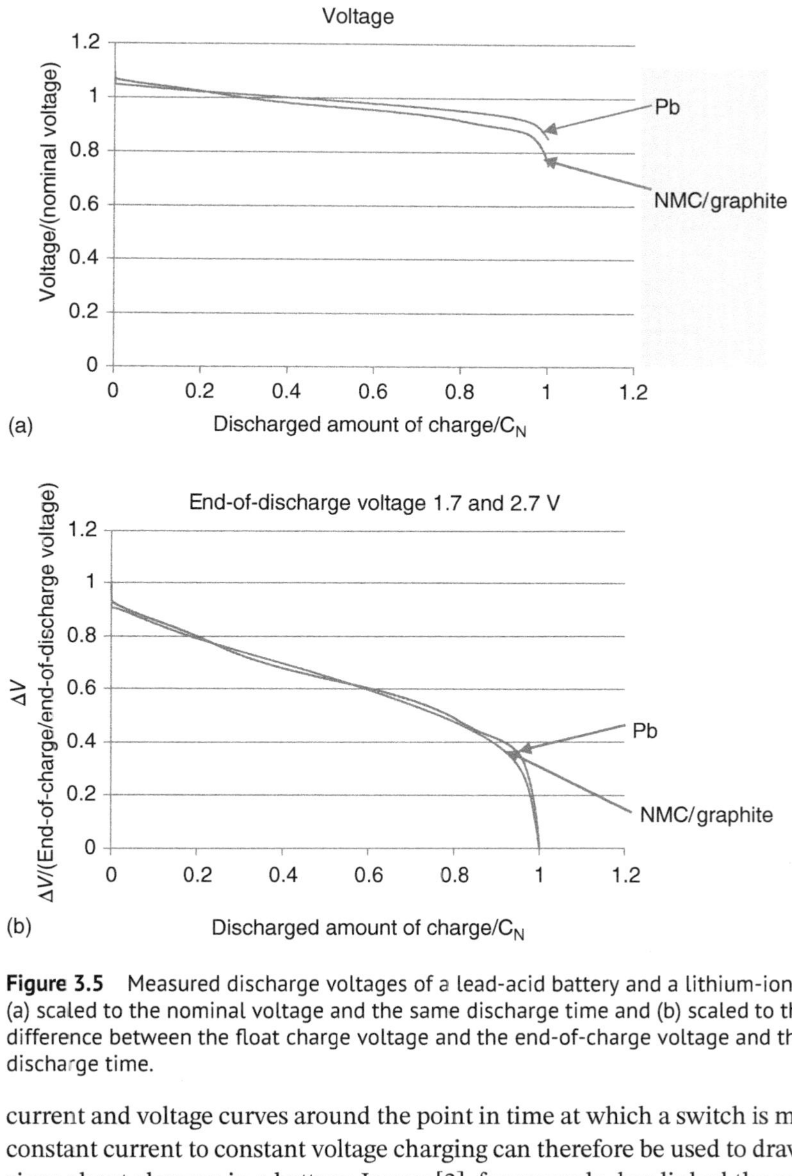

Figure 3.5 Measured discharge voltages of a lead-acid battery and a lithium-ion battery: (a) scaled to the nominal voltage and the same discharge time and (b) scaled to the voltage difference between the float charge voltage and the end-of-charge voltage and the same discharge time.

current and voltage curves around the point in time at which a switch is made from constant current to constant voltage charging can therefore be used to draw conclusions about changes in a battery. Jossen [2], for example, has linked the gradient of the current change immediately after the start of a constant voltage charge with the state of charge of the battery at this point in time.

3.3.4 Current and Voltage Curve of Batteries

Figure 3.6 shows a discharge followed by a charge with a CC/CV characteristic curve for a lead-acid battery. For better understanding, the voltage is shown in relation to the nominal voltage and the discharge and charge currents are converted to nominal currents.

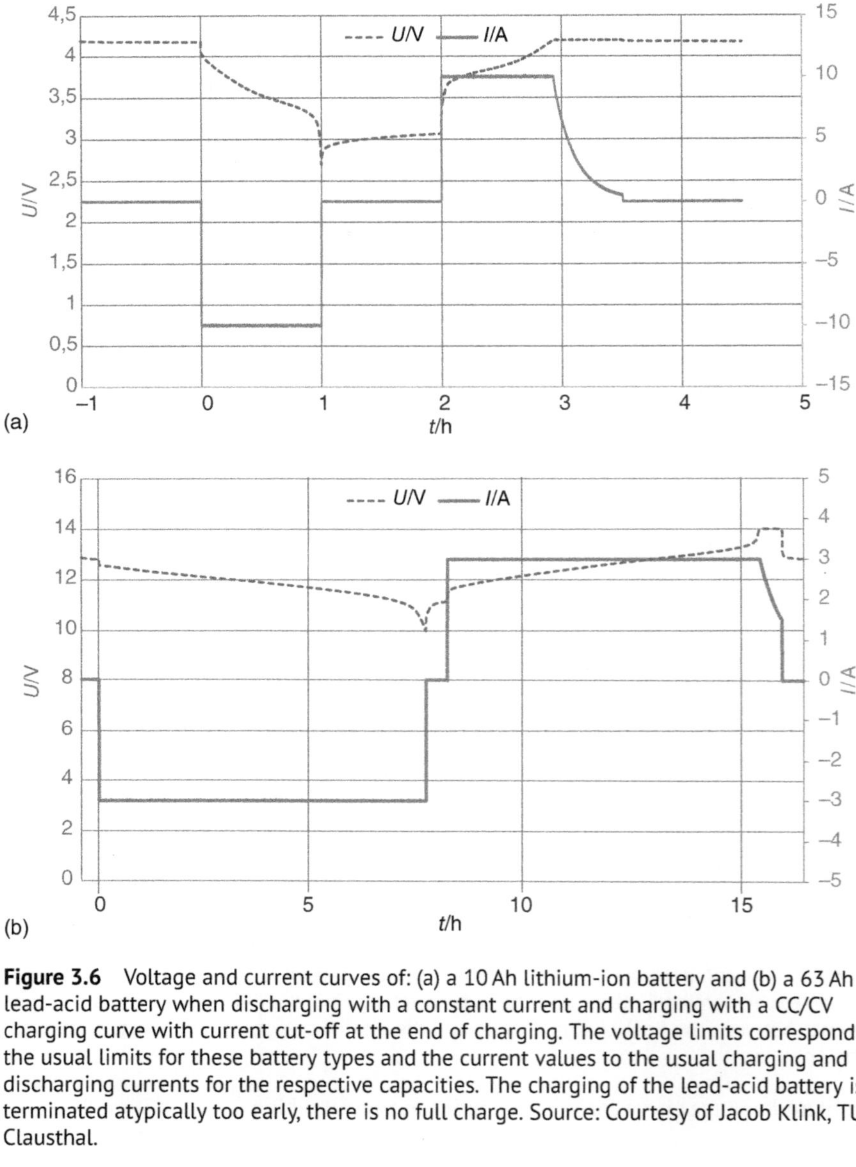

Figure 3.6 Voltage and current curves of: (a) a 10 Ah lithium-ion battery and (b) a 63 Ah lead-acid battery when discharging with a constant current and charging with a CC/CV charging curve with current cut-off at the end of charging. The voltage limits correspond to the usual limits for these battery types and the current values to the usual charging and discharging currents for the respective capacities. The charging of the lead-acid battery is terminated atypically too early, there is no full charge. Source: Courtesy of Jacob Klink, TU Clausthal.

A comparison of the measurement curves shown in Figure 3.5 with the curves for the electrode voltage curve shown in Figure 3.4 resulting from the BV equation when reactants are depleted, and the exponential drop in current at constant electrode voltage (see Eq. 3.15) clearly shows that the general form of the discharge and charge characteristics results from the BV equation alone – despite the considerable simplifications that have been made: constant equilibrium voltage and no change in electrical conductivities due to the discharging and charging process. Only the beginning of the discharge does not correspond to the calculated curve because battery-specific effects are superimposed here (see chapter 12 for lead-acid batteries and chapter 13 for lithium-ion batteries).

The similarity of the characteristic curves with the curves resulting from the analysis of the BV equation under simplifying assumptions is an indication that the change in the charge transfer voltage or the BV resistance is the dominant influence on the behavior of batteries during discharging and charging. This is again relevant when discussing battery models.

3.4 Series Connection of Electrodes – Interactions Between Electrodes

The value for the potential of an electrode relative to the reference electrode and the value for the current flowing through it are coupled via the Nernst voltage and the BV equation. When two electrodes are connected in series to form a cell, the current through both electrodes will be the same, and the cell voltage is the difference of both electrode potentials. If the property of one electrode changes, for example, due to a loss of active mass corresponding to a reduction in the exchange current, its current-voltage point will change and this will have an effect on the other electrode. Figure 3.7 shows this schematically for the case of new electrodes and for the case in which one electrode has a shifted Tafel line due to a reduced exchange current (reduced effective surface area, loss of available active material). If the current through the cell remains constant, the cell voltage and the potential of the electrode with a smaller effective surface area change. If the voltage between the electrodes is kept constant, then not only the current changes, but also the potential of both electrodes in relation to the reference electrode.

It is technically relevant that changes in one electrode due to aging effects can change the potential of the other electrode in such a way that it is operated outside the optimum voltage range and ages, possibly even faster than the electrode that triggered the voltage change.

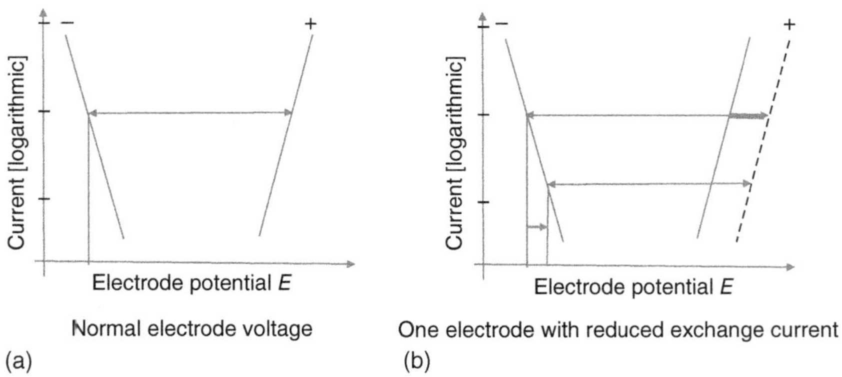

Figure 3.7 Current-voltage characteristics of two electrodes of a cell: (a) normal electrode voltages and (b) electrode with reduced exchange current. A change in current at constant cell voltage has an effect on the potential of the electrode, which has not changed. With constant current, on the other hand, the cell voltage and the potential of the electrode with a lower effective surface area (marked) change.

3.5 Discharging and Charging Electrodes in a Cell

The properties of the electrodes in a cell must be matched because the same current flows through both. For a cell to function, only electrodes with precisely defined states of charge and matched capacities are used in production. It is obvious that a cell consisting of a discharged and a charged electrode cannot conduct current because one of the electrodes always has a very high BV resistance due to the unavailability of reactants.

The voltage and current curves of matched electrodes correspond to the sum of the curves expected for individual electrodes. The aim must be to achieve that:

1) Both electrodes reach their respective electrode-specific end-of-discharge voltage at very similar times during discharging. The sum of the two voltages, the end-of-discharge voltage of the cell, is used to end the discharge process.
2) During charging, the maximum permissible electrode voltage of both electrodes is reached almost simultaneously. The sum of these two voltages is the end-of-charge voltage, which is used to control the charging current. In the subsequent charging phase with constant cell voltage, the two electrode voltages remain largely constant due to the same characteristic current control.

Minor deviations are unavoidable and are taken into account when designing the electrode sizes and determining the voltage thresholds.

Differences resulting from differences in the production process and electrode-specific aging mean that the electrode-specific voltage limits cannot always be adhered to during both discharging and charging. During discharging, one electrode is discharged beyond its voltage limit, while the discharge of the other electrode is terminated prematurely, and during charging, one electrode is operated at a high charging voltage, while the other does not reach its final charging voltage. Technical problems or safety risks only occur in rare cases when discharging or charging a single cell. With several cells in series, however, different electrode properties lead to massive problems and dangers.

Different electrode or cell properties become particularly apparent at the end of discharging or charging.

3.5.1 Significance of Side Reactions in Series Connection

Similar to matched electrodes of a cell, cells must have the same state of charge and capacity at the time of connection in series. Cells connected in series can be charged and discharged and have almost identical voltages at all times. To achieve this, individual cells are sometimes measured and selected individually before connection in series for applications with very high requirements. Even after a long period of operation, the voltage differences between the cells are only a few mV in each operating state. For reasons of safety and reliability, however, precautions must be taken to avoid ever-increasing differences in cell characteristics.

A significant difference between different electrochemical systems, which affects safety, reliability, and service life and which results in the need for battery management systems, is the presence of high ampacity side reactions causing little or no degradation. A distinction must therefore be made between cells with corresponding side reactions at both electrodes and cells without side reactions or destructive side reactions at both electrodes.[4]

In the following, the equivalent circuit diagram for mixed potentials shown in section 2.5 is used for cells with side reactions for each of the two electrodes, and the equivalent circuit diagrams shown in section 2.4 are used for cells without side reactions. In the following, it is assumed that of n cells in the series connection $n-1$ cells have the same properties, referred to here as "good cells," and only one cell has different properties, especially much lower capacity, and is referred to here as "defective cell." The discussion is limited to defective cells in which either one or both electrodes have a lower available active mass or effective surface area than the other cells and whose BV resistance therefore increases much faster than that of the other cells at the end of discharging or charging. The defective cell therefore becomes highly resistive at the end of charging and discharging before the good cells.

A distinction is made between the following technically relevant cases:

1) Discharging without side reactions,
2) Discharging with side reactions,
3) Charging with side reactions, and
4) Charging without side reactions.

In the following description, it is assumed that all cells were fully charged at the start of the discharge and that the same amount of charge was removed from all cells from the fully charged state during a normal discharge before charging.

The equivalent circuit diagrams used for the discussion are shown for all four cases in Figure 3.8.

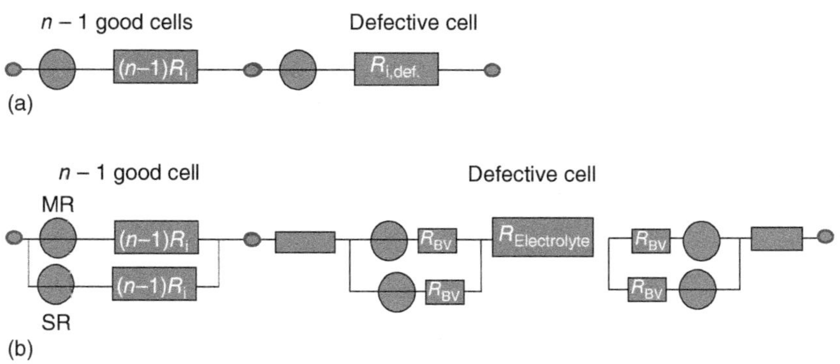

Figure 3.8 Equivalent circuit diagram of a battery with n cells in series with $n-1$ good cells and one defective cell: (a) without side reactions and (b) with side reactions.

4 The special case where only one electrode has appropriate side reactions, but the other electrode of the cell does not, is not relevant for the battery systems considered in this book and is therefore not discussed further.

3.5.2 Discharging Cells Without Side Reactions in Series Connection[5]

Reactions that lead to self-discharge, age-related loss of capacity, and corrosion are not considered below, even if, strictly speaking, these can also be electrochemical side reactions.

Soon after the start of discharging but at the latest when the amount of charge drawn from the battery is equivalent to the capacity of the defective cell, the BV resistance of the defective cell will increase because of the lack of charged material. The voltage of the defective cell is then significantly lower than that of the other cells. It is not possible to tell from the voltage behavior of the defective cell which of the two electrodes or whether both are the cause.[6]

The voltage of the $n-1$ good cells leads to a discharge current through the load and the defective cell, even if the voltage of the defective cell continues to drop and approaches 0 V. The discharge must then be stopped. However, if this does not happen and the discharge of the battery continues, the defective cell must be considered as an additional load in addition to the actual load. The BV resistance of the defective cell continues to increase, which leads to a massive build-up of heat in the cell. As the process continues, the cell becomes so highly resistive that the discharge current becomes lower and lower until no discharge current can flow. When the discharge current is nearly zero as a result, the voltage of the good cells will return to an equilibrium voltage associated with their state of charge. There will be no voltage drops across a resistive load because no current flows, and the total voltage of the $n-1$ good cells, possibly more than 100 V, drops across the defective cell as long as the circuit is closed (Kirchhoff's law). The more cells are connected in series and the higher their equilibrium voltage, the greater the voltage drop across the defective cell. The terminal voltage of the defective cell can assume high negative values. In Task 3.2, the voltage and the voltage curve are calculated.

This description is not realistic insofar as the defective cell exhibits uncontrolled behavior (fire, explosion, etc.) before reaching a very low terminal voltage. There are no combinations of electrode and electrolyte that are stable against such voltages. The reactions that take place at low voltages are chemical reactions and/or electrochemical side reactions, but they are little understood and highly destructive.

To avoid these safety risks, safety precautions must be taken:

a) Safety measures in a cell, which interrupt or reduce the current flow at high temperatures; and
b) Battery management systems that recognize voltage differences of a defective cell and stop the discharge.

Normally, individual cell voltage measurements are carried out for cells with a larger capacity, or a model-based attempt is taken to detect the existence of defective cells in the battery string.

A very simple monitoring procedure is: an additional voltage measuring point after $n/2$ cells can be used which allows to compare voltages of two parts of the

---

5 This case is relevant for all lithium-ion systems, for example.

6 A high cell resistance caused by electrochemically passive components in the cell will lead to a constant voltage difference throughout a constant current discharge. Such a voltage difference is less critical, except for the resulting heat generation.

string with equal number of cells and thus the detection of a high-resistance cell if very similar behavior can be assumed for all other cells.

The discharge of all cells can be interrupted using any of these safety precautions as soon as the defective cell falls below a voltage threshold. The defective cell thus limits the discharge of the string and the energy stored in the $n - 1$ good cells can no longer be used.

Task 3.3 shows the development of the voltages of the good cells and the defective cell during discharging.

3.5.3 Discharging Cells with Side Reactions in Series Connection[7]

The voltage of the overall system and the defective cell corresponds to the curve described above for cells without side reactions at the beginning of the discharge: Sooner or later, the capacity limitation of the defective cell causes the voltage to drop sharply due to the ever-increasing BV resistance of the main reaction. However, secondary reactions ensure that both the positive and negative electrodes can allow current to flow through the cell. For a better understanding, the main and secondary reactions as well as their equilibrium voltages under normal concentration conditions are listed below using the example of lead-acid batteries for both the positive and negative electrodes:

Positive electrode:

Main reaction, $E_{0,\,\mathrm{MR},\,+} \approx 1.7\,\mathrm{V}$:

$$PbO_2 + H_2SO_4 + 2H^+ + 2e^- \rightleftarrows PbSO_4 + 2H_2O. \tag{3.16}$$

Side reaction, $E_{0,\,\mathrm{SR},\,+} \approx -0.4\,\mathrm{V}$:

$$PbSO_4 + 2\,H^+ + 2\,e^- \rightleftarrows Pb + H_2SO_4. \tag{3.17}$$

Negative electrode:

Main reaction, $E_{0,\,\mathrm{MR},\,-} \approx -0.4\,\mathrm{V}$:

$$Pb + H_2SO_4 \rightleftarrows PbSO_4 + 2H^+ + 2e^-. \tag{3.18}$$

Side reaction, $E_{0,\,\mathrm{SR},\,-} \approx 1.7\,\mathrm{V}$:

$$PbSO_4 + 2H_2O \rightleftarrows PbO_2 + H_2SO_4 + 2H^+ + 2e^-. \tag{3.19}$$

It should be noted that the side reaction of the positive electrode corresponds to the main reaction of the negative electrode and the side reaction of the negative electrode corresponds to the main reaction of the positive electrode. If differences resulting from different material structures and additives of the positive and negative electrodes are neglected, the Nernst voltages and the parameters of the BV equation of the side reactions are identical to those of the respective main reactions.

The insufficient capacity of the defective cell causes the difference between the electrode voltage and the equilibrium voltage of the main reaction, the BV resistance, to increase. As soon as the electrode voltage reaches the equilibrium voltage of the side reaction, the current can also flow via the side reaction path in addition to the main reaction path – the electrode is overdischarged. The BV resistance of the side reaction is very low because enough "charged" material (= discharged material from the main reaction) has been generated beforehand.

7 This case is relevant for many lead-acid battery systems, for example.

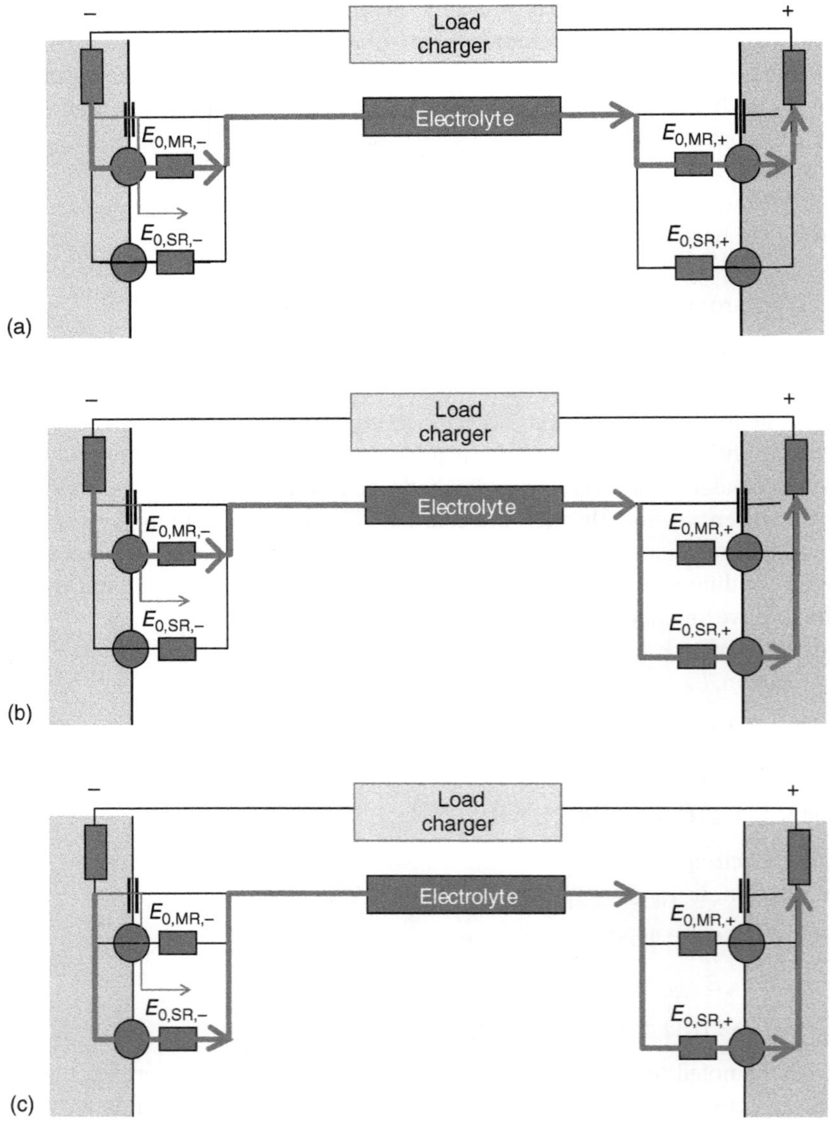

Figure 3.9 Current paths through the defective cell in different phases during discharging: (a) normal discharge, (b) one of the two electrodes is overdischarged, (c) both electrodes are overdischarged and the total current flows via the side reactions.

Figure 3.9 shows the current paths of the defective cell for the three different cases:

1) Normal discharge via both main reactions (A).
   Depending on the state of charge and current amplitude, the cell voltage is in the range of approx. 2 V.
2) One electrode is overdischarged, the other is discharged normally (B).
   The cell consists of two electrodes at which the same electrochemical processes take place, but a discharging process takes place at one electrode and a charging process at the other. The cell voltage is approx. 0 V.

3) Both electrodes are overcharged (C).
   The cell behaves like a cell installed with the wrong polarity, which is charged by the discharge current of the $n-1$ good cells. The cell voltage is approx. −2.2 to −2.4 V.

Figure 3.10 shows the voltage curve of a lead-acid telecommunications battery with a 2 V drop resulting from a cell with 0 V due to an electrode that is overdischarged. It cannot be concluded from this data whether the positive or negative electrode of the cell has failed. The discharge process of the entire battery continues normally, and the functionality of the battery system is largely maintained despite the cell defect. Overdischarge is a low-resistance fault whose impact on the entire battery system decreases as the number of cells increases.

When recharging, the process is reversed. First, the overdischarged electrode is charged until the normal charging process can begin. The mechanical stress on the electrode and its microstructure is very high during this process. Some field experience shows that after a full charge, the time at which an electrode is overdischarged occurs earlier and earlier. The cell must be replaced at the earliest possible time.

Lead-acid batteries have side reactions for both discharging and charging. In Task 3.5, it is shown what the voltage curve of a defective cell with a comparatively very low capacity looks like compared to good cells during discharge.

3.5.4 Charging Cells with Side Reactions in Series Connection[8]

All electrochemical systems with aqueous electrolytes have side reactions that always take place in competition with the main reactions if the cell voltage is sufficiently high.

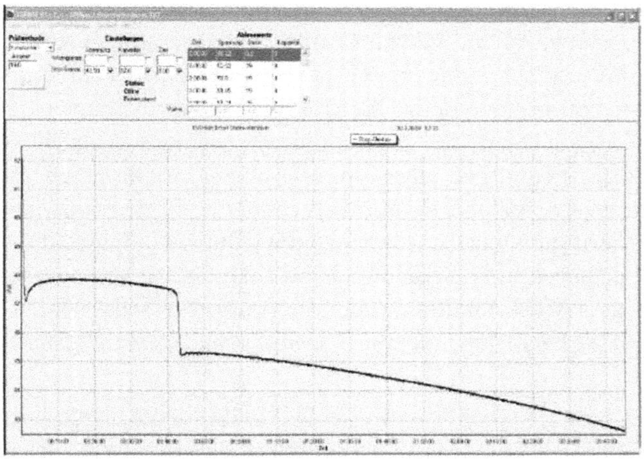

Figure 3.10 Measured discharge curve of a 48 V lead-acid battery of a telecommunications system, in which one electrode of a cell is overdischarged after approx. 45 min.

8 This case affects all batteries with aqueous electrolyte, including lead-acid batteries.

Positive electrode: formation of oxygen at 1.23 V compared to the hydrogen reference electrode:

$$H_2O \rightarrow \frac{1}{2}O_2 + 2H^+ + 2e^-. \tag{3.20}$$

Negative electrode: formation of hydrogen at 0 V relative to the hydrogen reference electrode:

$$2H^+ + 2e^- \rightarrow H_2. \tag{3.21}$$

The reactions of these two electrodes are the electrolysis of water to hydrogen and oxygen. Both have a high current capability at high voltages. The equilibrium voltages of the two side reactions are lower than the equilibrium voltages of the main reactions of lead-acid batteries (approx. 1.7 and −0.4 V respectively). Electrically, main reaction and side reaction on each electrode are two voltage sources connected in parallel where the voltage source with the higher voltage forces current through the other voltage source. However, the side reactions are so strongly inhibited (low exchange current density, high BV resistance) that the voltage of the main reactions (3.20) and (3.18) only leads to very small discharge currents and the voltages of the lead-acid electrode remain stable. The higher the cell voltage, the greater the currents according to the BV equation.

The current supplied by a charger increasingly flows through the side reactions of both electrodes, the more discharged active material has been converted and thus the greater the BV resistances of the main reactions become.

The equivalent circuit diagram shown in Figure 3.9 can also be used to represent the charging process with side reactions and the current paths shown through the cell in different phases also apply here. However, current paths A and B are only an approximation that describes the start of a charge because the side reactions (3.20) and (3.21) always take place in competition with the main reaction. Current path C corresponds to the situation in a fully charged cell because there is no more discharged material, no more current can flow via the main reaction path.

Figure 3.11 shows a measurement of the individual voltages of 22 series-connected cells of a nickel-cadmium battery from space travel [3] which is typical for cells with aqueous electrolytes before optimization of the charging characteristic. Chapter 12 shows a similar curve for lead-acid batteries. Due to the side reactions in aqueous electrolytes, the cell voltage is equalized after complete chemical conversion of the discharged active masses. Comparable experiments cannot be carried out with lithium-ion batteries, and corresponding data have not been published. Figure 3.11b shows the improvement achieved by intervening in the charging process with a narrow voltage tolerance, as is typical for lithium-ion batteries and must be ensured for safety reasons by use of the balancing function of a battery management system (BMS).

When charging cells in series, the BV resistance of different cells increases at a similar rate, and the differences are small at the beginning. Significant differences only become visible at the end. The end-of-charge voltage of a few cells is exceeded at a time when the voltage of most other cells is still below the end-of-charge voltage. Main and side reactions take place in parallel at each electrode in different

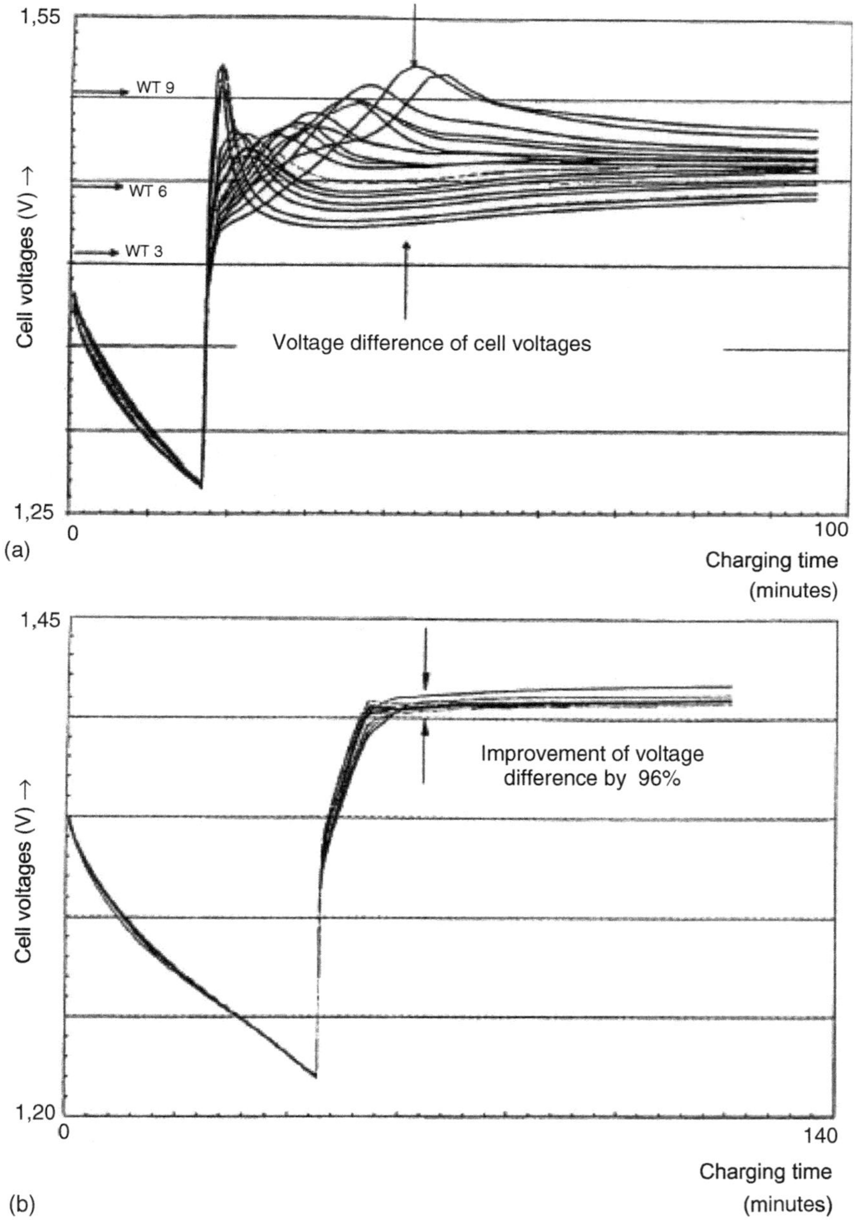

Figure 3.11 Individual cell voltages of a 22-cell nickel-cadmium battery for space applications (a) before and (b) after optimization of the charging method. Source: courtesy of the Jet Propulsion Laboratory, Pasadena, California.

proportions. When the end-of-charge voltage of the entire battery is reached, the current is reduced so that the overall voltage remains constant. For some cells, this means a further charge with an already impermissible voltage, which further increases the BV resistance and increases the cell voltage and heat generation in the cells, while for other cells, the current is already reduced at a time when they are not yet fully charged and their voltage may even drop. All cell voltages then slowly approach a similar value, the entire charging current flows at both electrodes exclusively via the side reaction paths. Figure 12.3 in chapter 12 shows another striking example of this behavior.

3.5.5 Charging Cells in Series Without Side Reactions[9]

The initial charging process is similar to that shown in Figure 3.11b. The BV resistance of a defective cell increases faster than that of the good cells. However, there are no side reactions that can take up the charging current within safe voltage limits. Despite maintaining a constant battery voltage, the cell voltage of the defective cell continues to rise.

Assuming that the defective cell is stable against the ever-increasing cell voltage, it becomes so highly resistive that the charging current tends toward 0 A. The voltages of the good cells then approach their state-of-charge-dependent equilibrium voltage. The difference between the voltage of the charger and the equilibrium voltage of all good cells then corresponds to the voltage of the defective cell. The greater the number of cells, the greater the voltage. The good cells can no longer be fully charged. In Task 3.4, the voltage of a defective cell that can occur is calculated – for a battery with 360 V nominal voltage and an end-of-charge voltage of 420 V possibly in the range of 20 V.

The electrodes and electrolytes are not stable against such high cell voltages. The processes that start as a result of the instability of the materials cause considerable safety risks. For this reason, precautions must be taken, similar to those taken when discharging cells without side reactions. In the case of high-capacity cells, this necessarily leads to single-cell voltage measurement or model-based algorithms and control of the charger via a battery management system. The charging current is controlled in such a way that the defective cell does not receive an impermissibly high voltage, which means that the good cells cannot be fully charged unless additional components, for example, bypasses, are installed (see more detailed information on battery management systems in chapter 13).

In cells without side reactions, the capacity that can be drawn is limited by a cell with a low capacity; when charging, recharging is limited by the cell that reaches the end-of-charge voltage first. A full charge can then only be achieved by the so-called cell balancing, that is, a controlled equalization of the cells.

It should be noted that, if the electrodes in a cell are very different, the term "full charge" can only refer to the full charge of one electrode or, due to the interaction of the two electrode voltages, causes overcharging of one electrode and undercharging of the other electrode. The second electrode cannot be fully charged.

---

9 This in turn affects many lithium-ion batteries.

3.6 Effects of a Short Circuit in a Cell Connected in Series

Figure 3.12a shows the equivalent circuit diagram of a short-circuited cell, and Figure 3.12b shows the equivalent circuit diagram of a series connection of cells with a short-circuited cell.

A short circuit[10] is an electron-conducting connection between the electrodes (contact between the active masses and/or the current collectors inside the cell), but also an electron-conducting path between the terminals outside the cell housing, for example, due to contamination or unintentional contact of the terminals. The equivalent circuit diagram does not distinguish between a high-impedance short circuit, through which only a few mA flow, and a low-impedance short circuit, through which very high currents can flow and which can completely discharge the cell in a very short time.

Short circuits in a cell lead to the release of energy in the cell and/or in the current-conducting path outside the cell container, even at rest. These cells can heat up very quickly, and the cell voltage drops to values close to 0 extremely quickly in case of a low-resistance short circuit. High-resistance short circuits will only lead to a slow self-discharge. Charging currents increase heat generation because, in addition to the short circuit current that increases with increasing cell voltage, the charging current also feeds the short circuit and contributes to heating. For cells in series, the charging current increases with a given charging voltage because the cell with a short circuit leads to higher charging voltages for all other cells due to the lower voltage of the short-circuited cell. Cells without a short circuit are overcharged and may be damaged as a result. The fewer cells in series and the lower the resistance of the cell short circuit, the more seriously the cells without a short circuit will be overcharged. If a constant charging current is used, there is no effect on the cells without a short circuit, but the shorted cell cannot be fully charged.

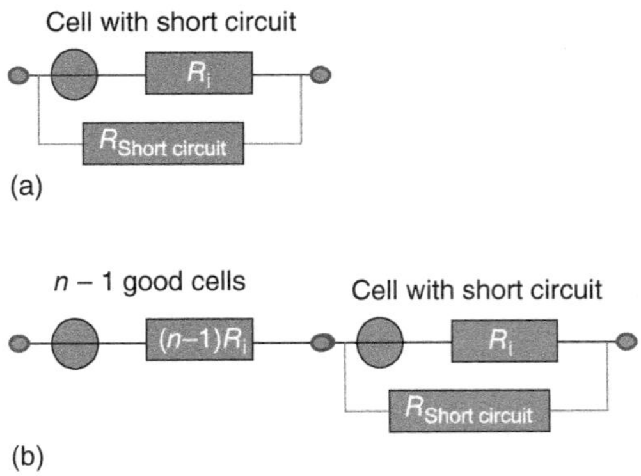

Figure 3.12 Equivalent circuit diagram of (a) a short-circuited cell and (b) a series connection of cells with a short-circuited cell.

10 The term "short circuit" is used here in line with electrical engineering, i.e. the connection between the electrodes may have either a high resistance resulting in only a few mA of current, or a very small resistance that lead to short circuit currents in excess of 10,000 A for high-capacity cells.

3.7 Fault Propagation, Parallel Battery Strings, and More

In the following, further issues arising from unequal cell properties that have technical relevance are listed. Some of these are dealt with in more detail in further chapters.

1) It should be noted that a cell with different properties (capacity, internal resistance, and self-discharge rate) has an effect on how other cells are operated, both when connected in series and in parallel.
2) With high-resistance cells, these effects increase with the number of cells in series and decrease with the number of battery strings connected in parallel.
3) There are differences as to whether the charging process at the end of the charge is current or voltage controlled. A fixed charging current only leads to undesirable voltages in the defective cell, a fixed charging voltage of the battery string also has an effect on the other electrode and other cells.

References

1 Wenzl, H., Benger, R., and Hauer, I. (2024). *State of Charge; Encyclopedia of Electrochemical Power Sources*, 2e. Elsevier.

2 Jossen, A. (1994). *Operational Management of Photovoltaic Systems with Energy Storage Systems* (Original German title: *Betriebsführung Photovoltaischer Anlagen mit Energiespeichern*). Dissertation. University of Stuttgart.

3 Blosiu, J.O., Deligiannis, F., and Di Stefano, S. (1996). *Nickel-Cadmium Battery Operation Management Optimization Using Robust Design*. Jet Propulsion Laboratory cl #96-0541. http://hdl.handle.net/2014/24727.

Tasks

Task 3.1

A fully charged lead-acid battery, with a nominal capacity of 40 Ah and rated voltage of 12 V, can be discharged at a rated current of 2 A for exactly 17.5 hours until the final discharge voltage specified by the manufacturer is reached. During a subsequent charging process, 36 Ah are recharged into the battery. Calculate

a) the measured (i.e. actual) capacity of the battery;
b) the *SoH* of the battery (in relation to the nominal capacity);
c) the *SoC* of the battery (in relation to the nominal capacity) at the start of discharge if it was not known at that time whether the battery was fully charged or not; and
d) the specific energy of the battery, which weighs 8 kg including the housing.

Task 3.2

Draw the curves of current, voltage, and amount of charge for the following charging methods:

a) IUa
b) CC/CV or IU
c) WIa
d) UOU ( )$U_2 < U_1$

Note: If there is a termination criterion for the charging procedure, this should be considered as time-controlled, that is, charging should be completed at time t_a, even if other criteria, for example, threshold value for current, exist.

Task 3.3

Discharging a lithium-ion battery

A lithium-ion battery with 100 cells in series, a nominal voltage of 360 V, and a nominal capacity of 18 Ah contains a defective cell that has only half the capacity of the other cells. Discharging takes place via a load resistor with 20 Ω.

Assume that the voltage of the defective cell corresponds approximately to the voltage of the good cells for the first 25 minutes, only then does it show a noticeable voltage deviation and after 35 minutes it has become so high resistance that no more current flows or the current can be neglected.

Assume that the equilibrium voltage of the good cells is 3.5 V after removing approx. 50% of the capacity. For the voltage curve of the good cells, use data sheet information for cells with a nominal voltage of 3.6 V. Assume that the defective cell remains stable in relation to the voltage that will develop.

a) Using the nominal voltage, calculate the current flowing through the load at the start of discharge. Plot the course of the current.
b) Draw the curve of the battery voltage and the voltage curve of a good cell and the defective cell.
c) Calculate the voltage that drops across the defective cell when the current has become 0.
d) Draw qualitatively a graph of the heat generation in the defective cell and make an estimate of the maximum expected value of heat generation in the defective cell.

Task 3.4

Charging a lithium-ion battery

In a lithium-ion battery with 100 cells with a maximum charging voltage of 4.2 V/cell and a maximum charger voltage of 420 V, one defective cell reaches the end-of-charge voltage after only 50% of the normal charging time. The charger continues to charge at the end-of-charge voltage until the voltage of the entire battery reaches 420 V. Only then is the charging current reduced to maintain a constant voltage of 420 V.

Assume that the good cells have reached a voltage of 4.1 V when the end-of-charge voltage of 420 V for the overall battery is reached.

a) Draw the voltage curve of a good cell and the defective cell up to the time at which the entire battery has reached the end-of-charge voltage.
b) What is the terminal voltage of the defective cell at this time?
c) At the end of charging, assuming that the defective cell is still intact, the charging current has dropped to 0 A. Calculate the terminal voltage of the defective cell, assuming that the state-of-charge-dependent equilibrium voltage of the good cells is 4.0 V.
d) Qualitatively estimate the course of heat generation in the defective cell.

Task 3.5

Overdischarging a lead-acid battery

When discharging a battery with a cell that has only 20% of the intended active mass on both electrodes due to a production error, this cell is overdischarged during discharging. The discharged active material formed at the beginning of the discharge is converted into charged active material.

a) What happens if the entire battery is completely discharged,
   i) after approx. 20% capacity of the good cells has been removed, and
   ii) after approx. 40% capacity of the good cells has been removed and the overdischarge process of the defective cell stops due to a lack of discharged active material?
b) Qualitatively draw the voltage curve of the defective cell.

4

Structure of Electrodes, Design of Cells, and Complete Battery Systems

The equilibrium voltage, or open-circuit voltage, after a rest period is determined exclusively by the electrochemical reactions taking place and the concentration of the reactants. It is not dependent on the design, size, or the number of electrodes connected in parallel. On the other hand, the performance of electrochemical systems and their service life depend to a large extent on the materials used, the properties of the active masses, and the design of the electrodes and cells. The basic considerations that apply to all electrochemical systems regarding the structure of active masses and the design of electrodes, cells, and complete battery systems are presented in the following.

Electrochemical reactions can only take place where reactants, electron-conducting, and ion-conducting structures exist at the same location, the so-called three-phase boundary. A high number or large area of three-phase boundaries in electrodes is created by an open-pore electrode structure, whose electron conductivity have to be improved by the addition of electron-conductive materials. The electrode structure must be maintained during discharging and charging. In these conversion processes, a distinction must be made between reconstitution mechanisms and intercalation mechanisms.

To improve mechanical strength and to conduct electrical current to the terminals, current collectors are used, whose surfaces are covered with active mass.

Usually, a cell consists of a negative electrode, followed by a separator, followed by a positive electrode, followed by a separator, another negative electrode, etc. All negative and positive electrodes are electrically connected in parallel so that the cell voltage remains the same regardless of the number of electrodes, but the capacity increases with the number and size of the electrodes.

To minimize electrical resistance and improve current homogeneity, the dimensions of the current collectors and the positions of the current collector tabs are of great importance.

(Continued)

Battery Technology: Fundamentals of Battery Electrochemistry, Systems and Applications,
First Edition. Alexander Börger and Heinz Wenzl.

Companion website: www.wiley.com/go/Börger/Battery_Technology

(Continued)

Depending on the application and for production reasons, two long electrodes can be wound to form a coil, or many individual electrodes can be joined together to form a stack. To increase the voltage and capacity, individual cells are connected together as modules, which are then connected in series or parallel to form a battery system.

The specific energy of a cell is often only approx. 25–35% of the theoretical specific energy due to the various additional components and the limited use of the electrochemically active materials and is even less for complete battery systems.

4.1 Electrochemical Requirements for the Structure of Active Masses

4.1.1 General Requirements

For electrochemical reactions to take place, the following conditions must always be met at the reaction site:

a) Conductivity for electrons:
Many active materials in rechargeable batteries are sufficiently electrically conductive so that the electrons released (on the anode) or the required electrons (on the cathode) can be conducted away from or toward the active charged and discharged materials. The active materials often contain additives (e.g. graphite) or conductive fibers that improve the conductivity for electrons. These additives must be electrochemically passive so that a long service life can be achieved.
b) Conductivity for ions:
Ion-conducting structures must be present at the reaction site to supply the ions required on the cathode and transport the ions generated on the anode away from the reaction site. Therefore, the electrode structure of the active materials must always have an open-pored structure with the pores filled with electrolyte.
c) The negative and positive electrodes must always be separated by means of an open-pored separator so that the electrodes never touch. Otherwise, the stored energy would be released uncontrollably inside the cell (short circuit).
d) Availability of reactants:
The availability of reactants must be guaranteed both during discharging and charging. This means that the discharged material
   - is deposited in such a way that it has little or no effect on the availability of the charged active materials, or
   - must be removed from the reaction site by mass transport processes or diffusion. During the charging process, however, the discharged material must be present at the site of the reaction so that an electron-conducting and open-pored ion-conducting structure, which can again be discharged, can be recreated.

e) For some systems, the presence of a suitable (electro-) catalyst: in the case of electrochemical reactions that require catalysts (e.g. air cathodes of metal-air batteries), a suitable catalyst must also be present at the reaction site.

The points at which the three conditions:

1) electron conductivity,
2) ionic conductivity, and
3) availability of reactants

are fulfilled are called three-phase boundaries. Figure 4.1 shows schematically the conditions at a reaction site and the structure of the active mass:

- the material of the current collectors can be in direct contact with the electrolyte in some areas, and
- the ohmic resistance for electrons from the reaction site to the current collector and for ions from the reaction site to the free electrolyte space between the electrodes is different at different points in the electrode.

It is easier to explain the electrode structure using a schematic diagram than with microscopic images. Points of different resistance have slightly different electrochemical potentials, which can lead to technically relevant effects in certain operating phases.

The conditions of the three-phase boundary are fulfilled for most electrochemical systems due to the sufficient electron conductivity of the active masses. To achieve the largest possible effective surface area for the reaction to take place, the electrode must have a very large microscopic surface area, for example, the material consists of very small particles. This property must exist during the entire charging and discharging process.

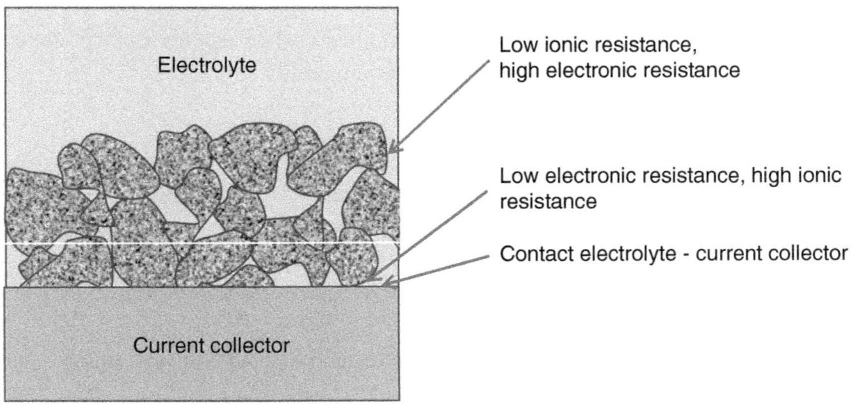

Figure 4.1 Schematic diagram of the structure of the active material. Points with different ratios of electron and ion resistance are marked. All surfaces that are in contact with the electrolyte and are connected to the current collector via other active mass particles are three-phase boundaries.

A distinction is made between active masses with reconstitution mechanisms (lead-acid batteries) and active masses with intercalation mechanisms (most lithium-ion batteries).

a) Reconstitution mechanisms
   In electrochemical systems with reconstitution mechanisms, the reactants and products of the reaction are different materials with possibly completely different properties (electrical and thermal conductivity, heat capacity, specific density, crystal structure, solubility in the electrolyte, reactivity toward other components of the cell, in particular current collectors and separators, etc.). As a result, the properties of the electrodes change significantly during charging and discharging. In the fully charged state, apart from auxiliary materials, only charged material is present; in the discharged state, the electrode consists of a mixture of charged and discharged material. For technical reasons, a complete conversion of the charged material is almost never possible. The properties of the charged and discharged material influence the macroscopic electrode properties. These changes in turn significantly influence the charging and discharging behavior and are technology dependent.
   In principle, it is possible to recreate an optimum structure of the electrodes during charging, but in practice, the structures slowly deteriorate with each discharge/charge cycle.
b) Intercalation mechanisms
   In electrodes with intercalation mechanisms, the ions required for the electrochemical reactions are located in a host material that has space for ions at certain sites. In most lithium-ion batteries, both the positive and negative electrodes consist of such materials. The ions are removed from high-potential sites and deposited in low-potential sites on the counter electrode. The changes in the electrode properties (metallic conductivity, volume, and size of the pores for the electrolyte) are often much smaller than in reconstitution mechanisms, but they can also play an important role.

The following explains how the basic requirements for the existence of large areas with three-phase boundaries are technically implemented.

4.1.2 Availability of Reactants

The reactions that take place during discharging come to an end once at least one of the required reactants is no longer available, or the reaction products inhibit the reaction. A lower availability of the reactants results in a higher overvoltage for the same current (see chapter 3). Therefore, favorable conditions for the availability of reactants must exist at all times.

To enable the discharged material to be converted back during recharging, the discharged material must be available for the electrochemical reactions. In many battery systems, the discharged material remains in the immediate vicinity of the reaction site and is then available for charging. In other battery systems, the discharged material is soluble in the electrolyte and then precipitates – often into chemically or electrochemically inert crystals (non-rechargeable battery systems).

High discharging and charging currents are only possible in the long term with a very large effective electrochemically active surface of the charged and discharged material. In addition, however, mass transport restrictions must be small so that rapid transport of the reactants to the three-phase boundary is possible.

Mass transportation is subject to Fick's laws.[1]

$$\frac{dc}{dt} = D\frac{d^2c}{dx^2}, \tag{4.1}$$

with D as the temperature-dependent diffusion constant and c as the concentration of the reactants. The higher the temperature and the greater the concentration gradient, the faster the concentration at the electrode-electrolyte interface approaches that of the concentration of the electrolyte in the pores of the electrodes and the electrolyte volume between the electrodes. The higher the temperature and the more educts present during the reaction, the lower the mass transport restrictions. With regard to the effects on the equilibrium voltage and the overvoltage, please refer to chapter 3.

4.1.2.1 Mass Transport Restrictions in the Electrolyte

Low mass transport restrictions in electrolytes are the result of high specific ionic conductivity and a high concentration of the ionic species required for the reaction. In batteries in which the electrolyte not only has an ion-conducting function but is also part of the reactions taking place, there must be sufficient electrolyte material between the electrodes and in their pores. The movement of ions in the open-pored electrode structures is restricted (great effective length) and leads to a local depletion of ions at very high currents and low temperatures, which is technically relevant. In batteries where no electrolyte is consumed, there are no requirements regarding the amount of electrolyte in a cell. The distance between the electrodes is chosen to be as small as technically possible.

4.1.3 Ionic and Electronic Conductivity of Electrodes and Cells

With electrodes and cells, a distinction must always be made between ion and electron conductivity.

4.1.3.1 Ionic Conductivity of the Electrolyte

The availability of ions at the three-phase boundary is achieved by the open-pored structure of the active masses. High specific conductivity of the electrolyte is important.[2] The ionic conductivity within the electrodes depends significantly on the pore structure or the number and diameter of connections between the pores filled with electrolyte. The active material adjacent to closed pores is electrochemically inert.

4.1.3.2 Electrical Conductivity of the Charged and Discharged Active Materials

For discharging and charging to take place, the porous active mass structure of electrodes must always have a certain minimum conductivity for electrons in all

---

1 The additional influence of migration, i.e. the movement of ions in the electric field, is not discussed further here.

2 Ionic conductivity is always low compared to the electronic conductivity of active materials.

permissible states of charge. A distinction must be made between the conductivity of the active material itself and the conductivity of a porous structure. The specific conductivity of an active mass as a compact material is irrelevant. Only the conductivity of the porous material, which depends on the number of contact points and contact areas between active mass particles, is important. Changes in the specific conductivity can be expected during discharging and charging, and these changes are particularly severe in the case of reconstitution mechanisms. To support the electron conductivity of electrodes, a conductive auxiliary structure made of electrochemically inert materials is used in many batteries to ensure good contact between the individual particles of the active mass and the current collectors.

4.1.4 Mechanical Stress on the Electrodes

The porous, electrolyte-filled active mass structure of electrodes has only limited mechanical stability. The high mechanical strength required during production and operation, especially for high-capacity electrodes, is primarily provided by the current collectors for conducting the electron current to the cell terminals.

4.1.4.1 Impact of Production Processes and Use of Battery

During production and also during subsequent use, electrodes are subject to mechanical stress that must not lead to damage (breaking of contacts between particles, cracks, or flaking) to the porous structure. The following measures are taken to absorb mechanical stress and prevent damage to the active mass structure, ensuring both mechanical stability and electrical conductivity:

1) Use of current collectors which, in addition to the primary function of conducting the electron current, stabilize the mechanical structure of the porous active mass, for example, grids, usually in lead-acid batteries, or foils with or without holes.
2) Additional use of woven gauntlets that exert permanent pressure on the active material and prevent loss of material.
3) Use of fleece covering on the separators, which prevent mass particles from detaching.
4) Tight assembly into the cell container and stabilization of the active mass structure by external pressure.[3]
5) Addition of additives such as binders.

4.1.4.2 Volume Changes Due to the Discharging and Charging Process

The volume of all active materials changes during charging and discharging. Although part of the volume change is at the expense of the pore volume, which can lead to a displacement of the electrolyte, especially with liquid electrolyte, there is always a macroscopic volume change of the electrode. The differences can be very large (5–400%).

The change in volume of the active materials during charging and discharging is one of the main reasons for the limited number of cycles of batteries and the increase

---

3 For batteries that may be subject to strong mechanical vibrations (e.g. in forestry and mining machines) or that are used in earthquake-prone installations, numerous additional measures are taken, e.g. bonding together electrode stacks sets, etc.

in the number of cycles with decreasing depth of discharge. Current collectors with high mechanical strength and a mechanically tight installation of the electrodes in a cell container prevent the loss of contact between the active mass and the current collector and have a positive effect on the service life.

4.2 Structure of Cells

4.2.1 General Information

The design of cells is determined by the need to achieve low internal resistance and high current homogeneity.

The most conductive electrolyte in conventional battery systems (aqueous sulfuric acid in lead-acid batteries) has a specific conductivity that is approx. 10^5 lower than that of copper at the same temperature; organic electrolytes are even less conductive. The resistance of the electrolyte depends on the distance between the reaction sites on the cathode and anode and the area (Ohm's law). The resistance of the electrolyte can therefore be minimized by a short length (thin electrodes closely spaced) and by a large macroscopic electrode surface. It should therefore not be concluded from the very high specific resistance of the electrolyte that the electrolyte also has the largest share of the total ohmic resistance of a cell.

There must never be an electron-conducting connection between the positive and negative electrodes within a cell. Otherwise, a discharge current will flow within the cell, and the energy released cannot be used by an external load. The electrodes must therefore always be separated from each other by a separator that only conducts ions but not electrons.[4] Damage to the separator and an electron-conducting connection between the electrodes within a cell not only leads to an unwanted discharge but can also lead to considerable safety problems, especially with lithium-ion batteries (thermal runaway, see chapter 13).

When designing and installing the electrodes, an uneven distance between the electrodes must also be avoided. This would otherwise lead to greatly differing resistances between the various sections of the electrode and to considerable current density inhomogeneities. For this reason, most batteries are designed for mechanically tight assembly by means of cell containers with high dimensional stability. In Task 4.1, the effect of different distances between electrodes is calculated.

4.2.2 Bipolar Plate Design

A possible solution for the requirements discussed above is a so-called bipolar design, which minimizes the total electrical resistance and current inhomogeneity. A bipolar design (see Figure 4.2) consists of the following components:

1) A bipolar electrode, that is, a very thin, electrically conductive plate that is coated on one side with negative active material and on the other side with positive active material. The plate must separate the two layers of active material in an

4 In the case of batteries with solid electrolytes, the solid electrolyte can also fulfill the role of the separator under certain circumstances.

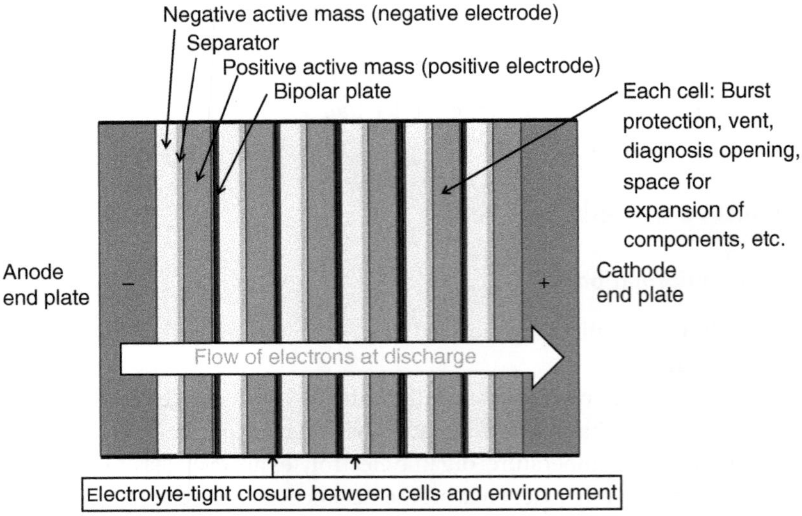

Figure 4.2 Schematic diagram of a bipolar plate design similar to those used in fuel cell systems.

electrolyte-tight manner and be electrochemically stable toward both the positive and negative active materials, if necessary by means of hole-free coatings.
2) Separators for permanently safe separation of the bipolar plates.

The more layers there are on top of each other, the higher the total voltage with unchanged capacity. So-called anode and cathode plates are on the outside which only contain active material on one side and must be mechanically stable to hold the electrode stack together. The power cables for charging and discharging are attached to the outside of the end plates.

Bipolar plate systems are limited in terms of electrode surface area, so the capacity of a cell will remain relatively small. A major problem is the permanent separation between the electrolyte of adjacent bipolar plates to avoid electrical short circuit between bipolar plates which would cause self-discharge processes, and worse, may cause corrosion. As of 2024, bipolar batteries have no significant commercial importance. However, the bipolar design is the standard design for fuel cells.

4.2.3 Stacked and Wound Cells

To produce high-capacity cells, a design consisting of a sequence of negative and positive electrodes separated by separators is generally used (see Figures 4.3 and 4.4). The current collectors must have a so-called terminal lug on one side for connection to the terminals of the cell. All negative electrodes of a cell are electrically connected in parallel via these terminal lugs. In lithium-ion batteries, where the electrodes and separators are very thin (in the range of 20–100 μm depending on the design), this is achieved by slight mechanical deformation of the terminal lugs and welding using industrial joining processes. The electrodes in lead-acid batteries are much thicker (approx. 0.8 mm for starter batteries and up to 2 mm for certain industrial batteries).

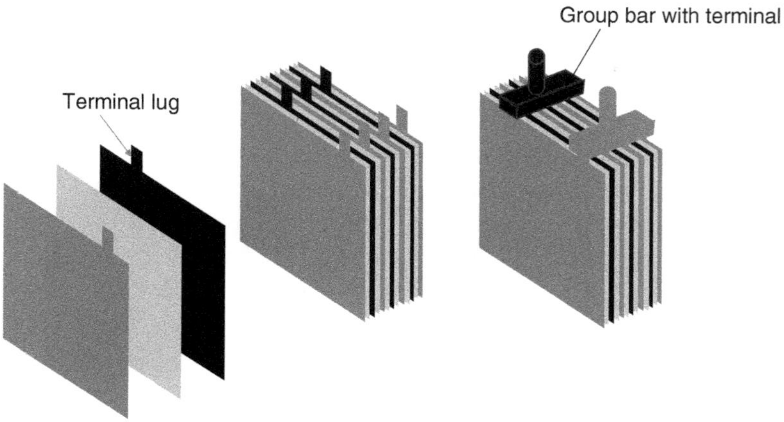

Figure 4.3 Schematic diagram of stacked cells. The connection of all negative and positive electrodes via a connection with a terminal, shown on the right, is used for lead-acid batteries but not for lithium-ion batteries.

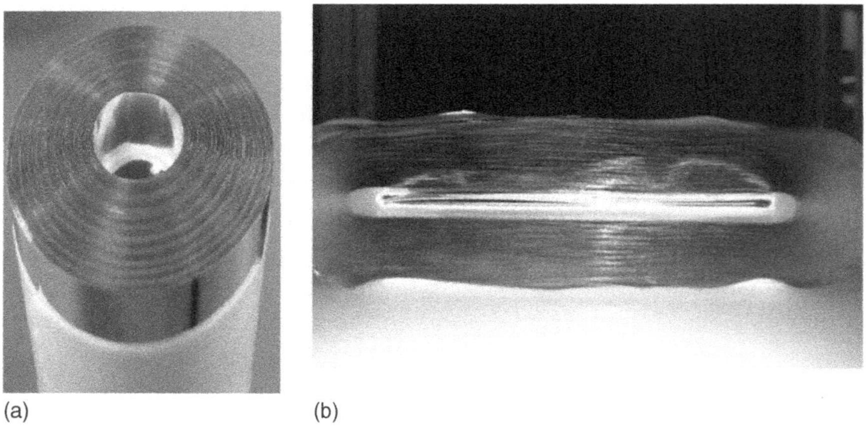

Figure 4.4 (a) Wound round cell and (b) wound prismatic cell. Source: Courtesy of Clarios Power Solutions GmbH.

As a result, deformation is not possible, and the electrodes are joined to each other via a so-called group bar and thus electrically connected in parallel. The group bar in lead-acid batteries is directly connected to the terminals of the cell. The more electrodes connected in parallel, the greater the capacity of the cell, but the voltage remains the same.

In so-called prismatic cells, the electrode stack is often made using many individual electrodes (Figure 4.3), up to approx. 21 individual electrodes, depending on the battery technology and application. For production reasons, the sequence of positive and negative electrodes, separated by a separator, can also be wound around a cuboidal core (prismatic wound cell) (Figure 4.4b). So-called round cells are wound around a cylindrical core (Figure 4.4a). The basic structure, that is, the sequence of positive and negative electrodes, is identical to that of stacked cells.

Due to their core, wound cells generally have a lower energy density (Wh/l) and specific energy (Wh/kg), but they have higher mechanical stability, especially in the case of round cells, and require a less complex production process.

Due to the highly conductive materials used for the current collectors and the low specific conductivity of the electrolyte, the direction of the electron current is always parallel to the current collector, and the direction of the ion current is perpendicular to the electrode surface. This leads to the lowest total resistance of the cell. Electrode areas that are directly opposite each other therefore always have the same charge throughput.

By design, the current density of bipolar plates is very homogeneous. In contrast, cells with current collectors with terminal lugs cause a voltage drop depending on the length, thickness, specific conductivity of the current collector, and the position of the terminal lugs (see, e.g. Refs. [1, 2]).

Current collectors, whose function for ensuring mechanical stability of the porous active masses has already been mentioned, must be optimized with regard to the ohmic voltage drop. The specific conductivity of the materials used must be high. The electron conductivity of the active masses is always lower in comparison but must not be neglected particularly for electrodes with a thick active mass coating (see Task 4.2).

The resistance of the current collectors only contributes little to the total internal resistance of the cell, but causes the current density along the current collector to decrease – the higher the current amplitude, the more. Figure 4.5 shows the equivalent circuit diagram of two electrodes, for simplification without the electrochemical voltage sources to represent the reactions and the resistance of the separator, with the electrodes divided into five segments. According to Kirchhoff's laws, the voltage drop must be the same on each current path between the terminals. It is easy to see that the total resistance from terminal to terminal on the top path is lower than the total resistance on the bottom path.

The equivalent circuit diagram shown in Figure 4.6 is equivalent to that shown in Figure 4.5. In the equivalent circuit diagram shown in Figure 4.6, the resistances of one electrode is added to the resistances of the other electrode. Figure 4.6 shows that an improvement in the conductivity of one electrode has a positive effect on the current homogeneity between the electrodes (see task 4.3).

The current inhomogeneity along the electrode caused by the grid resistances is difficult to measure. However, simulations show that the difference between the highest and lowest current density can vary by up to a factor of 10 [1] and depends on the length of the electrode, the current amplitude, and variations of the local state of charge caused by the discharge. The areas with the highest and lowest local current density change during discharging. The decisive parameter here is the ratio of the grid resistance to the resistance of the current between the electrodes (electrolyte and BV resistance). Due to the current dependence of the BV resistance and the low voltage drop across the grid at low currents, the current density distribution is most uniform at very low currents and most nonuniform at very high currents [3–6].

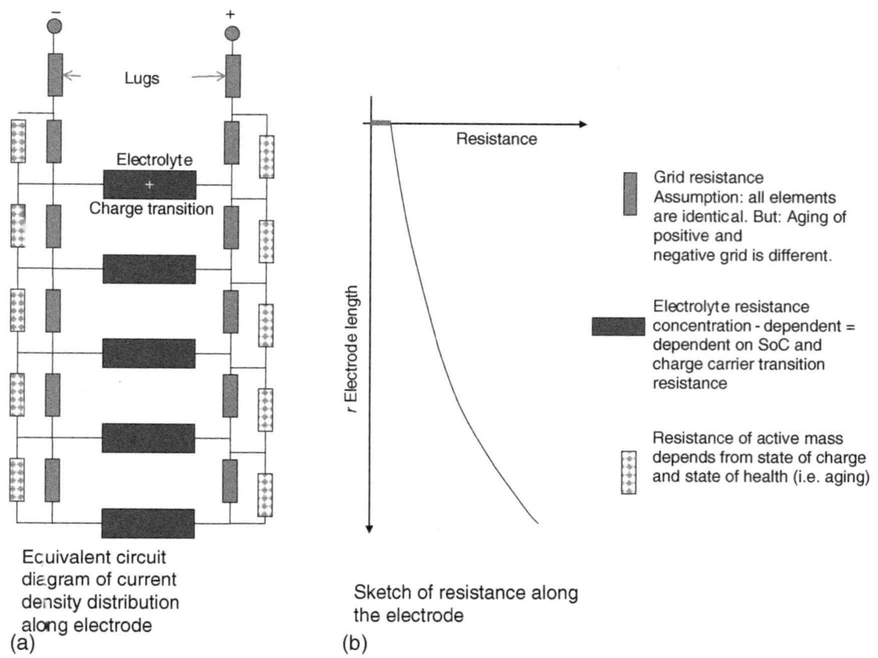

Figure 4.5 (a) Equivalent circuit diagram showing the resistance of a cell along the length of a current collector with active mass. (Voltage sources to represent the electrochemical reactions and the resistance of the separator can be neglected for this discussion.) (b) Diagram of the length-dependent resistance.

To avoid very large current inhomogeneities, it is necessary to avoid very long electrodes with terminal lugs on the same side. The position of the terminal lugs of the current collectors is of great importance for the current homogeneity. Figure 4.7 shows two different positions of the terminal lugs of the two electrodes. The positions in Figure 4.7a are the most commonly used, whereas the arrangement in Figure 4.7b is never used for lead-acid batteries and is more common for lithium-ion batteries. Brief considerations using Kirchhoff's laws show that the current homogeneity must be significantly higher for the positions shown in Figure 4.7b.

The best solution in principle is the position of the terminal lugs on different sides of the electrodes, as this leads to very low current inhomogeneities. In addition, the materials and collector thicknesses should be selected appropriately, and the thicknesses and conductivities of the active masses are similar.

Terminals on different sides of the cell are often considered unsuitable for handling cells and building modules. Two different cell designs are used to improve current homogeneity:

- Electrically conducting cell containers, which are connected to one of the electrodes at the bottom. The terminals of the cell are then arranged next to each other at the top of the cell. Cells must not touch each other when assembled into a module. Small cylindrical cells are often built in this manner.
- The lugs of the cell are on opposite sides, but inside the cell container, there is a connection from the lugs at the bottom to the top, and the terminals are next to each other on the top of the cell.

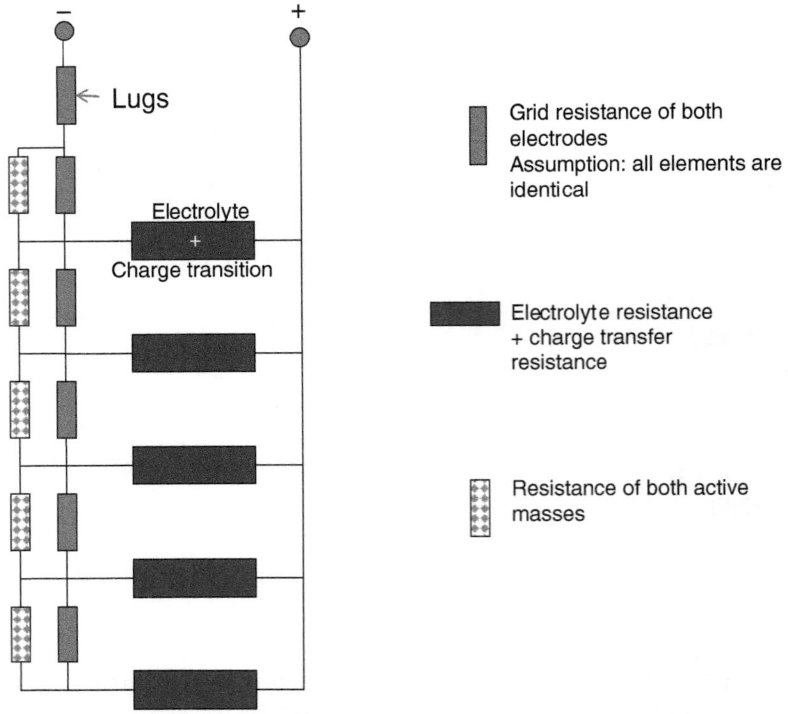

Figure 4.6 Equivalent circuit diagram of a cell consisting of two electrodes to illustrate the influence of the grid resistance of one of the electrodes on current homogeneity. By combining the grid resistance and active mass resistance into one resistance, a further simplification is possible, which is completely equivalent to the equivalent circuit diagram shown in Figure 4.5.

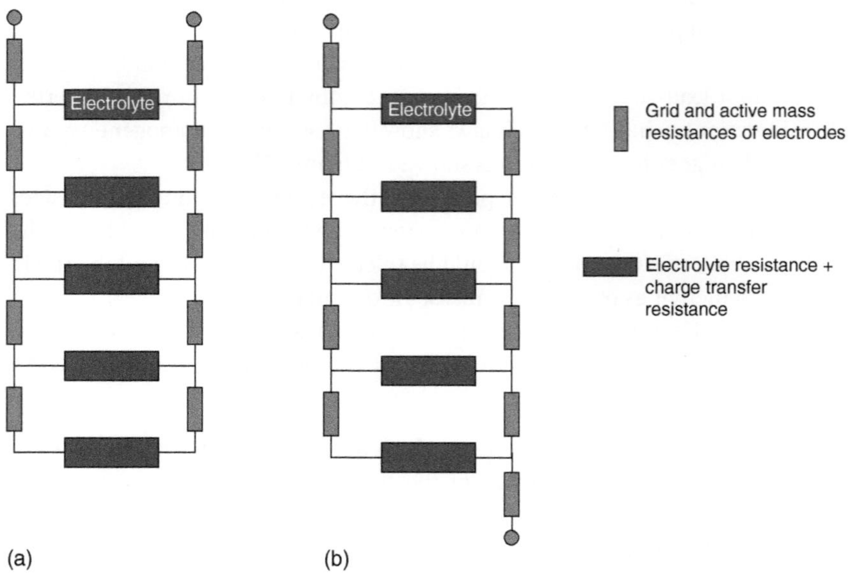

Figure 4.7 Equivalent circuit diagram to explain the differences in current homogeneity for connection of the electrodes on (a) the same side of the electrodes and (b) on opposite sides.

4.3 Combined Ion and Electron Conductivity of the Electrodes

Macroscopically, the electrode must be conductive for both electrons and ions. This is ensured by the open-pored, electrolyte-filled structure of the electron-conducting active masses.

On the path from the positive to the negative terminals, charges flow via the current collectors, the active masses, and the electrolyte between the electrodes. The electrochemical reactions take place in the electrodes at the three-phase boundaries. As the same voltage drop must exist along all different paths, the sum of the voltage drops (ohmic resistance of the active masses, current-dependent BV resistance, and ohmic resistance of the electrolyte in the pores of the active masses) must be the same along all current paths. Due to the significantly lower specific resistance of the active mass compared to the specific resistance of the electrolyte, the conversion processes take place predominantly on the outside of the electrode adjacent to the electrolyte. The reaction density (local distribution of the reaction sites with homogeneous distribution of the three-phase boundaries) in the electrode is therefore uneven. A homogeneous reaction density would only be possible if the specific resistance of the active material and the electrolyte (each in relation to the porous structures) were the same. The higher reaction current density on the outside of the electrodes is a function of the total current, and the effects (change in the resistances of the active material and the electrolyte, volume expansion and change in porosity) are technology-specific and also change the current density distribution in the electrodes. This can become technically relevant with thick electrodes. Figures 4.8 and 4.9 show the relationships in the form of an equivalent circuit diagram and a schematic diagram of the ratio between electron and ion current. To simplify the diagram, it is assumed that the electron current in the active masses only flows perpendicular to the current collector.

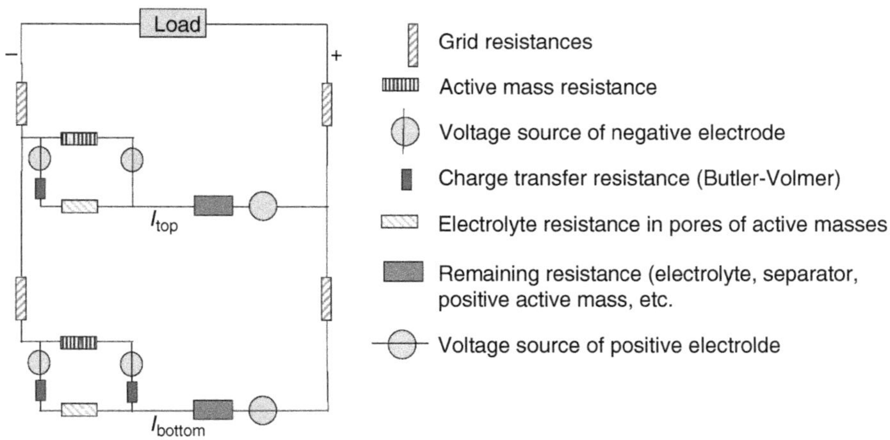

Figure 4.8 Equivalent circuit diagram to illustrate the current density distribution in an electrode perpendicular to the electrode surface.

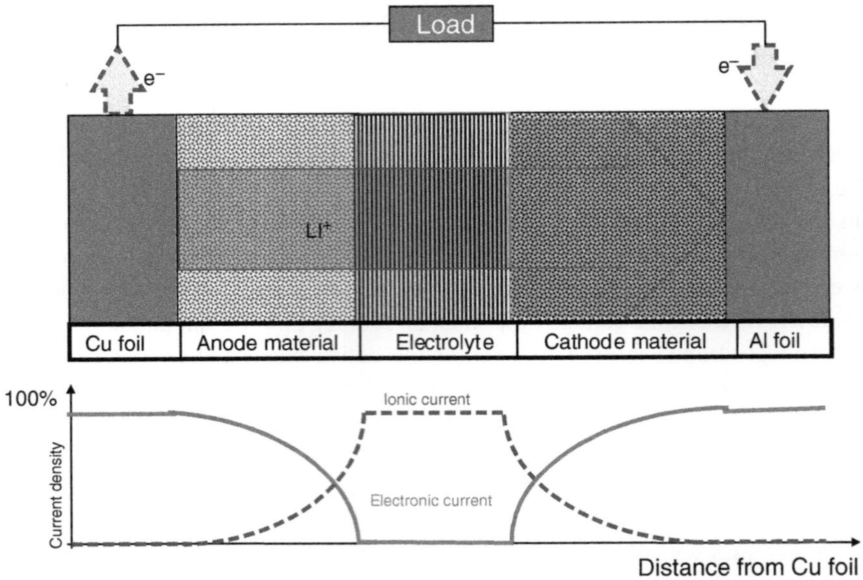

Figure 4.9 Schematic diagram of the ratio of electron and ion current in a cell between the positive and one negative electrode. The reaction current density is highest where the electron or ion current changes most rapidly.

The curve shown in Figure 4.9 depends on the exact ratio of the specific resistances of the active mass and electrolyte and, due to the current dependence of the BV resistance, on the current density. There is no generally valid curve.

4.4 Cell Containers and Battery Systems

4.4.1 General Requirements

The electrodes must be inserted into a container that tightly encloses the stack or coil to ensure that mechanical requirements and the required distance between electrodes are met. In addition, a seal must be used to protect against environmental influences and to protect the environment from the substances contained in the cell, as electrolytes and the materials used are usually toxic and chemically aggressive.

The following requirements generally exist:

- All feedthroughs of the cell container and the joint between the lid and the cell container must be leak-proof against the electrolyte.
- Controlled pressure relief.

It must be noted that electrochemical systems generate gases either in normal operation (always with aqueous electrolytes) or in the event of a fault (for most electrochemical systems). To prevent uncontrolled pressure build-up in the cell or deformation, there are openings that are sometimes closed by resealable valves and/or a rupture valve that allow pressure to be released safely if the internal pressure exceeds a threshold.

These and other requirements (low corrosion, diagnostic and maintenance openings in the cell container) regarding the design are so technology-specific that they are dealt with in more detail in chapters 12 and 13 for the respective battery systems.

To provide a sufficiently high voltage for an application, cells are almost always connected in series to form a string, and strings are also connected in parallel to provide sufficient capacity and energy. For characterization purposes, the term "xsyp" is used with x for the number of cells connected in series and y for the number of strings connected in parallel.

To increase operational safety and reliability, additional components are often required, the number and complexity of which are high and technology-specific. They are also dealt with in chapters 12 and 13 and additionally, if application-specific, in the respective chapters in the last part of the book.

4.4.2 Specific Energy of Cells, Modules, and Battery Systems

It is clear from the explanations in this chapter that the technical utilization of electrochemical reactions requires numerous components that are not involved in the electrochemical conversion. The theoretical specific capacity and energy of the electrochemical reactions described in chapter 2 are thus considerably reduced, by 60–80% depending on the technology. When cells are interconnected to form modules and batteries, additional components are required, which further reduces the specific energy. Figure 4.10 shows the weight distribution of the components of different types of lithium-ion cells.

When giving the specific energy of a battery system, it must therefore be carefully considered whether the information refers to the cells or the entire battery with all components required for safe and reliable operation and, if applicable, also to the energy that can be supplied during operation for reasons of service life or operational

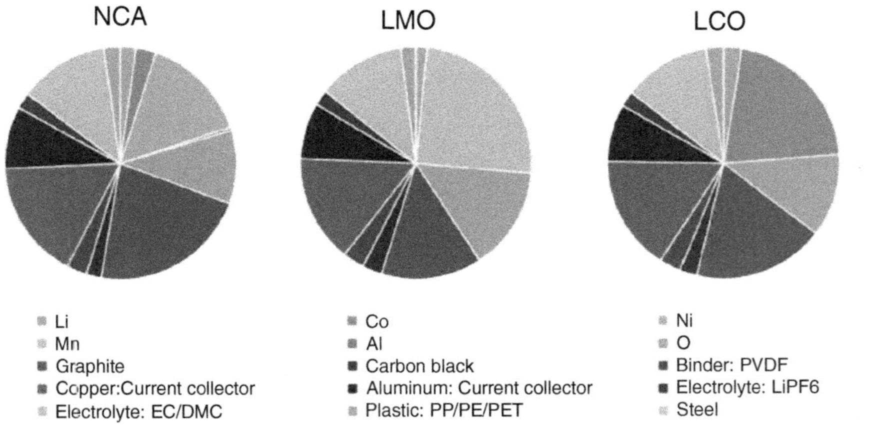

Figure 4.10 Weight distribution of the components of a 2.1 Ah lithium-ion high-performance cell. Source: Adapted from Ref. [7].

constraints. The specific energy can then be reduced significantly further compared to a cell and may then only be 15–25% of the theoretical value. For further details, see Ref. [8].

References

1 Král, P., Krivac, P., Baca, P. et al. (2002). Current distribution over the electrode surface in a lead acid cell during discharging. *Journal of Power Sources* 105: 35–44.

2 Krivac, P., Baca, P., Milan, C. et al. (2006). Current distribution over the electrode surface in a cylindrical VRLA cell during discharging. *Journal of Power Sources* 154: 518–522.

3 Arunachala, R., Oarthasarathy, C., Jossen, A., and Garche, J. (2016). Inhomogeneities in large format lithium-ion cells a study by battery modeling approach. *ECS Transactions* 73 (1): 201–212.

4 Euler, J. and Horn, L. (1965). Current distribution over the length of accumulator plates made of tubular elements (Original German title: Stromverteilung über die Länge von Akkumulatorenplatten aus rohrförmigen Elementen). *Archive for Electrical Engineering* 50: 85–90.

5 Euler, J. (1971). On the determination of the effective electrical resistance of the grids in lead accumulators (Original German title: Zur Ermittlung des effektiven elektrischen Widerstands der Gitter in Bleiakkumulatoren). *Archive for Electrical Engineering* 54: 122–126.

6 Kindermann, F.M., Oswald, P.J., Ehlert, G. et al. (2017). Reducing inhomogeneous current density distribution in graphite electrodes by design variation. *Journal of the Electrochemical Society* 164 (11): E3105–E3113.

7 Marshal, J., Gastol, D., Sommerville, R. et al. (2020). Disassembly of Li ion cells, characterization and safety considerations of a recycling scheme. *Metals* 10 (6): 773.

8 Wenzl, H., Benger, R., and Hauer, I. (2024). *State of Charge; Encyclopedia of Electrochemical Power Sources*, 2e. Elsevier.

9 Wenzl, H., Benger, R., and Hauer, I. (2025). *Energy; Encyclopedia of Electrochemical Power Sources*, 2e. Elsevier.

Tasks

Task 4.1

Prismatic cells are a frequently used type of lithium-ion battery, cells in which the electrodes are wound around a flat core (see Figure 4.5). In contrast to wound round cells, prismatic cells lead to better space utilization and facilitate production compared to cells that consist of a stack of individual electrodes. When winding the electrodes around the core, the mechanical stress regarding bending radius and tension must remain within safe limits. The distance between the electrodes in the bending area is smaller than in the area without bending due to the production process.

In the following, disregard all effects that can lead to different potentials and current densities, except for the distance between the electrodes.

Calculate the current density (A/cm^2) for two areas assuming an electrode distance of 25 and 50 μm. The specific conductivity of the electrolyte is 8.5 mS/cm, the specific capacity of the electrodes is 2.0 mAh/cm^2, which can be drawn until the end of the discharge. Calculate the differences in current density for a current density of 0.01 and 0.002 A/cm^2, i. e. a current rate of ca. 1C. These current values apply to the bending section with an electrode distance of 25 μm. Draw an equivalent circuit diagram with which the task can be solved.

Assuming a constant discharge current over the discharge duration: How much capacity does the area with the larger distance still have at the end of the complete discharge of the area with a smaller distance between the electrodes? Is such an extrapolation permissible under the assumption of a constant current discharge? Plot how the current densities change over the discharge time.

Task 4.2

Calculate the resistance per centimeter of length for the anode of a lithium-ion battery consisting of a copper foil with a thickness of 25 μm and an active mass layer of 25 μm on both sides. How much does the active mass layer contribute to the reduction in resistance?

Data:

- Specific conductivity of copper: $1.7 \cdot 10^{-8} (\Omega\,m)^{-1}$, and
- Specific conductivity of the anode material: $10^3 (\Omega\,m)^{-1}$.

Task 4.3

Using Kirchhoff's laws, show that the equivalent circuit diagrams shown in Figures 4.6 and 4.7 are equivalent.

5

Thermal Properties of Cells and Batteries

This chapter discusses the heat generation in a cell and its temperature change under normal operating conditions as well as the impact of temperature changes on cell properties. However, extreme situations that can occur in the event of faults and system malfunctions, such as very high temperatures, fires, and thermal runaway (uncontrolled self-heating process of a cell), will be dealt with later.

> The temperature of an electrochemical system influences its behavior during use, and in turn temperature changes caused by use influence the electrical properties. The temperature of cells has a significant effect on aging and safety and must therefore be kept within a narrow range for reasons of service life.
>
> Three different heat sources must be distinguished:
>
> - ohmic losses, proportional to the square of the current amplitude,
> - losses due to the charge transfer voltage and other overvoltages, increasing with the current amplitude in a complex manner, and
> - reversible heat, proportional to the current, which is caused by the thermodynamic data of the reaction $T\Delta S$ and can lead to both heating and cooling of a cell.
>
> The heat generated in a cell is a complex function of the current amplitude and current direction.
>
> The inhomogeneous temperature distribution due to locally varying heat generation and dissipation poses a fundamental problem when analyzing discharging and charging characteristics. Especially at very low temperatures and very high currents, a constant temperature of the cell and all its components can never be assumed.
>
> The structure of cells leads to inhomogeneous heat capacity and anisotropic thermal conductivity. Significant temperature differences are to be expected between the inside of a cell and the cell wall as well as different areas of the cell wall. The anisotropic thermal conductivity must be taken into account in cooling systems.

Battery Technology: Fundamentals of Battery Electrochemistry, Systems and Applications, First Edition. Alexander Börger and Heinz Wenzl.

Companion website: www.wiley.com/go/Börger/Battery_Technology

5.1 Inhomogeneous Heat Capacity and Anisotropic Heat Conduction

Due to the large differences in the thermal properties of the various battery materials, a cell is to be expected to have inhomogeneous heat capacity and anisotropic thermal conductivity. Figure 5.1 shows this schematically.

The heat capacity of a cell is largely determined by the amount of electrolyte and the plastic parts used for the cell; metallic components, on the other hand, have a comparatively low specific heat capacity. The heat capacity of the cell can be estimated if the weight of the cell components and their individual specific heat capacities are known, or it can be determined experimentally, for example, by measurements in a calorimeter. The values differ greatly between different cell types; 1 kJ/(kg K) can be used as a rough estimate.

The heat generated in the cell is almost exclusively transferred to the cell wall by thermal conduction. With high currents, temperature differences of several Kelvin must be assumed between the outer wall of the cell and the interior. Convection in the cell requires free air volumes or fluids that can circulate between cooler and warmer areas within the cell. This aspect is usually neglected as is heat transport within the cell by radiation. Even if only heat conduction in the cell is used to equalize the temperature, an analytical calculation is not possible, and a simulation is difficult. For a detailed analysis, a simulation using models capable of spatial resolution must be carried out.

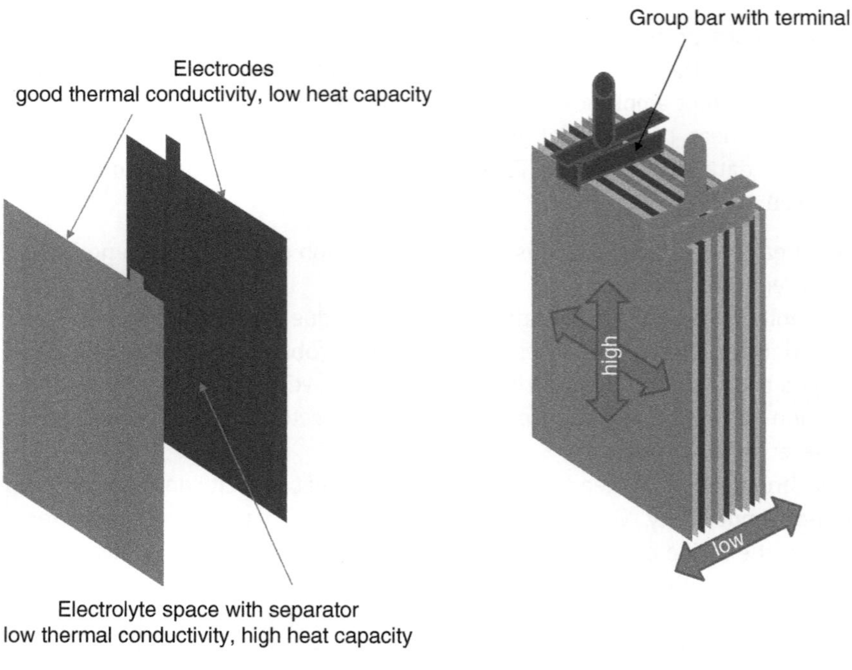

Figure 5.1 Diagram showing schematically the anisotropy of the thermal conductivity and heat capacity of cells.

Heat conduction from inside the cell to the cell wall is determined primarily by the orientation of the current collectors. It can therefore be expected that heat conduction to the cell wall is better in the direction of the current collectors than perpendicular to them. In the case of prismatic cells, differences in thermal conductivity λ of more than 300% have been measured in different directions, along with an additional pressure dependence of approx. 200%.

This anisotropic heat conduction must be taken into account when planning battery cooling systems. It also leads experimentally to the problem of where measuring points should be placed for temperature measurement.

It is important to remember these inhomogeneities and anisotropies and the inhomogeneity of the current density in the following discussion about heat generation and heat dissipation. For example, temperature differences of more than 100 K were measured at the cell wall during internal short circuits in lithium-ion batteries with very high local heat generation [1, 2].

5.2 Heat Generation

5.2.1 Types of Heat Generation

The heat generation in a cell depends on the current amplitude and its direction. It is necessary to distinguish under normal operating conditions between heating due to:

- ohmic resistances,
- charge transfer resistance, and
- the electrochemical reactions themselves (reversible heat).

The enthalpy of mixing due to reaction products and intercalation processes is generally not considered. These effects are small and disappear when considering the entire discharging and charging cycle. Furthermore, losses attributable to the dielectric properties of the electrolyte are not included.

When calculating the temperature of the cell wall and estimating the temperature inside cells, a homogeneous heat source density is often assumed, that is, current inhomogeneities and the different locations of heat generation are not considered. The maximum temperature inside a cell is inferred from the heating of the cell wall and the heat dissipation. When considering large cells, however, it is often necessary to take into account differences in the local current density. Temperature sensors inside cells are only used for special investigations in the laboratory. Estimates of the internal temperature using impedance spectroscopy are possible [3]. Temperature sensors at different points on the cell wall show differences of several Kelvin. For battery modules, it is important to choose the location of the measurement carefully so that the measured value is meaningful.

The materials used in cells can react with each other at very high temperatures, releasing large amounts of heat. In lithium-ion batteries in particular, this leads to an extremely steep rise in temperature at temperatures well above the permissible operating temperature, which can no longer be stopped from the outside[1] (see Thermal Runaway, chapter 13). Because this heat generation is only significant outside normal operating conditions, it will not be discussed further in this chapter.

5.2.2 Resistance Losses – Ohmic Heating

Current flow, regardless of whether it is ions or electrons, causes heat development in the respective components. Not only the electrochemically passive components, such as current collectors, group bars, and terminals, must be considered but also electrochemically active components, such as the active masses of the electrodes.

The heat generation Q_Ω in the individual components depends on the local current amplitude and the local resistance:

$$Q_\Omega = I^2 \cdot R_\Omega = I^2 \cdot \sum \rho_i d_i / A_i, \tag{5.1}$$

with R_Ω as the resistance of all components i, ρ as the specific resistance of the respective material, d as the length, and A as the area of the resistance i.

The following should be noted with regard to ohmic resistances:

- The greater the current inhomogeneities in a conductor, the greater the losses even if the specific resistance is uniform.
- The specific resistance ρ does not depend on the current amplitude but on the temperature and, in the case of electrochemically active components, on the state of charge, but changes only slowly during charging or discharging.
- The specific resistance of electrolytes for ions is usually several orders of magnitude higher than that of electron-conducting electrodes. In conventional designs, however, the path for the ions is very short compared to the path of the electrons, and the area through which the current flows is very large. The resistance of the electrolyte can therefore be in a similar range to the resistance of the electrodes and does not necessarily dominate the total ohmic resistance of a cell (see Figure 5.1).
- The resistance R_i of a component is in principle independent of the current. However, due to the dependence of the current distribution on the current amplitude, the effective length through which the current flows changes [4, 5] and the area of the current collectors and the electrolyte changes with the current amplitude so that the resistance R can have a small current dependence. For some problems (e.g. starting currents), it may make sense to use an effective length and area for calculating the resistance of a component.
- At very high frequencies and very steep and large current changes, the current displacement (so-called skin effect) must be taken into account. Depending on the geometry and materials (e.g. nickel and iron compounds with very high magnetic permeability), the effect can be observed above a frequency of a few kHz. The

1 Thermal runaway can occur not only in lithium-ion batteries, but also in other battery technologies. However, it is of particular importance for lithium-ion batteries due to the high amount of energy released.

temperature of cells rises as a result of this effect at very high alternating current loads [6]. For the vast majority of operating conditions, however, this effect is negligible.

- As the temperature rises, the resistance of the electron-conducting materials increases slightly and the resistance of the electrolyte decreases.

5.2.3 Charge Transfer Resistance

The difference between the charge transfer voltage and the equilibrium voltage corresponds to an energy loss that releases heat. The charge transfer voltage as a function of the current is described by the BV equation, which can also be written as a current-dependent resistance (see chapter 2).

$$Q_{\eta_T} = I^2 \cdot (E - E_o) = I^2 \cdot R_{BV}, \tag{5.2}$$

with E_o as the thermodynamic equilibrium voltage of the reaction, E as the electrode voltage, and R_{BV} as the resistance calculated from the BV equation. If there are several parallel reactions (main and side reactions) taking place simultaneously, the contributions of all reactions must be added.

$E - E_o$, the charge transfer voltage η_T, and I are either negative (discharge direction) or positive (charge direction), so that heat is generated regardless of the current direction. $E - E_o$ can differ considerably from the voltage measured on the terminals, especially at high currents.

In the range of the Tafel approximation (approx. $(E - E_o) > 0.05\,V$), the heat generation becomes:

$$Q_{\eta_T} = a \cdot I + b \cdot \ln(I) \cdot I \tag{5.3}$$

because of

$$R_{BV} = a/I + b \cdot \ln(I)/I, \tag{5.4}$$

with the constants a and b (see section 2.3.3).

Thus, Q_{η_T} increases disproportionately with the current but less than the ohmic heating Q_Ω. Because the two electrodes have different exchange currents, the heat generation of the two electrodes is normally different.

The temperature dependence of heat generation Q_{η_T} is large and complex. Equation (5.2) shows that at constant voltage, the heat generation will approximately double for every 10 K, because the current increases according to Arrhenius's law, but at constant current, the charge transfer voltage $(E - E_o)$ will decrease significantly (see chapter 2), as the temperature is explicitly in the denominator of the exponent of the BV equation. Additionally, the equilibrium voltage E_o is temperature dependent. The temperature dependence of the reaction constants, which changes the exchange current density, must in principle also be considered.

If more than one electrochemical reaction has to be taken into account, it is necessary to estimate how the terminal current is divided between the different reaction paths. The respective heat generation must be calculated for each reaction. Under otherwise identical conditions, the ratio of main to side reaction currents changes with temperature.

5.2.4 Reversible Heat of Reaction

In all chemical reactions, there is a temperature change caused by the entropy change of the reaction products $T\Delta S$ (see chapter 2), which can be positive (heat is released and the system releases heat to the environment – temperature increase) or negative (heat is required and removed from the system – temperature decrease), depending on the direction of the reaction. The reversible heat is proportional to the current and changes sign when the direction of the current changes due to the reversibility of the reactions.

$$Q_{rev} \sim I \cdot T\Delta S. \tag{5.5}$$

As the temperature increases, heat generation or heat consumption increases in proportion to the temperature. For a charge-discharge cycle with the same initial and final state, the integration over the entire heat generation results in a value of 0.

Just as the equilibrium voltage of the reaction can be calculated from the Gibbs free enthalpy ΔG by dividing by zF, a voltage value can be calculated from $T\Delta S$ by dividing by zF, which is explicitly used for heat considerations in the float charging mode of fully charged lead-acid batteries. Multiplying this term by the current gives the heat released or absorbed by the reaction:

$$U_{rev} = T\Delta S/zF. \tag{5.6}$$

U_{rev} may only be used for heat calculations, not for voltage calculations.

Each electrochemical reaction has a different reversible heat. In lead-acid batteries, for example, the positive electrode cools down slightly and the negative electrode heats up slightly during discharge. This effect is technically irrelevant in lead-acid batteries and is very difficult to measure. Figure 2.2 shows the relatively small influence of reversible heat on a lithium-ion battery. In some alkaline nickel batteries, however, the effect can be clearly observed at high currents and is technically used to control a charging characteristic.

In the case of electrochemical reactions running in parallel, the reversible heat of each individual reaction must be considered, and the distribution of the total current to the various reaction paths must be estimated.

5.2.5 Comparison of Heat Generation Terms

The heat generation processes associated with electrochemical reactions take place at the interface between the active masses of the electrode and the electrolyte. In contrast, ohmic heat generation is distributed over all components in the cell, depending on the resistances and the local current densities. The combination of

a) locally distributed different heat generation, that is, a heat generation density resulting from the combination of the location of the physical effect and an inhomogeneous current density distribution, and
b) the inhomogeneity of heat capacity and the anisotropy of heat conduction lead to an inhomogeneous temperature distribution in a cell.

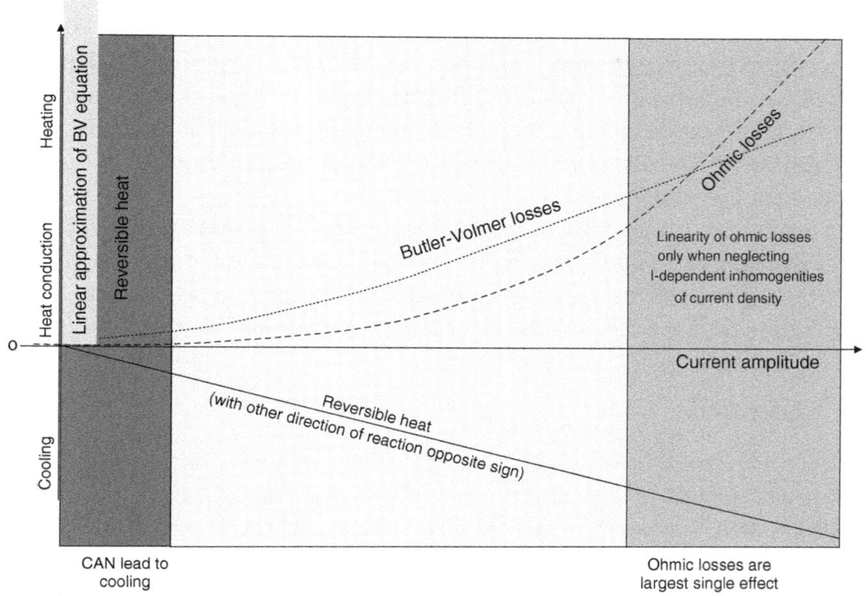

Figure 5.2 Diagram showing schematically the proportions of the various effects on heat generation as a function of the current amplitude. If the reversible heat has a negative sign, the system can cool down during charging or discharging.

The different dependence of the above three terms on the current (ohmic component quadratic, charge transfer disproportionate $[I + I \cdot \ln I]$, and reversible heat proportional to current) means that their contribution to the total heat generation changes with the current amplitude. At very low currents, charge transfer dominates due to the high BV resistance close to the equilibrium voltage. As the current increases, a phase may follow in which the reversible heat effect predominates. In this range, it is possible to observe cooling of a cell during charging or discharging despite the flow of current. This cooling effect is shown in Figure 5.2, as it is counterintuitive. At very high currents, ohmic heat generation dominates due to the quadratic increase with increasing current amplitude, although in systems with very low ohmic resistance, this range can often not be reached in normal operation.

The various dependencies of the three heat generation types described above on current are shown in Figure 5.2.

5.3 Heat Exchange with the Environment

Heat is exchanged with the environment by convection, conduction, and radiation. Large surfaces are decisive for convection and radiation. The greater the heat generation for a given surface area, the more difficult it is to achieve a high degree of heat exchange with the environment.

5.3.1 Heat Conductivity

Heat conductivity through cables and between cells of a string can be largely neglected. The temperature difference between cells is only relevant in the event of a fault, and the heat conduction through the normally relatively long connection cables from the terminals to the loads or the charger is small due to the usually small temperature differences.

Heat conductivity from cells to the bottom or sides of a container or cooling surfaces depends to a large extent on the thermal conductivity of the cell in the respective direction and the exact conditions of heat transfer. Contact of a large heat sink with the side of the cell with the highest thermal conductivity is very effective.

5.3.2 Convection

Convection requires a fluid flow (air, water, coolant, etc.) with high velocity and heat transfer coefficient. Depending on the surface and geometry, a distinction must be made between laminar and turbulent flow. Natural convection is often sufficient for cells to dissipate the heat generated during normal operation. In multicell batteries, natural convection requires design features, for example, a distance of a few millimeters between cells, to ensure that a sufficiently high fluid flow can develop, increasing the volume and reducing the energy density. Using technical convection, such as fans, smaller distances are possible, although the energy required to generate a sufficiently high fluid flow must be taken into account. Most lead-acid batteries only have natural convection; lithium-ion batteries often have small ventilators or liquid-based cooling systems due to installation conditions.

5.3.3 Radiation

The heat exchange via radiation between the cell and the environment can be relevant if the ambient temperature differs significantly from the cell temperature. Cell containers with high emissivity, for example, made of plastic or corresponding coatings, are then advantageous.

5.3.4 Heat Balance

Bernardi proposed the following approach for determining the heat balance [7]:

$$\frac{\mathrm{d}T}{\mathrm{d}t} = \left[I_{\mathrm{MR}}\left(U - U^0_{\mathrm{MR}}\right) + I_{\mathrm{SR}}\left(U - U^0_{\mathrm{SR}}\right) + I_{\mathrm{MR}}\frac{T\Delta S_{\mathrm{NR}}}{n_{\mathrm{MR}}F} + I_{\mathrm{SR}}\frac{T\Delta S_{\mathrm{NR}}}{n_{\mathrm{SR}}F} - \alpha A\Delta T - \sigma \in A\Delta T^4\right] \cdot \frac{1}{m_{\mathrm{B}}C_P}, \tag{5.7}$$

with

T: battery temperature,
ΔT: difference between ambient and battery temperature,
t: time,

$I_{MR, SR}$: current of the main reaction (MR) or side reaction (SR),
$E_{o, MR, SR}$: equilibrium voltage of the main or side reaction,
$\Delta S_{MR, SR}$: entropy of the main or side reaction,
$n_{MR, SR}$: number of electrons transferred in the main or side reaction,
F: Faraday constant,
α: heat transfer coefficient,
ε: radiation coefficient,
σ: Stefan-Boltzmann constant,
m_B: battery mass,
C_P: thermal capacity of the battery, and
A: surface.

The equation does not include heat sources or sinks based on other electrochemical side reactions, such as corrosion processes, chemical processes, enthalpies of mixing, or heat of intercalation and deintercalation, provided these are distinguished from chemical processes. Technically, these effects are normally negligible.

The explanations presented here show that heat generation is a complicated function of the current amplitude and the state of charge and that the parameters for heat dissipation and the temperature differences to a heat sink are difficult to calculate. A solution of the equation is only possible numerically and is often not necessary. In the chapters on lead-acid and lithium-ion batteries, the requirements of battery systems for cooling and heating systems are discussed.

5.4 Effect on Temperature

The temperature resulting from the inhomogeneous heat sources, heat dissipation, and inhomogeneous properties of a cell is a function of the heat capacity (approx. 1 kJ/(kg K) ± 20% for many cells). The temperature varies locally not only inside the cell but also on the wall of cell containers. Following is an example of a commercial battery module with 12 cells connected in series.

The following temperature differences were measured at high loads within the permissible operating conditions without cooling (see Figure 5.3):

a) 4.2 K maximum temperature difference between the bottom and top of a cell,
b) 9.6 K maximum temperature difference between the top of the cells,
c) 10.9 K maximum temperature difference between the bottom of the cells, and
d) 18.7 K maximum temperature difference between the lowest temperature of the bottom of a cell and the maximum temperature on the bus bar above the cell with the highest temperature.

However, the question of which position for a temperature sensor might be most suitable for measuring the average temperature of the cells depends not only on the load profile, but also on the duration of the load. In addition, it must be ensured that rapid changes in the ambient temperature (e.g. driving out of a heated underground car park in very cold weather) do not distort the measurement too much. Determining the best position and number of temperature sensors is therefore a difficult task.

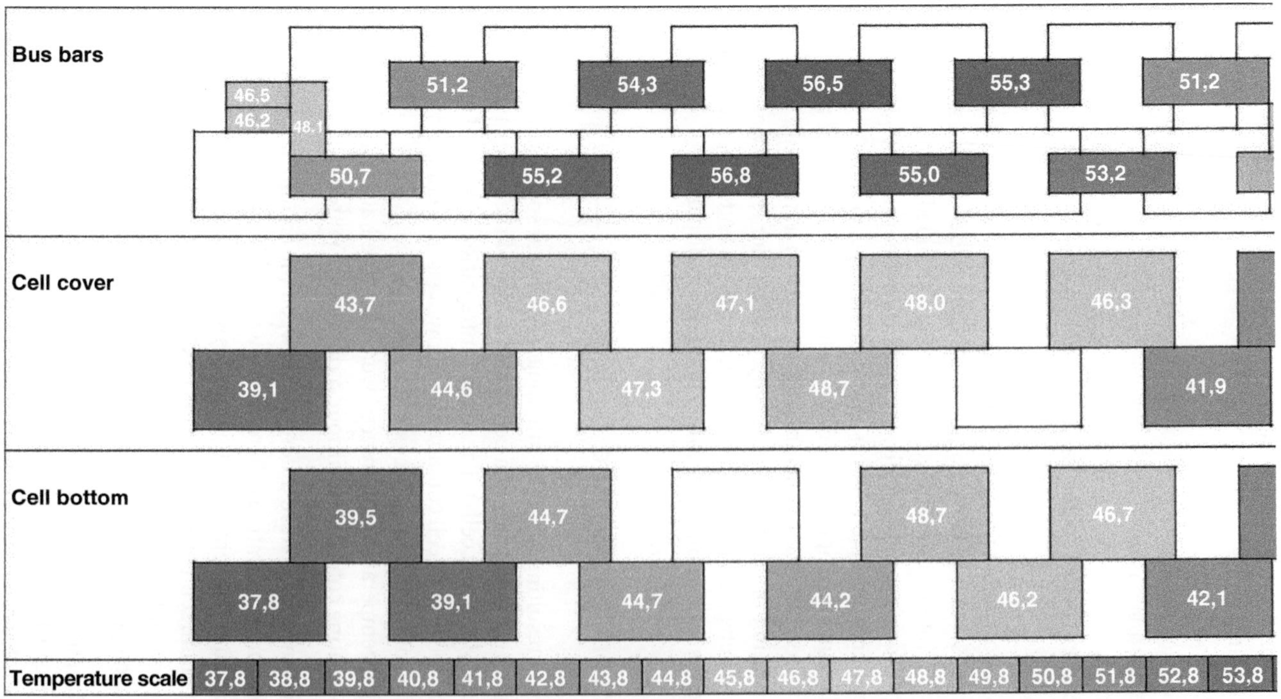

Figure 5.3 Temperature measurements of a lithium-ion battery. Source: Courtesy of A. Schmieder.

5.5 Determination of Thermal Parameters

Various methods can be used to determine the thermal properties of a battery. One possibility is to start from the thermal properties of the individual components. However, for many batteries, data (e.g. heat capacity of the materials and weight) are not available or not available with sufficient accuracy.

In recent years, thermal tests have therefore increasingly been carried out on complete cells or even batteries. For example, accelerated rate calorimetry, for which standardized laboratory devices are commercially available, is used to heat a cell with a given temperature rate and measure the heat input. At a certain temperature, irreversible decomposition reactions start, the heat quantities of which are measured. From this, conclusions can be drawn about the total amount of energy that can be released in the event of an internal short circuit, from which the necessary protective measures can be derived.

References

1 Grabow, J., Klink, J., Orazov, N. et al. (2023). Triggering and characterisation of realistic internal short short-circuits in lithium-ion pouch cells – a new approach using precise needle penetration. *Batteries* 9 (10): 496.

2 Klink, J. (2021), *Oral Communication.*

3 Haußmann, P. and Melbert, J. (2017). *SAE International Journal of Alternative Powertrains* 2: 6.

4 Euler, J. and Horn, L. (1965). Current distribution over the length of accumulator plates made of tubular elements. *Archive for Electrical Engineering* 50: 85–90.

5 Meiwes, J. (1985). Investigation of current distribution in batteries using a model with one-dimensional distribution (Original German title: Untersuchung zur Stromverteilung in Batterien anhand eines Modells mit eindimensionaler Verteilung). *etzArchiv* 7 (12): 381–387.

6 Benger, R. (2012). *Dynamic Behavior of Inverter-Fed Energy Storage Systems (Original German title: Dynamisches Verhalten von Umrichtergespeisten Energiespeichersystemen).* Dissertation. TU Clausthal.

7 Bernardi, D., Pawlikowski, E., and Newman, J. (1985). A general energy balance for battery systems. *Journal of the Electrochemical Society* 132: 5.

Task

Task 5.1

Design of a battery system

Based on the information provided so far, design a grid-connected, multicell battery system (including charger) that must be installed outdoors in a continental climate (very hot summers and very cold winters) and must supply loads with high power for short periods. The requirement for avoiding temperatures above approx. 30 °C must be fulfilled.

6

Ageing Processes and Service Life of Batteries and Cells

Aging processes in different battery systems differ greatly. However, for the following chapters on battery states, models, parameter determination, and battery diagnostics, it is useful to present general aspects of aging and lifetime together.

Aging processes are essentially based on the mechanical stress on the active masses caused by discharging and charging processes as well as on chemical and electrochemical processes. These processes are often the result of coupled anodic and cathodic reactions at one electrode. It is therefore always expected that the number of cycles and the cycle depth will have an effect on the service life, and that the aging processes, which are independent of the charge throughput, are dependent on temperature and time.

To investigate aging processes in batteries, test conditions are selected that either

- cause particularly the aging processes to be studied, while limiting all other aging processes as much as possible, or
- consist only of the continuous repetition of application-specific load profiles, for example, precisely defined cycles (depth of discharge [*DoD*], average state of charge, current amplitude, and temperature) or precisely defined periods without current (state of charge and temperature).

The tests must be carried out until a predetermined end-of-life criterion, usually a reduction in capacity or an increase in internal resistance. When investigating aging processes at the component level, material characterizations are often carried out at the end of the test.

To estimate the lifetime, the period for which a battery can fulfill a certain functional requirement, it is necessary to select test conditions based on the respective operating conditions of the application (generally a combination of calendar aging at different temperatures and states of charge and cyclic aging with many different cycles). Such tests determine the effects of all aging processes together on the properties of the battery. Identical aging processes

(Continued)

Battery Technology: Fundamentals of Battery Electrochemistry, Systems and Applications,
First Edition. Alexander Börger and Heinz Wenzl.

Companion website: www.wiley.com/go/Börger/Battery_Technology

(Continued)

proceeding at identical rates (e.g. corrosion rates) may lead to very different lifetimes depending on the design of the cells or battery.

As batteries generally show only a slow deterioration of their properties, it is important to precisely define the end of their service life.

6.1 Classification of Aging Processes

A brief overview of aging processes is required for the following chapters. A rough distinction can be made between three groups of aging processes:

1) Electrochemical processes
   In many cases, these processes involve coupled anodic and cathodic reactions at one electrode, similar to those described in section 2.5 on mixed potentials. In particular, the high or low voltages at the end of discharging and charging often cause significant degradation.
2) Mechanical processes
   During discharging and charging, the volume of the active masses and/or the crystal structure of the active masses changes. This volume expansion and contraction leads to long-term damage to the macro- or microstructure of the electrodes. The *DoD* as the difference between the initial and final state of charge in a cycle and the frequency of cycles play a major role in these aging processes.
   In addition, external mechanical influences (e.g. external pressure, shocks, vibrations, etc.) can affect the battery properties and lifetime.
3) Chemical processes
   Regardless of the potential of the active masses in relation to the electrolyte and other passive components of a cell, chemical reactions can take place, for example, the oxidation of separators and other materials by cathode materials, which are often strong oxidizing agents.

When investigating aging processes in cells without subsequent analyses of materials and components (see chapter 10), it is worth pointing out that the results of such investigations usually only state the number of cycles or the duration until a previously defined end-of-life criterion is reached, but not data on the actual aging processes, for example, changes of materials due to corrosion (e.g. weight loss per area and time).

A frequently used relationship between the number of cycles or equivalent full cycles[1] and *DoD* is shown in Figure 6.1. The decrease in the number of equivalent full cycles is primarily attributed to the mechanical stress on the structure of active materials.

---

1 An equivalent full cycle is calculated by dividing the charge throughput of a cell by its nominal capacity regardless of the cycles (*DoD* and average state of charge) that were carried out.

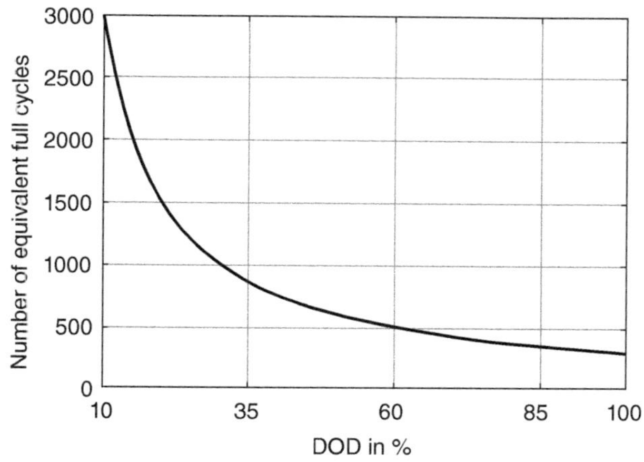

Figure 6.1 Number of equivalent full cycles of a lithium-ion battery cell as a function of the depth of discharge at constant current amplitude, room temperature, and an initial state of charge of 100% Courtesy of E. Tchoupou Lando.

It should also be noted:

- It is almost never possible to study a single aging process in cells alone. All other aging processes take place simultaneously, but at a reduced rate. Therefore, the results of aging studies in cells always refer to the complex interaction of all aging processes under the selected test conditions.
- When test conditions consist of continuously repeated load profiles (e.g. a precisely defined cycle with a fixed current, average state of charge, *DoD*, and temperature), transferring the results to complex load profiles with a range of different cycles and periods of rest is difficult and may be misleading.

Aging processes in different battery technologies differ so greatly that it does not make sense to discuss them further here. These processes are described in more detail in the chapters on the individual battery technologies.

The effects of all individual aging processes together on the ability of batteries to fulfill operational functional requirements are important for their use. The terms "end-of-life" and "lifetime estimation," which are independent of battery technology, are explained in detail in the following because they are important for the precise definition of some of the state variables described in chapter 7.

6.2 Service Life

6.2.1 Definition of End of Life

The usual criterion for the end of a battery's service life is the reduction of capacity under precisely defined conditions (often nominal conditions) to a lower value, for example, 80% of the nominal capacity. In many cases, this criterion may be sufficient from a general point of view but is often inadequate. A very obvious example is hybrid vehicles, where only a small proportion of the capacity is used at a medium state of charge, but the internal resistance at this state of charge is much more important than the capacity. In such cases, it is therefore of minor importance how high the capacity is, as long as the internal resistance is still low enough. For this reason, the term "end of life" is analyzed in detail in the following. The discussion is based on terms used in standards and by industry associations. The International Electrotechnical Commission (IEC) manages an International Electrotechnical Vocabulary (IEV), which is available under IIEC 60050 or www.electropedia.org for further details. Translations and definitions are available in many languages.

The service life of technical components, here a battery, is the time of use of a product which can no longer be repaired between commissioning and failure, that is, the inability to fulfill a required function. The service life of a product therefore depends on the ability to carry out repairs and replace defective parts technically within the time and cost restrictions of the application.

DIN EN 60896-22 defines "performance" or "performance level" as the degree to which an intended function can be fulfilled or as a characteristic value that indicates the ability of a piece of equipment (battery) to perform a specified function. If the performance of the battery is no longer sufficient to fulfill an intended function, it has failed, and the end of service life has been reached. Therefore, it is necessary to define which functions the battery must fulfill. If even one required function can no longer be fulfilled, the battery has reached the end of its service life for that particular application. Further use in other applications with other functional requirements may still be possible. This is sometimes referred to as the "second life" of a battery.

The most important functions and requirements are

- sufficient capacity,
- sufficiently low internal resistance, and
- ability of the battery to absorb or release high power for short periods of time (for braking, accelerating, or starting combustion engines) within the range of permissible operating conditions.

Permissible operating conditions include, for example, the lowest temperature and the lowest expected state of charge for which the overall system has been designed.

However, there are other requirements for defining the end of life. Therefore, the end of service life should not be understood as a scalar value but as a multidimensional vector, in which each vector represents an indispensable function of the battery. Not being able to fulfill one of these functions under normal operating conditions marks the end of the service life. The choice of capacity reduction as the sole criterion for the end of service life is inappropriate for many applications but may be justified by the fact that a decrease in capacity is often linked to the deterioration of other requirements.

Figure 6.2 shows a schematic diagram of a lifetime vector. In this example, dimensions of the lifetime vector include:

- capacity (if necessary, with further details such as temperature and discharge current);
- internal resistance at a given state of charge and temperature (in some circumstances, the low-temperature performance for starting combustion engines is a separate additional dimension);

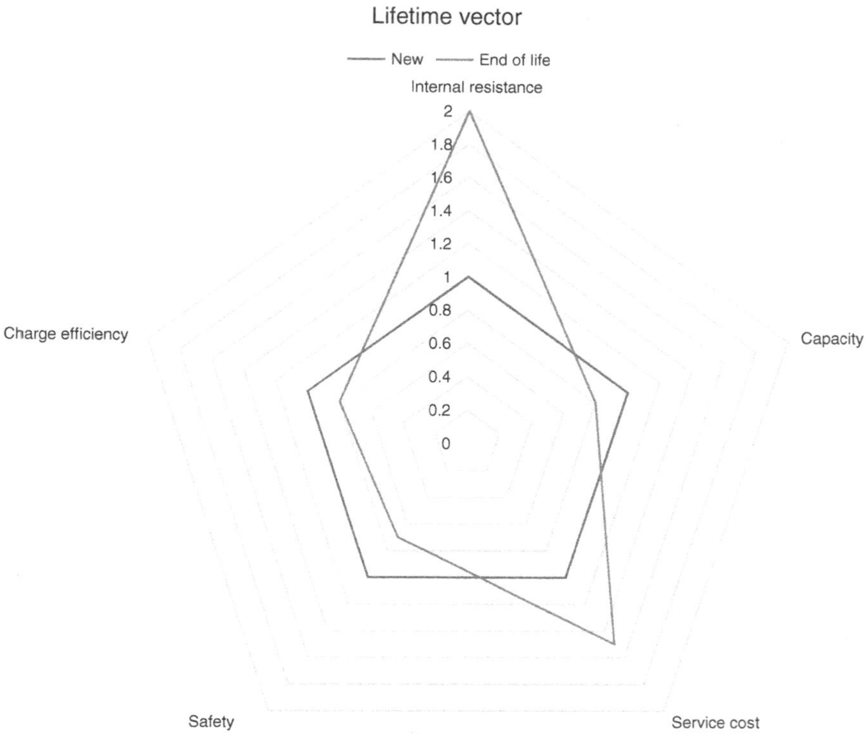

Figure 6.2 Schematic diagram of a lifetime vector.

- charging efficiency/charge acceptance;
- operating costs
  Costs for energy generation in diesel generator based remote autonomous power systems (RAPS) can become prohibitively high, because aged batteries often have a higher self-discharge rate and lower charge acceptance, which increases the costs of recharging, and maintenance costs for multicell battery systems increase dramatically once cells have to be replaced at each maintenance; and
- safety.

The definition of end of life chosen here is:

The end of life is reached as soon as one of the required functions of the battery can no longer be fulfilled within the permissible operating limits (temperature, state of charge, etc.).

6.2.2 Determining End of Life

The definition of end of life used here is conceptually very clear and unambiguous. Determining the exact point at which the end of life is reached is fairly straightforward in the laboratory but difficult, and sometimes impossible, in real applications. Studies of the actual service life of batteries in operation, covering more than just the replacement date, are rarely published.[2] The following considerations are intended to emphasize that data on the service life of batteries in real applications should be scrutinized with great care.

In the laboratory, continuously measuring the capacity and internal resistance under the chosen test conditions is possible without any problems, but it is not always easy to transfer the results to different functional requirements. What, for example, are the capacity or internal resistance values for determining the capability of a battery for starting an internal combustion engine under application-specific conditions of low temperature and low state of charge? The performance of a starter motor when starting a combustion engine depends on the oil and engine temperature, the viscosity and age of the engine oil, the time since the last engine start, the age and condition of the spark plugs, etc. The starting capability of a battery must be determined independently of such parameters, for example, by means of a standardized performance test. Knowledge of the internal resistance under nominal conditions, for example, 50% state of charge and room temperature may not necessarily help.

Test specifications from standards are not appropriate for this if they are designed as acceptance conditions for new batteries. For the starting of diesel engines for uninterruptible power supply (UPS) systems, for example, the standard DIN 6280-13:1994-12 1 [2] requires that it must be possible to start the engine three times at 5 °C, each time for 10 seconds with a 5 seconds rest. This requirement could be implemented in a test cycle that could be used in the laboratory to determine for how long it will still be possible to start a combustion engine safely and reliably even under difficult conditions. However, no generally applicable performance profile for testing a battery is specified in this standard.

In practice, the end of service life is determined in the worst case by the fact that a required function can no longer be fulfilled, and this can be attributed to the battery.[3] The engine no longer starts or the uninterruptible power supply system for production systems or data centers fails. In these cases, however, a meaningful analysis of the achieved service life of batteries is only possible if the batteries are tested after replacement and an estimate is made of when the battery replacement should have taken place to avoid such failures.

The best way to replace a battery without a previous functional failure is to carry out measurements during operation and determine if the measured values have fallen below values that experience has shown are associated with the end of the battery's service life. Examples of this for lead–acid batteries in UPS systems are:

- strong fluctuations in the cell voltages during float charge mode, and
- internal resistance exceeding threshold values (by means of AC resistance measurement, see IEC TR 62060 [3]).

---

2 The degree to which a battery fulfills functional requirements at the time of replacement, that is, capacity test after a previous full charge, internal resistance at the specified conditions regarding temperature and state of charge, must be carried out.

3 In many cases, the failure to start is due to faults in the alternator or an unfavorable energy balance of the vehicle, rather than a fault in the battery.

In lithium-ion batteries, increasing self-discharge rates are a strong indication for a near end of service date.

6.2.3 Changes in Properties During Use

Depending on the battery technology and application, battery properties change differently over time. A particularly instructive curve for this discussion of changes of capacity and internal resistance is shown in Figure 6.3. The capacity of the battery increases after commissioning and remains at a constant level for a long time, while the internal resistance initially decreases and then also remains at a constant level. Only after approx. 70–80% of the service life does these two values, which are relevant to the functionality of the battery, start to deteriorate. Unchanged battery characteristics therefore do not mean that no aging processes have taken place, but only that the effects of all aging processes have not yet had any impact on the functional parameters of the battery.

The development of important battery properties over time must be tested experimentally with the respective application-specific load in each individual case.

The changes in properties often follow exponential functions (e.g. calendar aging of some lithium-ion batteries). A slight change in a threshold value, which defines the end of life, can thus lead to a significant change in the expected service life. Figure 6.4 shows the typical change in capacity and internal resistance of lithium-ion batteries. This clearly shows that, although the decrease in capacity is associated with the increase in internal resistance, there is no simple mathematical relationship between the two values. The lines drawn for 80% and 75% state of charge show that the expected service life would increase by approx. 30% when the threshold values are changed.

The curves in Figure 6.4 clearly show that the measurement of battery properties during operation does not allow any direct statement to be made about the remaining

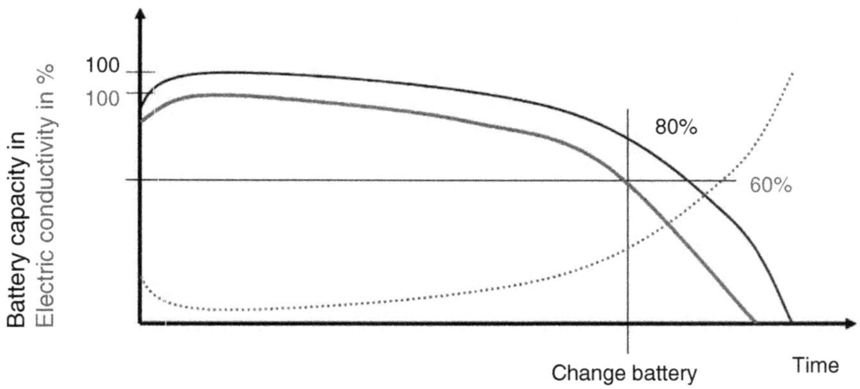

Figure 6.3 Diagram showing schematically the development of internal resistance (alternating current measurement) and capacity of a fully charged lead–acid battery in the nominal temperature range. The values initially improve slightly at the beginning of service life and only deteriorate toward the end of the service life.

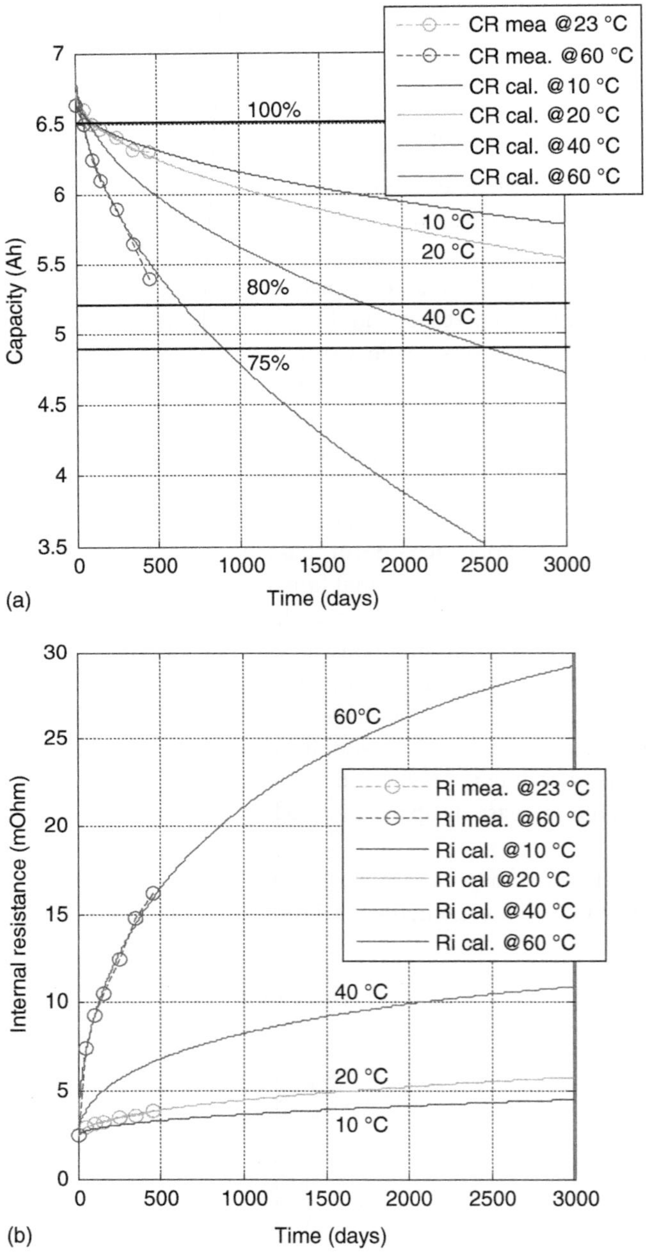

Figure 6.4 Measured and simulated curves of (a) capacitance and (b) internal resistance under calendar aging conditions. The threshold values for 75% and 80% capacity are shown.

service life. Therefore, further methods must be developed to predict the service life of a battery as a function of its use. It should also be noted that the end of a battery's service life is often determined by the weakest cell. The point at which experience shows that the cell properties start to drift apart plays an important role in the service life of the battery as a whole.

6.3 Limits of the Service Life

6.3.1 Basic Limitation of the Service Life

The service life of commercial batteries is constantly increasing due to improvements in materials and manufacturing processes, as well as the optimization of designs and better operating strategies. This raises the question of whether there are limits to this development that will limit further improvements, such as thermodynamically unstable states with high kinetic inhibition.

Thermodynamically unstable states exist both in lead–acid batteries (corrosion of the positive grid, see chapter 12) and in lithium-ion batteries (solid electrolyte interface [SEI], see chapter 13). Such thermodynamically unstable states cannot be eliminated by improved designs or modifications of materials, but only fundamentally different materials, for example, by introducing other electrolytes and/or anode materials for lithium-ion batteries.

The lifetimes now achieved when using high-quality batteries adapted to the respective application are nevertheless very long. With well-controlled operating conditions and highly reproducible production, service life problems are limited, and the economic benefit of batteries with a longer expected service life at possibly higher prices is low.

6.3.2 Manufacturer's Information on the Expected Service Life

The above explanations have shown that the expected service life depends to a large extent on the operating conditions. A statement from battery manufacturers about the expected service life, possibly even coupled with warranty claims, would therefore only be possible with considerable limitations and restrictions regarding their use. To provide guidance on the use of batteries, the Association of European Automotive and Industrial Battery Manufacturers (EUROBAT) as the European industry association has agreed on definitions for various service life terms for batteries in stationary applications:

1) Use in theory: design life
   A value derived from durability tests, taking into account the design and construction of the individual components and the parameters limiting the service life.
   The term thus describes the lifetime of the battery that might be achieved theoretically.

2) Use in the laboratory: endurance
   Values determined under defined, partly standardized, and partly accelerated aging conditions.
   The term thus describes the lifetime achieved in the laboratory.
3) Use in practice: service Life
   A value determined on the basis of field experience under optimum conditions, describing the period in which a specified capacity or other performance levels will be available.

The term therefore describes the service life that can be achieved in practice under optimum application conditions, which must be specified. For any application, only the term "service life" is a reference point for technical and economic considerations.

6.4 Lifetime Prediction

For planning and design purposes, it is important to be able to estimate the expected service life of a battery under actual operating conditions. The current methods for predicting service life under complex load profiles can be divided into ampere-hour throughput methods and event-based methods [4, 5]. The values determined by both models are planning values based on average properties of the battery types used and do not allow for an exact service life prediction for a specific battery.

6.4.1 Weighted Ampere-Hour Throughput Methods

The basic assumption of this method is that a known amount of charge Q_{max} can be removed from a battery under standard conditions until the end of its service life, however defined. Normally, the amount of charge that is charged back into the battery is not counted. The ratio of the amount of charge withdrawn during the period under consideration Q_t in relation to the maximum amount of ampere hours that can be withdrawn until the end of the service life Q_{max} is referred to as the loss of lifetime LL_t of the battery:

$$LL(t) = Q_t/Q_{max}. \tag{6.1}$$

Since batteries lose capacity at different rates under different operating conditions and the total charge throughput can be less or more than under standard conditions, weighting factors are introduced. This approach, the so-called weighted ampere-hour model, has been used for some time. In each case, a reference value for the amount of charge that can be drawn from the battery under standard conditions must be determined empirically or calculated from detailed data sheet information on the number of cycles under standard conditions up to the end of service life. The weighted loss of service life during an ampere-hour throughput Q_i with the corresponding weighting factor w_i is then calculated as follows:

$$LL_i = (w_i \cdot Q_i)/Q_{max}, \tag{6.2}$$

where i is the number of different operating conditions, each with the same weighting factor.

The following approaches are available for determining weighting factors:

1) Averages, for example, using the average temperature and the average discharge current of the battery in an application. This is obviously a very rough but effective and easy-to-use method.
2) Model-based approach, for example, continuous calculation of a weighting factor from the operating conditions of the battery, as used, for example, by Schiffer et al. [6].

Common to all ampere-hour throughput models is the inability to handle rest periods (calendar aging) and the difficulty of determining the weighting factors experimentally. The weighting factors are usually fitted [6] and then assumed to be generally applicable values.

The total percentage lifetime loss LL_t in the period under consideration, in which i different phases with different weighting factors occur, is:

$$LL_t = \sum_i LL_i = \sum_i (w_i \cdot Q_i)/Q_{\max}. \tag{6.3}$$

The end of life is reached when $LL_t = 1$ is reached.

6.4.2 Event-Based Lifetime Prediction Methods

In event-based lifetime prediction methods, the use of the battery is divided into specific events – short sections of the operating conditions. This is comparable to the division into phases of equal weighting in the ampere-hour throughput method. When defining events, for example, cycles with a certain current amplitude, *DoD*, average state of charge, and temperature, it is important that they enable aging tests on batteries in which the batteries are subjected only to a continuous repetition of one event. It must be possible to allocate different events to the operating conditions throughout the service life without exception. Events must include both cycles (average state of charge, *DoD*, current amplitudes, and temperature) and rest periods (state of charge, temperature, and duration).

An event-based method has great similarities with methods used in fatigue strength theory in mechanical engineering for estimating the service life of components under mechanical stress. Analogies between concepts from mechanical engineering and concepts from electrical engineering and electrochemistry suggest that the methods of fatigue strength theory can be applied to batteries in principle. The method of crack propagation [7–9] used in fatigue strength theory has already been successfully used to quantify the electrode degradation of lithium-ion batteries, and the number of cycles as a function of the *DoD*, which is frequently used for batteries, corresponds to the so-called Wöhler curves used in fatigue strength theory, albeit with a different axis scaling.

Each event i is assigned a loss of service life per event from tests in which a battery is subjected to one event only until the end of life is reached:

$$LL_i = 1/N_i, \tag{6.4}$$

with N_i as the number of cycles until the end of service life is reached. If the battery is subjected only to calendar aging at a constant state of charge and temperature, then the percentage loss of service life per day is as follows:

$$LL_i = 1/T_i, \tag{6.5}$$

with T_i as the service life (in days) until the end of the service life.

The analysis of operating data or data from simulations is used to determine the number of events n_i under cyclical load or the dwell time per event t_i under calendar aging that occur within the observation period. The lifetime loss in the observation period t is then:

$$LL_t = \sum_i n_i/N_i + \sum_i t_i/T_i. \tag{6.6}$$

The summation of the loss of lifetime per event for all the different events regardless of the sequence and type of events without weighting factors is called linear damage accumulation in fatigue strength theory. Tchoupou Lando has experimentally validated the concept of linear damage accumulation for the cyclic operation of lithium-ion batteries. This means that, in contrast to ampere-hour throughput models, all parameters (N_i and T_i for all events i) can be determined experimentally for lifetime prediction.

As each event can be assigned an exact duration, the time can be calculated that corresponds to the total lifetime loss calculated in this way. As with the ampere-hour throughput model, the end of service life is reached when the value calculated for loss of life becomes one.

6.4.3 Forecast of the Capacitance and Internal Resistance Curve

Lifetime predictions are not able to predict the development of functional properties such as capacity and internal resistance during aging, but only the point at which the end-of-lifetime criteria are exceeded. However, it is possible for the event-based lifetime model to predict capacity and internal resistance changes over time under any load profiles if the calculated loss of lifetime is matched to measured values of capacity and internal resistance changes.

References

1 Tchoupou Lando, E. (2021). *Event-based lifetime model for lithium-ion cells using linear damage accumulation (Original German title: Ereignisbasiertes Lebensdauermodell für Lithiumionenzellen mittels linearer Schadensakkumulation).* Dissertation. Clausthal University of Technology.

2 DIN 6280-13:1994-12 (1994). *Generating Sets – Reciprocating Internal Combustion Engine Driven Generating Sets – Part 13: For Emergency Power Supply in Hospitals and in Structures Designed to Accommodate Large Numbers of People.* Berlin: Beuth.

3 IEC TR 62060:2001 (2001). *Secondary Cells and Batteries – Monitoring of Lead Acid Stationary Batteries – User Guide.* International Electrotechnical Commission.

4 Sauer, D.U. and Wenzl, H. (2008). Comparison of different approaches for lifetime prediction of electrochemical systems – using lead-acid battery ice as example. *Journal of Power Sources* 1176: 534–546.

5 Wenzl, H., Baring-Gould, I., Kaiser, R. et al. (2004). Life prediction of batteries for selecting the technically most suitable and cost effective battery. *Journal of Power Sources* 144 (2): 373–384. https://doi.org/10.1016/j.jpowsour.2004.11.045.

6 Schiffer, J., Sauer, D.-U., Bindner, H. et al. (2007). Model prediction for ranking lead-acid batteries according to expected lifetime in renewable energy systems and autonomous power-supply systems. *Journal of Power Sources* 168 (1): 66–78. https://doi.org/10.1016/j.jpowsour.2006.11.092.

7 Deshpande, M., Verbrugge, M.W., Cheng, Y.-T. et al. (2012). Battery cycle life prediction with coupled chemical degradation and fatigue mechanics. *Journal of the Electrochemical Society* 159 (10): A1730–A1738.

8 Kotak, N., Barai, P., Verma, A. et al. (2018). Electrochemistry-mechanics coupling in intercalation electrodes. *Journal of the Electrochemical Society* 165 (5): A1064–A1083. https://doi.org/10.1149/2.0621805jes.

9 Lin, C., Tang, A., Wu, N., and Xing, J. (2016). Analysis for mechanical failure of DISs with graphite anode in Lithium ion batteries for electric vehicles. *Nanomaterials and Nanotechnology* 6: 184798041668080. https://doi.org/10.1177/1847980416680806.

Tasks

Task 6.1

Describe the different service life terms as defined in ZVEI data sheet no. 23 using the following examples:

a) Starter battery in a vehicle with a combustion engine.
b) Electric traction battery for a purely electric vehicle.
c) Stationary battery for uninterruptible power supply.

Task 6.2

What is the difference between calendar aging and cyclical aging, especially with low current amplitudes, during which not only aging processes take place through discharging and charging processes but also aging processes that are time-dependent only?

Task 6.3

Figure 6.1 shows the number of cycles to the end of service life as a function of the *DoD*. What do 500 cycles with a *DoD* of 100% mean if the end of service life is defined as 80% of the initial capacity and therefore less and less capacity is removed per cycle?

7

State of X Definitions and Calculations

Various terms, sometimes referred to as state of X, are required to describe the condition or status of a battery:

- State of charge (*SoC*)
  The amount of charge that can still be removed from a battery at any given point in time in relation to the amount of charge that can be removed from a fully charged battery.
- Depth of discharge (*DoD*)
- State of energy (*SoE*)
  The amount of energy that is still available in relation to the energy of a fully charged battery.
- State of health (*SoH*)
  The ratio of important performance characteristics of an aged battery, in particular capacity and internal resistance, in relation to the performance characteristics of a new, fully charged battery. The *SoH* describes the overall effects of all aging processes on important battery characteristics.
- State of function (*SoF*)
  The performance characteristics of an aged battery to fulfill the specified functional requirements at permissible operating temperature and *SoC*, compared to the minimum required performance characteristics of a new battery.
- State of safety (*SoS*)
  Concepts are still under discussion to quantify safety levels of batteries.

Capacity and internal resistance of a battery are not state variables, but rather properties of the battery.

7.1 Background

The properties of batteries are the result of the materials used, design, and production processes. An overview of the most important parameters can be derived

Battery Technology: Fundamentals of Battery Electrochemistry, Systems and Applications,
First Edition. Alexander Börger and Heinz Wenzl.

Companion website: www.wiley.com/go/Börger/Battery_Technology

from international standards. In addition to capacity under nominal conditions and internal resistance, or the capability to provide or absorb power pulses under standardized conditions, the most frequently referred to properties are:

- Low-temperature capacity and starting currents at low temperatures,
- Self-discharge rate,
- Charge acceptance,
- Electrolyte tightness,
- Vibration resistance, and
- Service life under cyclic and calendar aging.

These properties are determined in the laboratory under defined conditions, such as *SoC* and temperature. The methods for this are explained in detail in chapter 9.

When using a battery, some properties change quickly, and it is important to know the values of the properties or states exactly for safe and reliable operation. Methods for determining states therefore refer to batteries that are in operation, and it must be possible to determine them during operation. It has been found that the battery condition can be described sufficiently well with the following parameters[1]:

- State of charge (*SoC*),
- Depth of discharge (*DoD*),
- State of energy (*SoE*),
- State of health (*SoH*),
- State of function (*SoF*), and
- State of safety (*SoS*).

These state variables are defined in detail in the following, and methods for determining them are explained. In addition to these state variables, others are occasionally used, but these are generally closely related to the state variables discussed here.

Technical properties of batteries, such as changes in capacity and internal resistance, can be estimated using these state variables.

7.2 State of Charge, Depth of Discharge, State of Energy

The term *DoD* refers to either $1 - SoC$ or a difference in *SoC*, ΔSoC. To understand the term, it is necessary to first explain the term *SoC*, which is the most important parameter for controlling operation and for battery management systems. It should be pointed out that energy, not charge, is far more important, but historically *SoC* and not *SoE* has become the most widely used term. For further details, see Ref. [1].

1 The change in some properties during use is either indeterminable or typically does not occur (e.g. expected service life, vibration resistance).

7.2.1 Definition of State of Charge

SoC is the ratio of the amount of charge that can be drawn from a battery until the end-of-discharge voltage is reached, to the amount of charge that can be drawn from a fully charged battery. The general definition, using the nominal capacity as the reference value and using the sign convention that discharge currents have a negative sign, is therefore:

$$SoC(t) = 1 + \frac{1}{C_N} \cdot \int I_{\mathrm{MR}} \mathrm{d}t, \tag{7.1}$$

whereby the battery must be fully charged at the start of integration (time t_0), ($SoC = 1$), or

$$SoC(t) = SoC(t_0) + \frac{1}{C_N} \cdot \int I_{\mathrm{MR}} \mathrm{d}t, \tag{7.2}$$

where the integral is calculated from time t_0 and *SoC* $SoC(t_0)$ must be known exactly at time t_0. If side reactions can be neglected, the battery current I_{Batt} can also be used for the calculation instead of the main reaction current I_{MR}.

A fully charged battery therefore has an *SoC* of 1 or 100%, and a completely discharged battery has an *SoC* of 0. Despite this simple and precise definition, it is always necessary to pay close attention to what is meant. For a precise definition, questions regarding the capacity, estimation of the main reaction current, current measurement, and definition of full charge or the *SoC* at time t_0 must be answered (see Ref. [2]).

The capacity $C(t)$ of a battery depends on age, current, and temperature. When calculating the *SoC*, the nominal capacity is often selected as the reference point, and rather than the operating capacity, which depends on the application-specific discharge current and temperature. This definition has the following advantages:

- The *SoC* does not change simply because the temperature of the battery changes.
- During discharge, no assumptions need to be made about the previous and future use of the battery that influence the operating capacity.
- There is no need to estimate or measure the current battery capacity to calculate the *SoC*.

A *SOC*-dependent operating strategy leads to restrictions on use at the end of discharging or loads being switched off at the end of discharging. The following consequences arise when selecting the nominal capacity as a reference for calculating *SoC*:

+: If the battery capacity is greater than the nominal capacity, the battery can provide as much charge as it has been designed for, but is not discharged to its limit. The service life of all battery systems becomes shorter the deeper the discharge. A discharge to the end-of-discharge voltage that exceeds the system's requirements is perceived as less important than the advantage of increased service life.

−: If the battery capacity is lower than the nominal capacity under operating conditions or due to aging, the battery will reach its voltage limit unexpectedly early, although the *SoC* is still sufficient according to the calculation.

It should therefore be noted that knowing the *SoC* according to the above definition is misleading if the capacity under nominal conditions is not known at the same time. A calculated *SoC* of 50% for a battery that has only half of its capacity left due to aging means something different than the same value for a new battery.

The *SoC* can take on negative values when using the nominal capacity as a reference if the battery capacity is greater than the nominal capacity and the operation is controlled using only the voltage.

7.2.2 Main Reaction Current

When using the term *SoC* for batteries, no differentiation is made between the two electrodes. However, it is a fundamental difficulty of the above definition that the term "main reaction current" can only ever refer to one electrode. On each electrode, the battery current I_{Batt} has to be divided into partial currents, the main reaction current I_{MR}, which leads to charging of the battery and storage of energy, and side reaction currents I_{SR_i}:

$$I_{\mathrm{Batt}} = I_{\mathrm{MR},+} + \sum_i I_{\mathrm{SR}_i,+} = I_{\mathrm{MR},-} + \sum I_{\mathrm{SR}_i,-}, \tag{7.3}$$

with the indices + and − for the positive and negative electrodes, and *i* for the case of several parallel side reaction currents. It cannot be expected that

$$\sum I_{\mathrm{SR}_i,+} = \sum I_{\mathrm{SR}_i,-}, \tag{7.4}$$

so that the *SoC* of the two electrodes will normally be different, especially at the end of charging when side reactions increase due to the higher voltage. The same applies if the two electrodes have different capacities due to aging, design reasons, or production deviations. A full charge of both electrodes is only possible if, as with lead–acid batteries, there is a phase of overcharging with side reactions that cause little degradation. For lead–acid batteries, special algorithms must be introduced to calculate the *SoC*, with which the main reaction current can be estimated as a proportion of the total battery current, possibly separately for the negative and positive electrodes.

With lithium-ion batteries, side reactions can usually be neglected. Differences in the *SoC* of the electrodes resulting from aging processes of the two electrodes can no longer be compensated for (see chapter 3). Neglecting side reactions when calculating the *SoC* during overcharging tests of lithium-ion batteries results in values such as 200% *SoC*, which can no longer be linked to the amount of charge that can be removed later.

7.2.3 Measuring the Battery Current

Calculating the *SoC* solely by measuring the battery current is associated with the following difficulties: even without the further complication of side reactions:

a) Precision of the current measurement
   Depending on the application, discharge currents of significantly more than 100 A and charging currents of a similar magnitude (at least for very short

operating phases) must be expected for a 100 Ah battery. With a very good measuring device, an accuracy of 0.01% of the maximum value can be expected for the entire measuring range, that is, an average error of 10 mA. Using a measuring range of 100 A, the accumulated difference caused by measuring errors is already 7.2 Ah after 30 days and 86.4 Ah after one year.
At the same time, very low currents must also be measured precisely, for example, for the supply of monitoring electronics in an on-board power supply system with a power consumption of just a few watts. Due to long periods without any current, measurement errors add up and lead to further problems.

b) Self-discharging effects cannot be recorded via current measurement.
c) Analog-to-digital conversion (AD conversion) leads to an accuracy of only 0.1% when using a standard 10 bit AD converter.

Even with battery technologies without any side reactions, it is not possible to calculate the *SoC* solely by measuring the charge throughput (Coulomb counting) over a longer period of time (weeks or longer). For very short periods of time, for example, in applications with a daily full charge, estimating the *SoC* using Coulomb counting is, however, possible.

7.2.4 Calculating the State of Charge

Methods for determining the *SoC* therefore use combinations of several measured values and algorithms with stored discharge and charge characteristics. The most important are:

1) Voltage measurement
The voltage of a battery changes during charging and discharging and can therefore be compared with measured voltage values and used to estimate the *SoC*.
Measurement difficulties arise when the discharging and charging currents are unknown and fluctuate greatly. With knowledge of the expected current ranges, an *SoC* calculation can be achieved with very simple means, which has led to commercially successful products for lead–acid batteries used in industrial trucks.
When no current flows, there are voltage differences depending on the previous direction of the current and the length of time without current flow. Depending on the measurement method, the accuracy of determining the *SoC* using open-circuit voltage measurement is only approx. ±5% (Figure 7.1). Aged lithium-ion batteries also show significant deviations in voltage both in phases without current and under load, so the accuracy of the *SoC* determination for lithium-ion cells is problematic with pure voltage measurement. Lithium-ion batteries with iron phosphate as cathode material have a very flat voltage curve, so a purely voltage-based calculation of *SoC* is impossible. With lead–acid batteries, the problems with pure voltage measurement are even greater due to acid stratification (chapter 12).[2]

---

2 With other battery systems, there may be even greater difficulties due to very flat characteristic curves (e.g. lithium iron phosphate cathodes, lithium titanate anodes), hysteresis effects (alkaline nickel batteries), or downright pathological discharge curves (e.g. lithium sulfur batteries).

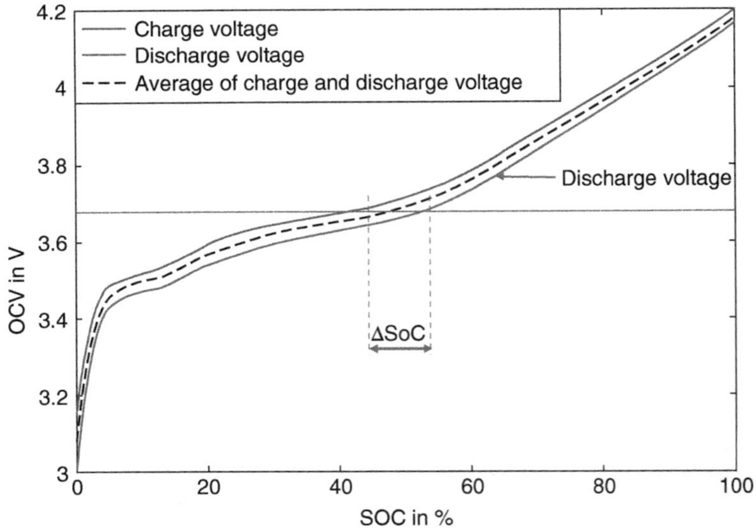

Figure 7.1 Open-circuit voltage measurement of a lithium-ion battery in the charging and discharging direction with C/20 current. Source: Courtesy of E. Tchoupou Lando.

2) Combined voltage and current measurement
   When measuring current and voltage at the same time, it is possible to evaluate the voltage value more accurately if the corresponding characteristic curves are stored (see also Figure 2.10, U–I characteristic curve). The additional measurement of the amount of charge, which is then easily possible, enables an even better calculation of the *SoC*. This combination is used in many uninterruptible power supply (UPS) systems and industrial trucks (see the descriptions in the last part of the book) and, in addition to accurate measurement, also allows for estimating the capacity reduction due to aging.
3) Rule-based algorithms
   The combination of voltage, current, and charge throughput measurement is further improved by applying rules. The calculated *SoC* value is adjusted using reference points. Almost all methods for determining the *SoC* use different reference points, generally the full *SoC*, at which the calculated *SoC* value is reset to 1. There are numerous other rules that are used [3]. One particularly easy-to-understand rule is to assign a specific value to the *SoC* within which the calculated value must lie (approx. 90% for most battery systems depending on the charging current) to the time at which the end of charge voltage is reached.

The need for recalibrating the result of any *SoC* calculation arises from the problems of the algorithms for calculating *SoC* and measuring the battery current. Criteria are necessary for identifying when "full charge" has been reached during charging. Without such recalibration, there is a risk of accumulating errors from any of the factors described so far. Full charge, $SoC = 1$, is reached in a lead–acid battery when there are no longer any lead sulfate molecules in either electrode. In lithium-ion batteries, the full *SoC* is reached when the degree of intercalation has reached the levels specified by the manufacturer. But how can this point be determined during charging?

In lead–acid batteries, float charge conditions, in which the current remains constant during constant voltage charging, can only exist after all discharged active materials have been reconverted, or when lead sulfate crystals in the active material are no longer accessible for charging (see chapter 12 for further details). When this point is detected, the battery is fully charged, and the calculated value of *SoC* can be reset to 1.

For lithium-ion batteries, the point of full charge has been reached when the charging current during constant voltage charging falls below a predetermined value, that is, 1/20 C under standard charging conditions. At this point, the calculated value of any algorithm can be reset to 1. Using this criterion but disregarding the condition of "standard charging" leads to misinterpretations. In chapter 3, it was stressed that at the end of charging, diffusion processes play a major role. Lithium ions must reach a uniform distribution in the host particles for a full charge (see chapter 13 for further details). Using fast or ultrafast charging methods (e.g. charging current of 10 C), the charging current will fall below the threshold value a few minutes after the start of charging, and the "full charge criterion" is reached. However, the required intercalation levels have not been reached, and the battery has a reduced capacity for the next discharge.

7.2.5 Yazami Theorem

Yazami has proposed a universal *SoC* relationship for various battery types and validated it empirically [4]. In chapter 2, it was pointed out that by measuring the equilibrium voltage E_0, the Gibbs free enthalpy ΔG and the temperature dependence of the equilibrium voltage dE_0/dT can be used to determine the change in entropy ΔS and thus the change in enthalpy ΔH of the overall reaction of both electrodes. The changes in the electrochemical system that occur during charging and discharging can therefore be measured. Yazami has established that the *SoC* and the thermodynamic variables are linked via an empirical relationship (the so-called universal battery theorem).

$$SoC = \alpha + \beta\Delta S + \gamma\Delta H. \tag{7.5}$$

Microchips that utilize this approach are now commercially available.

7.2.6 Experimental Determination of the State of Charge

The following procedure is required to determine the *SoC* experimentally, for example, to check devices and their algorithms:

1) Remove the battery from the application at any time (preferably when an *SoC* is shown that is in a particularly relevant range for the application).
2) Bring the battery to the nominal temperature or select the time when the battery is removed from the application so that the battery is already at the nominal temperature.

3) Determine the residual capacity by discharging with nominal current up to the end-of-discharge voltage.
4) Charge the battery fully according to the manufacturer's specifications.
5) Carry out a capacity test under nominal conditions.

Without determining the capacity (step 5), the measured residual capacity and the calculated *SoC* cannot be evaluated.

7.2.7 Depth of Discharge

Experimentally, only the fully charged state of a battery can be clearly determined, but not a completely discharged state, unless reaching the end-of-discharge voltage after a previous discharge under defined conditions is used as a reference point.Due to mass transport phenomena, a battery can continue to be discharged after reaching the end-of-discharge voltage and a short rest period until the end-of-discharge voltage is reached again. Using the fully charged state as a reference point ($SoC = 1, DoD = 0$), the *DoD* corresponds to a value of $1 - SoC$. Both terms are synonymous.

However, the term *DoD* is now mostly used to describe the amount of charge of a cycle in relation to the nominal capacity. A discharge of 20% *DoD* thus corresponds to a change in the *SoC* by 20%, and DoD is equivalent to ΔSoC. Information on the *DoD* of a cycle is therefore insufficient to describe a cycle without information on the average *SoC*.

7.2.8 State of Energy

For some time now, the *SoE* has been used in addition to or occasionally even as an alternative to the state variable *SoC*. The *SoC* provides information on the amount of charge that can still be drawn from the battery. For most applications, however, it is more important to know the amount of energy that can still be removed. The definition of *SoE* corresponds to the definition of Eqs. (7.1) and (7.2), but using energy and power as reference values.

$$SoE(t) = SoE(t_0) + \frac{1}{U_{\mathrm{Av}} C_N} \cdot \int U(t) I_{\mathrm{MR}}(t) \mathrm{d}t, \tag{7.6}$$

with U_{Av} as the average discharge voltage (data sheet value can be determined, for example, from the integral over the discharge characteristic curve) or $C_N \cdot U_{\mathrm{Av}}$ as the energy that can be drawn under nominal conditions and the integration over the power instead of the current.

All the above questions about defining and measuring the term *SoC* apply accordingly to the *SoE* with the additional difficulty of the required voltage measurement, which is normally easier and more precise than the current measurement.

In addition, it might be useful to have a term that describes the amount of energy provided during a discharge cycle in analogy to the term *DoD*. However, such a term has not yet been used or considered necessary in the scientific literature.

7.3 State of Health and State of Function

7.3.1 Definition of Terms

The performance of batteries, that is, the ability to fulfill intended functions within specified voltage and temperature limits, changes during use and as the battery ages. To predict whether the battery can fulfill a required function and to control the overall system, it is very important to know changes in performance levels. To this end, two terms, *SoH* and *SoF*, are introduced to describe the state of the battery.

The two terms are defined as follows:

- State of health
  The ratio of important performance characteristics, such as capacity and internal resistance under nominal conditions, to those of the new battery.
- State of function
  The ability of the battery to fulfill a required function in its present conditions regarding *SoC*, temperature, and aging.

For further details, see Ref. [1].

7.3.2 Differentiation and Discussion of the Terms State of Function and State of Health

The frequently used term *SoH* is often used without a more precise definition and usually only refers to capacity. The two terms *SoH* and the much less frequently used term *SoF* are treated together to understand them and their differences better.

SoH: How good is the battery now compared to when it was new?

For the comparison, the aged battery must be in a precisely defined state, for example, fully charged at nominal temperature, and the test conditions must be the same as those for a new battery, for example, discharging at nominal current and temperature.

SoF: Is the battery in its current condition concerning *SoC*, *SoH* and temperature, which must lie within the permissible operating limits, capable of performing all the required functions?

In chapter 6, the term "end of life" was introduced as a multidimensional vector to mark the time when the battery is no longer capable of fulfilling all functional requirements. End of life is therefore reached when the *SoF* is insufficient and not when the *SoH* falls below a predetermined value.

Under this definition, *SoF* and the state variables *SoC*, *SoE*, and *DoD* vary quickly during operation, whereas *SoH* changes only very slowly during its lifetime, between one at the beginning of lifetime and its lower threshold value, which depends on the exact definition of the term.

7.3.2.1 *SoH* and *SoF* in Relation to Capacity

The following definition of SoH_C, where the index C stands for reference to the capacity, has become established:

$$SoH_C = C/C_N, \tag{7.7}$$

with C_{act} as measured actual capacity under nominal conditions and C_N as nominal capacity.

If the end of the battery's service life has been defined as a drop in capacity to 80% of its nominal capacity, then a SoH_C of 0.8 or 80% means the end of service life has been reached. The value of SoH_C according to this definition fluctuates between 0.8 (end of life) and a value slightly above 1, as the capacity of a battery can be higher than the nominal capacity during part of its life.

An (albeit rarely used) alternative definition (here: SoH_M) from [5] is:

$$SoH_M = (C - C_{\min})/(C_N - C_{\min}). \tag{7.8}$$

Here, $C_{\min}$ is the capacity that must still be available under nominal conditions at the end of the service life for the battery to still be considered usable. With this definition, SoH_M has the value 0 at the end of the service life.

For SoF_C, the definition is analogous to Eq. (7.8):

$$SoF_C = (C_{\text{aged},SoC,T} - C_{\min,SoC,T})/(C_{\text{new},SoC,T} - C_{\min,SoC,T}), \tag{7.9}$$

where the indices SoC and T represent the most unfavorable battery condition within the permissible operating limits under which the available capacity must still have at least the value $C_{\min,SoC,T}$. $C_{\text{aged},SoC,T}$ is the capacity of the aged battery available under the existing SoC and temperature conditions. $C_{\text{aged},SoC,T}$, the capacity that is available under the most unfavorable but still permissible operating conditions, must be determined to calculate the SoF. The value of SoF_C varies between values significantly above 1 and 0. A negative value indicates that the capacity that can still be removed is no longer sufficient to fulfill the required function (provision of a certain amount of charge at a low temperature and SoC) and that the end of service life has therefore been reached.

An example of the use of SoF_C are batteries in renewable energy systems, where a full charge is rarely possible due to the weather and the batteries can reach low temperatures. It is therefore not the capacity under nominal conditions, SoH_C, but the capacity at low temperatures and an SoC that is likely to exist, SoF_C, that determines the end of service life.

7.3.2.2 *SoH* and *SoF* in Relation to the Internal Resistance

The definition of SoH_R, where the index R stands for the internal resistance of the battery, was proposed in Ref. [6] as follows:

$$SoH_R = R_{i,\text{aged}}/R_{i,\text{new}}, \tag{7.10}$$

with $R_{i,\text{aged}}$ as the measured internal resistance of the aged battery and $R_{i,\text{new}}$ as the internal resistance of the new battery, each measured under the same conditions. The SoH_{Ri} of a new battery has the value 1, the value of an aged battery has a higher value. Tchoupou Lando chose this definition so that the development of internal resistance and are parallel, analogous to the development of capacity and SoH_C.

For SoF_R, the definition analogous to Eq. (7.8) is used:

$$SoF_R = (R_{i,\text{aged},SoC,T} - R_{i,\min,SoC,T})/(R_{i,\text{new},SoC,T} - R_{i,\min,SoC,T}), \tag{7.11}$$

where the indices SoC and T, as in the definition of SoF_C, describe the most unfavorable conditions relevant for the application, at which the internal resistance must still be at least.$R_{i,\min,SoC,T}$.

An example of the use of SoH_R and SoF_R is the starting of combustion engines or batteries in hybrid vehicles, where the value of the capacity and its age-related decrease do not play a decisive role with regard to the end of service life. More decisive is the internal resistance, especially at low temperatures, and the SoC expected under unfavorable conditions but still permissible. The assumption that it is possible to deduce values of the SoH_R and SoF_R from the age-related decrease in the capacity of the battery should not be made for a detailed analysis.

7.3.3 Measurement of *SoH* and *SoF*

To estimate the SoH_C, the battery must be subjected to a capacity test under nominal conditions. Alternatively, the capacity under nominal conditions can be estimated from the behavior of the battery at other temperatures and with currents of different and variable amplitude using models (see chapter 8 on models). The same is required for measuring the SoH_R.

To calculate the SoF according to the above definitions, it must first be determined what the amount of charge and internal resistance of a battery are acceptable under the most unfavorable conditions within the permissible SoC and lowest expected temperature of the application. Only when these values have been determined is a model-based approach feasible (see chapter 8). Measurements during operation can be used to modify the model parameters to the parameters of an aged battery.

7.4 State of Safety

A standardized definition of a *SoS* of a battery has not yet been established in the literature. As the term is most commonly used in connection with lithium-ion batteries, section 13.20 provides an overview of the current discussion. For further details, see also Ref. [1].

References

1 Wenzl, H., Benger, R., and Hauer, I. (2024). State of charge. In: *Encyclopedia of Electrochemical Power Sources*, 2e. Elsevier.

2 Sauer, D.-U., Bopp, G., Jossen, A. et al. (1999). *State of Charge – What Do We Really Speak About? Proceedings: INTELEC'99*. Copenhagen: INTELEC.

3 Kaiser, R.J. (2005). *Optimization of a battery management system for use in photovoltaic power supply systems (Original German title: Optimierung eines Batteriemanagementsystems zum Einsatz in photovoltaischen Stromversorgungssystemen)*. Dissertation. University of Ulm.

4 Yazami, R. (2019). *The Universal Battery SOC Law*. The Battery Intelligence Company, Company Information.

5 Meissner, E. and Richter, G. (2003). Battery monitoring and electrical energy management: precondition for future vehicle electric power systems. *Journal of Power Sources* 116: 79–98.

6 Tchoupou Lando, E. (2021). *Event-based lifetime model for lithium ion cells using linear damage accumulation (Original German title: Ereignisbassiertes Lebensdauermodell für Lithiumionenzellen mittels linearer Schadensakkumulation).* Dissertation: Clausthal University of Technology.

Task

Task 7.1
What methods are there for determining the state of charge, and what are their advantages and disadvantages?

8

Battery Models

In the previous chapters, models based on equivalent circuit diagrams have already been introduced to illustrate electrochemical relationships and design issues. This chapter explains battery models that are frequently used in practice. These models improve the understanding of the behavior of batteries, and they are important for determining battery properties. The focus is on equivalent circuit diagram models because they can be intuitively understood and allow simple calculations. The most frequently used models are:

- Thévenet or Shepherd model, consisting of a voltage source and a resistor, which correspond to the frequently used "look-up tables," and
- Randles model, consisting of a voltage source, an inductance, a resistor, and two RC circuits.

Multicell battery systems are usually treated only as the sum of single-cell models.

8.1 Classification, Use, and Limitations of Models

8.1.1 The Concept of Battery Modeling

Battery models are used to better understand the complex interaction of many different processes during discharging, charging, and aging. However, they are also technically required to predict the behavior of batteries. Model-based algorithms offer potential for optimization, particularly in hybrid systems with several power generators, where the power demand can be distributed to individual power generators depending on their current status. When using energy management systems capable of controlling the power profiles of loads, the model-based predictions of battery performance play a major role.

Basically, battery models can be categorized according to their scale (atomic/microscopic, mesoscopic, and macroscopic) and according to the use of physicochemical information (the so-called analytical models) or the absence of such information (the so-called empirical models).

Battery Technology: Fundamentals of Battery Electrochemistry, Systems and Applications,
First Edition. Alexander Börger and Heinz Wenzl.

Companion website: www.wiley.com/go/Börger/Battery_Technology

In most battery models, the focus is on individual cells only. The behavior of an overall battery, consisting of several cells in series and several strings in parallel, is often extrapolated on the basis of identical cell properties. However, when modeling temperature, aging, and faults, it is often necessary to consider cells as having different properties and their interactions (see, e.g. the analysis of cells in series with a high-resistance cell in chapter 3).

This chapter deals with models that describe the electrical behavior of cells. Models for developing materials at particle level (single-particle model) and for investigating abuse conditions are not included in this chapter. Models for heat generation due to current flow and external heat sources or sinks (see chapter 5) and for a detailed description of mass transport phenomena, unless they can be represented by RC circuits with long time constants, must be coupled with the equivalent circuit-based models presented here, but are not the focus of this chapter.

Historically, model developments have focused on the voltage curve during discharging. Now, however, the development of models for issues relating to short and high-power pulses and charging processes has become equally important.

To set up a model, it is important to precisely specify the question to be clarified by the model and then select the simplest model for the task at hand.

8.1.2 Use of Models

The following questions are at the forefront of technical systems:

1) Voltage curve for power consumption or output
   During high power consumption for a limited period of time (e.g. starting combustion engines, accelerating vehicles, providing short-circuit power, and inrush currents of electrical motors), the voltage must remain above a voltage threshold so that cells are not damaged and/or loads or converters are still within a permissible voltage range for their operation. During a high-power input (e.g. regenerative braking) for a short time, the voltage must not exceed a voltage limit that could damage cells or destroy system components connected to the battery.
2) Voltage and temperature curve for a specified power profile over a longer period of time
   The voltage and temperature curve during discharging and charging under normal operating conditions is technically important for estimating the amount of charge or energy that can still be supplied or absorbed. It must be possible to calculate the voltage and temperature curve not only for different initial temperatures and states of charge over a longer period of time, but also for different states of health.
3) Diagnosis and estimating battery states
   When changes in model parameters result in a better fit between model and measurements, they may allow quantifying changes in battery properties (e.g. internal resistance and capacity) and aging processes.
   Additional requirements for batteries, such as a long time between maintenance and low self-discharge, are rarely investigated using models, although they could

play an important role for the operating strategy and the service life of batteries in some applications.

8.1.3 Limits When Using Models

Battery models are simplifications, and their interpretation must always be carried out carefully.

- Accuracy of the model parameters (see chapter 9):
  Model parameters are often difficult to determine, especially for complex models. When calculating cell properties, uncertainties in model parameters can quickly lead to considerable misinterpretations, so quantitative statements must be carefully checked for their validity.
- Interpretation of model parameters:
  Changes in model parameters, for example, during aging tests and changes in environmental or operating conditions, often cannot be assigned to physical processes and components, or only roughly.
- Observing model limits:
  Obviously, only the phenomena that are represented in the selected model can be investigated. This applies in particular to aging processes such as corrosion or the growth of the solid electrolyte interface (SEI) layer (see chapter 13), as well as temperature development. Equivalent circuit diagram-based models are generally not suitable for describing the heat sources or sinks of the battery because the reversible heat cannot be represented as electrical resistance.

8.2 Equivalent Circuit Diagram Models

8.2.1 Fundamentals

This chapter focuses on the description of electrical properties of cells. An electrode or cell is considered as a component with homogeneous properties or, if current inhomogeneities are to be investigated, the cell is divided into different but relatively few volume segments in which all physical and chemical properties are considered homogeneous for the respective segment. There are then discontinuities at the boundaries of the segments, which become smaller the smaller the segments are defined, but are always remain in principle. In finite element models, on the other hand, there are no discontinuities between volume segments, but the models become much more complex and less transparent.

Using equivalent circuit diagrams, it is often possible to immediately recognize why and at which points on the electrodes voltages and currents are higher or lower relative to other points without calculating the voltages and currents. This makes it easy to understand batteries and important aspects of their behavior and detailed calculations with knowledge of the exact model parameters are then not necessary. This is an important reason for choosing equivalent circuit diagram models.

In addition, algorithms for battery or energy management systems are often implicitly based on equivalent circuit diagrams. When using models to calculate battery properties, however, the model parameters must be accurately known (see chapter 9).

Finite element models, models based on neural networks or fuzzy logic, as well as other empirical and analytical models, are not covered, although they are a powerful tool for clarifying detailed questions. Please refer to further literature, for example, Ref. [1], for an overview.

8.2.2 Structure of Equivalent Circuit Diagram Models

The information already presented in chapter 2 is expanded in the following. The individual components of an equivalent circuit diagram for a battery correspond to basic physical principles.

- The two electrodes are represented as spatially separated ideal voltage sources that are completely independent of each other and – in the case of current flow – are coupled by the same total current. If mass transport effects are neglected, the voltage of the voltage sources correspond to the equilibrium voltage of the electrochemical reaction it describes (section 2.2).
- The voltage sources lead to a charge carrier separation at the electrode–electrolyte interface (Helmholtz layer), which corresponds to a capacitor and is therefore referred to as a double-layer capacitor.
  When a constant current can be assumed, it is possible to remove the capacitors that describe the double layer of the electrodes from the equivalent circuit diagram (see Figure 2.12).
- Current flow leads in all real voltage sources to a change in voltage. This is modeled using an internal resistance that is always associated with the ideal voltage source. The BV equation (see section 2.4) describes the difference between the equilibrium voltage and the electrode voltage $E - E_0$, the charge transfer voltage and can be converted to the internal resistance of the voltage source.

A battery therefore reacts to changes in current or voltage with a time delay described by the time constant of the RC circuit.

- An ohmic resistance is assigned to each of the current (ion or electron) conducting components. Because the individual resistance components can differ depending on temperature, aging, and state of charge, they are often not combined into one resistance but shown as separate resistances.
- All electrical systems show some inductive behavior in case of very large current changes. Depending on the use of the model, however, the inductance can be neglected in many cases.
- The positive and negative electrodes, which are separated by the electrolyte as a dielectric and have different potentials versus a reference potential respectively the electrolyte, form a "capacitor." Figure 8.1 shows the discharge of a battery above its equilibrium voltage and the subsequent increase in voltage that can be explained by the discharge of this additional capacitor. For further details, see Ref. [2].

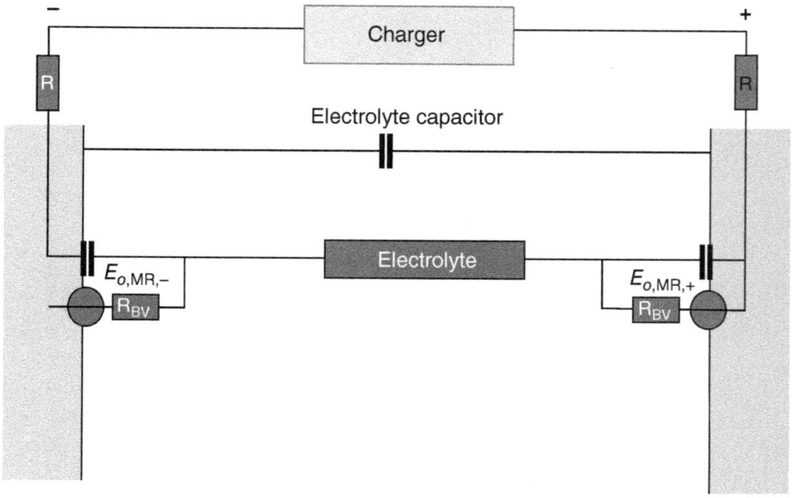

Figure 8.1 Complete equivalent circuit diagram of a cell without side reactions including the capacitor formed by the positive and negative electrodes with the electrolyte as dielectric. This capacitor was not yet included in chapter 2. This capacitor is shown as an ideal capacitor without self-discharge.

The equivalent circuit diagram (if the inductance and mass transport phenomena are neglected) of a battery that is capable of explaining the electrical behavior of a battery above and below the equilibrium voltage is shown in Figure 8.1. The equivalent circuit diagram can be significantly simplified depending on how the model is to be used. At constant current, it can be reduced to a voltage source and an internal resistance (Thévenet model) as a fully valid equivalent circuit diagram (see chapter 2 and Figure 2.12). Equivalent circuit diagrams cannot be used to calculate temperature changes as the reversible heat of reaction, which leads to either cooling or heating of a battery cannot be modeled using a resistance.

8.2.3 Capacitance of a Battery

Figure 8.2 shows the voltage curve of a 12 V lead–acid battery in the voltage range between 14 and 13 V, that is, above the equilibrium voltage, when discharging with small currents. The current and energy supplied by the battery cannot be explained by the main reaction because the battery voltage is above the equilibrium voltage determined by the concentration of the electrolyte. The main reaction therefore cannot supply any energy. After the end of the discharge, which was terminated above the equilibrium voltage, the battery voltage rises again. The amount of charge and energy supplied by the battery can only be provided by a capacitor, which is in parallel with the double-layer capacitors at the electrode/electrolyte interface as a result of electrochemical reactions and the separation of charge carriers caused by the reactions. This additional capacitor has properties similar to those caused by the pseudo capacitance of conventional ultracapacitors (electrochemical double-layer capacitors [EDLC]). From the amount of charge removed and the voltage difference, it can

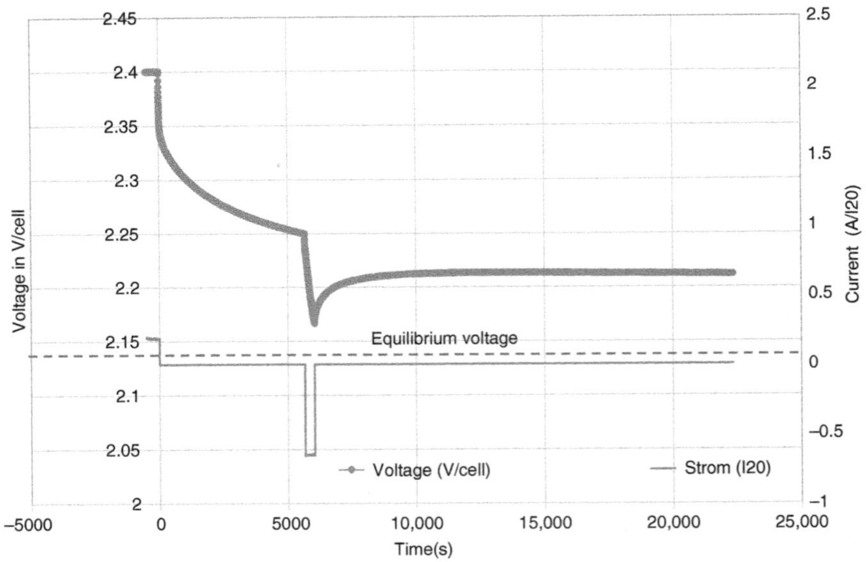

Figure 8.2 Voltage curve of a 12 V lead–acid battery with a nominal capacity of 65 Ah when discharged with 50 mA above the battery's equilibrium voltage.

be estimated that the capacitance of this capacitor is in the range of 1000 F/cell and that approx. 0.6 Wh can be provided to a load. An exact estimate is difficult and not necessary for practical applications.

It is important to note that, due to the capacitance property of fully charged batteries, discharge and charge currents during float voltage operation are not necessarily based on conversion of active materials, that is, not Faraday currents – as long as the electrode potentials are above their equilibrium potentials.

The relationship shown here for lead–acid batteries can be transferred to other batteries, including lithium-ion batteries. However, a comparable evaluation is not possible for lithium-ion batteries because the equilibrium voltage in the fully charged state is not known with similar certainty.

8.2.4 Integration of Mass Transport and Temperature

Temperature effects and mass transport phenomena that change the values of the voltage sources and resistances can only be described by coupling models that convert the heat generated by current flow into a temperature change and concentration changes of reactants to changes of voltage and resistances (see section 8.5).

Mass transport phenomena follow Fick's laws. When a constant concentration ($\mathrm{d}c/\mathrm{d}t = 0$) is changed by current flow, the local concentration changes exponentially and also the Nernst voltage and the concentration-dependent exchange current, which leads to an exponential change in electrode voltages. The variation of the voltage difference over time can be simulated using RC circuits with a large time constant. In particular, when using the Randles model (see section 8.4) and impedance spectroscopy (see chapter 10 and appendix D) to determine model parameters, RC circuits are used for modeling both diffusion and migration effects.

8.2.5 Locally Resolved Equivalent Circuit Diagram Models

Equivalent circuit diagrams with local resolution must be used for investigations of current inhomogeneities, which can arise due to the design (see chapter 4) or due to inhomogeneous temperatures (see chapter 5). These model extensions are briefly explained here to show that they can be used to achieve qualitative approximations to finite element models or other models, for example, computational fluid dynamics (CFD), and because they are used for some specific problems.

One or both electrodes are divided into several volume segments or spatially separated current paths through the cell. A simplified equivalent circuit diagram with four current paths per electrode (top and bottom, inside [near the current collector] and outside [near the electrolyte]) is shown for constant currents (capacitors are omitted) in Figure 8.3.

The electrolyte between the electrodes is divided into two volume segments that electrically couple the outer volume segments of both electrodes. These two electrolyte segments are also electrically connected to each other. The resistance values are assigned from the center of the volume segment to the center of the adjacent volume segment. Differences in material properties are permitted at the boundary between the volume segments.

As the specific conductivity of the electrolyte is much lower than that of the other materials in all batteries, it is immediately clear that the charge throughput is greatest in the segments at the top and outside and smallest in the segments at the bottom

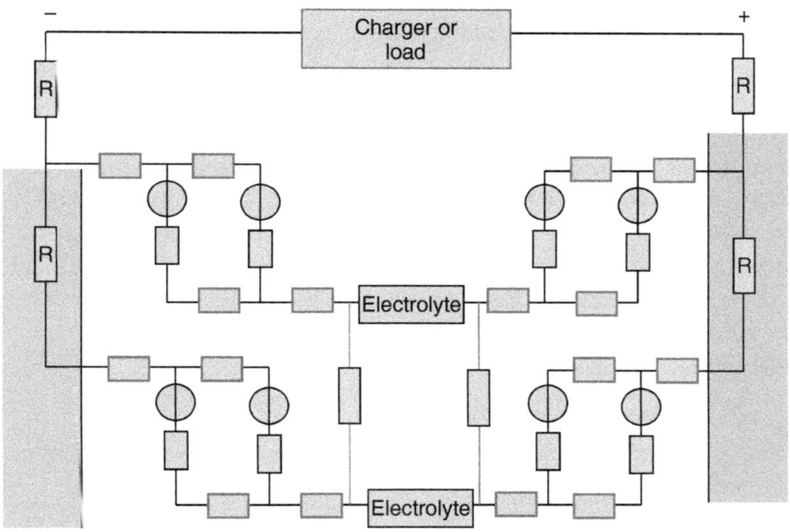

Figure 8.3 Equivalent circuit diagram of a cell showing current inhomogeneities.

and inside.[1] This results in differences in temperature and state of charge, which changes the local equilibrium voltages. After the end of an external current flow, there are areas with different equilibrium voltages at both electrodes. As the voltage sources are connected in parallel, the area with the lower voltage is recharged by discharging the area with the higher voltage. A constant voltage can only be measured at the terminals once these processes have equalized.

The slow voltage change after the current has been switched off which can be easily explained on the basis of this simple equivalent circuit model should therefore be described as the effect of equalizing currents and not diffusion. Genuine diffusion effects also contribute to the slow voltage change, but they are not modeled in this equivalent circuit diagram.

8.2.6 Relaxation Processes

In the previous sections, various processes were explained using equivalent circuit diagrams that show a change in voltage over time during current flow (difference between the electrode voltage and equilibrium voltage, mass transport phenomena, equalizing currents due to locally different equilibrium voltages, and effect of cell capacitance). Only one of these is diffusion-driven. The information presented here is intended to illustrate that internal equalizing currents can occur without an external current flow and that the cause of an external current flow is not exclusively due to electrochemical reactions.

8.3 Models with Parameters Independent of State of Charge: The Shepherd Model

The Shepherd model described in the following [3, 4] is not only one of the first models to describe the discharge process of batteries but also the only one whose model parameters are independent of the state of charge. The voltage curve during discharging and charging up to the end-of-charge voltage can be well represented with the Shepherd model with only four constant parameters to be determined experimentally for any currents (except for high current loads) and, unlike most other models, no state of charge-dependent parameters need to be determined. In its original version, the charging process is not treated. The importance of the Shepherd model results from the frequent, albeit only implicit, use of this model concept for many of the battery and energy management systems used today.

The model consists of four components:

1) Open-circuit voltage U_0 when fully charged
   The voltage of the two electrodes is combined with one voltage source and has a constant value.

---

1 It is also immediately clear that the charge throughput in opposite segments of the electrodes will be the same.

2) Reduction of the open-circuit voltage at a lower state of charge
   The reduction in the open-circuit voltage is assumed to be proportional to the amount of charge removed Q_t respectively the decrease in the state of charge over the entire range of state of charge under consideration. Electrically this is a voltage source with opposite polarity to the open-circuit voltage source.
3) Voltage drop proportional to the current according to Ohm's law
   The resistance r is considered to be independent of the current and state of charge.
4) Voltage drop, which depends on the reduced availability of the reactants (see section 3.2).
   The amount of charge in the fully charged state Q_o is compared with the amount of charge still available $(Q_o - Q_t)$ and treated as resistance I

$$U_Z = U_o - G \cdot Q_t + r \cdot I + k \cdot I \cdot Q_o / (Q_o - Q_t). \tag{8.1}$$

By measuring discharge curves, the cell voltage can be calculated for any discharge currents using four variables to be determined experimentally (G, r, k, and Q_o). Figure 8.4 shows the Shepherd model in the form of an equivalent circuit diagram.

Like all other simple equivalent circuit diagrams, the Shepherd model is not capable of describing diffusion and equalization processes after changes in current. The voltage only depends on the current and the charge throughput.

Shepherd specified a method for analytically calculating the parameters of the model from two measured discharge curves with different currents by cleverly selecting the measuring points from four pairs of U–I values. Today, the parameters are determined as a parameter fit from at least two, but often also several discharge curves with different currents. When parameterizing the model, physically unrealistic values may occur as parameters for the best fit, for example, a negative resistance value for r. It should be noted that Q_o does not correspond to the capacity of the battery, but is one of the parameters to be determined from measurements. Q_o also does not correspond to the fully available active mass. It is worth noting the fundamental similarity of the fourth term with the term derived in section 3.2 for the general form of the discharge curve using only the BV equation.

The original Shepherd model was only presented for discharging and did not include temperature and aging effects. It has now been extended and can be used for both the discharging and charging direction (until the end-of-charge voltage or

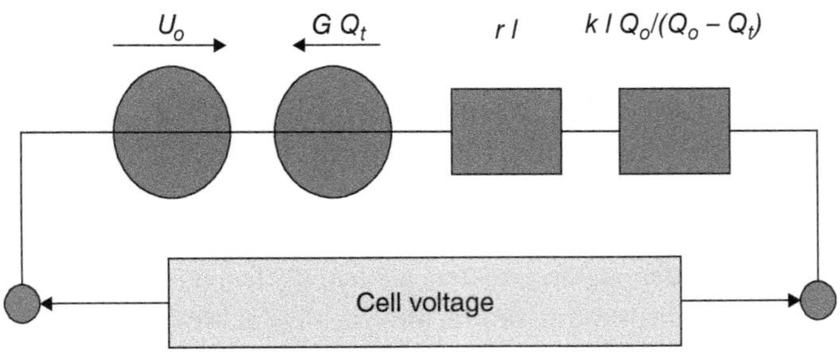

Figure 8.4 Shepherd model as an equivalent circuit diagram.

a significant proportion of side reaction current in the battery current is reached). It can also be used for other temperatures by adjusting the parameters:

a) r is corrected for the temperature-dependent change in resistance of the electrically conducting components.
b) Q_0 is regarded as a value that must be changed in proportion to the temperature-dependent capacity of the battery and reduced in proportion to the decrease in capacity due to aging.
c) The dependence of the open-circuit voltage value on temperature does not have to be taken into account because the effect can usually be neglected.
d) The four parameters to be determined experimentally must be determined separately for the discharging and charging direction and labeled with the appropriate indices.

Equation (8.2) shows the enlarged model for the discharge direction (d), changed temperatures, and the resistance and capacity value changed by aging (ag).

$$U_Z(T, \text{Alterung}) = U_{o,d} - G_d \cdot Q_t + r_{d(T,\text{aged})} \cdot I + k \cdot I \cdot Q_{o(T,\text{aged})} / (Q_{o(T,\text{aged})} - Q_t). \quad (8.2)$$

To calculate the current curve at constant voltage at the end of charging, the current value I must be calculated such a way that the cell voltage remains constant.

Despite the many simplifications, the agreement between the cell voltage and the measured data is very high, except immediately after current changes and at a very high state of charge (approx. 100–90%), because special effects sometimes play a role. Extrapolation of the data outside the current values used for determining the parameters is possible up to a certain point, but not for very high discharge currents (starting processes, short circuits).

8.4 Models with State of Charge-Dependent Parameters

The most frequently used and practically applied battery models require state of charge-dependent model parameters. Without changing the parameters when the state of charge changes, the model can only be used around the operating point, the state of charge for which the model parameters were determined. Section 8.5 describes the simulation process for calculating current and voltage curves over the entire state of charge range and for taking temperature and mass transport effects into account.

8.4.1 Thévenet Model

The Thévenet model consists of a voltage source (open-circuit voltage) and a resistor (internal resistance of the battery, see chapter 2). It is electrically completely equivalent to the equivalent circuit diagram shown in Figure 8.1 derived from electrochemical principles (see also Figure 2.12) for a quasi-stationary state (constant current, neglecting the state of charge changes and mass transport phenomena as well as temperature changes). The model parameters are usually listed in tables as a function of the state of charge and the temperature, and possibly also of the current (the so-called look-up tables), which are used for many battery management systems. An important application is the determination of the internal resistance under high current load (chapter 9).

The option of using parameters of the Shepherd model in the tables is not normally used but the similarity to the Shepherd model is obvious.

8.4.2 Randles Model

Another frequently used model was first established by Randles. The cell is represented as a series connection of a voltage source, two RC circuits, a resistor, and an inductance, see Figure 8.5. The model is suitable for characterizing the battery properties at the temperature and a state of charge range at which the parameters were measured. To calculate discharge and charge curves over the entire state of charge range, the model parameters must be adjusted accordingly.

Apart from including an inductance, the model is a formal equivalent to the equivalent circuit diagram shown in Figure 2.12. The voltage sources of both electrodes and all ohmic resistances are each combined in one component of the equivalent circuit diagram. In the Randles model, however, the first RC circuit is used to represent the behavior of the double-layer capacitors of both electrodes, and the second RC circuit is used to describe the long-lasting diffusion and equalization effects of both electrodes together.

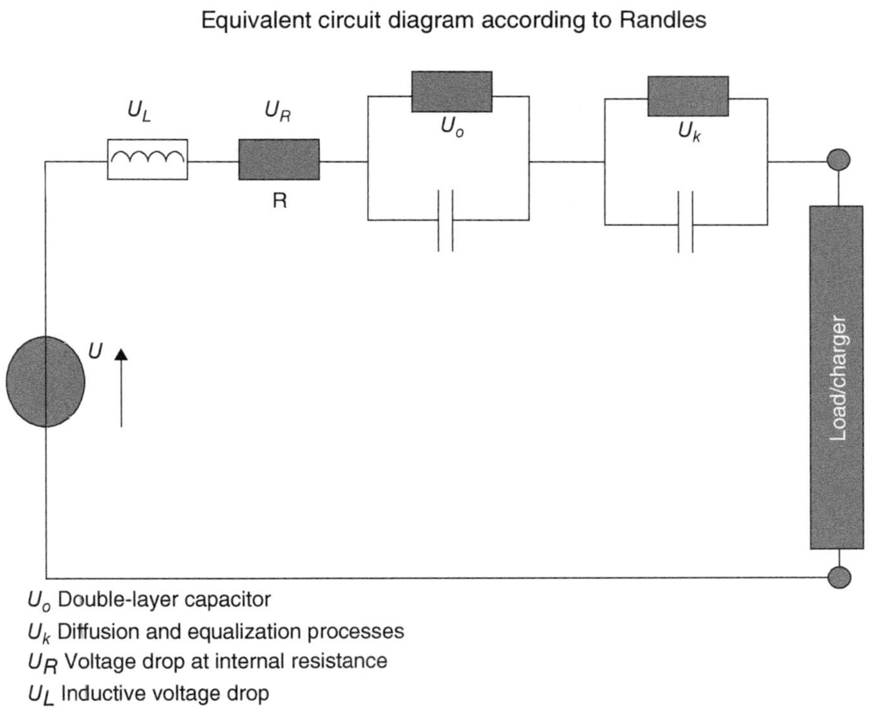

Figure 8.5 Randles model.

The time constant of the first RC element is usually in the range of a second, whereas the time constant of the second RC element ranges from a few minutes to many hours. Depending on the time interval under consideration, either one or the other RC circuit can therefore be neglected.

The importance of this equivalent circuit diagram results from its frequent use, and it is also often used as a basis for model extensions, for example, by adding one or more further RC circuits or other elements.

As a rule of thumb, batteries have an inductance in the range of μH, an internal resistance in the range of mΩ and a capacitive behavior in the range of a few F. The second RC circuit in the Randles model often has such a high time constant that its contribution is negligible in many applications.

8.5 Simulation Process

The Shepherd model can be used to directly simulate the voltage curve over time with variable current amplitudes because the model parameters are the same for the entire discharging and charging process. With all other models, the model parameters change with the state of charge. The model parameters must therefore be measured as a function of the state of charge and constantly changed in the event of major changes in the state of charge. The resulting process with simultaneous consideration of temperature, mass transport, and aging effects is described in the following.

When using models for a detailed discussion of charging and discharging processes, a current or power profile is necessary, and separate models for heat or temperature development and mass transport phenomena have to be integrated. In addition, models can be used to calculate aging effects which lead to changes in the model parameters on the basis of temperature, charge throughput, and voltage. The initial state of charge and temperature must be defined and a time step for the calculation, for example, one second or one minute, must be chosen. The calculation is carried out in the following incremental steps:

1) Conversion of a power profile into a current profile:
   For this purpose, the model with its initial voltage value derived from the state of charge and temperature is used.
2) Calculation of the incremental change in voltage, heat generation, and the resulting temperature change as well as the change in state of charge.
3) Adaptation of the model parameters (Nernst voltage, BV resistance, etc.) to the new state of charge and new temperature. With a locally resolved model, the adjustment must be made for each volume segment of the cell.
4) Calculation of changes due to aging processes.
5) Recalculation of the current from the power profile for the next step and proceed with step 2 for the next time interval.

This process is carried out iterative for the specified current or power profile until voltage limits are reached which must not be exceeded. As soon as voltage limits are reached during charging, either the simulation is terminated or the specified current

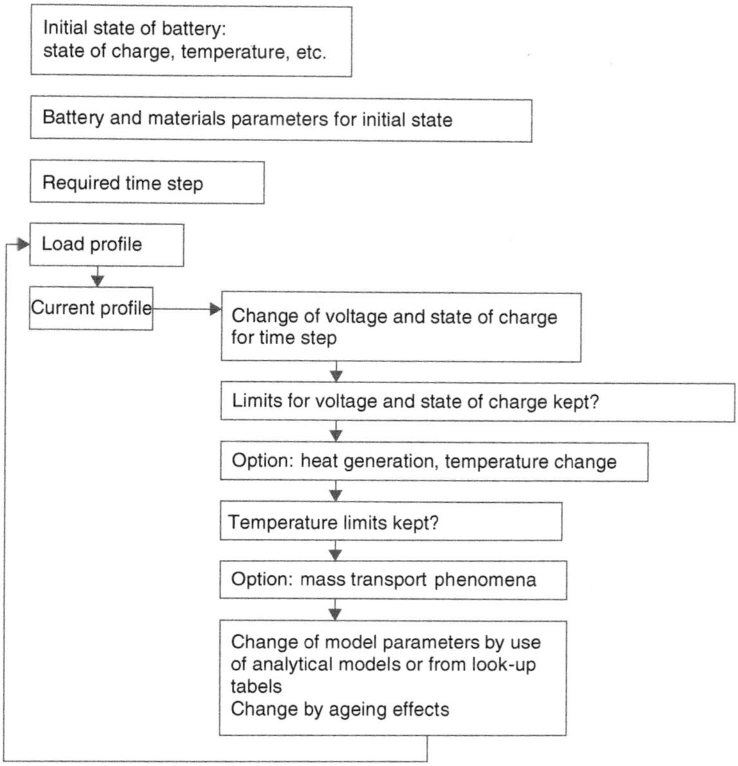

Figure 8.6 Diagram showing schematically the simulation process to calculate the current, temperature, and voltage curve of a battery including aging processes.

or power profile must be changed so that the voltage at the terminals corresponds to the specified voltage limits. Figure 8.6 shows the procedure schematically.

When a cell is divided into several volume segments, inhomogeneous current densities, locally different states of charge, equilibrium voltages, and temperatures result, even if the material properties were homogeneous everywhere at the start of the analysis. The locally different equilibrium voltages then lead to internal currents, even if no external current is present, and to slow changes in the cell voltage even if there is no more current.

8.6 Comparison of Models

Figure 8.7 shows the comparison of three models

1) Voltage source and resistor.
2) Voltage source, resistor and one RC circuit.
3) Voltage source, resistor and two RC circuits.

The parameters were determined on the basis of the same measurement data, which were carried out using the current step method (see chapter 9), and then fitted

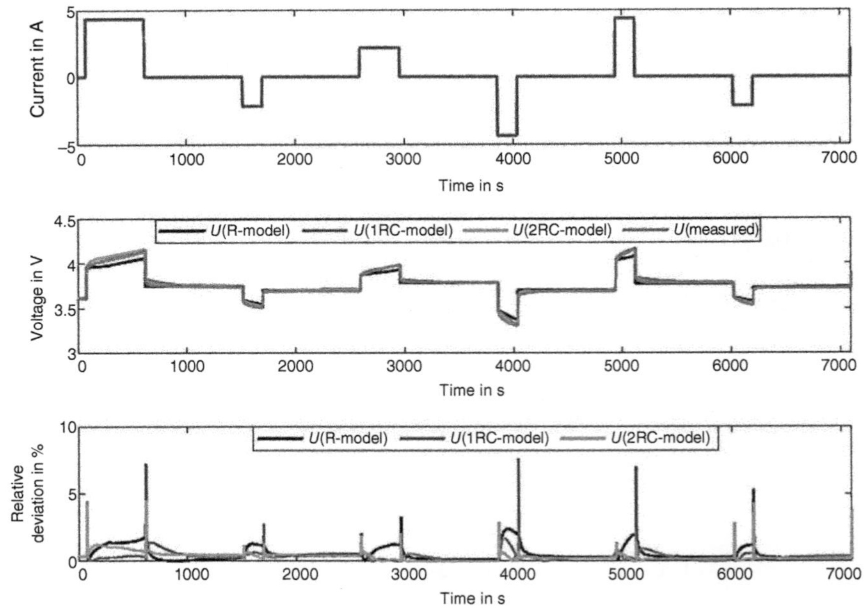

Figure 8.7 Comparison of the terminal voltage of measured voltage of a lithium-ion cell at 30 °C with simulation results using equivalent circuit diagram models (R only, R + RC, and R + 2RC). The cell was subjected to a dynamic current profile. Source: Courtesy of E. Tchoupou Lando.

using standard procedures. Obviously, the error is greatest immediately after current changes when using the Thévenet model. However, it is questionable whether the additional effort for significantly more complex models is worthwhile if the model is only used to simulate the voltage curve from a current or power profile.

8.7 Modeling Larger Systems

Modeling and simulation of battery modules and complete systems are frequently used tools, especially in the automotive sector, both in product development and for characterizing system performance. For larger systems, in particular, they are used for

- evaluation of design requirements and their limitations,
- cost-effective investigation of a large number of different possible scenarios, and
- identification of worst-case conditions for ensuring system performance.

Special attention is paid to the safety behavior and here in particular to thermal propagation issues. Simulations are essential, especially in the context of a complete safety assessment with regard to thermal propagation in lithium-ion batteries, because:

- experiments at module, system, or even vehicle level are far too expensive and complex for experimental data alone to be sufficient,
- regulatory requirements already include safety verifications in accordance with current standards, which also require simulation or interaction between simulation and experiment [5],
- test capacities for all conceivable propagation tests are not available, and
- sometimes simulations help to verify hypotheses from test results or steer development in new directions.

The modeling carried out here uses electrical models, as explained at the beginning of the chapter, but also includes the behavior under abuse conditions. However, at the core of the models and simulations are thermal models (see chapter 6) and mechanical models (with regard to vibrations and mechanical stress). Details are given in chapter 13.

References

1 Cugnet, M., Dubarry, M., and Liaw, B.Y. (2009). Modeling. In: *Encyclopedia of Electrochemical Power Sources*, vol. I (ed. J. Garche), 816–828. Amsterdam: Elsevier.

2 Wenzl, H., Benger, R., and Hauer, I. (2024). Capacitance. In: *Encyclopedia of Electrochemical Power Sources*, 2e. Elsevier.

3 Shepherd, C.M. (1965). Design of primary and secondary cells. I. Effect of the polarization and resistance on cell characteristics. *Journal of the Electrochemical Society* 112: 252–257.

4 Shepherd, C.M. (1965). Design of primary and secondary cells. II. An equation describing battery discharge. *Journal of the Electrochemical Society* 112: 657–664.

5 Global Technical Regulations (GTR) (2018). *No. 20 (Electric Vehicle Safety (EVS)).* http://unece.org/transport/standards/transport/vehicle-regulations-wp29/global-technical-regulations-gtrs (last accessed 27 May 2022).

Further Reading

Abada, S., Marlair, G., Lecocq, A. et al. (2016). Safety focused modeling of lithium-ion batteries: a review. *Journal of Power Sources* 306: 178–192.

Benger, R. (2012). *Dynamic behavior of inverter-fed energy storage systems (Original German title: Dynamisches Verhalten von umrichtergespeisten Energiespeichersystemen).* Dissertation. TU Clausthal.

Krewer, U., Röder, F., Harinath, E. et al. (2018). Dynamic models of li-ion batteries for diagnosis and operation: a review and perspective. *Journal of the Electrochemical Society* 165 (16): A3656–A3673.

Lai, X., Zheng, Y., and Sun, T. (2018). A comparative study of different equivalent circuit models for estimating state-of-charge of lithium-ion batteries. *Electrochimica Acta* 259: 566–577.

Sauer, D.U. and Wenzl, H. (2008). Comparison of different approaches for lifetime prediction of electrochemical systems – using lead-acid batteries as example. *Journal of Power Sources* 1176: 534–546.

Turner, J. (2016). *Progress in Modeling and Simulation of Batteries*. SAE-Verlag. ISBN: 978-0-7680-8282-1.

Tasks

Task 8.1

Describe advantages and disadvantages of:
The Randles model, and
The Shepherd model.

Task 8.2

Draw a complete equivalent circuit diagram for a system with two different electrodes, two different phase boundaries, one electrolyte, one separator, and two different terminals. Why is such an equivalent circuit diagram difficult to parameterize and not very practical?

9

Determination of Parameters

This chapter describes methods that can be used to determine battery properties and model parameters under real and laboratory conditions.

The two most important properties of batteries are internal resistance (including short-circuit current) and capacity. Both are functions of temperature, current, and aging. The internal resistance can be determined by pulse measurements (the most common case) or by alternating current resistance measurement (impedance spectroscopy). In the case of capacity, a distinction must be made between nominal, residual, initial, and practical capacity, which are needed to define the terms "state of charge" (SoC) and "state of health" (SoH) or "aging."

To predict the voltage of batteries with variable load profiles, it is necessary to determine model parameters that go beyond knowing the internal resistance and capacity. For this purpose, step discharges and charges, also known as the current interrupt technique (CIT), in the time domain (accurate measurement of voltage changes over time is required) are used, or the battery behavior in the frequency domain (measurement of the impedance spectrum) is evaluated.

9.1 Definitions

A qualitative discussion of battery behavior, which is the main focus of this textbook using equivalent circuit diagrams, does not require precise knowledge of battery properties and model parameters. However, when operating conditions such as current and temperature vary and quantitative information for operational control is required, it is necessary to determine suitable characteristics and parameters. The accuracy requirements depend on the use of the results and must be weighed against the length of time and cost to determine them. The most important electrical parameters of batteries are capacity as a function of temperature and load current, and the internal resistance, for which SoC and the direction of current must also be taken into account.

Battery Technology: Fundamentals of Battery Electrochemistry, Systems and Applications,
First Edition. Alexander Börger and Heinz Wenzl.

Companion website: www.wiley.com/go/Börger/Battery_Technology

Different methods are used to determine battery properties and model parameters. A distinction must be made between the parameters that result from physicochemical conditions of the system and those that can only be determined by appropriate experiments.

9.2 Determination by Physicochemical Methods

When parameters can be determined from properties of materials, thermodynamic and kinetic properties, the data required are design specifications (e.g. dimensions of the electrodes) and many material properties including values required to calculate the Nernst voltage and the charge transfer voltage (e.g. electrochemically active surface, exchange current density, specific resistances, etc.). If necessary, the temperature- and *SoC*-dependent changes in the material properties must also be known. Under laboratory conditions, material properties can be determined fairly accurately. When using material parameters, however, the question often arises whether the results of laboratory tests can be transferred to the use of the materials in batteries. The measurement of the specific resistance of graphite powder, for example, cannot be directly converted into the resistance of an electrode material characterized by binders and an open-pore structure. Another example is the corrosion resistance of current collector materials to the electrolyte – with or without an active mass coating. In models with local resolution, temperature and mass transport phenomena can be represented by analytical systems of equations consisting of several coupled differential equations. Their numerical solution is usually carried out using finite element models, and the results are integrated via coupled models to calculate local differences in material parameters and temperatures. Specialist literature must be consulted for further illustrations and questions regarding how such results can be transferred and interpreted for use in complete cells.

However, the properties of complete cells and entire battery systems can only be determined in this way at the expense of considerable time and cost, and may, even then, not be successful.

9.2.1 Experimental Determination

The most important methods for the experimental parameter determination rely on charge and discharge curves with

- constant current or constant power for the determination of capacity and open-circuit voltage values,
- pulse current loads[1] to determine values for the open-circuit voltage and the internal resistance as well as all variables of the equivalent circuit diagram of the Randles model (see chapter 8) except inductance, and

1 This method is sometimes also referred to as the CIT method, where CIT stands for current interrupt technique in the discharge direction and current injection technique in the charging direction.

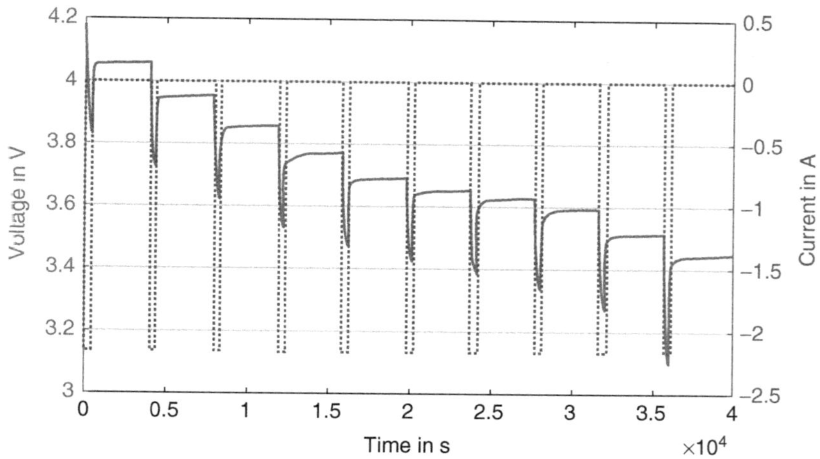

Figure 9.1 Voltage curve of a lithium-ion cell during discharging. The cell was discharged at 2.25 A from 100% *SoC* to 0% *SoC* with 10% *DoD* steps (6 min) including a 30 min rest period after each discharge step. Source: Courtesy of E. Tchoupou Lando.

- AC resistance measurements (impedance measurement or impedance spectroscopy at different frequencies).

An example of the use of pulse loads is shown in Figure 9.1. Starting from a full charge according to the data sheet and a defined rest phase after the end of the charge, the battery is discharged in steps, usually with

- discharging 10% of the capacity during constant current phases with rated current or the current amplitude of particular interest for the application, followed by
- rest periods until the voltage has reached a largely stable value (15–30 minutes for many lithium-ion batteries, in the range of hours for lead–acid batteries).

For a complete and precise determination of parameters and model parameters, the method should also be used in the charging direction at different temperatures and, if necessary, different currents.

Figure 9.2 shows schematically the relaxation behavior of a cell in the charging and discharging direction with short rest periods. The points are labeled that are used to determine model parameters. When using these methods to determine the parameters of RC circuits for the Randles model (see chapter 8), the voltage after each current change must be measured with high time resolution (e.g. <1 ms).

9.2.2 Capacity Measurement

The actual capacity and the initial capacity of a new battery are always measured under precisely defined conditions (full charge, rest phase after end of charge, constant current, and temperature). The practical capacity is usually determined by discharging with application-specific current and temperature, sometimes by using the connected loads. The final discharge voltage may be determined according to the application and may deviate from the datasheet value.

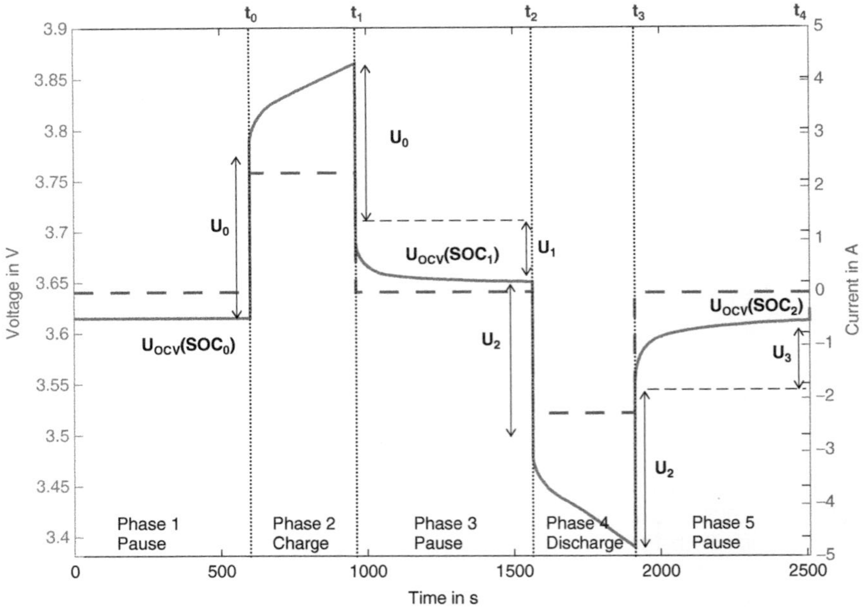

Figure 9.2 Voltage curve for charging and discharging pulses. Source: Courtesy of E. Tchoupou Lando.

9.2.3 Temperature and Current Dependence of Capacity

Measurements are required for both the temperature and current dependence of the capacity. A uniform temperature of the cells must be ensured, especially at high currents. The unavoidable temperature inhomogeneity may lead to an increase in voltage during discharge caused by the heating inside the cell at very low temperatures but with high currents. The results may be affected by the temperature and charging current before the test.

An important empirical parameter of batteries for characterizing the current dependence is the so-called Peukert exponent. Empirically, it has been shown that the decreasing capacity with increasing discharge current can be described by the following equation, the so-called Peukert equation. Knowledge of the Peukert coefficient allows a quick estimate of the capacity at other current amplitudes:

$$I^n \cdot t_I = K. \tag{9.1}$$

where I is the discharge current, n is the empirical Peukert exponent (which is approx. the same for all discharge currents), t_I is the discharge time associated with the discharge current I, and K is an empirical constant. For lead–acid batteries, the Peukert exponent is often between 1.15 and 1.25; for lithium-ion batteries, it is usually much closer to 1 (1.02–1.04). As a result of the Peukert equation, it should be observed that the current related to a specific capacity value cannot be converted directly to another capacity value, for example, $4I_{20} \neq I_5$. Due to the Peukert exponent being much closer to 1 for lithium-ion batteries, capacity differences are significantly less relevant for these batteries.

A coefficient $\Delta C/\Delta T$ is often introduced for the decrease in capacity ΔC at temperatures below the nominal temperature,[2] for example, approx. −0.6 %/K for lead–acid batteries in the temperature range up to 0 °C, and approx. −1 %/K at lower temperatures. Equivalent specifications for lithium-ion batteries depend greatly on the materials used.

9.2.4 Cold Temperature Capacity and Cranking Current

The capacity measurement at very low temperatures and high currents is an acceptance and test criterion for starter batteries. In the case of lead–acid starter batteries, for example, the cold start performance is in reality a capacity test and not an internal resistance test. The performance is determined according to various standards, for example, [1], which requires discharging at a low temperature (−18°C in Europe, −29°C in other world regions) with a cold-test current assigned to the battery by the manufacturer (data sheet information) until a cut-off voltage (often 6 V, i.e. significantly lower than the normal end-of-discharge voltage) is reached. Then, after a short rest period and without recharging in the meantime, the battery is discharged again at 60% of the cold-test current. The resulting amount of charge that can be removed is referred to as the cryogenic capacity.

9.2.5 Bridging Time at Constant Power

Capacity tests are generally carried out with a constant current, which corresponds to a constant rate of electrochemical reactions. For many applications, however, the behavior at constant power for supplying loads is more important. The time of the discharge is the bridging time, and the amount of energy supplied is simply the product of power and discharge time. For batteries with a high current load and stationary applications, discharge curves with constant power or the bridging time as a function of the constant power drawn and the respective end-of-discharge voltage are usually specified.

9.3 Open-Circuit Voltage Curves

Knowledge of the open-circuit voltage is particularly important for the measurement of the *SoC* or the calibration of algorithms to calculate *SoC*. Two methods are used to determine this curve:

1) Constant current discharge with very low current:
   Starting from a fully charged battery, the battery is discharged with a current of $1/20\,C$, that is, a discharge time of 20 h, sometimes also with a current of $1/100\,C$. The voltage values as a function of the amount of charge removed are referred to as open-circuit voltage despite the current. To improve accuracy, the voltage

2 The increase in capacity above the nominal temperature is low and usually ignored.

curve is often also recorded during charging with the same current amplitude and the mean value of the two voltages is used as the open-circuit voltage value. Figure 7.1 shows such an open-circuit voltage curve.

2) The voltage values for a pulse discharge, as shown in Figure 9.1, are used at the end of the respective rest period before the next discharge phase.
   Measuring the open-circuit voltage is important for lithium-ion batteries, as a precise calculation using the Nernst law is not possible. The methods of incremental capacity analysis (ICA) and differential voltage analysis (DVA) are used for a more precise analysis of the voltage curve, which is also dependent on aging (see chapter 10). For lead–acid batteries, on the other hand, the measurement of the open-circuit voltage curve is unusual because the open-circuit voltage can be easily calculated from technical design data (the amount of electrolyte, initial concentration) – provided there are no inhomogeneities (see the explanations on the so-called acid stratification in chapter 12).

9.4 Internal Resistance Determination with Current or Voltage Pulses

An important parameter of a battery is the power that can be delivered or absorbed within the permissible voltage limits, for example, for electric vehicles during acceleration or braking. Measurements are usually carried out with current or voltage pulses. Alternatively, the internal resistance is measured using alternating currents or impedance spectroscopy see also appendix D. AC methods use in most cases only one frequency, whereas the term “impedance spectroscopy” refers to many different frequencies over a wide frequency range.

When using current pulses or CIT methods, cells are subjected to high currents (charging and discharging direction) at different states of charge and temperatures in the laboratory, and the associated voltage curve is measured after a few seconds (e.g. 10 seconds). The measurement corresponds to the determination of the total ohmic resistance and the resistance of the first RC circuit of the Randles model (see chapter 8) by means of a step response. The capacitance of the first RC circuit in the Randles model has been fully discharged by then and therefore can be neglected.

The voltage curve during this time plays no role for calculating the internal resistance. The appropriate simplified equivalent circuit diagram is therefore a voltage source (sum of both electrode voltages) with a resistor (Thévenet model, see chapter 8). It was shown in chapter 2 that such an equivalent circuit diagram at constant current is completely equivalent to equivalent circuit diagrams with RC elements if current changes can be neglected. Figure 9.3 shows the measuring points and measurements of a lithium-ion cell with a constant resistance load instead of a technically much more complex constant current load.

No inductive voltage peaks can be seen in the measurements in the figure due to an insufficiently high rate of current change. With reference to the definition of the

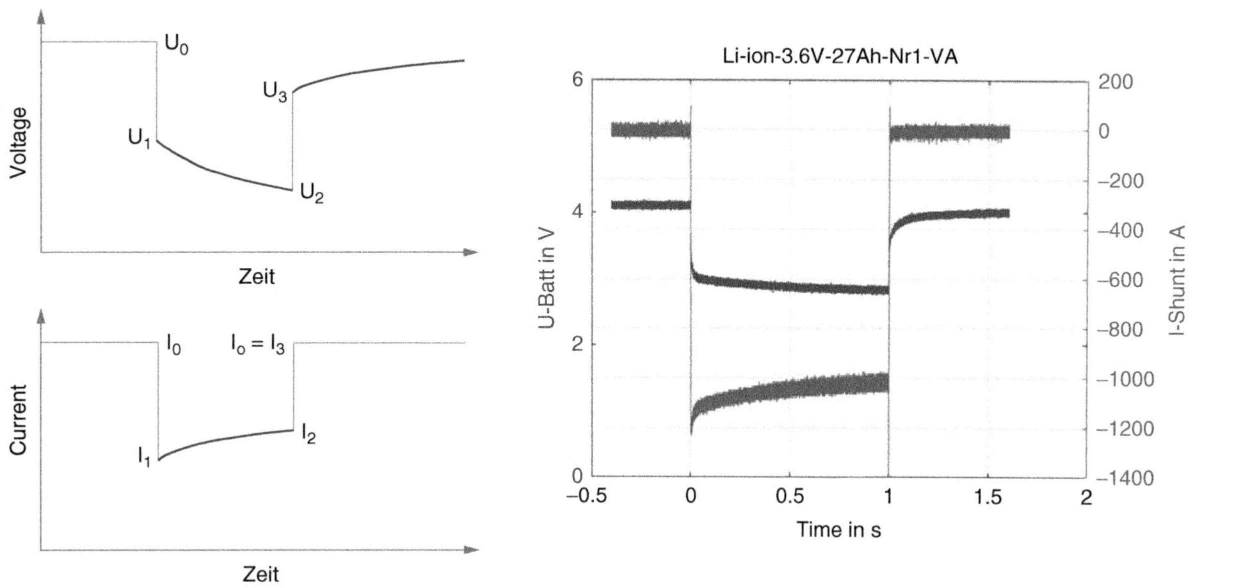

Figure 9.3 Diagram showing schematically the measuring points for calculating the internal resistance and exemplary measurement curves for a constant resistance discharge.

various voltage and current points marked in Figure 9.3, the following results[3]:

$$R_\Omega = \frac{U_0 - U_1}{I_0 - I_1} \tag{9.2}$$

respectively

$$R_\Omega = \frac{U_2 - U_3}{I_2 - I_3}. \tag{9.3}$$

However, these values can generally only be measured using an oscilloscope or measuring equipment with high time resolution and are not suitable for predicting performance. The internal resistance is therefore usually measured without looking at the voltage curve, that is, according to the following formula:

$$R_i = \frac{U_0 - U_2}{I_0 - I_2}. \tag{9.4}$$

Using these measuring points does not require a complex measuring equipment. Usual time intervals for the measurement are either 1, 3, or 10 seconds. A time interval of 10 seconds corresponds to the maximum expected duration of acceleration or braking. The measured values can therefore be used to predict the performance of the battery for electric vehicles. The data must be measured for different states of charge and temperatures. It is also important to choose the load current appropriately because the resistance depends significantly on the current-dependent BV resistance and the increasing current inhomogeneity at very high currents.

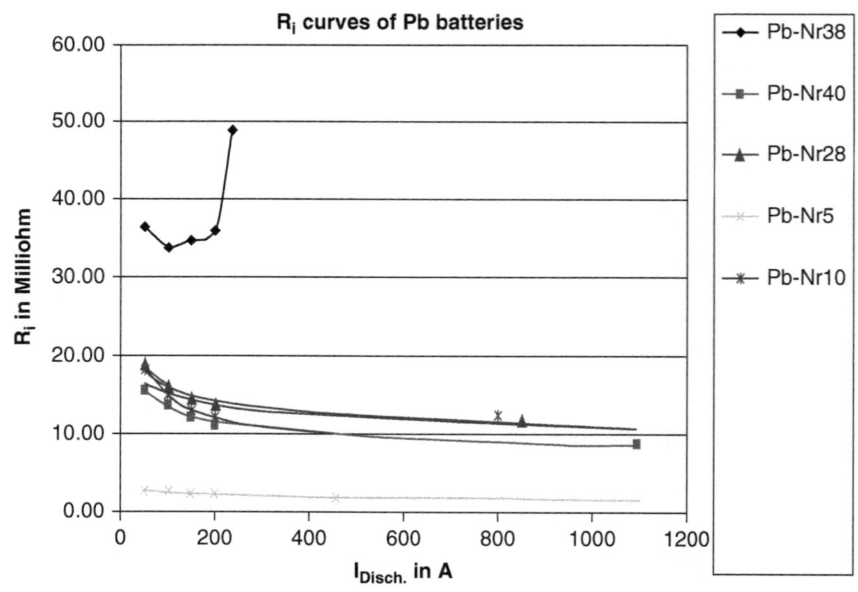

Figure 9.4 Resistance determination at different measuring points.

3 Again, the sign convention must be observed, that is, discharge currents must have a negative sign.

For this reason, it is often more appropriate to use pulse current methods to determine the internal resistance than impedance spectroscopy discussed in the following, as the values determined using pulses correspond more closely to the applications.

Figure 9.4 shows measurements of five lead–acid batteries with different currents. There is a clear decrease in the internal resistance by almost half with increasing current, but at the same time also an atypical behavior in the case of a defective battery with very high internal resistance.

Internal resistance measurement methods are also used to diagnose the *SoH* of the battery. The sensitivity of the method depends on the pulse duration and the load current (see chapter 12).

9.5 Short-Circuit Current

The short-circuit current of batteries, which must be known for the design of fuses and switches, can be several 10,000 A for high-capacity cells and can flow for many seconds. Measuring this current or the internal battery resistance for an extremely small resistance is technically very complex. In Ref. [2], the following method was therefore proposed for lead–acid batteries for stationary applications for new and fully charged batteries at nominal temperature:

- Discharging of a cell with a current in the range from 4 to $6I_{10}$ for max. 25 seconds, measuring the current and voltage values 20 seconds after the start of discharge.
- After a rest period of 2–5 minutes without recharging, a discharge is carried out with a current in the range of 20–$40I_{10}$. After 5 seconds, the current and voltage values are measured.
- Both values are marked in a U–I diagram, and a straight line is drawn through both points. The intersection of this straight line with the *I*-axis at the value $U = 0$ is the estimate for the short-circuit current (see Figure 9.5).

The internal resistance is calculated as the differential resistance, as shown in Eq. (9.5), and the short-circuit current I_K is then calculated as follows:

$$I_K = (U_1 \cdot I_2 - U_2 \cdot I_1)/(U_1 - U_2). \tag{9.5}$$

In Figure 2.10, a U–I characteristic curve for a lithium-ion battery was shown corresponding to the schematic diagram here.

The short-circuit current can also be estimated using the internal resistance of the battery measured by other methods.

$$I_K = U_N/R_i. \tag{9.6}$$

The nominal voltage is often used as basis for this calculation. Due to the dependence of the internal resistance on the current amplitude, the internal resistance must be determined in an appropriate manner.

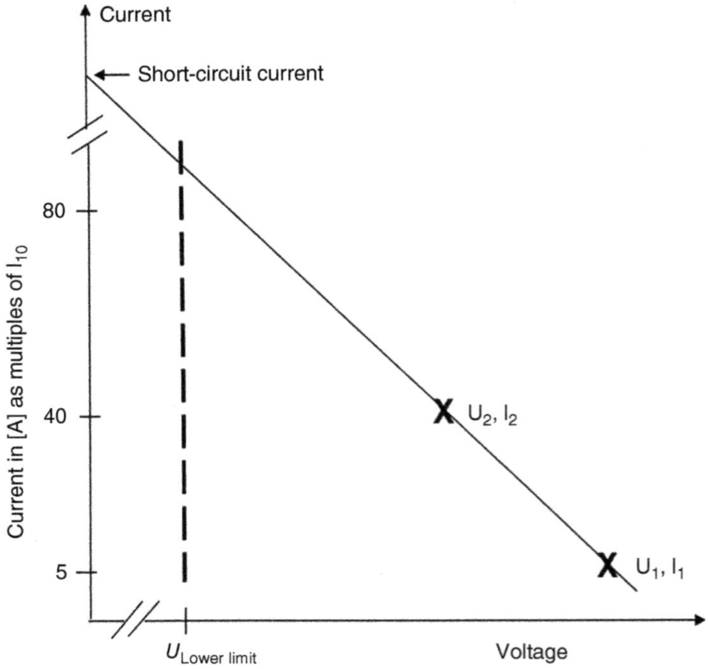

Figure 9.5 Diagram showing schematically the calculation of the internal resistance and short-circuit current according to the specifications of DIN EN 60896-11:2003.

As a rule of thumb, a short-circuit current of approx. 250–300 times the rated current I_{10} for stationary lead–acid batteries can be assumed for new, fully charged batteries at nominal temperature. In the case of lithium-ion batteries, the short-circuit current is even about 5–10 times higher than for lead–acid or other battery types.

9.6 Parameterization for the Randles Model Using Pulse Currents (Measurement in the Time Domain)

To determine the model parameters, the batteries are discharged with current pulses (see Figure 9.1), and the voltage curve is measured with high time resolution (e.g. 1 kHz) from the beginning to the end of the discharge. It is possible to calculate the expected voltage curves for a current pulse analytically for equivalent circuit diagrams with one or two RC elements (see chapter 8). By adjusting the model parameters, the greatest possible agreement between the calculated and measured values is achieved. The quality of the agreement or parameter determination depends on the initial values, the programs used, and their settings (e.g. convergence criteria) in addition to the precision, time resolution (1 ms or better), and low noise measurement.

9.7 Parameter Determination by Measuring the Impedance Spectrum (Measurement in the Frequency Domain)

Electrochemical impedance spectroscopy is a measurement method that requires sophisticated equipment and considerable experience to use and interpret the results. The method is explained in detail in appendix D. The method is only described here to the extent that it is required to determine the internal resistance of a battery and the model parameters for the Randles model.

Impedance is the complex alternating current resistance (amplitude and phase angle) of an electrical system. Impedance measurement is the determination of the AC resistance at a specific frequency, while impedance spectroscopy involves determining it over an entire frequency range (for batteries, usually in the range from 1 mHz to 100 kHz). When a sinusoidal current with the frequency ω and small amplitude flows through a cell, a sinusoidal but phase-shifted voltage response occurs. The same happens when a voltage signal is generated and causes a current response. The impedance is:

$$U(t) = U_0(\sin \omega t) \sim I_0 \sin(\omega t + \vartheta). \tag{9.7}$$

with U_0 as the amplitude of the voltage signal, for example, a few mV, and I_0 as the amplitude of the current response and ϑ as phase shift angle.

If the system has capacitive properties at the selected frequency, then ϑ is negative and the current follows the voltage. If the system has inductive properties, then ϑ is positive and the current leads the voltage signal. The same applies if a sinusoidal current signal with sufficiently low amplitude is used. Figure 9.6 schematically shows the input signal and the response signal.

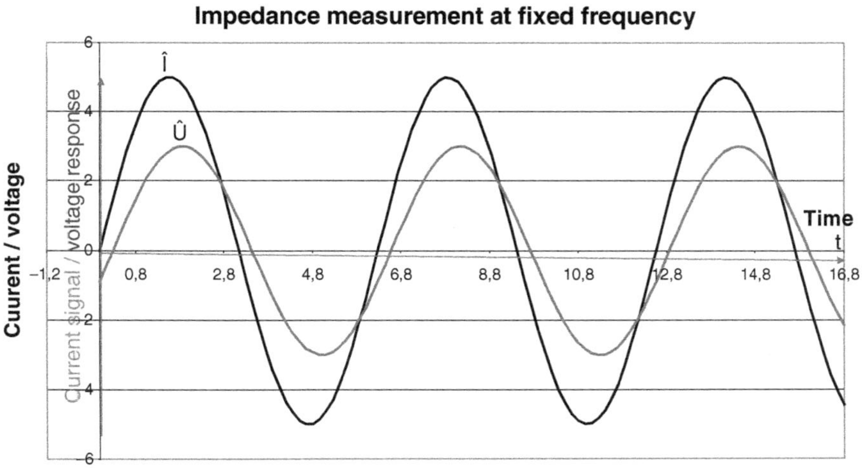

Figure 9.6 Measurement principle of impedance spectroscopy.

The AC resistance is characterized by two variables at each frequency:

- Amplitude ratio of the peak value of voltage U to current I, and
- Phase angle.

Electrochemical systems always have an inductive behavior at very high frequencies and a capacitive behavior at low frequencies. Typically in the range from approx. 100 to 1000 Hz, the phase angle is very small or the transition from the inductive to the capacitive range is in this range. The amplitude ratio at the point $\vartheta = 0$ corresponds to the ohmic resistance of the of the overall system (see appendix D).

In addition to simply determining the ohmic resistance – by measuring at a well-chosen frequency – all parameters of the equivalent circuit diagram shown in Figure 8.5 can be determined, in particular the values of the capacitors and resistances of the two RC circuits. For this purpose, the value of the AC resistance, the impedance, is determined for many frequencies within the selected frequency range, for example, 10 mHz to 10 kHz, whereby the same number of frequencies, usually 3 to 5, are usually used for each decade of the frequency range.

The impedance for any frequency can be calculated analytically for an equivalent circuit diagram with one or two RC circuits if the values of all components of the equivalent circuit diagram are known. However, calculating the values of the individual components or the equivalent circuit diagram from measurements requires a complex mathematical procedure to determine frequency-independent values.

It is immediately clear that the *SoC* and temperature change the model parameters of the Randles model or other models used to interpret the results. Impedance spectroscopy can therefore also be used, in principle, to measure the *SoC*, temperature, and *SoH*.

9.8 Measurement of the AC Resistance

For the diagnosis of lead–acid batteries for vehicles and stationary applications, internal resistance or conductance measuring devices are often used. They perform a very simplified impedance measurement or measurement of the AC resistance according to section 9.7 or a DC resistance measurement according to section 9.4, but with low current amplitudes and very short pulse durations. In a technical report by the IEC [3], this method is explicitly proposed for testing lead–acid batteries in stationary applications and is widely used. Numerous publications have demonstrated a close relationship between the decrease in the internal resistance measured in this way and the battery capacity. For lithium-ion batteries, in contrast, such measuring systems are rarely used because of the existence of battery management systems that are required to ensure safe and reliable operation. See chapter 13 for further details.

Commercial devices do not usually generate a sinusoidal signal, but rather a pulse-shaped discharge of the battery with a frequency of, for example, 22 Hz

(Midtronics/Franklin© measuring device for stationary applications) or 1000 Hz (the most commonly used value), occasionally several frequencies in succession. The resistance values calculated by different measuring devices differ because the signal amplitudes, sometimes also frequencies, shape of signal and above all the calculation algorithms are different. In many applications, changes in the measured values during use are used for diagnosing batteries and not the absolute value. DC and AC resistance values also differ; comparing them or asking which value is better or more correct is not very meaningful.

Commercial devices are therefore rarely used to characterize batteries, but primarily for monitoring tasks, for example, inspection of deliveries or diagnosis of battery properties during use.

9.9 Parameterization of the Randles Model Across All Operating States

The methods for determining the model parameters for the Randles model in sections 9.6 and 9.7 only provide data sets for a specific *SoC* and temperature. Both methods must therefore be used over the whole range of interest. Determining all parameters over the required range of *SoC* and temperature therefore requires that either

- an impedance spectroscopic measurement is carried out at the end of each rest phase (measurement in the frequency domain), or
- the voltage curve is measured with high time resolution (measurement in the time domain).

These measurements must be carried out for all temperatures for which a calculation of the current and voltage curve is required.

The time to collect the data differs and is significantly lower for the pulse current method.

It is analytically possible to calculate the expected voltage curves of a battery under any load profile with either of the two methods. The quality of the agreement or parameter determination depends not only on an accurate and noise-free measurement but also on the initial values, the programs used, and their settings (e.g. convergence criteria).

Figure 9.7 shows the comparison of a current profile and the comparison of the measured voltage response with the calculated voltages for the Randles model, whereby the parameters were determined once using step discharge (CIT) with a recording rate of 1 ms and once by means of impedance spectroscopy and parameter fit using commercial software. It is clear from this that the generally higher effort required for parameterization using impedance spectroscopy does not necessarily provide better results. The quality of the simulation is very high in both cases, and the differences are too small to conclude that one method is generally superior to the other. Table 9.1 shows the comparison in detail.

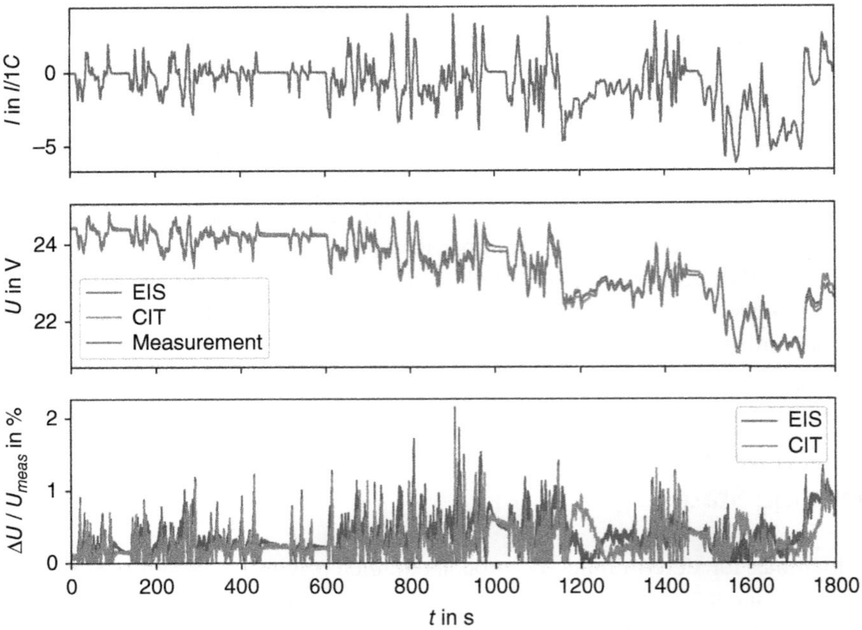

Figure 9.7 Comparison of measured and simulated voltage values. The simulation was carried out using the Randles model, which was parameterized once with a step discharge method and once using impedance spectroscopy. The current curve was converted from the WLTP cycle (speed curve) to a voltage curve (module with six cells, 6s1p, 6.5 Ah pouch cells). WLTP: worldwide harmonized light vehicles test procedure. Source: Courtesy of J. Klink.

Table 9.1 Comparison of the simulation results.

Evaluation of differences	EIS	CIT	Explanation
Average error	0.337	0.254	The mean value of the relative error in %
Standard deviation	0.232	0.191	The standard deviation of the relative error in %
Maximum relative deviation	2159	1691	The maximum relative deviation in %
RMSE	0.096	0.075	The root mean square error of the simulation error in V

References

1 DIN EN 50342–1:2019 (2019). *Lead-Acid Starter Batteries: Part 1: General Requirements and Tests*. Berlin: Beuth.

2 DIN EN 60896–11:2003 (2003). *Stationary Lead-Acid Batteries – Part 11: Sealed Batteries; General Requirements and Test Methods*. Berlin: Beuth.

3 IEC TR 62060:2001 (2001). *Secondary Cells and Batteries – Monitoring of Lead-Acid Stationary Batteries – User Guide*.

Further Reading

Schmidt, J.P. (2013). *Methods for the characterization and modelling of lithium-ion cells.* Dissertation. Karlsruhe Institute of Technology. ISBN: 1868-1603 Available online at digbib.ubka.uni-karlsruhe.de/volltexte/1000036622.

Schweiger, H.-G., Obeidi, O., Komesker, O. et al. (2010). Comparison of several methods for determining the internal resistance of lithium ion cells. *Sensors* 10 (6): 5604–5625. (Basel, Switzerland).

Tasks

Task 9.1

Why are current/voltage pulses used much more frequently for parameterization than impedance measurements, that is, measurements with alternating current that do not cause degradation due to charge throughput?

Task 9.2

Compare the parameterization methods described in the text with regard to:

1) Accuracy that can be achieved,
2) Complexity,
3) Options for extending the model, and
4) Physical plausibility.

Describe the time required, as well as the time and cost for the measurement if both the capacity and the internal resistance are to be measured at several states of charge after every 100 full cycles during lifetime tests.

Task 9.3

Estimate how many measurements are required to determine the necessary parameters of a simple Randles model if the parameters are to provide reasonably useful results in the temperature range from −10 °C to +40 °C and at states of charge from 0% to 100%. Assume a required resolution of 5 K and 10%, and for simplification, assume a linear temperature and state of charge dependency of all parameters.

10

Battery Diagnostics and Analytics

Battery diagnostics and analysis refer to the use of methods to obtain more precise information about the battery and its materials. Many methods do not require battery models. A distinction must be made between:

1) Determination and evaluation of changes in battery parameters as described in chapter 9,
2) Electrochemical analysis methods, subdivided into stationary, quasi-stationary, and nonstationary,
3) Chemical, spectroscopic, diffractometric, and microscopic analysis methods after dismantling the battery (i.e. postmortem methods), and
4) In situ methods.

10.1 Overview of Methods

The term "battery diagnostics" is used for collecting detailed information during operation or to identify the causes of a fault without the use of complex analytical techniques. All other methods are referred to as battery analytics, but the boundaries between diagnostics and analytics are blurred. The methods include the determination of state variables, but also information about processes on a microscopic, molecular, or atomic level, as well as processes during the production of the battery (which will not be discussed in detail in this book). Due to this general approach and the diversity of battery types and their fields of application, there is a wide range of methods for battery analysis, which can be divided into four groups, which in turn are subdivided into various subgroups:

1) Determination and evaluation of changes in battery parameters (properties and model parameters according to chapter 9, mostly diagnostic procedures).
2) Electrochemical analysis methods, which in turn can be divided into stationary, quasi-stationary, and nonstationary methods.
3) Chemical, spectroscopic, diffractometric, and microscopic postmortem analysis methods (mostly destructive methods).

Battery Technology: Fundamentals of Battery Electrochemistry, Systems and Applications,
First Edition. Alexander Börger and Heinz Wenzl.

Companion website: www.wiley.com/go/Börger/Battery_Technology

Table 10.1 Overview of battery diagnostics and analysis methods.

Method	Goal
Determination of electrical parameters	Determination of the operating parameters as a function of aging
Stationary electrochemical analysis methods	Investigation of the properties of electrochemical reactions
	ICA/DVA: Investigation of the energy levels of the intercalation stages and loss of lithium through solid bonds
Quasi-stationary electrochemical analysis methods	Cyclovoltammetry: Investigation of the existing electrochemical reactions
	Impedance spectroscopy: Determination of the internal resistance and model parameters, as well as the relaxation and diffusion behavior
Nonstationary electrochemical analysis methods	Dynamic behavior, internal resistance, diffusion properties
Chemical and physical processes	Determination of the atomic and molecular composition of the samples
	Determination of the microscopic surfaces (BET)
Microscopic procedures	Structure elucidation
Spectroscopic methods	Local determination of atomic and molecular composition
Diffraction methods	Crystal structures, determination of stoichiometry

4) In situ analytics (i.e. probe methods that provide information beyond current and voltage during battery operation).

Table 10.1 provides an overview of the aims of the methods.

10.2 Evaluation of Changes in Electrical Parameters

Changes in the properties of batteries, in particular internal resistance and capacity, but also other model parameters, are important for battery diagnostics. All methods described in chapter 9 for characterizing batteries can and are used for this purpose. For special applications of batteries, other parameters can be interesting and important; for lead starter batteries, the cold start performance and the trickle charge current for stationary applications used in emergency power systems. These aspects are technology- and application-specific and are therefore not described further here.

The determination of capacity has already been described in chapter 9 and always requires the entire system to be taken out of operation for a short time, because during the capacity test, the amount of energy that can be supplied to a load will be reduced. Other parameters, such as cell or block voltages, internal resistances, or temperatures, are immediately available via permanently installed sensors that continuously store the data in battery management or monitoring systems. Alternatively, appropriate measuring equipment can be used as long as measuring points were provided when the battery was installed. Special attention should be paid to

- Current sensors must be accurate over the entire range of expected current amplitudes, for example, several 100 A for vehicles when starting or accelerating, and a few mA when the vehicle is stationary. Current sensors used in the automotive industry are so sensitive that they can also be used for testing loads below 1 W power consumption.
- When using permanently installed equipment for measuring the voltage of cells in batteries with a high nominal voltage, care must be taken to ensure a safe arrangement of the measuring leads and, if necessary, individual galvanic isolation of each measuring point.

However, the estimate of actual or residual capacity and internal resistance can also be model-based. For this purpose, the measured voltage curve is compared with the voltage curve calculated using known model parameters.

10.3 Electrochemical Analysis Methods

Electrochemical analysis methods are usually divided into three groups [1]:

a) Stationary analysis methods, that is, those that leave important parameters for battery operation at a constant value without changing the battery condition. These include, in particular, potentiostatic and galvanostatic methods.
b) Quasi-stationary analysis methods, that is, those in which battery operating parameters are changed sufficiently slowly so that quasi-stationary states can be expected. These include, in particular, various types of alternating current methods. Examples are electrochemical impedance spectroscopy (EIS) and cyclic voltammetry (CV). The boundaries with the methods under c) are blurred.
c) Nonstationary analysis methods, that is, those in which the state of the battery is deliberately and abruptly changed, which applies in particular to measurements with current or voltage pulses.

To obtain more precise information about a cell, electrochemical analysis methods often use a third electrode, the so-called reference electrode. The reference electrode (which should have a very high resistance to avoid current flow and thus avoid an additional voltage drop in the electrodes under investigation) has a potential relative to the electrolyte that is precisely known and independent of all changes in the system under investigation. This makes it possible to assign separate potentials to the anode and the cathode.

10.3.1 Stationary Electrochemical Analysis Methods

The term "stationary electrochemical analysis methods" is used for methods that do not or only insignificantly change the state of the battery or the electrode under investigation. The methods are referred to as the galvanostatic method if constant current is used or the potentiostatic method if constant voltage is used. Commercial devices exist for both methods. Ultimately, the aim is to determine the current-voltage characteristics of the battery or the test system, that is, diagrams already shown in the discussion of the BV equation (see chapter 2).

10.3.2 Quasi-stationary Electrochemical Analysis Methods

The two most important quasi-stationary electrochemical analysis methods are CV and impedance spectroscopy.

In CV (also known as delta voltage method), the voltage is increased with a linear rate and the current increase is measured. After reaching a maximum value, the voltage is reduced linearly back to the initial value. Among other things, diffusion coefficients can be determined from the cyclic voltammograms. The relevant literature must be consulted for further details and the corresponding equations.

10.3.2.1 Impedance Spectroscopy

EIS, which has already been introduced in chapter 9 and is described in more detail in Appendix D, is not only used to determine the internal resistance and model parameters of the Randles model (chapter 9), but is also an important method for characterizing aging processes and determining complex material properties. However, measurements are difficult to interpret. Many research institutes use impedance spectroscopy for comprehensive investigations. Common research objectives are the determination of internal resistance, electrolyte resistance, charge transfer resistance, mass transport phenomena, and diffusion coefficients, as well as their changes as a function of aging, state of charge, and temperature.

To analyze the measurement data, in addition to the physically easily interpretable elements of the equivalent circuit diagram according to chapter 8, further elements that cannot be directly interpreted physically are introduced: Warburg impedance and/or a constant phase element (CPE). Equivalent circuit diagrams with more than two RC circuits are also often used, which tend to improve the agreement between model and measurement but make the interpretation and assignment of physicochemical phenomena of measurements more difficult. The aim should not be to achieve an optimum match between model and model parameters, but rather a detailed understanding of the physicochemical processes.

When using reference electrodes separate impedance, spectra can be measured for each of the two electrodes.

The frequency range, which is sufficient for determining internal resistance and model parameters, is extended toward both lower and higher frequencies for more in-depth investigations. When using very low frequencies (e.g. <0.1 mHz, i.e. approx. 3 h for a single sinusoidal excitation, which is often repeated at least once to increase measurement accuracy), the stability of the measuring device and the question of whether the cell under investigation does not suffer any changes during the measurement (e.g. change in ambient temperature) are important. At very high frequencies, cable inductance plays an increasingly important role, so a limit of 100 kHz is usually set. A four-point measurement[1] is always required, that is, the cell voltage must be measured at the terminals, and the ohmic and inductive voltage drops of the current cables can be disregarded.

It is important to strictly adhere to reproducible measurement conditions and a reproducible evaluation of the measurement results to increase the potential of impedance spectroscopy.

For more details on this important but complex method, see the appendix and further literature.

In general, the potential of impedance spectroscopy in the field of battery analysis and diagnosis has not yet been fully exploited. Among other things, this is due to difficulties measuring in the megahertz range where measurements can be very fast, but cable inductances dominate the signal of the test specimen if the measuring cables are arranged unfavorably. In the range of a few millihertz and lower frequencies, where diffusion processes often dominate, measurements take several minutes to hours, which is too slow for practical purposes.

10.3.3 Nonstationary Procedures

Pulse methods described in chapter 9 for determining model parameters can also be used for basic chemical investigations. The objectives of the investigation, measurement accuracy, and preparation of test specimen must be planned carefully, for example, with reference electrodes and the high capacity of the counter electrode when concentrating on the properties of only one electrode.

Particularly important methods are the so-called switch-on or pulse methods, where a voltage pulse and its effect on the current (chronoamperometry) or a current pulse and its effect on the voltage (chronopotentiometry) are examined. Modern developments of these techniques are the potentiostatic intermittent titration technique (PITT) and the corresponding galvanostatic intermittent titration technique (GITT). In the latter method, a voltage (PITT) or current pulse (GITT) is applied sequentially until approx. 2–5% of the total capacity of an electrode (e.g. an intercalation material) is discharged. After each pulse, there is a rest period until equilibrium is reached again. The titration techniques are used in particular to

1 Four-point measurement: Four cables are connected from the measuring device (or a charger) to the battery: two for power supply and two for voltage measurement. The cables for voltage measurement must be attached as close as possible to the battery, for example, directly to the terminals and not to clamps used to attach the power cables. The voltage measurement is carried out with as high a resistance as possible so that the current through the measuring cables can be neglected, and therefore, there is no voltage drop due to the cable resistances. When measuring impedance spectra at high frequencies, the cables must also be twisted to reduce the effects of cable inductance. Impedance spectrum measurements without a four-point measurement are of little value.

determine phase boundaries and chemical diffusion coefficients, which in turn can be used to calculate internal resistances.

Another group of transient methods are incremental capacity analysis (ICA) and differential voltage analysis (DVA).

When the open-circuit voltage curve of a battery is determined during charging or discharging with a small constant current amplitude, the different phases of the electrochemical processes can show up in steps and plateaus of the open-circuit voltage curve. However, the stages and plateaus are often not recognizable clearly. Methods such as ICA or DVA are used to identify steps and plateaus in the voltage curve which then appear as peaks or troughs. ICA and DVA also provide considerable information about the aging processes [2] and the *SoH* of the battery, in particular for lithium-ion cells [3].

ICA is calculated by dividing an incremental change in capacity (charging or discharging) of a battery through the corresponding voltage change, so that plateaus are highlighted in the form of peaks:

$$\mathrm{ICA}_{SoC} = \left(\frac{\mathrm{d}C}{\mathrm{d}U}\right)_{SoC}, \tag{10.1}$$

with the dimension Ah/V.[2] To form this quotient from measurement data, not only a precise voltage measurement technique is required, but also suitable smoothing and averaging methods due to the very slow change in the voltage value. Figure 10.1 shows corresponding measurements.

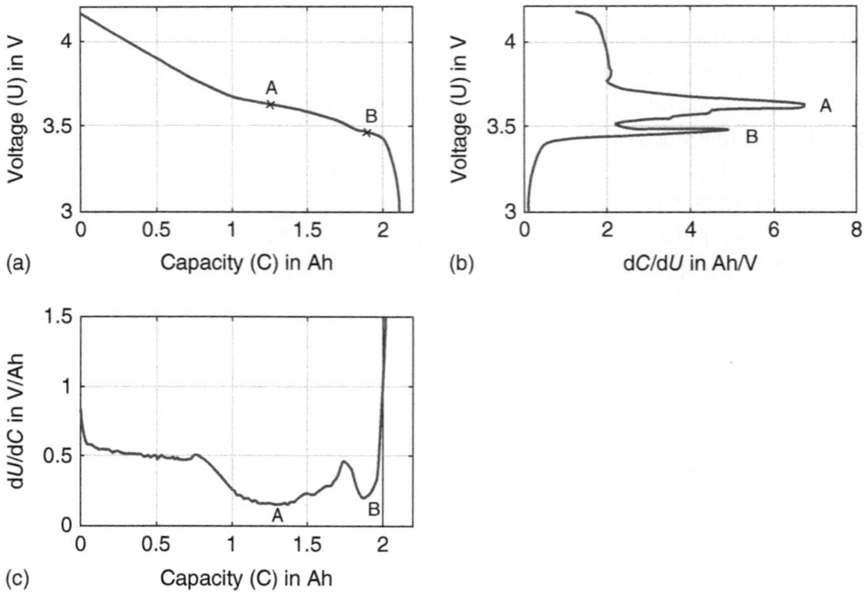

Figure 10.1 Different methods of capacity analysis. (a) Open-circuit voltage measured during a discharge process with a constant current amplitude of 1/20 C. (b) Incremental capacity analysis and (c) differential voltage analysis. Source: Courtesy of E. Tchoupou Lando.

2 Ah/V is analogous to As/V, the unit of capacitance of electrical capacitors.

In contrast to ICA, DVA converts the plateaus into troughs. The larger the peaks, the smaller the troughs.

$$DVA_{SoC} = \left(\frac{dU}{dC}\right)_{SoC}, \tag{10.2}$$

where DVA_{SoC} is the differential capacity analysis in V/Ah.

In both cases, it should be noted that the denominator contains a unit that only changes very slowly. The change dC can be calculated from as the current is constant, but the voltage change must be measured and leads to significant problems, especially when using ICA if the voltage measurement is not accurate and stable and the measurement data is not averaged and filtered beforehand, see, for example, Refs. [4, 5].

10.4 Chemical and Spectroscopic Methods – *Postmortem* Analysis Methods

10.4.1 General

Considerable information about manufacturing or aging processes of batteries can best or only be obtained by dismantling the batteries in question and subjecting the components to more detailed analysis. Depending on the battery chemistry, these analyses can be carried out in air and on fully charged batteries (lead–acid batteries, other aqueous systems) or only in an inert gas atmosphere and preferably on discharged batteries (lithium-ion batteries).

A visual inspection alone often provides valuable information about the reasons for specific battery performance or early failure. For example, short circuits or dendrites can sometimes be identified visually, possibly also by discoloration of the surrounding battery materials. Deformations of the separator, insufficiently stable current collectors, etc. can also be easily recognized. The nature of the active masses also provides clues: in lead–acid batteries, for example, parts with soft active mass with poor adhesion to the grid are generally an indication of heavy cyclic use, while shiny metallic areas on graphite electrodes are an indication of the deposition of metallic lithium.

Complementary to visual inspection, various chemical, microscopic, spectroscopic, diffractometric, and elemental analytical techniques can be used. For a detailed description of these methods, reference must be made to further literature [6]. The explanations here are intended to provide some indications of where the use of these techniques could be worthwhile and help to better understand conclusions made on the basis of such analyses.

In all cases, more or less complex sample preparation is required (washing and drying of cells with liquid electrolytes, separating separators from electrodes, cutting or sawing of group bars and terminals, preparation of metallographic sections, etc.). These processes are not explained further here, nor are the problems of changes to the sample caused by the sample preparation, for example, damage to separators that are to be examined for holes and cracks.

10.4.2 Chemical Techniques Including Separation Processes and Characterization Methods for Surfaces and Grain Sizes

Basic techniques of analytical chemistry are the determination of concentrations by gravimetry and titration techniques. As these methods can only be automated to a limited extent, their importance in battery laboratories has diminished over time. However, chemical techniques are still the method of choice for certain issues and are a frequently used technique, such as when determining the sulfate content of an electrode in a lead–acid battery, which is generally determined chelatometrically, that is, using a complexing agent.

Certain separation techniques are also based on chemical interactions. This applies, for example, to chromatographic processes in which a mobile phase (liquid, gas, or supercritical fluid) and a stationary phase interact with each other, resulting in different flow times for the individual dissolved components. For example, ions can be separated by ion chromatography, but the individual additives in the electrolyte solution of a lithium-ion battery can often only be separated by high-performance liquid chromatography (HPLC). The exact component is then usually identified using coupled spectroscopic techniques.

Finally, different methods for characterizing surfaces and grain sizes are also based on chemical (or physical) interactions between different components. In the production process, such processes are used before the electrolytes are added; for aging investigations and failure analyses, the electrolyte must be washed out beforehand. The BET method[3] for surface area determination uses the pressure dependence of the adsorption of gas molecules (often nitrogen, but argon, helium and other gases can also be used) to determine the available inner surface area of a sample. The electrochemically active surface area can be calculated from the space requirement of a gas molecule and the amount that has been absorbed. In addition, such investigations are carried out at low temperatures. Another method is mercury porosimetry in which liquid mercury is forced into a sample under pressure.

10.4.3 Microscopic Techniques

Classical light microscopic methods help, for example, when investigating separators (e.g. with regard to the existence of holes) or when evaluating the extent of dendrite formation. However, the true potential of advanced microscopic methods is only unlocked by electron microscopy and scanning probe microscopy. Electron microscopy mainly exists in its two versions, scanning electron microscopy (SEM) and transmission electron microscopy (TEM). In SEM, the surface of a sample (e.g. a grid or sections of active mass) is systematically scanned, whereas TEM requires a sample so thin that electrons can penetrate the sample. This places special demands on sample preparation, which significantly limits the range of applications of TEM. In contrast, SEM only requires a conductive sample (usually not a problem with battery materials; if necessary, an insufficiently conductive sample such as a separator material can also be made conductive by sputtering with gold or similar); the

3 Brunauer-Emmett-Teller theory = BET theory.

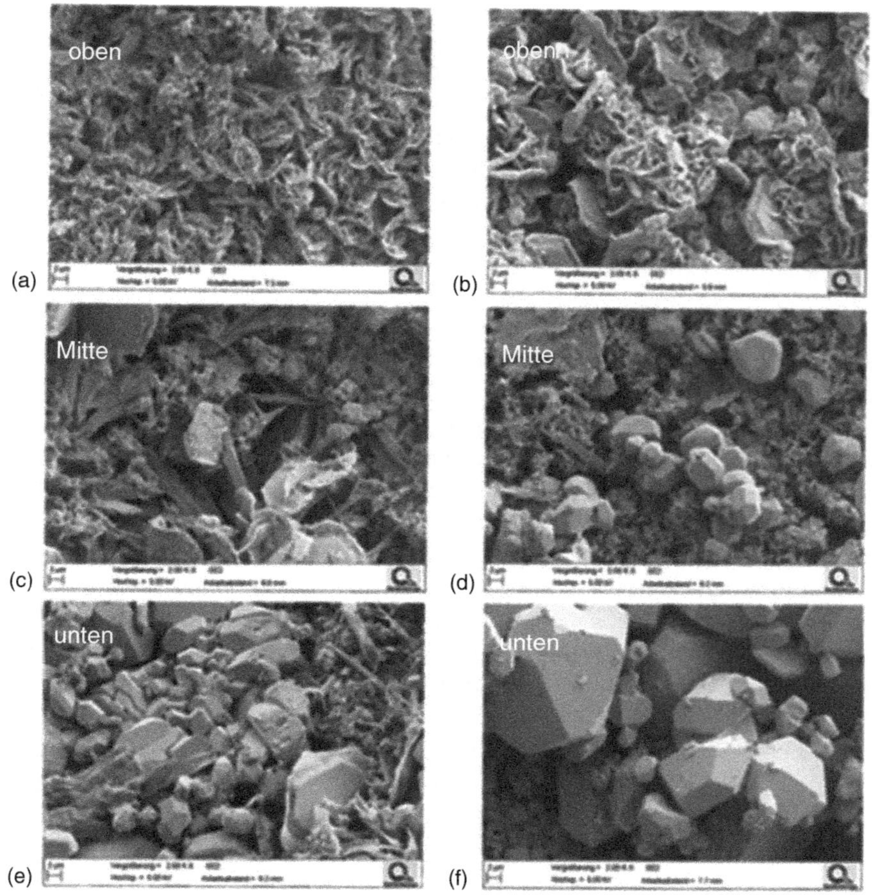

Figure 10.2 (a)–(f) SEM images of different surface segments (top-center-bottom) of the negative active mass of two lead–acid batteries with different designs. Source: Reproduced with permission from Ref. [7]/John Wiley & Sons.

disadvantage of SEM, however, is a lower resolution. An example of an application can be found in Figure 10.2. In addition to areas with very fine-grained structure (e.g. Figure 10.2a), large lead sulfate crystals can also be recognized, particularly in Figure 10.2f.

In recent decades, the range of microscopic methods has drastically expanded by scanning probe techniques. The principle is always the same: a thin, very narrow, and focused beam scans the surface of interest. Depending on whether the tunneling current, mechanical adhesion force, optical properties with fiber optics, or other physical effects are used, the terms scanning tunneling microscopy (STM), atomic force microscopy (AFM), or scanning near-field optical microscopy (SNOM) are used.

Another method is scanning electrochemical microscopy (SECM), in which surfaces are scanned with a microelectrode to determine local thermodynamic and kinetic data [8].

10.4.4 Spectroscopic Techniques

With spectroscopic techniques, a distinction must first be made between atom-specific and molecule-specific techniques. While the former provide information about atomic composition, the latter also provide information about chemical bonds. The basis for quantitative determination using spectroscopic methods is Lambert-Beer's law, which describes the relationship between the absorption of electromagnetic radiation, the concentration of the absorbing species c, and the layer thickness d through which the radiation passes:

$$I = I_0 \cdot e^{-\varepsilon c d}, \tag{10.3}$$

where I is the intensity after passing through the sample, I_0 is the initial intensity, and ε is the extinction coefficient.

Among the atomic spectroscopic methods, atomic absorption spectroscopy (AAS) and atomic emission spectroscopy (optical emission spectroscopy, AES, or OES) are most important. Both methods are usually carried out in solutions, although solid AAS devices have been commercially available for some time.

In contrast, X-ray fluorescence spectroscopy (XRF) works directly with solid samples. This technique is normally used in addition to electron microscopy, in which characteristic X-ray radiation emitted during ionization of samples in the electron beam (energy-dispersive X-ray diffraction, EDX) can be used to obtain information about the atomic composition of the sample of interest. An application can be found in Figure 10.3.

Information about chemical bonds can be obtained using spectroscopic methods that work with light in the visible or infrared part of the electromagnetic spectrum. While absorption of solutions in the UV or visible range (the so-called UV/Vis spectroscopy) can normally only provide relatively unspecific information about the presence or absence of certain chromophoric (i.e. color-bearing) groups, vibrational spectroscopic techniques have become very important, namely FTIR spectroscopy (Fourier transform infrared spectroscopy) and Raman spectroscopy. Both methods are based on the excitation of molecular or solid-state vibrations by absorbing electromagnetic radiation of a suitable wavelength. FTIR spectroscopy leads directly to the excitation of vibrations by absorption. Raman spectroscopy is an indirect method where inelastically (i.e. with a change in energy) scattered light is observed – the energy loss (or, more rarely, gain) is due to the excitation of an oscillation (or the absorption of energy from a previously thermally excited oscillation). Since inelastic scattering is a very rare process under normal conditions, high-energy light sources are required, usually lasers that operate in the visible or near-infrared part of the electromagnetic spectrum. The theory of Raman spectroscopy shows that there are different selection rules for FTIR and Raman spectroscopy, that is, the two techniques are often complementary to each other. Raman spectroscopy (Figure 10.4) has proven to be an indispensable method, particularly for the investigation of carbonaceous materials such as graphite anodes in lithium-ion batteries or conductive carbon black as an additive in lead–acid battery anodes.

The most precise information about chemical bonding states in analytical chemistry is possible using spectroscopic methods in the radio frequency range, so-called

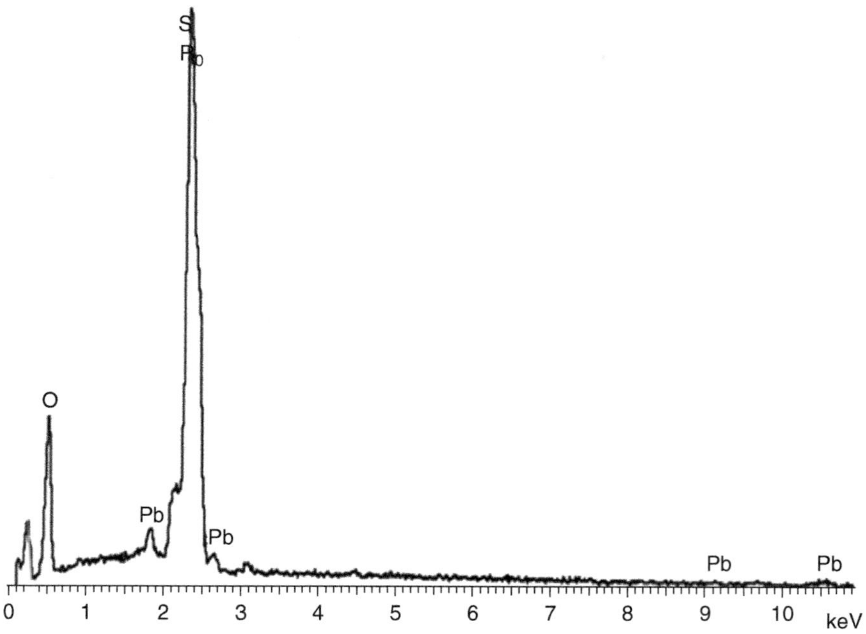

Figure 10.3 EDX spectrum of a surface segment in the middle of an electrode of a lead–acid starter battery. The ratio of lead to sulfur in damaged batteries can be used to determine the extent to which the battery is sulfated, that is, the extent to which lead has already been converted into electrochemically inactive lead sulfate. Source: From Reproduced with permission from Ref. [7]/John Wiley & Sons.

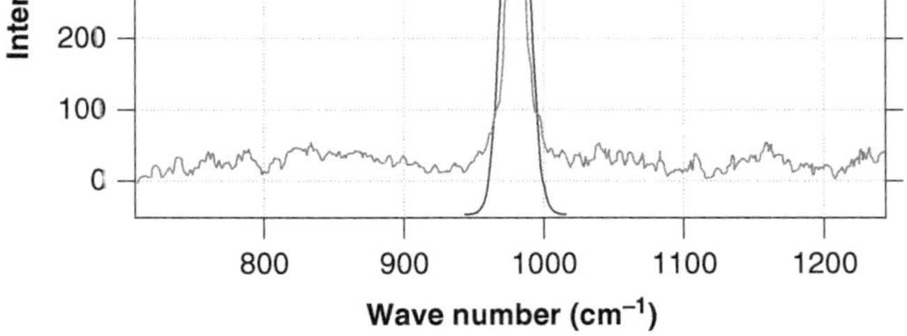

Figure 10.4 Example of a Raman spectrum for a sample containing sulfate. Source: Reproduced with permission from Ref. [9]/Elsevier.

NMR spectroscopy (nuclear magnetic resonance). In NMR spectroscopy, the sample under investigation is placed in a strong magnetic field in which the energy levels of atomic nuclei are split, provided they have a nonzero nuclear spin. Due to various couplings and the sensitivity of the nuclear spin energy levels for the exact chemical environment, very detailed knowledge about the chemical environment in a molecule can be obtained, which is used for structural analysis. Unfortunately, NMR spectroscopy often requires larger amounts of substance and is also primarily a technique that is used for solutions. Since metallic conductors, unpaired electrons, and magnetic substances interfere with measurements, NMR investigations are limited either to carefully prepared battery components after battery teardown or to special cell geometries that differ greatly from commercially available batteries.

Numerous other spectroscopic techniques have been developed that are used in solution, in solids, and on surfaces. The methods described above are also being continuously developed and refined.

10.4.5 Diffractometric Techniques

While spectroscopic techniques are always associated with the absorption or emission of electromagnetic radiation, diffractometric techniques, that is, diffraction methods, involve the constructive or destructive interference of the incoming radiation, which results in characteristic diffraction patterns. Due to the wave-particle duality known from quantum physics, diffraction phenomena can be observed not only with electromagnetic radiation but also with neutrons and electrons, which are usually considered to be particles.

All diffraction methods have in common that they only lead to sharp diffraction patterns for crystals. Liquids and amorphous solids such as glasses only produce uncharacteristic diffraction patterns with very broad bands. The basic equation for crystalline substances is the Bragg equation, which relates the distance between lattice planes in the crystal lattice, characterized by the Miller indices *hkl*, with the diffraction angle θ and the X-ray radiation of wavelength λ:

$$n\lambda = 2d_{hkl} \sin \theta, \tag{10.4}$$

n indicates the diffraction order.

Diffraction experiments with X-rays (XRD) make it possible to determine the crystalline phases present in solid samples, including their relative proportions and even the arrangement of the atoms that make up the crystal. In the battery sector, there are usually not single crystals, but polycrystalline materials, where it is much more difficult to determine the precise structure. The main field of application of XRD is therefore the analysis of crystalline phases involved in electrochemical processes and, if necessary, their relative chemical composition (in the case of nonstoichiometric compounds, the diffractogram, that is, the diffraction pattern often depends in a characteristic way on the exact atomic composition).

The scattering behavior of X-rays depends directly on the atomic number of the elements; light elements lead to comparatively low scattering and may therefore be difficult or impossible to detect in X-ray diffraction patterns. A complementary method is neutron diffraction, because the scattering coefficients for neutrons are nuclide-specific and not related to the atomic number. The disadvantage of neutron diffraction, which is also mostly to be carried out on powders (powder neutron diffraction [PND]), is the radiation source because neutron beams are only available at large research facilities, and the radioactive activation of the sample. However, it is often only the interaction of X-ray and neutron diffraction that enables exact information of structures or complex composition of different phases of the material.

The theory of electron diffraction is more complicated than that of X-ray and neutron diffraction because in the case of electron diffraction multiple scattering processes have to be taken into account. Due to the strong interaction of electrons with the surrounding matter, electron diffraction is a method requiring special equipment (e.g. vacuum), and is used in particular to examine surfaces and areas close to the surface.

10.5 In Situ Analysis Procedures

Ideally it should be possible to obtain further information about the chemical and physical processes in a battery directly during operation.[4] Unfortunately, commercial batteries are not suitable for most conventional non-electrochemical analysis methods, such as those presented in section 10.4, because the batteries are too large or too impermeable to electromagnetic radiation as used in spectroscopic, diffractometric, or microscopic methods. Electrochemical analysis methods such as impedance spectroscopy can be carried out under in situ conditions, but they only provide indirect information or require very precise knowledge of the system under investigation, which can only be obtained using complementary techniques.

Recently, however, in situ techniques have been developed and applied that can be used under real-time conditions. One example of this is neutron diffraction, which has now been used in a series of investigations. The disadvantage is that the radioactive activation of the examined battery in the neutron beam generally rules out further investigations on such batteries. However, Raman spectroscopy can also be used as an in situ technique, although modified battery designs often have to be used to allow the laser beam to penetrate into the battery interior or onto the electrode surface.

The range of in situ methods is still under development and it can be assumed that further complex measurement concepts will be presented in the future, which will enable a more detailed understanding of the processes taking place in batteries under real-time conditions. A recent advance in microscopic methods, for example, is the use of tomographs to generate entire 3D images of batteries.

4 In situ does not mean testing the battery at the installation site, but testing a complete, operational battery, which may have to be specially prepared.

Table 10.2 Summary of some characterization methods for batteries.

Question	Methods	Notes
Total atomic composition	AAS, ICP-OES, RFA	Not suitable in situ
Atomic composition locally and on surfaces as well as stoichiometric ratios	EDX, SECM	Not suitable in situ
Crystal structure	XRD, PND	Limited suitability in situ
Crystal structure	Microscopic methods (SEM, TEM, …)	Not suitable in situ
Particle size distributions	XRD, BET, screening process	Not suitable in situ
Concentration gradients (lithium intercalation, acid stratification, …)	Raman spectroscopy, electrochemical methods	Suitable in situ with appropriate design of the battery to be tested
Local temperatures	EIS	Complex evaluation

10.6 Summary

Table 10.2 gives a summary of the methods briefly described can be used. The experimental effort required to carry out the tests and the experience required to interpret the results should not be underestimated.

References

1 Hamann, C.H. and Vielstich, W. (2005). *Electrochemistry*, 4e. Weinheim: Wiley-VCH.

2 Barai, A., Uddin, K., Dubarry, M. et al. (2019). A comparison of methodologies for the non-invasive characterization of commercial Li-ion cells. *Progress in Energy and Combustion Science* 72: 1–31.

3 Tian, J., Xiong, R., and Yu, Q. (2018). Fractional order model based incremental capacity analysis for degradation state recognition of lithium-ion batteries. *IEEE Transactions on Industrial Electronics* 25: 49–56.

4 Grabow, J., Benger, R., Beushausen, L., and Deblon, F. (2019). *Advanced Filtering for Incremental Capacity Analysis Applied to Various Lithium-Ion Batteries*, Kraftwerk Batterie, Aachen (April 02–04, 2019). Clausthal University of Technology, Institute of Institute of Electrical Power Engineering and Energy Systems.

5 Li, Y., Abdel-Monem, M., Gopalakrishnan, R. et al. (2018). A quick on-line state of health estimation method for Li-ion battery with incremental capacity curves processed by Gaussian filters. *Journal of Power Sources* 373: 40–53.

6 Schwedt, G. (2008). *Analytical Chemistry. Fundamentals, Theory and Practice*, 2e. Weinheim: Wiley-VCH.

7 Ebner, E., Metge, I., Keil, S.N., and Börger, A. (2013). Electrolyte convection in lead-acid wet batteries – testing of passive mixing elements in vehicles. *Chemie Ingenieur Technik* 85: 1294–1301.

8 Krueger, B., Balboa, L., Dohmann, J.F. et al. (2020). Solid electrolyte interphase evolution on lithium metal electrodes followed by scanning electrochemical microscopy under realistic battery cycling current densities. *ChemElectroChem* 7: 1–8. https://doi.org/10.1002/celc.202000441.

9 Börger, A., Ebner, E., Rühl, E. et al. (2017). On the use of Raman microscopy for sulfation analysis in lead-acid battery research. *Journal of Energy Storage* 12: 305–310.

Tasks

Task 10.1

Draw a schematic diagram of the current, voltage, and *SoC curves* over time for a simple lithium battery test (two 25 Ah cells connected in series), which should consist of the following sequence:

a) Triple capacity test with rated current,
b) Determination of the 5 C internal resistance (duration of pulses for discharging and charging: 30 seconds each) at 50% *SoC*,
c) Ten consecutive charging and discharging cycles at 10 C each without a rest period between individual cycles, whereby the charging should take place according to a CC/CV characteristic curve,
d) Repetition of the capacity test as in a), and
e) Repetition of the internal resistance test as in b).

The voltage limits should be: 4.1 V when charging and 2.8 V when discharging. How many hours will this take, and how many working days?

(Note: The exact shape of the charging and discharging curves is not relevant).

Task 10.2

Name and describe three different electrochemical test methods from different classes that are suitable for determining the intercalation behavior of a lithium titanate anode. Give a brief description of each method (including classification).

11

Overview of Battery Systems

Lead–acid and lithium-ion batteries are the two most technically and economically important battery systems because their overall properties make them superior to other battery systems in most applications. However, for certain applications, many other battery systems, due to their specific properties, are superior to lead–acid and lithium-ion batteries. It is expected that these battery systems will remain commercially important, possibly in very small, niche applications.

11.1 Physicochemical Data and Characteristics

The most technically and economically significant battery systems are lead–acid batteries and lithium-ion batteries, which are discussed in detail in chapters 12 and 13. Additional battery systems are discussed in chapter 14, either because they continue to have technically and economically interesting properties for small areas of application (e.g. very good low-temperature properties) or because they offer attractive development potential. Chapter 14 also briefly mentions the properties of ultracapacitors, also called electrochemical double-layer capacitors (EDLC).[1] The information presented in chapters 2–10 applies to all battery technologies. The bibliography lists books in which numerous other battery systems are described.

Table 11.1 provides an overview.

The specific energy that cells have in practice is approx. 20–25% of the theoretical value; only in the case of metal–air batteries does it reach up to 40% because the cathode material is air, and its weight does not have to be included. When discharged, however, the cells become heavier because the absorbed air is now chemically bound in the cell.

Comparative values for the specific energy achievable in practice are problematic. The main reason is that in some cases meaningful data can only be provided for the

1 In this book, the term "ultracapacitor" is used to avoid confusion with the electrode–electrolyte interface of batteries which is called a double layer, and its electrical properties are those of a double-layer capacitor.

Battery Technology: Fundamentals of Battery Electrochemistry, Systems and Applications,
First Edition. Alexander Börger and Heinz Wenzl.

Companion website: www.wiley.com/go/Börger/Battery_Technology

Table 11.1 Overview of important battery systems.

System	Positive electrode	Ion for transporting charge in electrolyte	Negative electrode	Theoretical specific energy[a)] (Wh/kg)
Lead–acid (chapter 12)	PbO_2	H^+ Aqueous sulfuric acid	Pb	160
Lithium-ion (chapter 13)	Various metal oxides, iron phosphate	Li^+ Different solvents with conductive salts	Mostly graphite (C), rarely lithium titanate	1001 ($LiMnO_2$/Grafit)
Alkaline nickel systems (section 14.1)	Nickel oxide hydroxide NiOOH	OH^- Aqueous caustic potash solution	Metal hydride, zinc, iron, cadmium	278 (NiMH)
Zinc–air (section 14.2)	Oxygen from the air	OH^- Aqueous caustic potash solution	Zinc	1086
Vanadium redox flow batteries (section 14.3)	VO_2^+	H^+ Aqueous sulfuric acid	V^{2+}	25
High-temperature batteries (section 14.4)	Sulfur (S) or $NiCl_2$	Na^+ Aluminum oxide ceramic at approx. 350 °C	Sodium	787 ($NaNiCl_2$) 792 (Na—S)
Lithium solid electrolyte (section 14.5)	As with lithium-ion batteries	Li^+ Solid material, e.g. ceramic	As with lithium-ion batteries, possibly also lithium-metal	1001 ($LiMnO_2$/Grafit)
Lithium-sulfur (section 14.6)	Sulfur	Li^+ Organic ethers	Lithium	2500
Lithium air (section 14.7)	Oxygen from the air	Li^+ Dimethyl sulfoxide	Lithium	3450
Sodium air (section 14.8)	Oxygen from the air	Na^+ Dimethyl sulfoxide	Sodium	1600

a) Information on the theoretical specific energy (see chapter 2 on terminology) refers to the values for theoretical specific capacity and cell voltage given in the *Handbook of Batteries* [1].

overall system and not for individual cells. In addition, different system boundaries must be considered. Therefore, a comparison is not carried out here.

The dominant role of lead–acid and lithium-ion batteries is due to the fact that for many purposes they are superior to all other battery technologies in terms of their overall properties. Table 11.1 shows that their importance cannot be due to theoretical specific energy or the practically achievable specific energy and energy density.

Lithium-ion batteries and lead–acid batteries are direct alternatives in some applications and are also used together in some applications, sometimes in direct parallel connection, because their different properties complement each other well in specific applications. For this reason alone, it is important to study both battery systems, even if the primary interest is only for one of the two technologies. In addition, knowledge of both battery systems beyond the information presented in chapters 2–10 will enhance understanding of their advantages and disadvantages. For instance, the positive and negative effects of side reactions in lead–acid batteries increase awareness of the need for battery management systems and single-cell voltage measurements in lithium-ion batteries.

Table 11.2 shows a comparison of the two systems. Lithium-ion batteries is a generic term that refers to different cathode and anode materials and different electrolytes. This must be taken into account when interpreting the table. A comparison to select the most suitable battery system on the basis of specific requirements and specifications shows that the number of applications in which the two battery systems are real alternatives is rather small and that general comparisons can contribute little to the decision-making process. This problem has already been discussed in more detail in chapter 1. The information presented in chapters 2–10 also makes clear that multi-cell lithium-ion batteries require a complex system environment (with the possible exception of low-capacity cells), whereas lead–acid batteries do not (apart from ventilation requirements). However, lead–acid batteries can benefit from simple battery management systems and heating and cooling systems in terms of reliability and service life. When comparing systems, lithium-ion batteries with a complex system environment should not be compared to the most basic lead–acid battery system, consisting of a series and possibly parallel connection of cells or blocks, but with lead–acid batteries that are equipped with commercially available and simple monitoring and temperature management systems.

A comparison with the other battery systems described in chapter 14 is not carried out due to its limited value.

The criteria used are taken from operational requirements in the application-oriented chapters.

In addition to differences in technical properties, other differences must be consider, some of which can be attributed to the fact that lead–acid batteries are based on more than 120 years of technological development. They represent a largely mature technology in which numerous innovations are still taking place.

Table 11.2 Comparison of the properties of lead–acid and lithium-ion batteries.

Criteria	Lead–acid batteries	Lithium-ion batteries
Specific energy and energy density	Approx. 20–40 Wh/kg for cells, blocks, and complete battery systems	Approx. 250 Wh/kg at cell level and >150 Wh/kg as a battery system Highest specific energy and energy density of all commercial rechargeable batteries (for comparison: commercial zinc-air batteries – non-rechargeable – exceed 400 Wh/kg)
Internal resistance or specific power	Approx. 2.25 mΩ/cell for approx. 15 Ah cells; for very large cells, a few μΩ/cell at room temperature	In relation to the same capacity or energy content, significantly lower internal resistance, but with a much greater increase at low temperatures
Cell sizes	Approx. 4.5–3000 Ah (stationary applications) or even larger for submarines	Approx. 0.1–ca. 200 Ah (2024) To achieve a very large energy content, the parallel connection of several strings is unavoidable
Series and parallel connection	Possible without special precautions, independent of rated capacity and rated voltage	Only for low capacity and nominal voltage is this easily possible. For all other cases: complex monitoring by means of BMS, individual cell voltage measurement. In addition, housing, cooling, and heating systems may be necessary
Self-discharge, storage time	Approx. 1–2% per month for new batteries; considerably higher for aged batteries; recharging required after a few months	Approx. 1% per month; occasional voltage check and recharging necessary
Stand-by losses without charger	Approx. 1–10 times the stored energy content per year	Low, essentially dependent on the energy consumption of the BMS
Cycle efficiency at full cycle (without charger)	Approx. 80% for flooded batteries and up to almost 90% for sealed batteries	Up to almost 100%
Safety	Explosion risk of individual cells; explosion of the entire battery is unlikely with proper ventilation	Thermal runaway of single cells and thermal propagation to neighboring cells causing high damage – a major issue

Temperature range	Increasing performance restrictions at very low temperatures and service life problems with increasing temperature	For many materials, significant performance limitations in terms of power output and rechargeability at low temperatures; service life and safety problems above approx. 60 °C
Operating restrictions	Frequent recharging is required for service life reasons For flooded batteries, refilling with distilled water is necessary in many applications	Charging at low temperatures and operation at very high temperatures must be avoided for reasons of service life and safety
Installation site	Ventilation requirement; coating of the floor against acid, if necessary	For larger batteries in closed rooms, certain fire protection classes must be observed
Service life	High number of cycles, with a charge throughput of well over 1000 times the nominal capacity achieved in some applications; calendar aging of up to 15 years is achieved in some stationary power systems	Very high number of cycles (e.g. more than 4000 full cycles); calendar aging is strongly dependent on temperature (approx. 10–15 years at room temperature and lower are easily achievable; less at higher temperatures)
Operating costs	Low; maintenance requirements depending on the details of cell types	Very low because the BMS contains an automatic monitoring function
Recycling	Well-organized and technically easy; recycling rate is well over 90%	Facilities are under construction; generally very much in need of development
Toxicity of ingredients in the event of accidents and recycling	Lead is a toxic heavy metal and must be prevented from entering the environment; sulfuric acid is corrosive	Very high risk of exposure to toxic substances during fires, especially due to the formation of fluorine and organic phosphorus compounds

11.2 Investment and Operating Costs

A key aspect when comparing the two battery systems is their investment costs.[2] Lithium-ion batteries currently cost around € 100 per kWh at the cell level. Added to this are the costs for a battery management system, possibly cooling, housing, etc., which cause additional costs, so that at system level, € 130 per kWh is to be expected. The extent to which the respective prices for lithium-ion batteries cover costs and the prices paid by major customers such as the automotive industry is naturally not known. The price level will also continue to change, whereby the material costs will play a special role, similar to lead–acid batteries. Lead–acid batteries are reported to cost less than € 50 per kWh, with the additional costs of building a complete battery system being negligible in most cases. Price changes are not expected unless the price of lead changes significantly and permanently.

However, the price range is very high. Lead–acid starter batteries are certainly available for well under € 40 per kWh in cases of high purchase volumes, while special lead–acid batteries for stationary applications with low production volumes tend to cost in the region of € 300 per kWh. Similar price differences can also be expected for lithium-ion batteries.

The operating costs for lithium-ion batteries are somewhat lower and a longer service life can be expected in many applications, although certainly not in all.

In many cases, lead–acid batteries continue to be used for cost reasons, although lithium-ion batteries can fulfill the technical requirements better and for longer. However, with amortization periods of a few years, which are required in many applications, lower lifetime costs over a period of 10 years or more are no longer relevant to purchasing decisions given the higher investment costs.

11.3 Market Structure

The system requirements for lithium-ion batteries require different market structures compared to those for lead–acid batteries. Lead–acid batteries for stationary applications and industrial trucks have very high values in terms of capital investment and energy content (in the gigawatt-hour range). They are commercial products for which chargers as well as monitoring and testing devices are available from specialized companies. For any application, these products can be purchased directly from many different companies and easily combined within certain limits as a result of extensive standardization. The same is not true for lithium-ion batteries due to their high system requirements and integration. Cells, BMS, housing, cooling, heating, etc. come from a single source. The replacement of a defective cell or battery by maintenance engineers from service companies or by the operator, which is common practice with lead–acid batteries, is not possible.

2 Ultimately, prices are only ever known to a company's purchasing department and therefore not to the authors. The values given below are estimates based on freely accessible sources on the internet (as of 2024). It should also be noted that great price differences can be expected depending on the exact type of cells and the number of cells needed for the battery system.

11.4 Availability of Information

In the case of lead–acid batteries, apart from details of the material composition and production process, a great deal is published and known. Examples include:

1) Technical reports (e.g. [4]) that can be used to evaluate the differences arising over the service life of cells in series connection.
2) Publications on the effects of different charging methods on capacity and service life.

Comparable information in relation to commercial products generally does not exist for lithium-ion batteries or is not available in the scientific literature.

In addition to the scientific literature (e.g. *Journal of Energy Storage*, *Journal of Power Sources*, etc.) and conferences (e.g. Advanced Battery Power, Advanced Automotive Battery Conference [AABC], European Lead Acid Battery [ELBC], Advanced Battery Power, etc.), important sources of information for both lead–acid and lithium-ion batteries include standards, leaflets from industry associations, and patents.

11.5 Level of Standardization

For lead–acid batteries, many specifications exist that enable the easy substitution of a product by products from other manufacturers, and there are extensive standards that cover many technical aspects, for example:

- Use of batteries in stationary applications, including test specifications and evaluation of the results.
- Ventilation requirements.
- Determination of the short-circuit current.
- Specifications for the design of containers for batteries.
- Test specifications concerning the requirements to be met by starter batteries.

The same level of standardization has not yet been reached for lithium-ion batteries, but numerous standardization specifications now exist for them. Rapid technological progress is a challenge for standardization. For example, minimum requirements for battery management systems for lithium-ion batteries still do not exist. Appendix B (Overview of Standards) contains a comprehensive overview of standards, technical reports from standardization organizations, leaflets from industry associations, etc., which is divided according to the areas of application presented in chapters 15–20.

A lack of standards can lead to significant costs and represent barriers to market entry if lithium-ion batteries are to replace existing battery solutions in complex, strictly regulated energy systems, such as for railways or aircraft.

Further Reading

Broad information about many different battery systems including primary systems

Beard, K.W. (2019). *Linden's Handbook of Batteries*, 5e. McGraw-Hill.

Crompton, T.R. (2000). *Battery Reference Book*, 3e. Newnes.

Garche, J. (2009). *Encyclopedia of Electrochemical Power Sources*, 2e. Elsevier.

IECTR 62060:2001 (2001). *Secondary Cells and Batteries – Monitoring of Lead Acid Stationary Batteries – User Guide.*

12

Lead–Acid Batteries

Lead–acid batteries are mature and robust products that serve as the best technical and economic energy storage solution in many applications. The scientific principles are very well understood, with current development activities largely focused on reducing production costs.

The active material of the anode consists of lead, while the cathode is made of lead dioxide. When discharging, the active materials are converted into lead sulfate, which has different physical properties compared to the charged active material. In particular, the electrical conductivity is approx. 10^{12} lower, and the specific volume is twice as big. The electrolyte used is aqueous sulfuric acid, which is consumed during discharge and is an electrochemically active material. The decrease in the concentration of aqueous sulfuric acid during discharge leads to numerous technically relevant changes in properties. The fully charged state – no lead sulfate – is clearly defined. The nominal voltage of all lead–acid batteries is 2 V/cell, with the difference between the usual end-of-charge voltages (approx. 2.4 V/cell) and end-of-discharge voltages (approx. 1.7 V/cell) is small in percentage terms compared to other battery systems.

There are two technical designs of lead–acid batteries: those with liquid sulfuric acid ("flooded batteries") and those with acid immobilized in either gel (gel batteries) or an absorptive glass mat (AGM) batteries. In the latter, an internal oxygen consumption cycle exists. There are side reactions with high current-carrying capability in both discharging and charging directions on both electrodes that are not very damaging and pose manageable safety risks. For this reason, many cells can be connected in series without the need for monitoring or cell balancing. Lead–acid batteries are very safe systems, even under abuse conditions. Apart from the general electrical hazards when handling voltage sources with high voltage and low internal resistance, the only danger is the formation of hydrogen gas, the explosion risk of which can be easily prevented.

(Continued)

Battery Technology: Fundamentals of Battery Electrochemistry, Systems and Applications,
First Edition. Alexander Börger and Heinz Wenzl.

Companion website: www.wiley.com/go/Börger/Battery_Technology

(Continued)

Overcharging reactions allow each electrode in each cell to be fully charged without cell balancing systems.

However, lead is a toxic heavy metal, so its use is generally restricted. A ban on lead–acid batteries in Europe is conceivable, although the collection and recycling rate for these batteries is very high.

12.1 Introduction and Economic Significance

As a toxic heavy metal, lead can only be used in products with an exemption, and its use is now banned in various products. It is also conceivable that lead–acid batteries will be banned in certain regions of the world in the near future. The current exemption for lead–acid batteries within the European Union was extended again in 2018. However, due to their combination of technical and economic advantages and their low environmental impact[1] thanks to a very high recycling rate, lead–acid batteries will remain an important industrial product for starting combustion engines, electric traction for industrial trucks and emergency power systems in many parts of the world in the long term.

The following explanations build on the chapters of the first part of this book and deepen and supplement them with technology-specific information.

12.2 Electrochemistry

The electrochemical principles for lead–acid batteries (equilibrium voltage, Nernst voltage, mixed potential, and Butler-Volmer equation) have been fully explained in chapter 2. Although the design details of different lead–acid batteries and the materials used for the respective applications are slightly different, the electrochemical processes that take place during discharging and charging as well as during aging processes are the same for all types of lead–acid batteries.

In the following sections, the active components – positive electrode, negative electrode, and aqueous sulfuric acid – as well as the main and side reactions are described. Relevant details are added to the information already presented in section 2. The electrode potential, or the equilibrium voltage with respect to the electrolyte, is measured with a hydrogen reference electrode (see section 2.2.2), whose measured values agree very well with the values calculated from thermodynamic data, is given.

1 Lead and lead compounds are usually insoluble or only very slightly soluble in water. In contrast to cadmium compounds, which are easily soluble in water and can lead to serious health problems when they enter the environment, lead therefore hardly ever finds its way into the food chain. Lead may therefore still be used for hunting and fishing.

12.2.1 Overview of Active Components

12.2.1.1 Electrolyte

The electrolyte in lead–acid batteries consists of aqueous sulfuric acid and can be either in liquid form (flooded batteries) or immobilized in a gel or fleece (AGM batteries, absorbent glass mat) (sealed batteries, VRLA batteries, and valve regulated lead–acid batteries). The density of the aqueous sulfuric acid used when fully charged (nominal density) is between 1.24 g/cm^3 (sealed stationary batteries) and up to approx. 1.34 g/cm^3 for batteries in which the sulfuric acid is immobilized in a gel or fleece.

In contrast to many other battery systems, the electrolyte is part of the reaction and is consumed during discharging. The acid density decreases measurably and, depending on the design, is approx. 20–50% of the acid density of the fully charged battery in a discharged battery. This changes the specific resistance and the specific volume to a technically relevant extent.[2] Some special properties of lead–acid batteries are due to the large change in electrolyte density during discharging and charging.

Mixing highly concentrated sulfuric acid with water leads to a violent exothermic reaction (enthalpy of mixing). In the relevant concentration range, approx. 80% of aqueous sulfuric acid dissociates to form HSO_4^- and H^+. The remaining proportion of SO_4^{2-} and H_2SO_4 varies and is of minor importance in terms of electrochemical effects. The use of the notations HSO_4^- and H^+ has the advantage that for reactions, the calculation of voltage values (Nernst voltage) and thermal properties using thermodynamic data is very precise and is therefore always used in the following. The explicit inclusion of the solvate shells surrounding both ions, which is reflected, for example, in the notation H_3O^+, is not used.

Further information on the electrolyte and its properties can be found in section 12.4.

12.2.1.2 Active Masses of the Positive and Negative Electrode

The material data given in this section are all taken from Sauer [1]. The active material of the positive electrode, lead dioxide (positive active material [PAM]), is predominantly present in a crystallographic form β - PbO_2 and consists of crystals with an average size (diameter) in the range of approx. 0.6 μm in the fully charged and new state. The specific electron conductivity of the charged active mass is approx. 10% of that of metallic lead and is ensured macroscopically by the contact between individual particles. Additives to improve conductivity are employed in many cases, but they are not necessary in principle.

The active material of the negative electrode (negative active material [NAM]) consists of highly porous lead (lead sponge) with a crystallite size, also in the new and fully charged state, of approx. 3 μm. Similarly, the electron conductivity is

---

2 In most other battery systems where the electrolyte is part of the reaction (such as some types of alkaline nickel batteries and metal–air batteries), water is consumed during discharging. However, water is present in high excess, so that the density and other properties of the electrolyte change only slightly.

produced by contact between individual particles without the need for additives. The macroscopic conductivity is approx. 50% of metallic lead. However, for certain applications, conductive carbon black or another special carbon additive is added to improve the conductivity properties, especially in the partially or fully discharged state.

During discharging, lead sulfate is formed in both active materials, and water is formed in the positive electrode. Discharging and charging is a reconstitution mechanism in which the structure of the discharged and charged active material is renewed each time. The discharged material lead sulfate is electrically non-conductive and, depending on the operating mode, can form large crystals with a significantly larger volume than the charged active material.

12.2.2 Overview of the Most Important Reactions at the Positive and Negative Electrode

Some of the reaction equations given in the following proceed via intermediate products, as seen in Refs. [2, 3]. However, for thermodynamic considerations, only the initial and final products of the reaction are important. Therefore, the details of the reactions are not further considered here. The formation of intermediate products does not change the Nernst voltage, but it does reduce the reaction rate and affects the charge transfer voltage via changes in the exchange current density.

Lead–acid batteries have powerful side reactions causing little degradation at both electrodes in the charging and discharging direction. These side reactions are of great technical importance and enable safe operation when cells are connected in series (see chapter 3). In contrast to lithium-ion batteries, it is not necessary to install individual cell voltage monitoring for lead–acid batteries, although this would be useful and extend the service life in certain cases. The side reactions in lead–acid batteries represent inherently existing cell balancing systems, which must be provided in lithium-ion batteries at an additional cost and complexity.

The complete equivalent circuit diagram for lead–acid batteries, which can be used to explain behavior during discharging and charging, is shown in Figure 12.1, but without the "capacitor properties" discussed in chapter 8, because their effects are rarely important. The oxygen consumption reaction that takes place in sealed batteries will be added later.

Table 12.1 contains all relevant reactions including the oxygen consumption reaction not shown in Figure 12.1, and the equilibrium voltages of the reactions at an electrolyte density of $1.28\,g/cm^3$.

The voltage and current of a cell can be used to calculate the respective BV resistances of the individual reactions and thus also which current paths are used to what extent in a particular operating state. It is not possible to directly measure the current for individual reactions in the battery, but it is possible to make calculations by estimating the parameters of the BV equations and Nernst equation of the individual reactions. This can be used in some algorithms for calculating the state of charge.

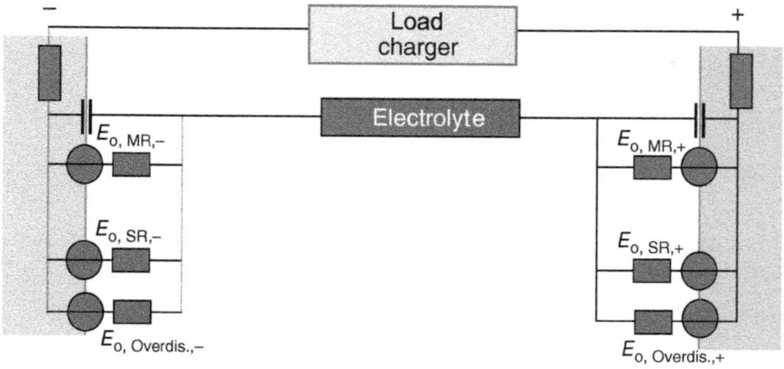

Figure 12.1 Complete equivalent circuit diagram of lead–acid batteries with the main and side reactions (MR and SR) relevant for discharging, charging, and overdischarging reactions $Ü_{entl}$.

Table 12.1 Overview of the relevant electrochemical reactions during discharging and charging, as well as internal oxygen consumption. The equilibrium voltages listed correspond to a nominal electrolyte density of 1.28 g/cm³.

	Positive electrode	**Negative electrode**
Main reaction	$PbO_2 + HSO_4^- + 3H^+ + 2e^- \leftrightarrow PbSO_4 + 2H_2O$ $E_o = 1.7\,V$ $Q_{rev}: -12\,kJ/mol$	$Pb + HSO_4^- \leftrightarrow PbSO_4 + H^+ + 2e^-$ $E_o = -0.43\,V$ $Q_{rev}: 25.2\,kJ/mol$
Overcharging reaction	$H_2O \rightarrow \frac{1}{2}O_2 + 2H^+ + 2e^-$ $E_o = 1.23\,V$ $Q_{rev}: 25.2\,kJ/mol$	$2H^+ + 2e^- \rightarrow H_2$ $E_o = 0\,V$ $Q_{rev}: -12\,kJ/mol$
Overdischarge reaction	$PbSO_4 + H^+ + 2e^- \leftrightarrow Pb + HSO_4^-$ $E_o = -0.43\,V$ $Q_{rev}: 25.2\,kJ/mol$	$PbSO_4 + 2H_2O \leftrightarrow PbO_2 + HSO_4^- + 3H^+ + 2e^-$ $E_o = 1.7\,V$ $Q_{rev}: -12\,kJ/mol$
Oxygen consumption reaction[a)]	$\frac{1}{2}O_2 + 2H^+ + 2e^- \rightarrow H_2O$ $E_o = 1.23\,V$ $Q_{rev}: -12\,kJ/mol$	$\frac{1}{2}O_2 + 2H^+ + 2e^- \rightarrow H_2O$ $E_o = 1.23\,V$ $Q_{rev}: -12\,kJ/mol$

a) The oxygen consumption reaction is not shown in Figure 12.1 because it is only important when discussing sealed batteries.

The following explanations can also be read without reference to chapters 2 and 3. They contain all the electrochemical information relevant to lead–acid batteries, but without the detailed description of the electrochemical principles. The thermodynamic data for these reactions are precisely known. All reactions are described taking into account the dissociated form of sulfuric acid.

12.2.3 Description of the Main Reactions

Main reaction at the positive electrode (cathode):

$$PbO_2 + HSO_4^- + 3H^+ + 2e^- \leftrightarrow PbSO_4 + 2H_2O. \quad (12.1)$$

Minimal differences in the crystal structure (α-PbO_2 β-PbO_2, etc.), stoichiometry ($PbO_{2-\delta}$ with δ in the range of 0.01 [3]), and particle shape, see [4], can be neglected with regard to the calculation of the equilibrium voltage. From the value of $T\Delta S$ in Table 12.1, it can be seen that the equilibrium voltage decreases with increasing temperature. The exchange current density is in the range of 0.5 – 0.01 A/m² [1].

Main reaction at the negative electrode (anode):

$$Pb + HSO_4^- \leftrightarrow PbSO_4 + H^+ + 2e^-. \quad (12.2)$$

The value of $T\Delta S$ in Table 12.1 shows that the equilibrium voltage increases with increasing temperature. The exchange current density is in the range of 0.5 – 0.01 A/m² [1].

The combined reaction of both main reactions is:

$$Pb + PbO_2 + 2HSO_4^- + 2H^+ \leftrightarrow 2PbSO_4 + 2H_2O. \quad (12.3)$$

The combined equilibrium voltage of both electrodes at an acid density of 1.28 g/cm³ is 2.13 V and the reversible heat is 13.2 kJ/mol. The reversible heat leads to a very small cooling of the cell during discharging and contributes to heating during charging, whereby the total heating is caused by the cooling of the positive and heating of the negative electrode if other effects are neglected. The release heat of the total reaction can only be measured experimentally with great difficulty and can generally be neglected, as can the enthalpy of mixing respectively changes in the degree of dissociation of sulfuric acid due to changes of the concentration during charging and discharging.

The changes in concentration of aqueous sulfuric acid during discharging and charging due to the consumption and formation of sulfuric acid, has the greatest influence on the equilibrium voltage, see Nernst's law in chapter 2. Table 12.2 gives the values according to Berndt [5].

Table 12.2 Equilibrium voltage as a function of concentration at 20 °C with a hydrogen reference electrode.

Density (g/cm³)	Concentration (mol/l)	Cell voltage (V)	Pos. electrode (V)	Neg. electrode (V)
1.05	0.825	1913	1619	−0.293
1.10	1652	1964	1665	−0.299
1.15	2507	2008	1701	−0.307
1.20	3391	2052	1734	−0.318
1.25	4310	2097	1768	−0.329
1.30	5259	2145	1805	−0.340
1.35	6229	2197	1845	−0.355

Source: Adapted from Ref. [5]/Research Studies Press.

When calculating the Nernst voltage, it can be assumed that the activities of the reactants involved do not change, with the exception of the hydrogen sulfate ion (HSO_4^-). The Nernst voltage (see Eq. 2.6 in section 2.2.2) can thus be simplified to

$$E_0 = E_{00} + (RT/zF) \cdot \ln a\left(HSO_4^-\right), \tag{12.4}$$

with E_{00} the equilibrium voltage at standard conditions (one molar concentration of aqueous sulfuric acid) and $a(HSO_4^-)$ the activity (see section 2.2) of the HSO_4^- ions at the existing concentration.

Linearization, equating the activity and concentration of the aqueous sulfuric acid, and adding the voltages of both electrodes, results in the frequently used rule of thumb for the dependence of the equilibrium voltage and acid concentration:

$$E_0 = 0.84 + \text{acid density (in g/cm}^3\text{)}. \tag{12.5}$$

This approximation formula is quite accurate except at very low states of charge and is often used. Additives to the electrolyte, such as H_3PO_4 and others, cause small changes [4].

12.2.4 Overdischarge Reactions During Discharging

Both the negative and positive electrodes offer an alternative reaction path if the main reactions become impossible during discharging due to a lack of available charged active material. The alternative, that is, side reactions taking place at the two electrodes are based on the availability of lead sulfate as a byproduct of the discharge reactions and aqueous sulfuric acid in the electrolyte.

Positive electrode:

$$PbSO_4 + H^+ + 2e^- \rightarrow Pb + HSO_4^-. \tag{12.6}$$

Negative electrode:

$$PbSO_4 + 2H_2O \rightarrow PbO_2 + HSO_4^- + 3H^+ + 2e^-. \tag{12.7}$$

These two reactions correspond to the main reactions. The overdischarge reaction of the positive electrode is identical to the discharge reaction of the negative electrode, but takes place in the charging direction. The positive electrode behaves like a negative electrode being charged during overdischarge. As a consequence, the voltage level of the cell is approx. 0 V. The same applies to the negative electrode, which behaves like a positive electrode during charging when overdischarged. If both the negative and positive electrodes of the same cell are overdischarged, the voltage is no longer approx. 2 V, but approx. −2 V (also known as polarity reversal or voltage reversal).

In chapter 3, Figure 3.6, the voltage curve of a 48 V battery during discharging was shown to illustrate the problems that occur toward the end of a discharge when cells are connected in series. The detection of an overdischarge process in a battery string is easily and clearly possible by means of voltage measurements. In many battery systems, measuring the voltage after $n/2$ cells and comparing the two partial voltages is used to detect an overdischarging process of an electrode (see Figure 12.2).

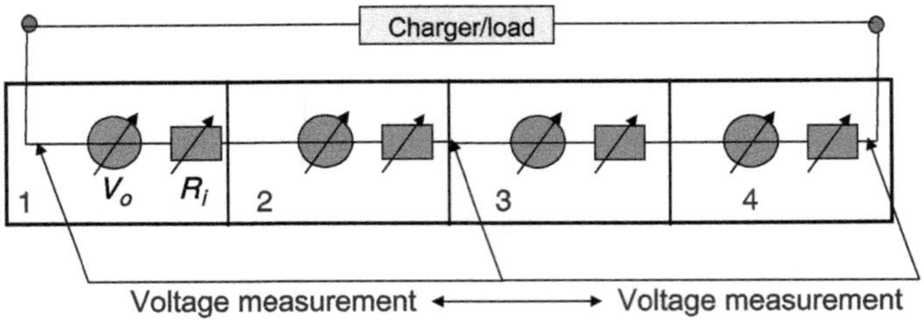

Figure 12.2 Equivalent circuit diagram for detecting overdischarge when cells are connected in series, with four cells in which the voltage and internal resistance are shown as variables. It is not possible to know from the voltage comparison of the two partial voltages whether it is the positive or negative electrode and which cell of the string is affected by the voltage drop.

The overdischarge reactions are electrochemically characterized precisely by the basic thermodynamic data and the exchange current density. Since the electrochemically active surface and the additives of the positive and negative electrodes are different, it is not expected that the total voltage of the cell will be exactly zero when one electrode of a cell is overdischarged.

Overdischarge is a low-resistance fault[3] that has hardly any effect on the functionality of the battery, especially in battery strings with a high nominal voltage. The voltage curve in Figure 3.6 shows this very clearly. Even after a cell has been overdischarged, a multicell battery still provides sufficient power.

When recharging a cell with overdischarged electrodes, the electrochemical processes take place in the opposite direction. The charging process first converts the "wrong" active material back into lead sulfate in the overdischarged electrode until the charging process of the original main reaction, the conversion of lead sulfate into active material, can begin. This double conversion process stresses the active materials and damages them further because the additives of electrodes and the design are only geared toward one discharge direction.[4] Cells with this behavior must be replaced quickly, the higher the demand on the battery, the faster.

12.2.5 Side Reactions of the Positive and Negative Electrodes During Overcharging

At the positive electrode, except for voltages below approx. 1.23 V/cell, the reaction[5]

$$H_2O \rightarrow \frac{1}{2}O_2 + 2H^+ + 2e^- \qquad (12.8)$$

3 A fault that does not cause a relevant voltage drop across the location of the fault when current flows and, unlike a high-impedance fault, does not interrupt the current flow.

4 Experience shows that the more frequently an electrode is overdischarged, the earlier the overdischarge of the cell begins.

5 The notation with the formation of half an oxygen molecule is chosen because the number of charge carriers corresponds to the number used in the notation of the main reaction.

takes place. The exchange current density is approx. 5×10^{-11} A/m^2 [1] and the equilibrium voltage is 1.23 V. The reversible heat of the reaction is −237 kJ/mol [5], leading to a non-negligible reduction in heat generation during charging due to water decomposition.

A further side reaction constantly takes place at the negative electrode:

$$2H^+ + 2e^- \rightarrow H_2. \tag{12.9}$$

The exchange current density is approx. 5×10^{-9} A/m^2 [1] and is strongly dependent on aging and material. The equilibrium voltage of this reaction is zero, as is the reversible heat.

12.2.5.1 Combined Reaction when Overcharging

The combination of both side reactions during overcharging is:

$$H_2O \rightarrow H_2 + \frac{1}{2} O_2. \tag{12.10}$$

It should be noted that both side reactions take place at different speeds and compete with the main reactions. The combined reaction formula can only be used for a cell in which both electrodes are fully charged (with no lead sulfate present). This combined reaction also shows that cells only lose water during charging, not sulfuric acid. Therefore, aqueous sulfuric acid must never be used to restore the required electrolyte level and purified ("distilled") water must be added instead so that no impurities are introduced which may act as catalysts to increase the reaction rate of the side reactions.

The reaction rate of both side reactions is described by the BV equation, where the difference between the equilibrium voltage and the electrode voltage is in the exponent. In trickle charge mode of a fully charged battery with a nominal electrolyte density of 1.28 g/cm^3, the cell voltage is then 2.23 V, and the difference between the equilibrium voltage and the electrode voltage for the side reaction at the positive electrode is approx. 0.53 V (1.76–1.23 V) and for the side reaction of the negative electrode approx. (−0.47 to 0 V). Despite the very high values in the exponent, the current associated with the reaction is very small due to the very low exchange current densities. The exchange current density is a function of lead alloys used in the current collectors and catalytic impurities, and it is about 7 to 9 orders of magnitude smaller than the exchange current density of the main reaction. The exchange current densities of the two side reactions differ by a factor of c. 100 when comparing them on the basis of the same effective surface area.

For new flooded batteries in stationary applications, IEC TR 62060:2001 gives values in the range of 10–30 mA per 100 Ah capacity in trickle charge mode. For batteries at the end of their service life, these values are approximately five to six times higher. Although the same current always flows through both electrodes, the distribution of the total voltage to the electrode voltages cannot be readily inferred due to different effective total surfaces (aging processes, effects of design) and age-related changes in catalyst properties.

During charging, the current required for the reaction is provided by the charger, which supplies both the current for the main reaction path and for the side reaction path. At high charging voltages from approx. 2.4 V/cell, the gassing voltage,[6] the reaction rate of the secondary reactions reaches such a high value that gas bubbles occasionally rise in cells with liquid electrolyte. During the recharging phase of batteries for industrial trucks, charging voltages in the range of approx. 2.7 V/cell with correspondingly high currents are reached. Many gas bubbles then rise, which can be seen very clearly. This will be discussed further in the discussion on aging and the removal of acid stratification (see section 12.9).

According to Faraday's law, which states the proportionality of the amount of current and the amount of substance converted, the water loss is 0.336 g of water per ampere hour of charge, but only if the charging current flows exclusively via the side reaction paths of both electrodes when neither of the two electrodes contain any more lead sulfate. This is the case when the battery is fully charged. A ratio of hydrogen to oxygen of 2 mol to 1 mol could therefore be regarded as proof that both electrodes are fully charged, but this method is not even used in the laboratory due to the complexity of measurement.

Under normal conditions (1 bar, 298 K), the amount of gas is 0.21 l of oxygen and 0.42 l of hydrogen per ampere hour. Data sheets provide the following values for the explosive effect of hydrogen:

Auto-ignition temperature	**853 K/580 °C**
Minimum ignition energy in air	0.019 mJ
Ignition limits in air	4–75% by volume
Flame temperature	2318 K/2045 °C
Detonation velocity in air	2.0 km/s
Detonation pressure	14.70 bar
Explosive energy	2.02 kg TNT/m^3

The effect of an explosion of hydrogen corresponding to 1 Ah is equivalent to an explosion of approximately 0.85 g TNT. The resulting safety problems have led to numerous requirements in standards that affect all areas of application (see the application-related chapters in the last part of the book).

The loss of water caused by the side reaction leads to maintenance requirements, regular topping up with purified water. Refilling cells is not necessary only if

- there is a large amount of electrolyte in the cell and the side reactions are suppressed by inhibitors to such an extent that the amount of water decomposed is small over the whole service life (e.g. in modern starter batteries),

6 The term is often incorrectly interpreted to mean that no gassing occurs below this voltage. However, the decomposition of water and the formation of hydrogen and oxygen gas take place during every charging process, whether at standstill or at any cell voltage above 1.23 V.

- an external recombination device ("recombination plug") is provided, which recombines the gas newly produced inside a cell in the a space above the cell back to water[7] and returns it to the cell, or
- there is an internal oxygen consumption cycle between the electrodes, which makes oxygen from the positive electrode available at the negative electrode and effectively reduces, but cannot completely prevent the production of hydrogen according to Eq. (12.9). The details of this process and the resulting consequences are described in section 12.3.

12.2.6 Side Reactions and Self-Discharge at Rest

Even without current respectively charging current, overcharging reactions take place because they have a cathodic or anodic counterreaction with the main reactions at both electrodes. The discharging reaction of the main reaction (Eqs. (12.1) and (12.2)) provides the electrons and energy required by the side reactions. The combined reaction for the positive electrode from Eqs. (12.1) and (12.8) is:

$$PbO_2 + HSO_4^- + H^+ \rightarrow PbSO_4 + \frac{1}{2} O_2 + H_2O \tag{12.11}$$

and for the negative electrode from Eqs. (12.2) and (12.9):

$$Pb + HSO_4^- + H^+ \rightarrow PbSO_4 + H_2. \tag{12.12}$$

The negative and positive electrodes are discharged with the formation of lead sulfate, hydrogen and oxygen. Both reactions are reversible self-discharge processes. Even very low continuous charging currents (a few mA per 100 Ah capacity for new batteries) can be sufficient to completely reverse the reactions on both electrodes. Self-discharge is similar on both electrodes and amounts to approx. 1–3% per month when new, which is only a problem during long storage periods without trickle charge. Battery manufacturers usually require the batteries to be recharged after approximately four to six months to prevent irreversible damage caused by the accumulated lead sulfate and other processes (see section 12.14 – Corrosion, microstructure).

It should be noted that the self-discharge of an electrode can differ greatly from the self-discharge of the other electrode in certain aging processes that only affect one electrode.

The self-discharge processes also take place during discharging up to the usual final discharge voltages. However, because the differential voltage between the electrode voltage and the equilibrium voltage always decreases during discharging, the self-discharge of the electrodes during discharging can always be neglected.

The information in Eqs. (12.12) and (12.13) shows that gas always develops in lead–acid batteries and that cells, in contrast to many other battery systems, must therefore never be completely gas-tight[8] and must always have an opening or valve.

---

7 Hydrogen (H_2) and oxygen (O_2) are converted to water on catalysts in contrast to the oxygen consumption reaction, in which oxygen (O_2) forms water with protons. In both cases, a great deal of heat is generated.

8 Flooded batteries with liquid electrolyte are sometimes supplied with sealing plugs for transportation purposes. Removing the plugs can lead to a significant release of gas mixed with an aerosol mist consisting of aqueous sulfuric acid.

12.2.7 Charging and Discharging Cells in Series

Chapter 3 described the processes and problems associated with a series connection of cells. Because lead–acid batteries have side reactions described earlier during charging and discharging, which can lead to large current amplitudes and cause relatively little degradation. As a result, lead–acid batteries are inherently protected against safety problems caused by connecting them in series.

All electrodes of all cells in series can reach a state of full charge, and the capacity of all cells can be used during discharge. Figure 12.3 shows the development of the voltages of series-connected 12 V blocks of a lead–acid battery during charging [6]. After largely similar voltages during charging until the end-of-charge voltage of the overall battery is reached, the individual voltages vary greatly. Blocks with a very high voltage lead to a rapid reduction in the charging current, so that the voltage of blocks with a low voltage can drop again as a result. After some time, the voltages become more and more similar, because once all the blocks are almost fully charged, the entire battery current only flows via the side reactions. This process applies to all battery technologies with aqueous electrolytes.

The highest block voltage is 16 V. Without further measurements, it is impossible to determine whether the voltage is evenly distributed across all six cells (approx. 2.67 V/cell) or whether only one electrode is charged, but then with approx. 1.6 V more than usual.

Although a battery management system with individual cell voltage measurement is not necessary [7, 8] for lead–acid batteries, it was shown that controlling the charging current depending on individual cell voltages and active charge equalization between cells can extend the service life. However, such systems have not established themselves due to their high costs.

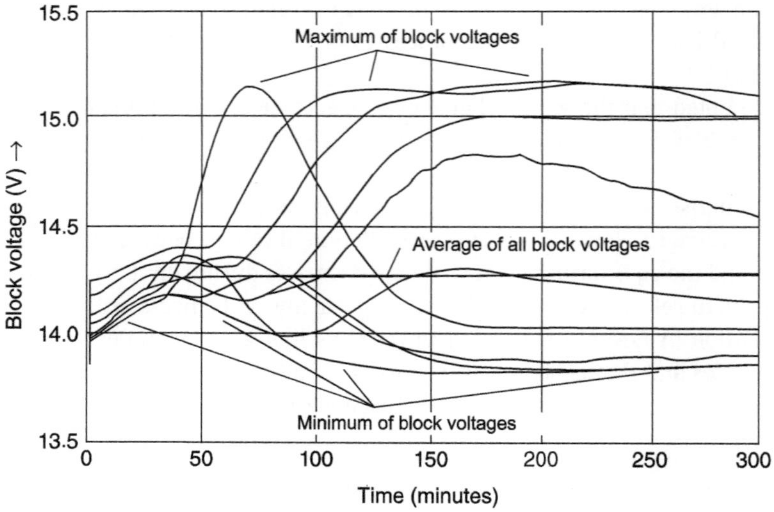

Figure 12.3 Development of individual block voltages of 12 V lead–acid batteries connected in series at the end of charging [6]. Source: VDE-Verlag GmbH.

12.3 Other Electrochemical Reactions

The reactions discussed earlier fully describe the behavior of lead–acid batteries in operation. The reactions that are important in the aging of lead–acid batteries (grid and terminal bridge corrosion) are discussed in section 12.14. To prevent water loss and the maintenance requirements to refill water, a further reaction path which was already included in Table 12.1 is available. This is of great practical importance.

12.3.1 Batteries with Internal Oxygen Consumption Cycle (Sealed Batteries, VRLA)

The loss of water from a cell can be almost completely prevented electrochemically by supplying oxygen, which is formed at the positive electrode during charging, to the negative electrode (oxygen cycle). Refilling with water during the entire service life of the batteries is no longer necessary and technically no longer possible. Batteries with these properties are referred to as sealed batteries[9] or VRLA batteries. VRLA stands for valve regulated lead acid.[10] Flooded batteries with liquid electrolyte are sometimes referred to as VLA batteries (vented lead–acid batteries). The decisive industrial advantage of VRLA batteries is that they are "maintenance-free"[11] (no water needs to be added).

In sealed batteries, the electrolyte is either immobilized in gel (gel batteries) or fleece (AGM batteries – absorbent glass mat [AGM]). Both gel (due to cracks in the gel) and fleece (no complete filling of all pores, filling level between 90 and just under 100%) enable rapid oxygen transport in a cell from the positive electrode to the negative electrode. The volume requirement of gel and fleece is compensated for by a higher electrolyte density so that a sufficient number of sulfate ions are available without increasing the electrolyte volume. This leads to a higher equilibrium voltage in the cells. In principle, the immobilization of the acid in the gel or fleece enables installation in any position and prevents sulfuric acid from leaking if the housing is damaged.

12.3.2 Electrochemistry

Figure 12.4 shows the electrical equivalent circuit diagram, which contains all the reactions to be considered for the discussion here. The other reactions shown in the equivalent circuit diagram in Figure 12.1 relate to the overdischarge reactions, which do not play a role in the consideration here, assuming a chemically fully charged battery.

---

9 The term "sealed battery" is sometimes incorrectly used by some users for batteries that do not offer the option of topping up with water. However, sealed (VRLA) batteries are actually batteries that have internal oxygen consumption reactions.

10 The term "recombination battery" is not used in this book because it is technically misleading.

11 The term "maintenance-free" is misleading because battery systems consisting of sealed cells still need to be maintained and checked. The time and cost required to accurately diagnose the functionality of sealed batteries is similar to that of flooded batteries.

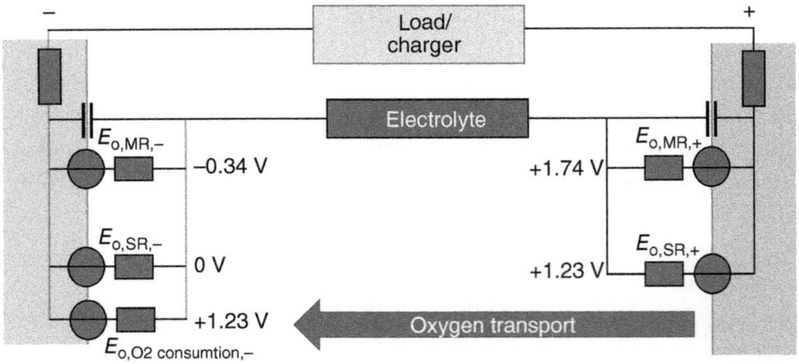

Figure 12.4 Equivalent circuit diagram of a cell to explain the oxygen consumption cycle.

In the fully charged state, no current flows through the main reactions in trickle charge mode because there is no more lead sulfate, that is, the two resistances of the main reactions $R_{BV, HR}$ are "infinitely" large. The entire charging current flows through the side reactions. At the negative electrode, these are the formation of H_2 and water with oxygen produced at the positive electrode consuming protons from the electrolyte. At the positive electrode, the side reaction to be considered for the functioning of sealed batteries is the formation of oxygen. The reactions required to understand sealed batteries are already listed in Table 12.2.

The reaction taking place at the negative electrode[12] is:

$$\frac{1}{2}O_2 + 2H^+ + 2e^- \rightarrow H_2O. \tag{12.13}$$

It corresponds to the oxygen-producing reaction (Eq. (12.8)), also with regard to equilibrium voltage and heat effect, but takes place in the opposite direction with consumption of oxygen at the negative electrode and is strongly exothermic.

The transport of oxygen from the positive electrode to the negative electrode, and thus the oxygen-consuming reaction, also takes place in lead–acid batteries with liquid electrolyte, because oxygen is soluble to a small extent in the aqueous sulfuric acid [9, 10]. The effect can be ignored concerning all technical properties of flooded batteries.

The formation of hydrogen (Eq. (12.8)) and the consumption of oxygen to form water (Eq. (12.13)) are competing reactions for charging currents, leading to a change in voltage at the negative electrode (see section 2.5). The equilibrium voltage of the negative electrode is determined by the main reaction and the acid concentration and is approx. 1.34 g/cm^3 at a density of –0.35 V and under trickle charge conditions, approx. –0.4 V. The difference between the electrode voltage and the equilibrium voltage for the oxygen consumption reaction is about 1.6 V (1.23 V + c. 0.4 V), but only about 0.4 V for the hydrogen formation reaction, because its equilibrium voltage is 0 V. It is therefore expected that the trickle

12 The reaction is often, but misleadingly, referred to as a recombination reaction. However, there is no recombination of hydrogen (H_2) and oxygen (O_2).

charge current I_{Batt} is divided very unequally between the two reactions, even if the exchange current densities and transfer coefficients are different.

$$I_{Batt} = I_{SR,H_2} + I_{O_2-Consumption}, \tag{12.14}$$

$$\text{with } I_{O_2-Consumption} \gg I_{SR,H_2}.$$

By selecting alloys to inhibit the hydrogen formation reaction and dimensioning the positive electrode so that it already produces oxygen at a high state of charge before hydrogen production begins, a ratio of $I_{O_2-Verzehr}$ to I_{NR,H_2} of up to 99 : 0.1 is achieved in batteries. However, the production of hydrogen cannot be completely suppressed. Safety precautions against hydrogen explosions must still be taken.

The reaction of hydrogen (H_2), which has already been formed, with oxygen (O_2) to form water is not possible at the negative electrodes of the battery.

It is of great practical importance that the voltage of the negative electrode changes at a constant float charge current compared to a flooded battery. Figure 12.5 shows this. On the left, the conditions without transport of oxygen through the electrolyte to the negative electrode are shown, and on the right with the transport of oxygen and the oxygen consumption reaction (Eq. (12.13)). The shift in the electrode voltage of the negative electrode is shown.

At the same trickle charge current, the voltage of the negative electrode of a sealed battery decreases compared to the negative electrode voltage of a flooded battery, but the voltage of the positive electrode remains unchanged. If the cell voltage remains the same, the float charge current increases and both electrode voltages change, although the electrochemistry of the positive electrode remains unchanged. However, the voltages of the two electrodes do not necessarily change by the same amount.

The voltage changes lead to differences in operation, corrosion behavior, and aging, which are explained in further sections.

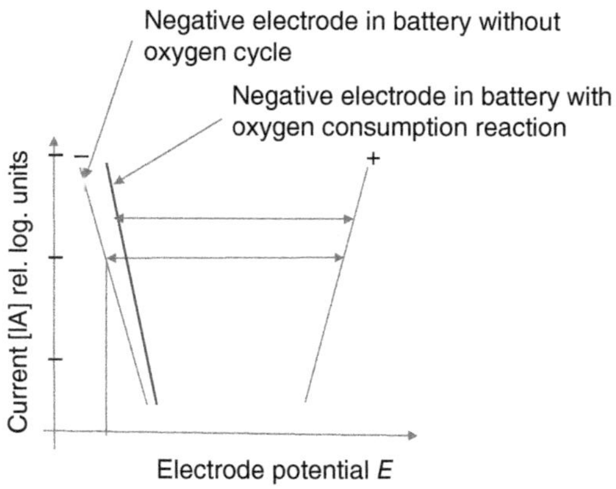

Figure 12.5 Voltages of the negative electrodes in flooded and sealed batteries. Shifting the negative potential leads to an increase in the total voltage and an increase in the electrode voltage of the positive electrode if the same current flow.

Due to the great importance of gel and AGM batteries, the most important common properties are summarized in the following:

1) The oxygen cycle only takes place properly in a gas-tight sealed cell. The vessel must therefore be sealed.
2) The formation of hydrogen gas is largely suppressed in favor of the formation of water via the oxygen cycle. As a result, the cells no longer need to and must not be refilled with water. There is no technical possibility of refilling.
3) The formation of hydrogen gas cannot be completely suppressed because the electrode voltage has a value that allows hydrogen formation.[13] The cells must therefore have a pressure relief valve that prevents the battery housing from being destroyed. This valve must close automatically. The overpressure at which the valve opens is set to a value in the range of approx. 50–150 mbar, depending on the type. Leaking cells reduce the rate of water formation, the electrolyte dries out and the cell becomes highly resistive. The quality of the valves is relevant for the service life. Some manufacturers even carry out tests for each individual valve despite the very high quantities. Figure 12.6 shows the pressure curve of two 12 V blocks connected in series. The internal pressure of one of the blocks constantly remains below the opening pressure of the valve, while the opening pressure of the other block is repeatedly exceeded, which leads to gas loss and the reduction in the formation of water at the negative electrode causes the electrolyte to dry out (increase in internal resistance).
4) Each cell of a block battery must have its own pressure relief valve and there must not be a common gas volume for several cells. Otherwise, the gas produced in one cell could be converted back into water in another cell with better oxygen consumption efficiency. One cell then dries out after a long time, while another is flooded.

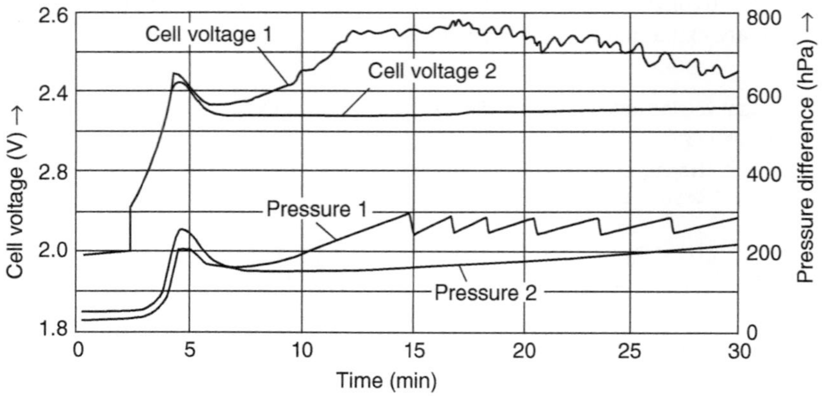

Figure 12.6 Pressure curve of two 12 V blocks in trickle charge mode with series connection. It can be clearly seen that the valve closes tightly again after opening [6]. Source: VDE-Verlag GmbH.

13 With other battery types such as alkaline nickel batteries, on the other hand, hydrogen formation can be completely suppressed. In this case, only a non-sealable safety valve needs to be provided for abuse and charger faults.

5) Special care is required when selecting additives for active materials and alloys. It is important for the oxygen cycle that oxygen production is low. If the oxygen production exceeds the transport capacity of the gel or AGM or the ability of the negative electrode to consume oxygen, the pressure increases and the oxygen escapes via the valve. The development of hydrogen must also be minimized by careful material selection. These conditions must also be observed as the batteries age. Alloys with antimony, which dissolves from the positive grid, deposits on the negative electrode, and acts as a catalyst for hydrogen evolution at the negative electrode, are therefore unsuitable and can no longer be used in many applications.
6) Design of the electrode sizes: During cyclic operation, it must be ensured that there is always sufficient oxygen at the negative electrode so that oxygen consumption can take place. For this reason, the positive electrodes are designed to be smaller than the negative electrodes. However, this makes it more difficult to fully charge the negative electrode (removal of all the lead sulfate previously formed by the discharge).
7) Acid stratification: The electrolyte immobilized in gel or AGM by capillary force has a lower vertical mobility than the liquid electrolyte. In contrast to flooded batteries, density differences that occur during discharging (see section 12.9 and Appendix) can no longer be reverted. Over long periods of time, acid stratification therefore also forms in sealed batteries [11]. The tendency for acid stratification is higher in AGM batteries than in gel batteries. For this reason, in some cases, the electrolyte that is used in AGM batteries contains gel-forming additives.
8) Electrolyte quantity or density: For the active masses to be converted, a certain minimum quantity of sulfate ions must be available. Increasing the distance between the electrodes is not appropriate due to the resulting increase in resistance. Instead, the electrolyte concentration is increased. The usual electrolyte density of sealed batteries is 1.30–1.34 g/cm^3, but products with up to 1.36 g/cm^3 are now also available on the market. The consequences are higher open-circuit voltage and different corrosion conditions (see section 12.15).
9) Internal resistance: Gel and AGM are not ion conductive so the ion conductivity of the electrolyte space decreases with otherwise identical dimensions. This can be counteracted by reducing the distance between the electrodes as AGM batteries do not require a separator in addition to the AGM material. In addition, due to the possibility of making better use of the battery housing, the internal resistance of both AGM and gel batteries is often somewhat lower than that of flooded batteries with the same capacity or size as the battery box. In flooded batteries, there must be a reservoir of free electrolyte space in the upper area to ensure that the plates are always covered with electrolytes even for aged batteries.
10) Wetting of the electrodes: In flooded batteries, there are no physical boundaries between the electrolyte and the electrodes; wetting is always guaranteed. In sealed batteries, however, the gel or AGM containing the electrolyte can become detached from the electrode. In the area of detachment, the electrical

connection of the active mass material to the electrolyte is very poor and the effectively usable amount of active mass decreases. A mechanically tight assembly of the electrodes in the battery box is therefore necessary, but without closing the open pores of the AGM materials. They must therefore have a permanent elastic compressibility.

11) Heat generation: The formation of water at the negative electrode is a highly exothermic process, which is largely thermally compensated by the formation of oxygen at the positive electrode. In contrast to heat loss through the generation of hydrogen and oxygen, there is no net heat loss in sealed batteries. Overall, the cells are significantly more thermally sensitive than flooded cells. In addition, the volume requirement of AGM or gel reduces the volume respectively amount of aqueous sulfuric acid and, as a result, the heat capacity of the cell decreases.

The production of sealed batteries is more complex and the batteries are therefore somewhat more expensive.

- Gel batteries: The electrolyte is mixed with additives, mainly SiO_2 (Aerosil), immediately before filling and then gels within a few minutes of being filled into the battery. After that, there is no way to compensate for production errors. At the end of production and after a short operating time, microcracks form in the gel, allowing oxygen to be transported from the positive electrode through the separator to the negative electrode. In the case of batteries for stationary applications (pure trickle charge operation), the process is only completed after approximately six months [12]. Until then, a constant increase in oxygen transport capacity through the electrolyte is expected.
- AGM batteries: In AGM batteries, the fleece is placed between the electrodes during production instead of the usual separator and then filled with electrolyte in situ using a vacuum process. The degree to which the AGM material is filled is usually between 90 and just under 100%. Here too, it can take some time for the oxygen transport capacity to reach a stable value.

12.4 Active Materials

Active masses of batteries must have good electron conductivity for every state of charge, a large total surface area in contact with the electrolyte, and an open-pored structure to absorb the electrolyte and transport ions (see chapter 4 for the basic requirements for the structure of active masses). This structure is created during the manufacture of the electrodes. An electrically conductive connection to the current collectors and avoiding passivation layers between active masses and current collectors is achieved by an appropriate choice of alloys for the current collectors and the design of electrodes.

The electrochemical conversion processes at both electrodes of lead–acid batteries are reconstitution mechanisms that lead to completely different properties of the charged and discharged materials. Table 12.3 shows the most important data on the active masses.

Table 12.3 Overview of properties of active masses (for both NAM and PAM in a fully charged state with usual porosity in the range of 50%) and lead sulfate. All data are strongly dependent on the aging condition of the battery.

	Negative active mass (NAM)	Positive active mass (PAM)	Lead sulfate
Electron conductivity (S/cm)	5.3×10^3	1.35×10^2	1×10^8
Effective surface (m^2/g)	2–5	1–2	Time- and age-dependent
Crystal size (diameter) (μm)	Approx. 0.6	Approx. 2.6	For heavily sulfated batteries up to approx. 100 μm
Molar volume (cm^3/mol)	18.25	25.72	48.22
Density (cm^3/g)	11.34	9.31	6.29

Source: Adapted from Ref. [1]/Sauer D U.

12.4.1 Electrical Conductivity of the Active Masses

The agglomerate of spheres model, which is also used in a similar form in powder metallurgy to explain the strength of pressed components, has proved useful for explaining the electron and ion conductivity of the electrodes [13–15]. The active material is considered to be a collection of microscopic spheres with a contact zone where they touch. This forms an open-pored structure and causes macroscopic ionic conductivity when the pores are filled with aqueous sulfuric acid. The porosity usually selected to ensure sufficiently high availability of sulfate ions without special transport restrictions is approx. 50%.

The number and cross-sections of the contact zones between individual particles determine the electron conductivity. Relevant factors are the macroscopic surface pressure, the temperature, the viscosity of the material, the surface tension against the aqueous sulfuric acid, and the electric field of the double-layer capacitor as a result of the electrochemical reactions. An analogy to the contact of two drops of a liquid with very high viscosity can be used to give a better idea. Figure 12.7 illustrates the agglomerate of spheres model.

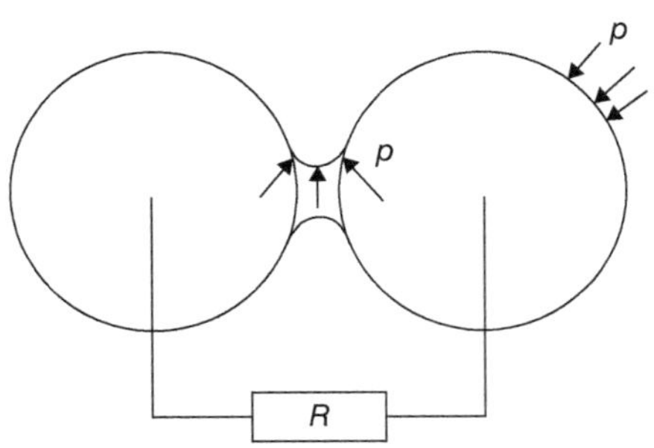

Figure 12.7 Diagram showing schematically the agglomerate of spheres model [14, 15]. Two particles, simplified as microscopic spheres, form a contact zone whose dimensions determine the electrical resistance R between the two particles. The arrows indicate the pressure on the active mass particles caused by surface tension.

The contact zone is reduced both when discharging, because the discharge occurs preferentially at the contact zone due to the local electric field, and when charging at a high charging voltage [14]. If the contact zone is destroyed and neighboring particles no longer touch each other, they are no longer electrically connected to the rest of the electrode, and the macroscopic electron conductivity is reduced. The mechanical connection of particles to the electrode is then also lost and particles can detach and fall to the bottom of the cell container. External mechanical pressure on the cells perpendicular to the electrode surface stabilizes the contact zones and increases the cyclic service life. The use of gauntlets usually woven from polypropylene for the positive electrode (so-called tubular plates, see section 12.6) and a tight assembly in a mechanically stable container are solutions that are widely used. Solutions to build up an additional external mechanical pressure on the housing that increases the mechanical strength of the battery housing [16], on the other hand, have not been established. All these measures also stabilize the active masses as they expand mechanically due to the formation of lead sulfate during discharging. For further details on the agglomerate of spheres model, please refer to the literature cited here.

The change of conductivity of such a structure consisting of conductive particles (lead or lead dioxide), in which nonconductive lead sulfate crystals accumulate during discharging, can be explained by the percolation theory[14] [17] or the effective path length model.[15] Both models predict the point at which the conductivity of the structure is no longer determined by the conductive particles but by the nonconductive particles. The electron conductivity collapses completely within a narrow ratio of charged and discharged material and corresponds to the conductivity of the nonconductive lead sulfate at a ratio of lead sulfate to the charged active material of approx. 55–45%. The electron conductivity of the electrode then decreases by approximately twelve magnitudes. In Ref. [1], a calculation of the development of the electrical conductivity was carried out on the basis of the effective path length model, which clearly shows this sudden decrease in conductivity, see Figure 12.8.

Lead and lead dioxide electrodes can only be discharged as long as the active mass structure is still conductive for electrons, that is, until approx. 55% of the active mass has been consumed [1]. For lifetime and voltage reasons, the discharge is usually limited so that only approx. 35–40% of the active mass is discharged. For this reason alone, the theoretical specific energy of the system and the practically achievable specific energy are significantly different.

To improve charging, conductive additives, mainly conductive carbon black or special carbons, are often added to the NAMs. The later addition of additives (i.e. after the battery has been put into operation) to improve the active compound structure is not recommended by most battery manufacturers.

---

14 The percolation theory describes the properties of two-phase mixtures with a statistical distribution of components.

15 With increasing nonconductive particles in a mixture of conductive and nonconductive particles, the effective length of a current path also increases. This is because there are fewer conductive particles present (effective medium theory).

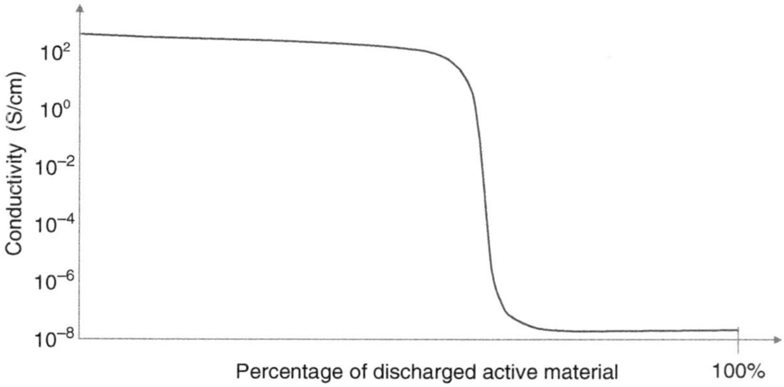

Figure 12.8 Conductivity of the active masses as a function of the percentage of discharged active material. Source: Adapted from Ref. [1]/Sauer D U.

12.4.2 Effective Surface Area and Microstructure of the Active Masses

In addition to maintaining a sufficiently high electrical conductivity of the active masses during discharging and charging, a large electrochemically active surface area of the active masses must also be maintained because the surface area has a direct effect on the charge transfer voltage via the BV equation. Due to the very different material properties of the charged and discharged active masses, maintaining a large effective surface area is not trivial.

The decrease of the effective surface area proportional to the amount of discharged active mass and therefore also the state of charge is characteristic for lead–acid batteries. However, it is important that after recharging, the active mass structure of both the positive and negative electrodes once again meets the specified requirements – a large effective surface area and high mechanical stability. Examples of measures to maintain the active mass structure are:

- Use of "expanders" (organic additives, e.g. based on lignin) which are added to the negative electrode to ensure that a fine-pored structure (lead sponge) is formed again during recharging. Without these additives, the electrode structure would become increasingly compact as the battery ages, its active lead would shrink and its effective surface area would decrease.
- For the PAM, additives [13] are used to ensure a fine crystalline structure during recharging.
- Barium sulfate serves as a crystallization nucleus at both the positive and negative electrodes to minimize the voltage dip which occurs at the beginning of a discharge (see section 12.4.4).
- Additions to the electrolyte (e.g. tin sulfate) to improve the particle structure have also been discussed.

12.4.3 Lead Sulfate

Sauer [1] has schematically described the process of formation and reconversion of lead sulfate as a coupled electrochemical and chemical process (Figure 12.9).

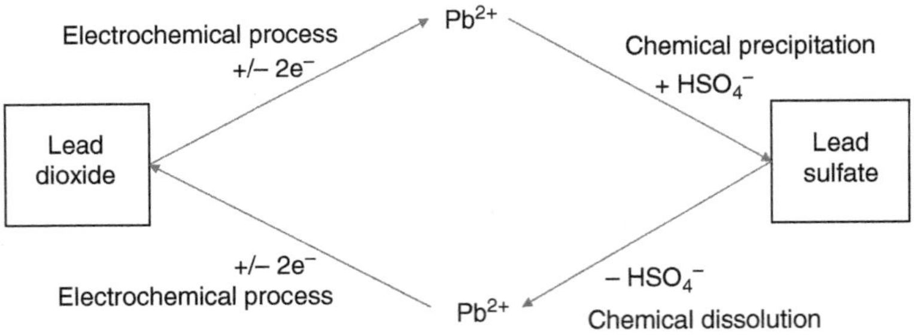

Figure 12.9 Deposition and dissolution of lead sulfate before and after the electrochemical reactions. Source: Adapted from Ref. [1]/Sauer D U.

The formation of lead sulfate crystals through a chemical deposition process and the provision of lead sulfate for redissolution are chemical processes that occur before or after the electrochemical processes. The effects of the crystallization processes are technically relevant, see the following section (voltage dip).

The reaction product of the positive and NAMs, lead sulfate, which leads to a significant increase in volume per mole compared to lead and lead dioxide (approx. factor of 2 due to the chemical bond with the large sulfate ion), is not electron-conducting and has a different crystal structure to lead or lead dioxide.

The formation of lead sulfate crystals in the highly porous lead and lead dioxide mass of the two electrodes thus leads to mechanical stress and changes in the electron conductivity of the active masses. The deeper the electrodes are discharged, the greater the mechanical stress. It is therefore expected that the service life depends on the depth and frequency of discharges. As the volume of lead or lead dioxide increases to lead sulfate, the pores in the electrodes become smaller so that less electrolyte volume is available. The effects of this process are more than compensated for by the decrease in concentration caused by discharging. The electrolyte level of sealed lead–acid batteries decreases during discharge and increases again during charging. Water may therefore only be added to a fully charged battery to prevent the cell from "overflowing" after full charging.

Lead sulfate has a low solubility in aqueous sulfuric acid, whereby the solubility increases with decreasing concentration. The minimum permissible concentration of the electrolyte at the end of the discharge (up to approx. 1.06 g/cm^3, see Ref. [18], depending on the design) is limited not only by the decreasing electrical conductivity of the aqueous sulfuric acid, but also by the increasing solubility of lead sulfate.

During the charging process, lead sulfate must be quickly and completely converted back into the active masses of the negative and positive electrodes. If this does not happen, the lead sulfate crystals change.

- Recrystallization takes place, which leads to a decrease in the number of crystals and increases the average crystal size (Ostwald ripening). This reduces the total surface area of the lead sulfate in contact with the electrolyte. The lower the acid concentration, the faster this process takes place due to the increased solubility of lead sulfate. Lead sulfate must dissolve prior to the charging process, which slows down the charging process at the end of charging if the surface area

Figure 12.10 Scanning electron microscope images of the active masses (surface) in different states. Source: Courtesy of D. Makar.

is smaller. Particularly at low temperatures, this reduces the charge uptake of lead–acid batteries (see hybrid vehicle applications) and increases the time it takes to fully charge them.

- If very large crystals form, they are very difficult or impossible to charge – the battery "sulfates,", also known as "hard" sulfate.

For reasons of service life, it is therefore important to keep lead–acid batteries in a fully charged state or to recharge them regularly. Leaving them in a partially discharged state and continuous operation in a partially discharged state without full charging are destructive for lead–acid batteries, as the Ostwald ripening process continues both during and without current flow.

Figure 12.10 shows scanning electron microscope images of active masses with different lead sulfate contents.

12.4.4 Voltage Dip at the Start of Discharging

After a full chemical charge[16] of a lead–acid battery, there is no lead sulfate on either the positive or negative electrode.

Lead sulfate forms crystal nuclei on the surface of the electrodes from a supersaturated concentration of lead sulfate molecules dissolved in the electrolyte. The crystal nucleus is stable once approximately ten molecules have formed a cluster on the surface of the active material as a result of thermal movement. Lead sulfate molecules formed at the beginning of discharging cover the surface of the active materials and lead to a reduction in the effective electrochemically active surface.

16 The term "chemical full charge" is chosen to distinguish it from technically achievable or practically acceptable full charge, where there is still discharged material left.

At a constant discharge current, this leads to an increase in the charge transfer voltage according to the BV equation due to the decrease in the effective surface area. Only from a certain concentration of lead sulfate on the surface, typically after approx. 1–2% capacity has been discharged (faster at high discharge currents), exist sufficiently many lead sulfate molecules so that stable crystal nuclei can form. As soon as this crystallization process begins, the coverage of the active mass with lead sulfate decreases and the effective surface area increases again. This causes a reduction in the charge transfer voltage via the BV equation, and the cell voltage rises again slightly. These phenomena are referred to as a voltage dip or coup de fouet, which is physically a crystallization overvoltage. A voltage dip has no technical significance in the use of lead–acid batteries except in the very rare cases where, due to low temperatures and high discharge currents, the voltage may fall below an unusually high setting of the end-of-discharge voltage.

The presence of a voltage dip indicates that there were no or only a small number of lead sulfate crystals at the start of the discharge and that the battery was therefore chemically fully charged. The absence of a voltage dip, on the other hand, indicates an incomplete charge. Observing the voltage dip is therefore part of a good battery diagnosis.

By adding additives (e.g. barium sulfate) to the active mass, crystallization nuclei for lead sulfate can be provided, which reduce the formation of a voltage dip.

A large number of lead sulfate crystals evenly distributed in the electrode volume is favorable for discharging. The higher the initial discharge current and thus the number of individual lead sulfate molecules on the electrode surface, the more crystal nuclei are formed. However, high discharge currents also lead to a current density distribution toward the edge of the electrode, so a high discharge current at the beginning of the discharge does not necessarily have to be part of a good operating strategy.

12.4.5 Manufacturing Process

The lead used to manufacture the active masses must be tested for correct compliance with purity because impurities in the ppm range can already cause quality problems, whereas desirable components can have a positive effect on service life.

In the first step, lead is converted into lead dust either in ball mills or by thermal processes and oxidizes in air. In the next step, the lead powder is mixed with additives depending on the application and later used either for positive or negative electrodes (paste production). Further processing into electrodes (plates) is described in the following.

12.5 Electrolyte

The electrolyte used is aqueous sulfuric acid, which is mainly dissociated into HSO_4^- and H^+. Depending on the application, the concentration (all data for 20 °C) in the fully charged state is between 1.24 g/cm^3 and approx. 1.34 g/cm^3, corresponding to an approx. 4.1 or 6.4 M solution. To improve the structure of the active masses, some phosphoric acid is sometimes added to the aqueous sulfuric acid (especially in gel batteries), and occasionally other materials as well.

The consumption of sulfuric acid during discharge and the new formation during charging change the density and electrical conductivity of the electrolyte as well as the equilibrium voltages at both electrodes significantly and unevenly across the electrode height. Many technically relevant properties of lead–acid batteries can be attributed to this.

The electrical conductivity of aqueous sulfuric acid depends on the temperature and density and has its maximum of approx. 22×10^{-2} S/cm at approx. 1.24 g/cm^3. As soon as the density falls below 1.24 g/cm^3 during discharging, the conductivity decreases continuously. In the discharged state, the average acid density is approx. 1.06–1.12 g/cm^3, depending on the design, and the conductivity decreases by approx. 30–40%. Figure 12.11 shows the conductivity as a function of temperature and density.

Aqueous sulfuric acid is the material in lead–acid batteries with the lowest specific conductivity. It is approx. 5×10^4 lower than porous lead and approx. 5×10^3 lower than porous lead dioxide. However, it is wrong to automatically conclude from this that aqueous sulfuric acid always has the largest share in the resistance of a cell, because the area to be considered is very large and the distance between the electrodes is very small (see Figure 12.1).

Aqueous sulfuric acid is hygroscopic, the density in thermodynamic equilibrium changes depending on the temperature and humidity.

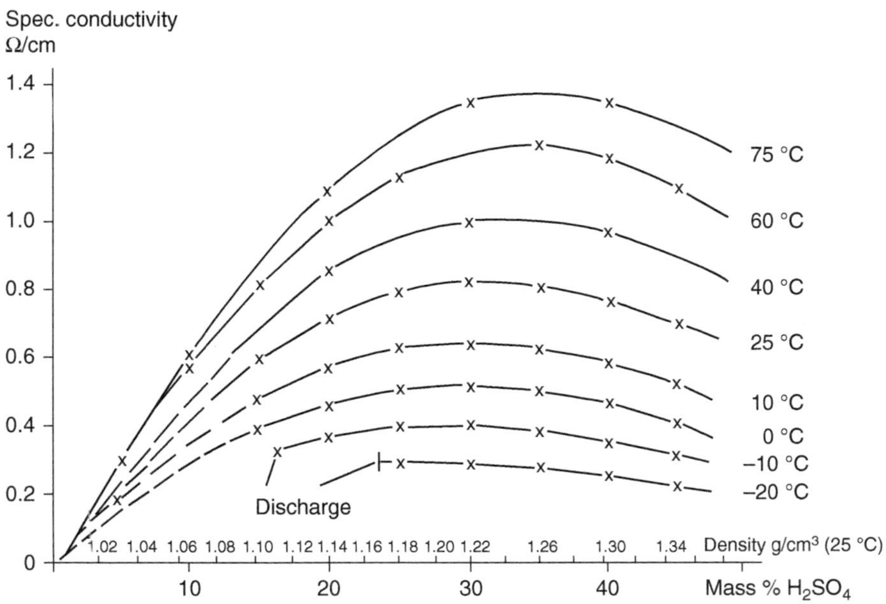

Figure 12.11 Specific conductivity of aqueous sulfuric acid as a function of temperature and concentration. Source: Adapted from Ref. [5]/Research Studies Press.

Different nominal concentrations for different applications are a compromise between different requirements:

1) The design of cells must ensure that sufficient sulfate ions respectively aqueous sulfuric acid is available for a complete discharge of the electrodes up to the specified voltage limit and that the electrodes are completely covered at all times.
2) Both distilled water and concentrated sulfuric acid are very poor ion conductors (see Figure 12.11). At the end of the discharge, the conductivity of the sulfuric acid must be sufficiently high for recharging.
3) To prevent the dissolution of lead sulfate in the electrolyte, the concentration of aqueous sulfuric acid must still be sufficiently high at the end of the discharge.

For these reasons, aqueous sulfuric acid must be present in excess. This reduces the specific energy of cells compared to the theoretical specific energy.

For low-temperature applications, it is important that aqueous sulfuric acid freezes only at low concentrations, corresponding to a very low state of charge, and at very low temperatures. The formation of ice crystals in the pores of the active masses is obviously a mechanical stress that damages the electrodes and must be prevented during operation.

To measure the acid density, either a simple manual hydrometer is used, with which the immersion depth of an evacuated glass body in the electrolyte is measured, or an automatic device that also measures the temperature at the same time. For accurate measurements and analysis, the temperature of the electrolyte must not be neglected when determining the density. In the case of sealed batteries, the density of the electrolyte immobilizes in the gel or AGM can only be measured in the laboratory.

Due to the side reactions explained in sections 12.2.5 and 12.2.7, water is decomposed in all flooded batteries and leaves the cell as hydrogen and oxygen. Depending on the operating mode, charging voltages, and temperature, this requires regular topping up with purified water (see Ref. [19] regarding the technical requirements). Contaminated water can promote side reactions in the cells and accelerate the self-discharge reaction. Cells then change their charging and discharging behavior and are damaged. In the case of intensively used flooded batteries for industrial trucks, water is often topped up daily after each charge, but at the latest after a week. For batteries in trickle charging mode for emergency power systems, it is sufficient to top up with water once a year. Modern starter batteries are built with a sufficiently large reservoir of electrolytes and additives or alloy components so that no water needs to be topped up over the entire service life.

It should be noted that sulfuric acid does not leave the cell through openings in the lid, or only in negligible quantities (leaking terminal feedthroughs, as an aerosol at high gassing rates, etc.). For maintenance reasons, it is never necessary to top up with aqueous sulfuric acid and battery manufacturers do not allow topping up with aqueous sulfuric acid in their maintenance instructions.

A little sodium or aluminum sulfate is often added to the electrolyte as protection against deep discharge. As lead sulfate has a certain solubility in pure water, the sulfate ions added by the salts reduce the solubility in accordance with the law of mass action and thus prevent or reduce the formation of short circuits due to dendrites and the like.

12.6 Current Collectors, Grids

It is necessary to increase the electrical conductivity of the electrodes, which have to conduct up to several hundred amperes per electrode at high loads, by means of additional conductive components, the current collectors, or grids. The current collector must offer the possibility of connecting several electrodes in parallel to increase the cell capacity and to conduct the current to the terminals of the cell. Furthermore, the mechanical stability of the open-pored, porous active masses is not sufficient to build and operate large electrodes without mechanical stabilization. Current collectors must also provide mechanical stability during production and operation.

Due to the corrosive conditions (sulfuric acid, electrical voltage), the current collectors must be made of particularly corrosion-resistant materials. So far there is no alternative to lead. In certain applications (e.g. submarines), a copper core with a dense lead coating can also be used on the negative electrode. Microcracks or coating defects on the current collector of the negative electrode are covered with lead when the battery is charged, similar to galvanic processes. With the positive electrode, on the other hand, alloy components can be released in the event of coating defects, made available for unwanted electrochemical processes, and destroy the current collector.

The usual form of a grid for both the positive and negative electrodes is shown in Figure 12.12. The design explained in chapter 4 with the terminal lugs of the positive and negative electrodes arranged on different sides for improving current homogeneity (see Figure 4.8) is not used for lead–acid batteries for production and handling reasons, with the exception of a few wound cells as special designs.

The properties of current collectors that are relevant to the functionality and service life of the battery are explained in more detail in the following:

- corrosion resistance,
- low electrical resistance,
- high mechanical stability, and
- good electrical contact between the collectors and the active masses.

12.6.1 Corrosion Resistance

12.6.1.1 Current Collector of the Negative Electrode

The lead of the negative current collector or grid is stable as soon as low charging currents shift the electrode voltage above the thermodynamic equilibrium voltage. A self-discharge reaction with conversion to lead sulfate, which affects not only the

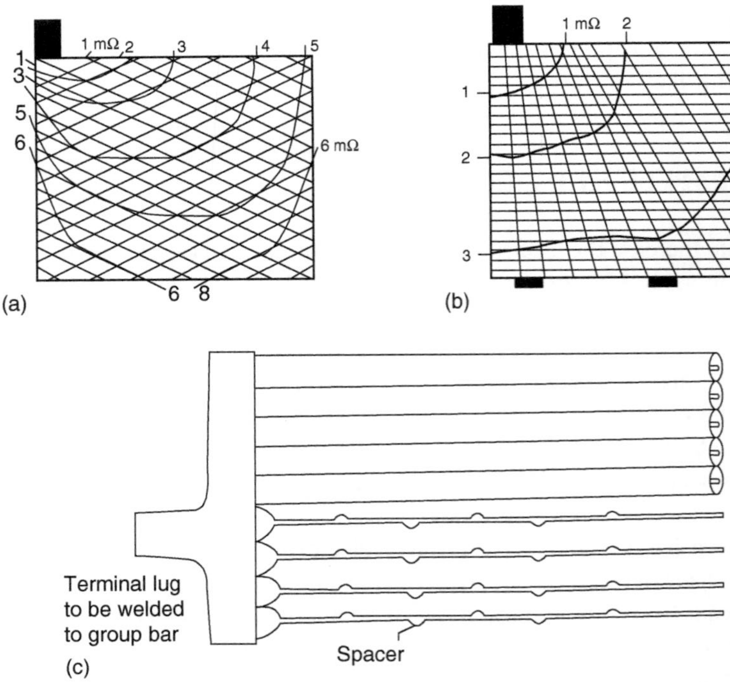

Figure 12.12 (a) Expanded metal grid and (b) cast grid with curves of equal resistance with respect to the point where grids are connected to the cell terminal. (c) Schematic diagram of a tubular plate electrode with textile gauntlets (only in the upper area for clarity). The individual rods have spacers to ensure that the individual rods are centered in the gauntlet.

active mass but also the lead of the current collector, cannot take place at such potentials. Corrosion attack on the negative current collector is therefore not observed. However, section 12.14.4 shows that corrosion can occur in the area of the group bars and terminals of sealed batteries due to small local voltage differences to the active masses.

12.6.1.2 Current Collector of the Positive Electrode

The lead of the positive current collector, on the other hand, is thermodynamically unstable and corrodes depending on the temperature and voltage level. This so-called grid corrosion determines the service life in many applications. Details are discussed in section 12.15.

Favorable operating conditions (low voltages in trickle charge mode, low temperatures), appropriate alloys for the grids, production processes and high production quality, and the thickness of the grids minimize the effects of this corrosion attack. Lead–acid batteries in stationary applications can thus achieve a service life of 15 years or more under favorable operating conditions. In particular, so-called large surface batteries (GroE) can achieve even longer lifetimes because the current collectors consist of solid, thick lead plates with only a thin layer of active material.

Conversely, cells with very thin grids have a short service life. Cells with extremely high specific power were developed under the brand name thin metal foil (TMF), but had a very short service life due to the high corrosion attack.

12.6.2 Electrical Resistance

The voltage drop from the terminal lug over the entire length of the grid should be as low as possible under volume and weight restrictions. Euler and Horn [20] have calculated the theoretically best shape for an ideal current collector, a shape tapering from the terminal lug of the current collector. The proposal also includes the use of a plate without any holes instead of a grid. The proposed ideal shape leads to different amounts of electrolyte between the electrodes in the upper and lower areas with a constant active mass layer and is therefore not practicable.

For production reasons, current collectors in the form of grids or, in applications with high cyclic loads, in the form of tubular plates with parallel rods have become established. The grid structure is often regular, but sometimes irregular to increase conductivity, with larger distances between the grid bars that are farthest away from the terminal lug. However, the electrical resistance of a grid must not be calculated without the active mass. Figure 4.6 shows the relationships using an equivalent circuit diagram. Although the specific resistance of the porous active masses is only a fraction of the specific resistance of the lead metal used for grids, the area of the active material that determines the value of its resistance is higher than that of the grid. As a result, the electrical conductivity of the electrode does not change much if the grid is broken or completely destroyed by corrosion, as long as the mechanical stability of the active masses still allows sufficient electrical conductivity. An example of this is calculated in Task 12.2.

The electrical resistance along the electrode is an important factor with regard to the homogeneity of the current density distribution and the formation of acid stratification (see section 12.8 and Appendix). When very long current collectors are used, the voltage drop along the electrodes is so great that the discharge of the active masses in the lower area hardly takes place. Lengths of more than 50 cm are therefore avoided or a so-called copper expanded metal grid is used for the negative electrode, which must be coated with lead, as is the case for submarine batteries, for example. Task 12.3 contains considerations on the arrangement of terminal lugs on current collectors with regard to the total resistance.

It is obvious that areas of active masses close to the grid have lower electrical resistance than areas that are further away from the grid. The resulting differences in the amount of charge throughput are not technically relevant.

12.6.3 Mechanical Stability

Lead is a soft metal. For production reasons, it is advantageous if the mechanical stability of the grids enables easy handling. Pure lead grids, which would be advantageous to use for corrosion reasons, are more complicated to process. The use of pure, unalloyed lead is therefore limited to special applications, that is, batteries for stationary applications with very high service life.

However, more important than technical production considerations is the dimensional stability of the grid during discharging and charging. The mechanical contact between the grid and the active mass is improved if the grid is hard and not bent by the mechanical forces that arise in the active mass due to the formation of lead sulfate. To improve mechanical stability, production processes are optimized and various lead alloys are used, mainly antimony and calcium. Cadmium, which would also be suitable, has not been allowed to be used for some time.

With a content of up to 4.5%, in the past up to 9%, antimony is mainly used for batteries subject to high cyclic loads. Over time, antimony goes into solution due to corrosion processes and is deposited on the negative electrode in a similar way to a galvanic process. There it increases the exchange current density of the hydrogen evolution reaction so that, all other things being equal, cells draw a higher float charge current and water loss increases (so-called antimony poisoning).

Calcium alloys (0.2% Ca content) are mainly used for starter batteries and sealed batteries, where hydrogen formation must be kept to a minimum.

12.6.4 Electrical Contact Between Grids and Active Masses

A good mechanical connection between the current collector and the active masses is an important prerequisite for good electrical contact, but it is not sufficient.

To avoid electrically insulating barrier layers between the grid and active materials, the grid must be alloyed with additional metals. Without suitable alloys, insulating barrier layers form after a few cycles, which significantly reduce the capacity. This effect is known as premature capacity loss (PCL) and was noticed when grid alloys were changed. Lead alloys with small additions of tin and other metals play an important role in modern batteries, so that PCL is no longer a technical problem under good operating conditions.

12.7 Manufacturing Process and Other Components for the Production of Cells or Blocks

The steps that lead to building cells or blocks (several cells in one housing) and other components required for this are described in the following.

12.7.1 Production of Current Collectors and Electrodes (Plates)

Grids are produced in various ways:

- Casting process:
  The liquid lead flows into a mold where it solidifies (drop casting) or, in the case of grids for tubular plates, is produced using a pressure casting process. In both cases, the grid must rest for some time after the casting process to harden for further processing.

- Punching process and expanded metal:
  Holes are punched out of a roll of lead sheet. In the case of expanded metal grids, a lead strip is cut lengthwise and then stretched to the desired width. This process is also used when copper is used as the core for large negative electrodes.
  In both cases, the strip of lead is then cut to the desired dimension, taking care to ensure a clean cut. Sharp edges can later destroy the separator and cause short circuits.

Galvanic processes to manufacture grids, which can be used to create composite materials consisting of different metal layers, have not established themselves [21].

When grids are used as the basis for complete electrodes or plates, the active masses are applied to both sides of the grid as a paste that is already mixed with aqueous sulfuric acid and other additives (pasting). It is important that the grid is completely covered with paste on both sides to reduce corrosion processes and ensure a uniform thickness. The total thickness of pasted grids can be approx. 0.8–1.2 mm for starter batteries and up to approx. 6 mm for batteries used in stationary and traction applications.

With tubular plates, the first step is to insert each rod into individual pockets of the gauntlet. The individual pockets are then filled with active materials, usually via a "paste injection," in which long cannulas are inserted into each pocket and paste is injected while the cannula is pulled out. The alternative process of filling the gauntlet with lead dust and then shaking the plate to fill each pocket completely is much more problematic for reasons of occupational health risks and is rarely used anymore. The pockets are closed off on one side against the appropriately shaped cast electrode and on the other side with plastic plugs. This type of plate (i.e. tubular plate or pocket plate) is only used for positive electrodes and offers high mechanical stability due to the mechanical pressure of the gauntlet on the active mass. Lead–acid batteries for industrial trucks in Europe which are cycled daily therefore almost always have these kinds of plates as positive electrodes with a diameter per tube of up to 8.4 mm. The use of a gauntlet leads to higher electrical resistance. The production costs are higher than for pasted grid plates. Sealed batteries with such plates are always built as gel batteries because AGM material cannot easily following the contour of a tubular plate.

In both designs, the terminal lugs of the current collectors are freed from adhering active masses and then matured for several days in special chambers at a controlled temperature and humidity (so-called curing process). This process results in chemical conversion of the active masses and better bonding of the active masses to the current collectors.

After curing, the plates are electrochemically formed either individually in a separate step in a tank filled with acid (tank formation) or, in the case of starter batteries, also in the fully assembled and filled battery (block formation).[17]

17 Formation refers to the process by which the active masses are electrochemically converted into the desired electrochemical form in terms of stoichiometry and crystal structure during production. This process may not be completed at the time of delivery, but will only be completed at the beginning of use.

12.7.2 Separator

Positive and negative electrodes must always be electrically connected by the electrolyte, but must not touch each other directly. This is the most important function of separators. In addition to this basic function, there are also the following requirements:

- High porosity so that the ionic conductivity is as high as possible.
- Corrosion resistance to PAM lead dioxide, a strong oxidizing agent (see section 12.14), which can damage the separator with the formation of oxidation products, for example, also, CO_2.
- Avoidance or reduction of dendrite growth (see section 12.14).
- Ensuring a constant distance between the electrodes to improve current homogeneity and creating a sufficiently large electrolyte volume so that sufficient sulfate ions are available during discharge – separators therefore often have a ribbed structure on the surface.
- Easy vertical movement of aqueous sulfuric acid to equalize acid concentrations (see Appendix and section 12.9) through appropriate design and orientation of the ribbed structure of the separators.

Separators are made of highly porous plastics (usually polyethylene, in the case of traction batteries also polyvinyl chloride-based) with a porosity in the region of 80%. In AGM batteries, the AGM material itself serves as a separator.

During operation, active mass particles can detach from the outside of the electrodes, which can lead to a loss of capacity and fall to the bottom of the cell housing (shedding, see section 12.14). To prevent this, separators with a fleece layer are often used to support the outside of the electrode. In starter batteries, so-called pocket separators are often used, which are sealed at the bottom and prevent loose active mass from causing a short circuit between neighboring positive and negative electrodes.

12.7.3 Production of Electrode Stacks

In the vast majority of cases, cells consist of several electrodes connected in parallel to increase the capacity. The production of individual electrodes with more than approx. 150 Ah is not possible due to the required dimensions, especially length – unless current grids with a copper expanded metal grid are used for the negative electrode.

Before assembling a stack of grids, all terminal lugs are freed from active mass particles. The stack is created by placing a negative electrode, separator, positive electrode, separator, negative electrode, etc. on top of each other, whereby a separator and a positive electrode form a pre-assembled unit when using pocket separation. For cost reasons, large industrial batteries have a negative electrode on the outside which only contains half of the active mass (half plate) because it only has a positive counter electrode on one side. The terminal lugs of the negative and positive electrodes are located on different sides of the current collectors.

For cells with special corrosion requirements, the plate stack is first immersed in a tin bath with the terminal lugs facing downwards to such an extent that the lugs receive a tin coating. This reduces corrosion attack on the group bar which connects all electrodes of the same polarity. The lugs of the plate set are then immersed from above in a mold filled with liquid lead, which also contains the terminal or, in the case of modules, intercell connectors, so that the lugs weld together. The individual electrodes are now connected to each other via a so-called group bar.

There is a high risk of corrosion in the area of these group bars (see section 12.14) if the individual lugs are not welded together without cracks or blowholes. The manufacturing processes and material selection must therefore be carried out with particular care.

12.7.4 Battery Housing and Lid

High-quality polypropylene housings and lids are now used for almost all battery types because of their very high chemical stability. For stationary batteries, styrene acrylonitrile (SAN) is also often used, which is transparent and allows for visual inspection of the cell. The housing must have high mechanical strength so that the distance between electrodes remains within desired limits, the volume expansion of the electrodes during discharge is compensated for and the transportation, and assembly of cells and blocks is possible.[18] After inserting the stack of plates, the lid and housing are electrolyte-tightly welded together using a hotplate welding process.[19]

In the case of block batteries or modules, that is, several cells are located in one housing, the cells must be connected to each other with so-called intercell connectors before the lid is welded on. In the first step, a hole is punched in the partition wall and then appropriately shaped attachments of the group bars are mechanically pressed through the hole and firmly connected to each other in a kind of spot welding process. To create a connection that is permanently electrolyte-tight, a great deal of effort is put into monitoring the contact pressure and the current profile during welding. The electrolyte space of neighboring cells must remain permanently separated from each other in an absolutely electrolyte-tight manner. The safety and reliability problems resulting from a leaky intercell connector are discussed in section 12.16. Figure 12.13 shows a block battery after manufacturing the intercell connectors and before welding the lid to the housing.

The terminal feedthroughs are made electrolyte-tight from the outside after welding the lid to the housing. Lead terminal inserts are already fitted in the lid for this purpose.

There are many other requirements for the lid and additional components. These are explained further in section 12.10. The valves for sealed batteries have already been described in section 12.4.

18 Large individual cells (e.g. 2500 Ah) and large starter batteries can weigh well over 50 kg.

19 For welding the lid and housing together, they are first pressed against a hot metal plate, which melts the material. The hot plate is then removed and the lid and housing are pressed against each other so that the melted materials weld together.

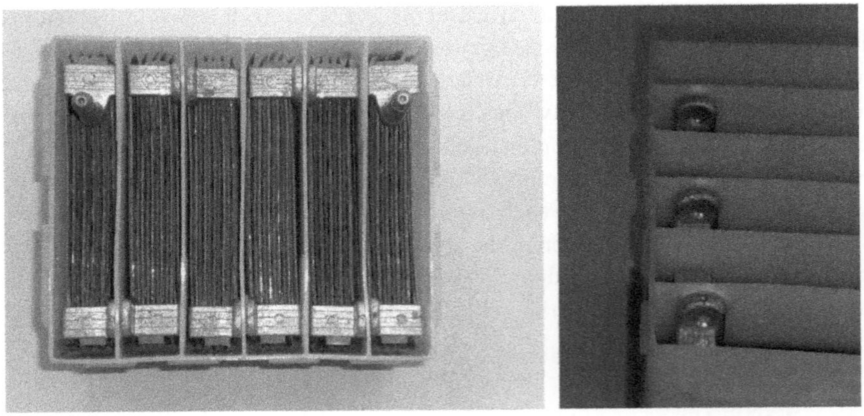

(a) (b)

Figure 12.13 Starter battery in block design without a lid. (a) From above and (b) section viewed from the side for showing the intercell connectors more clearly.

12.7.5 Cell Connector

If several cells or modules are connected in series to achieve the required system voltage, they must be connected to each other, to the loads, and to the charger.

- For starter batteries, lead terminals are usually used for connection cables to the vehicle electrical system and to other batteries, in the case of a higher system voltage.
- For small batteries with low power, crimp connectors are used which are molded into the terminals during production.
- In almost all other cases (batteries for stationary applications and industrial trucks) screw-on connections are used virtually exclusively in Europe.[20] For this purpose, there is a brass insert with a thread in the battery terminal for fastening a flexible, fully insulated copper connector with a terminal screw. The contact resistance between the terminal and the connecting cable is so low that the connectors only show a significant increase in temperature under prolonged use with the rated power range [22]. In normal operation without prolonged high current load, a measurable temperature increase indicates a fault at the terminal.

Battery manufacturers usually require the use of a torque wrench so that the terminals are not damaged, and often also the use of self-locking screws, which may only be used once.

---

20 In other parts of the world, lead connectors similar to those used in starter batteries are still used, which must be soldered to the terminals by trained personnel. These connectors have a higher resistance than flexible copper connectors and require more assembly work.

12.8 Current Inhomogeneity

The current inhomogeneity in electrodes, as explained in chapter 4, leads to significantly greater problems in lead–acid batteries than in lithium-ion batteries. Some of these problems are due to the often significantly larger capacity of individual cells in lead–acid batteries. Starter batteries with up to 225 Ah, traction batteries with up to 1200 Ah, up to 12,000 Ah for submarines, and stationary batteries for emergency power systems with up to 3000 Ah do not (yet) have a commercial equivalent in lithium-ion batteries. However, there are also current inhomogeneities in lithium-ion batteries that lead to local differences in the state of charge.

To discuss the current inhomogeneity, a spatially segmented equivalent circuit diagram is used, which divides the current path on both electrodes into three vertically separated current paths (top, middle, and bottom) along the current collector, and each of these three current paths into three segments (inside, middle, outside). Figure 12.14 shows this equivalent circuit diagram. Nine separate current paths from the cell terminal to the electrolyte exist for each electrode. The three horizontal current paths are coupled by electrolyte resistors in the free electrolyte so that ring currents can flow within an electrode. These electrolyte resistances are necessary to explain equalization processes in acid stratification. To simplify matters, only the main reaction without the double-layer capacitor is shown for each segment. The equivalent circuit diagram is based on Ref. [23]. The calculation of the currents and voltages per segment is easily possible with known values of the individual components using standard electrical engineering methods. The impact of the different current amplitudes per segment on temperature, state of charge, and acid density) must be estimated using coupled algorithms (chapter 8) and are fed back by changes in the resistance and voltage values per segment. Each segment is considered homogeneous with regard to all properties. Discontinuities in material

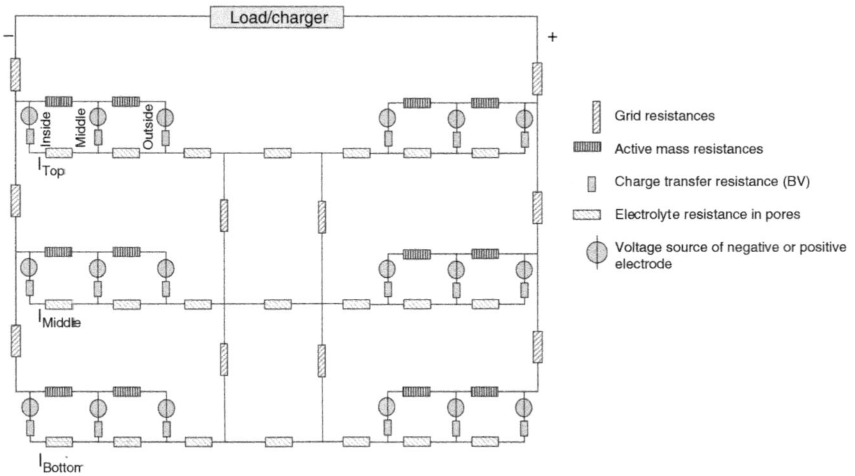

Figure 12.14 Spatially segmented equivalent circuit diagram (3 × 3). Source: Adapted from Ref. [23]/Elsevier.

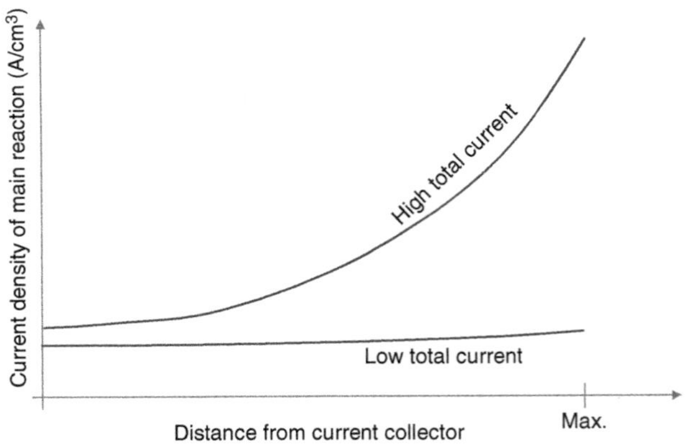

Figure 12.15 Schematic diagram of the main reaction current density in either the top, middle, or lower current path for a very low total current, where the BV resistance dominates in all current paths, and a high total current, where the high electrolyte resistance has a significant effect compared to the active mass resistance (see also Ref. [24]).

properties between segments are ignored. Despite this simplification, the model can provide good qualitative and, in some cases, quantitative predictions. A finer subdivision into further flow paths, which would then increasingly resemble finite element methods or analytical methods, is unusual and does not provide any further insight for the purpose of this book.

It is technically virtually impossible to measure the distribution of electron or ion current in electrodes perpendicular to the electrode surface. Figure 12.15 shows a schematic diagram of the main reaction current density for two different total currents. The figure illustrates that the main reaction current density is essentially constant at very low total currents because the resistance of all paths is mainly determined by the BV resistance. At very high total currents, the main reaction has its maximum at the interface of the active mass to the electrolyte, because the large difference in specific resistance between the active mass and the electrolyte means that current paths with a long path through the active mass and a short path through the electrolyte are preferred. This behavior is favored by the decrease in BV resistance with increasing current. The greater the current, the greater the displacement of the current into the outer segment. The formation of numerous nuclei for lead sulfate by high currents and a uniform distribution of nuclei throughout the electrode by low currents are therefore not compatible [25].

12.9 Acid Stratification

More technically significant than the uneven current distribution on the inside, center, and outside of the electrode is the current inhomogeneity in the vertical direction because it leads to the formation of acid stratification or an acid gradient (local differences in acid concentration) during discharging and recharging. Especially in flooded batteries, acid stratification, which is in principle reversible, leads to a reduction in performance and induces significant aging processes if it is not removed frequently.

The development of acid stratification depends on the operating conditions and has three causes:

- Height of the electrolyte level above the electrodes to compensate for water loss through electrolysis and to provide a sufficient amount of acid in the cell.
- Changes in acid concentration during discharging and charging between the electrodes. Driven by gravity, concentrated, heavier sulfuric acid above the electrodes moves downwards and less concentrated sulfuric acid moves upwards.
- Increase of the vertical concentration differences due to inhomogeneous current density distribution along the electrodes (chapter 4 and section 12.8).

Differences in acid concentration along the electrodes have an effect on the local state of charge of the electrodes and the local Nernst voltage. In lead–acid batteries, this has considerable consequences for the service life, operational management, and design. The differences in current density between the inner area of the electrode near the current collector and the outer area at the boundary to the free electrolyte volume described earlier in section 12.8 do not play a role in understanding acid stratification.

The steps for the formation of acid stratification are detailed in the Appendix. It is important to note that acid stratification develops most strongly toward the end of charging, shortly after the gassing voltage of 2.4 V/cell is reached, but before a full charge is achieved. The concentration of acid in the cell has the lowest value above the electrodes toward the end of charging and the highest value below the electrodes. Figure 12.16 shows measured values of a battery that was specially cycled to demonstrate acid stratification [26].

If acid stratification is not removed, the service life of the battery will be considerably shortened.

In addition to sufficiently long charging times and high charging voltages during normal charging processes, there are other methods for removing acid stratification:

- Short periods of charging with high voltage in the recharging phase, which leads to strong gassing reactions. The rising gas bubbles mix the electrolyte, especially if the structure and properties of the separators used support vertical mobility of the electrolyte. However, high voltages increase corrosion and water consumption.
- Mechanical electrolyte circulation: Air is pumped with low pressure and flow rate through a thin tube at one corner of the cell container, where it can bubble upwards unhindered. As a result, the maximum temperature during charging becomes lower, the service life increases and the water consumption and power requirement up to a full charge decrease. Many traction batteries for industrial trucks are operated with this type of equipment.
- Mammoth pump principle: The movement of the electrolyte level when the battery moves (curves, braking, and acceleration processes) is used to enable electrolyte to flow from the top to the bottom. Figure 12.17 shows the

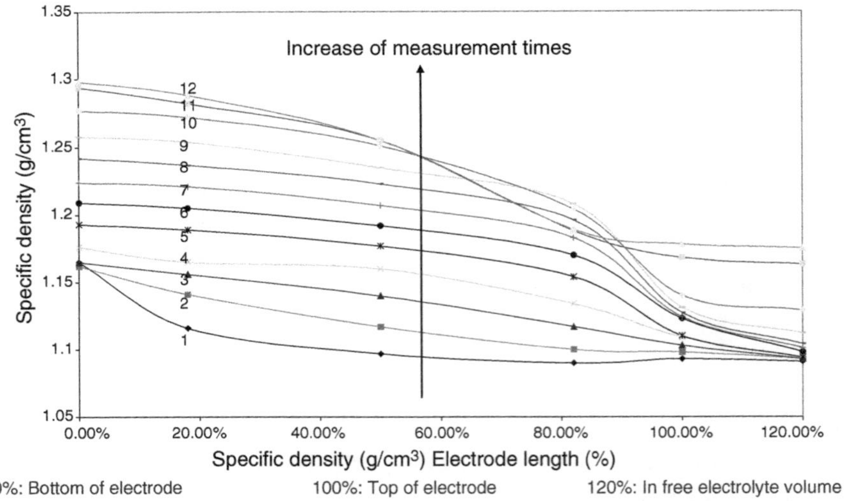

Figure 12.16 Development of acid concentration along the length of the electrode during charging. Source: Reproduced with permission from Ref. [26]/Elsevier. The measurement shows an acid concentration of almost 1.30 g/cm^3 toward the end of charging compared to the initial concentration of 1.26 g/cm^3 [27].

Figure 12.17 Electrolyte level in flooded starter batteries while driving. The batteries were placed in the trunk of a vehicle and illuminated from behind with a powerful lamp. For better visibility of the electrolyte level, the electrolyte was colored red [28]. Source: Courtesy of MOLL Batterien GmbH, Bad Staffelstein.

electrolyte level in a starter battery. Despite the differences in the electrolyte level shown there, movements of the battery alone do not lead to good mixing of the electrolyte, but must be supported by the installation of a mammoth pump device. The direction in which a starter battery is installed in a vehicle influences the effect.

If acid stratification is not completely removed before the start of the next discharge, subsequent discharging and charging processes lead to even greater acid stratification [26]. In addition, the active masses in the lower electrode area are discharged deeper and deeper, the charge throughput is reduced compared to electrode areas above and the acid concentration below the electrodes remains permanently above the nominal density of the electrolyte. In Ref. [1], this was demonstrated very impressively in experiments using three electrodes arranged one above the other in a common electrolyte reservoir. The consequences are massive sulfatation of the active masses in the lower electrode area, caused by reduced charging currents due to high local Nernst voltage, and high corrosion attack, also due to high local Nernst voltage (see section 12.15). Relatively frequent full charging of the cells to ensure a homogeneous acid concentration or other measures for electrolyte circulation (see section 12.10) are therefore necessary.

Acid stratification leads to problems for algorithms for the calculation of state of charge, because the terminal voltage is increased by areas with the highest acid concentration in the cell and the linearity between voltage, state of charge, and acid concentration no longer holds.

Acid stratification can be detected by changes in the relaxation behavior once current flow has been interrupted, by changes in the charging characteristics particularly when the constant current charging phase changes to the constant voltage curve and, when measuring acid concentration for monitoring purposes, by deviations between the measured and expected value.

The detailed description in the Appendix shows that the distance between electrodes and the electrolyte level, as well as the length of the electrodes and the resistance of the current collectors influence the development of acid stratification. The maximum acid difference in the cell and the speed at which it develops are therefore design-specific [26].

12.10 Design and Design Differences for Various Applications

The electrochemically active and passive components for lead–acid batteries were discussed in the previous sections. In doing so, points relating to the design of the battery and design features relating to different battery applications have already been mentioned in some cases. In the following, this information is supplemented, and design requirements for batteries in different applications are explained. The resulting effects on the specific energy of an entire battery are also presented.

12.10.1 Layout of Cells

The total amount of electrochemically active masses in a cell determines the capacity of the cell. Chapters 3 and 4 have already explained why the maximum utilization of the active masses on both electrodes can only be approx. 35–40%; the longer the planned service life, the lower the value. The weight used per amount of charge (g/mAh) is used as a parameter, for example, for service life specifications for different active masses during development. The amount of electrolyte that must be used so that the concentration is not too low at the end of the discharge must be determined using the molecular weight of the aqueous sulfuric acid. In Task 12.2, it is shown how the required active mass quantities can be calculated.

The theoretical specific capacity of a lead–acid battery of 80.7 Ah/kg is reduced to a value in the range of approx. 16 Ah/kg depending on the design, and the specific energy when using the nominal voltage for the calculation is thus reduced from 161.4 Wh/kg to approx. 32 Wh/kg. The following sub-chapters show the range in which the specific energy actually lies in various applications.

The weight of electrochemically passive components of a cell such as current collectors, intercell connectors, group bars, terminals, separators and the housing contribute little to the total weight in comparison to the weight of the required excess active masses and aqueous sulfuric acid. The same applies to all components for the production of a complete battery system (e.g. connectors, steel battery trays, etc.). The existence of side reactions during discharging and charging means that monitoring devices are almost never used to ensure a high level of safety and reliability.

12.10.2 Starter Batteries

The requirements for lead starter batteries are described, for example, in DIN EN 60342 or IEC 60095, and the area of application is dealt with in detail in chapter 16. The most important aspects for the design are:

- High starting currents even when partially discharged and at very low temperatures (test temperatures vary with locally applicable standards, typically around −20 °C).
- High corrosion resistance at high temperatures and long phases in float charge mode.
- Good charge acceptance.
- Low water loss so that refilling with water can be avoided over the entire service life ("maintenance-free according to DIN/EN").
- Controlled discharge of gases to allow installation inside the vehicle.
- Limited number of cycles during service life.

Very good starting capability from the smallest possible volume necessitates the use of thin electrodes (approx. 0.8–1.2 mm), whose low individual capacity requires electrode stacks consisting of many parallel electrodes. The grids must be thin and are therefore at risk of fast disintegration from corrosion attack, so the requirements for corrosion-resistant alloys and manufacturing processes are high. The low cycle stability associated with the use of thin grids is accepted. The desired low water consumption and high charge uptake also require special alloys and additives. Low water consumption allows reducing the amount of electrolytes, which must be available throughout the entire service life so that no water needs to be topped up. The use of pocket separators makes it possible to do with free space[21] below the electrodes. Both low water consumption and full expansion increase specific energy and energy density.

21 Area below the electrodes where active mass particles that have detached from the electrode can accumulate without causing a short circuit between the electrodes.

As deep discharges do not occur in normal operation, the electrolyte quantity is limited, and very low sulfuric acid concentrations (e.g. 1.06 g/cm^3) are acceptable at the end of discharge.

Despite low cyclic loads of starter batteries, acid stratification develops but can no longer be easily removed due to reduced side reactions (low gassing) that help reduce water consumption. Vehicle movements (see section 12.8) are not sufficient for this reason. Therefore, sealed batteries (mostly AGM batteries) are used, as they only develop negligible acid stratification compared to conventional sealed batteries, especially for start-stop functionality.

Sealed batteries do not require an electrolyte space above the electrodes, and AGM batteries do not require a separate separator. The electrodes can be larger than those of sealed batteries of the same dimension, which slightly reduces internal resistance and slightly increases energy density.

The production costs for sealed batteries are significantly higher than for flooded batteries. Enhanced flooded batteries (EFB) with improved charge acceptance and devices to eliminate acid stratification are therefore in competition with sealed starter batteries.

12.10.3 Traction Batteries for Industrial Trucks and Semi-Traction Batteries

Semi-traction batteries are used in wheelchairs, golf carts, mobile work platforms, construction machinery, etc., which are subject to low cyclical loads compared to industrial trucks. The energy required is often low (e.g. approx. 2 kWh corresponding to 12 V/160 Ah). Traction batteries outside of electromobility are used in industrial trucks for material handling and have a significantly higher energy content, from approx. 4.5 kWh (24 V/210 Ah) to approx. 72 kWh (80 V/900 Ah). The high weight of lead–acid batteries serves to improve the stability of the vehicles and is often a necessary property of the battery.[22] Further explanations are given in chapter 18.

The main requirement is a long cyclic service life, for example, 1200 full cycles for industrial trucks, which is clearly exceeded by many manufacturers in the laboratory. In Ref. [29], there are reports on batteries for electric buses in public transport (Mönchengladbach/Germany), which achieved 2300 full cycles with good thermal management, electrolyte circulation, and high charging currents.

The high service life expectancy under cyclic operation and high energy throughput with frequent, relatively deep discharges (limited to 20% state of charge for service life reasons) is achieved by using thick electrodes. For traction batteries in Europe, cells with tubular pocket plates are used almost exclusively for the positive electrode. Gauntlets enclose the active material and generate a contact pressure of the active masses against the lead rods of the current collector. The individual tubes usually have an outer diameter of 8 mm. In case of lower cyclic loads and shorter service life requirements,[23] semi-traction batteries and traction batteries have

22 For industrial trucks with low intensity of use, batteries with reduced capacity are often used, and the required counterweight is provided by lead plates or similar counterweights.

23 In the United States, traction batteries for industrial trucks are often manufactured with pasted grid plates, which have a much shorter expected service life.

thinner pasted grid plates in the range of approx. 2–6 mm as positive electrodes. The thinner the electrodes, the lower the internal resistance, but also the lower the cyclic service life at deep discharge operating conditions. The negative electrodes are always manufactured with pasted grid plates.

Constant cycling causes mechanical stress on the grids due to volume changes in the active materials. The grids must therefore be mechanically hard, which is achieved by alloy additives, especially antimony in the lead of the current collector of the positive electrode. In addition, the microstructure of the active masses is mechanically stressed by deep discharging and charging. Active material particles detached from the electrode sink to the bottom. This can be reduced by using fleece layers on the separators. To avoid a short circuit between positive and negative electrodes caused by active material falling to the bottom of the cell, the electrodes end a few millimeters above the bottom and form a so-called sludge space. There is no pocket separation as with starter batteries.

For handling reasons, cells for traction batteries are relatively long and narrow, which favors the formation of acid stratification. The cells are therefore massively overcharged at the end of charging (high water loss, temperature increase, corrosion), so that a strong gas development occurs and the rising gases mix the electrolyte, or equipped with so-called electrolyte circulation systems, in which the acid stratification is mechanically removed by air bubbles from the bottom. To compensate for the loss of water, the cells must contain a large amount of electrolyte above the electrodes, which induces the development of acid stratification, and are regularly topped up with water, often every day, but at least once a week. This is usually done semi-automatically or fully automatically using water refill systems on the battery.

The use of sealed cells eliminates the need for maintenance work to top up with water. Nevertheless, flooded cells are still predominantly used for traction batteries in intensive use because water refilling can be easily organized.

12.10.4 Batteries for Stationary Applications

The most important requirements for batteries in this area (see chapter 19) are:

- long service life in float charge mode,
- low maintenance requirements (once a year), and
- reliable monitoring as the cells are, in most applications, discharged only seldom and then often only slightly, but still have to be evaluated with regard to their reliability.

The batteries are operated in trickle charge mode for the majority of the time, usually at moderate temperatures. The effects of corrosion attack on grids, group bars, and terminals are reduced by using thick current collectors and plates and selecting appropriate materials. Where particularly high demands are placed on service life and reliability, tubular plates are used which minimize the shedding of active material on the positive electrode in float charge mode. A battery with a very long service life is a so-called GroE cell, whose current collectors consist of thick lead plates with ribs and comparatively little active mass. These cells have lifetimes in the range of 20–25 years but are very expensive.

Sealed lead–acid batteries are being used more and more frequently instead of flooded batteries.

For stationary systems with low energy requirements, low-capacity AGM batteries are usually used, which are similar to starter batteries but are optimized for corrosion resistance. However, the service life of burglar alarm and fire alarm systems, for example, is often only four years, or the batteries are then routinely replaced.

12.11 Properties

In the following, the typical and technically relevant properties of lead–acid batteries are explained and summarized on the basis of the information presented so far. Aging effects will be explained later.

1) Discharge behavior, capacity, Peukert coefficient.
2) Power output or uptake and internal resistance.

Charging and charging characteristics are dealt with in section 12.12.

Depending on the application, other properties are important, for example, water consumption, self-discharge, cycle stability, and vibration resistance, which are described in industry standards and must be demonstrated in acceptance tests. These are dealt with in the application-oriented chapters 15–20.

In the following, the usual current specification for lead–acid batteries in relation to the capacity:

$$I_n = C_n/n \tag{12.15}$$

is used, where n stands for the discharge duration in hours under nominal conditions.

12.11.1 Discharge Behavior and Capacity

Chapter 8 explained how to measure the capacity of a battery. Compared to lithium-ion batteries, the capacity of lead–acid batteries depends much more on the discharge current, but less on temperature.

A special feature of lead–acid batteries is the limitation on the use of the capacity required by the manufacturer to ensure the required service life. Table 12.4 shows the capacity that can be removed during discharge at different currents. The percentage of capacity that can be removed decreases as the current decreases, but the amount of charge that can be removed in ampere-hours is essentially the same, corresponding to equal utilization of the active masses regardless of the discharge current.

In chapter 3, the general shape of the voltage curve was already presented on the basis of a very simplified model, that is, the availability of charged active material of an electrode decreases proportionally to the amount of current drawn. This book does not provide a more detailed adaptation of this model to lead–acid batteries. The decrease in the effective surface area of the active mass structure, the change

Table 12.4 Discharge currents and capacity for a cell used in stationary applications. Taken from a data sheet of BAE Batterien GmbH.

Nominal capacity values	C_{20}	C_{10}	C_5	C_1	$C_{0.5}$	$C_{1/6}$
Capacity	270 Ah	229 Ah	203 Ah	161 Ah	135 Ah	103 Ah
Discharge current	I_{20}	I_{10}	I_5	I_1	$I_{0,5}$	$I_{1/6}$
Discharge current	13.5 A	22.9 A	40.6 A	161 A	270 A	618 A
Discharge time	20 h	10 h	5 h	1 h	0.5 h	10 min
End-of-discharge voltage	1.8 V	1.8 V	1.79 V	1.74 V	1.7 V	1.6 V
Limit of use of capacity	70%	80%	85%	90%	100%	100%
Discharge duration considering service life aspects	14 h	8 h	4.25 h	54 min	30 min	10 min

in the availability of sulfate ions due to diffusion effects and changes in the pore structure of the active mass, the effects of the local acid concentration on the Nernst voltage, and changes in the electrical conductivity of the active masses (electrodes and electrolyte) as a function of the amount of charge removed would have to be taken into account.

Table 12.5 gives an overview of all components and characteristics of a battery and their changes during charging and discharging.

Table 12.5 Overview of the components of a lead–acid battery and changes in properties during discharging and charging.

Component	Change during charging and discharging
Cell connector	No change
Terminals/group bars and intercell connectors	No change
Electrode grid	No change, but "effective" length changes due to increasing current inhomogeneity
Active materials	Conductivity of active materials is reduced due to the formation of nonconductive material and preferential conversion of the contact zones between active mass particles (agglomerate of spheres model)
Electrolyte	Strong concentration dependence of conductivity, diffusion effects influence high current behavior
Charge transfer resistance	Main reaction: current-dependent resistance, dependence on the surface, and the concentration of the electrolyte
Equilibrium voltage	Concentration of the electrolyte

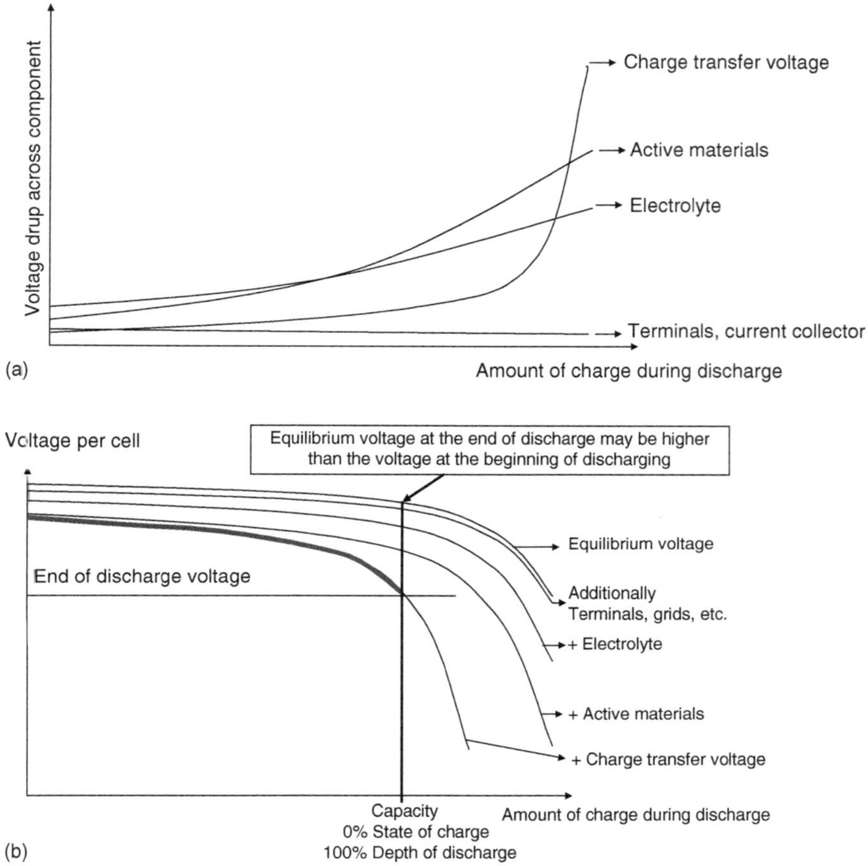

Figure 12.18 Schematic diagram of a voltage dip across the individual components of the current path in a cell and b the resulting cell voltage.

A schematic diagram showing the voltage drop across each component, including the BV resistance and the resulting cell voltage, is shown in Figure 12.18. The current only flows through the main reaction path of the two electrodes. In further discussion, the voltage curve in the event of an overdischarge will not be dealt with further as this is a cell fault.

The voltage drop at the end of the discharge is primarily determined by the charge transfer resistance. During a normal discharge, the point at which the active mass resistance becomes very high is not reached. After the end of discharging, the voltage rises for some time (several hours as opposed to approx. 15–30 min for lithium-ion batteries) to a constant value, which can be in the range of the voltage immediately after the start of discharging. This is a very clear indication that the voltage of a battery cannot be used to assess the state of a battery without knowing the current and the previous history. Depending on the design of the battery in

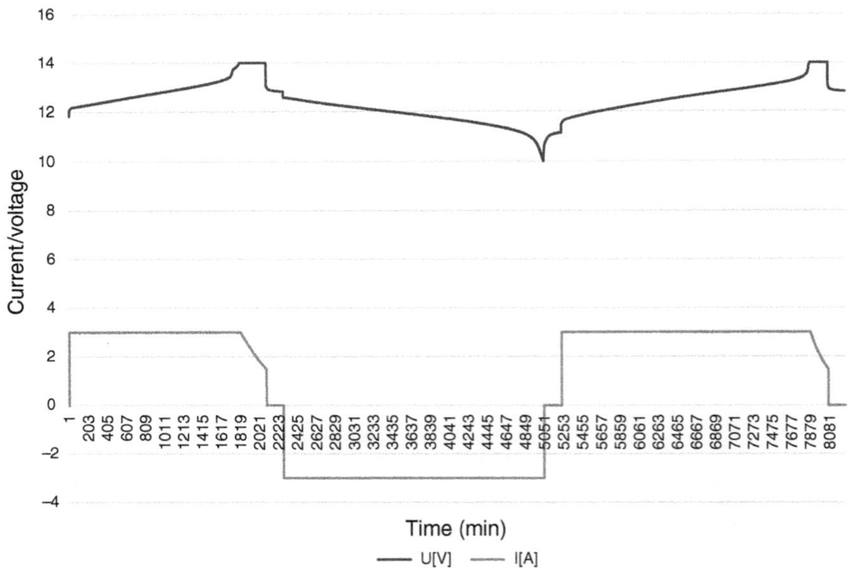

Figure 12.19 Discharging and charging curve of a cell with rest periods after the end of discharging and charging.

terms of electrolyte quantity and current amplitude, a fully discharged cell has an open-circuit voltage in the range of just over 1.90 to just under 2.0 V after a long rest period. Figure 12.19 shows a discharge curve of a battery with a subsequent rest period. Further discharges of the battery until the end-of-discharge voltage is reached again lead to a deep discharge and are not permitted for lifetime reasons. Also, they do not make sense due to the rapid voltage drop during additional discharging.

A voltage dip (see section 12.4) when a fully charged battery is discharged is not a quality defect and does not have to be prevented by adding additives as crystallization nuclei. On the contrary, the occurrence of such a voltage dip is a diagnostic tool for determining whether a battery is fully charged. The higher the discharge currents, the deeper the voltage dip. Technical limitations resulting from this are due to incorrectly selected, excessively high end-of-discharge voltages.[24]

The higher the discharge current, the faster the discharge voltage and the usable capacity drop. Under high current discharging, the current density distribution is concentrated on the electrode area in the immediate vicinity of the electrolyte and on the areas near the group bars. The material that is effectively available or used during discharging decreases as a result. Figure 12.20 shows a discharge curve at different current amplitudes resulting from these effects. The voltage curve is

24 One of the authors is aware of a case in which the end of discharge voltage of a 3000 Ah telecommunications battery was set to 2.00 V/cell. A few minutes after discharging began, discharging was interrupted because the cell voltage fell below the set end of discharge voltage due to a voltage dip.

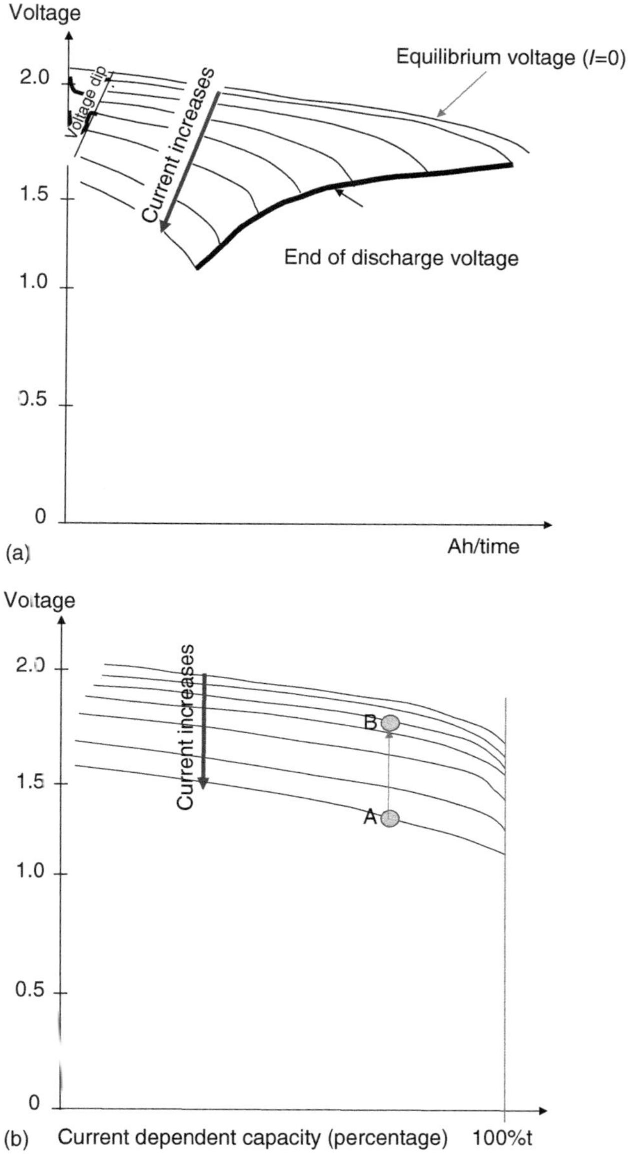

Figure 12.20 Voltage curve at constant currents as a function of (a) time and (b) usable capacity. Diagram (b) also shows how the voltage change as a result of a change of current, point A to B, can be easily estimated.

shown both as a function of the discharge time and as a function of the usable capacity, both at constant current. It should be noted that the voltage dip is not shown and the voltage curves are only shown after a short time interval after the start of discharging.

Figure 12.20 shows that, apart from an essentially constant voltage difference, the voltage curve is the same for all currents.

The end-of-discharge voltage is a function of the discharge current. If the discharge continues, the voltage drops very quickly below the minimum voltage required to operate the loads and would bring the discharge to an end.[25]

Constant current discharges rarely happen when batteries are used. If there are constant discharge conditions, they tend to be constant resistance discharges, which lead to a slow decrease of the discharge current due to slowly decreasing voltage (light bulbs, emergency lighting systems) or constant power discharges (see Figure 3.3). For batteries for stationary applications, such discharge curves or tables with the bridging time as a function of the power output are therefore often specified.

For batteries discharged with variable currents, it is difficult to determine the end of discharge using only voltage measurement, because the voltage during short high current loads can easily fall below fixed end-of-discharge voltages for a short time. On the other hand, very low discharge currents may not reach the fixed end-of-discharge voltage and lead to a very deep discharge. For this reason, state-of-charge-dependent or time-averaged voltage criteria are used for terminating a discharge.

The temperature dependence of the capacity, but also of the internal resistance, is approx. 0.6%/K in the nominal temperature range for lead–acid batteries, and approx. 1%/K at low temperatures.[26] Figure 12.21 shows a typical curve of the capacity values as a function of discharge current and temperature. Very large battery capacities must therefore be used for applications with very low temperatures and high energy consumption. An alternative to larger batteries would be the use of heating systems – an approach that is rarely used, however.

Further relevant information on capacity:

- Capacity as function of service life

  The capacity of new lead–acid batteries for industrial trucks is often only 90–95% of the nominal capacity. According to IEC 60245-1, the capacity must have reached

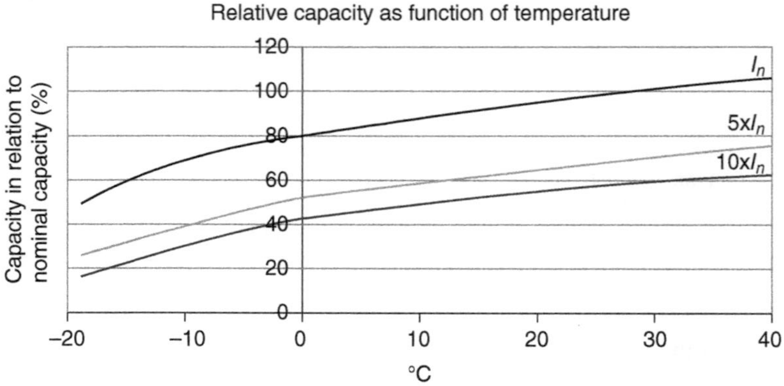

Figure 12.21 Capacity as a function of temperature and discharge current.

25 With the exception of pure resistance loads without special technical requirements.

26 Compared to many other battery technologies, the temperature dependence of the capacity is unproblematic.

full capacity after 10 cycles. The reason for the increase in capacity is that the active masses are not yet fully formed during production.

- During use, the capacity often increases, sometimes by a further approx. 20–25%, because the corrosion of the positive grid (see section 12.15) increases the available PAM at the expense of the positive grid. In most cases, the capacity decreases only toward the end of its service life.
- The capacity loss at the end of the service life is application-specific. Tests of batteries in photovoltaic systems have shown, for example, that the capacity curve shows a slow and continuous decrease [30]. The often shown rapid capacity collapse after approx. 20% capacity loss cannot generally be assumed.

- Heating of the battery during discharging
  At very low temperatures, the battery heats up during discharge to such an extent that the discharge voltage rises again during discharge, as is also the case with lithium-ion batteries. Discharge curves of batteries at very low temperatures therefore depend on how and whether the battery is cooled to ensure a constant discharge temperature.
- Fluctuating capacity values
  During capacity tests, the available capacity of an individual unit in a battery string fluctuates slightly even under otherwise identical discharge, temperature, and charging conditions. It is not always the same cell that reaches the end-of-discharge voltage first. The reason for this is the microscopic structure of the newly formed active mass which differs slightly after every recharging process.
  The capacity depends to a small extent on the conditions of the last discharge and more significantly on the charging conditions up to full charge.
- Pulsed discharge
  Symons [31] has shown that the capacity is lower with pulsed discharge than with discharges with a constant current of the same average amplitude.
- End-of-discharge voltage
  When discharging a battery with numerous cells and blocks in series, it is expected that the individual voltages will differ slightly at the end of discharge. Data from a battery for particularly safety-relevant applications, in which the cells were selected to have identical capacities, show that the end-of-discharge voltage of the individual cells only deviates from the average value (1.75 V) by a few mV – at least at the beginning of use. With cells that are not selected and after aging effects show up, however, the differences become greater and greater and the question of the end of discharge arises as regards differences in cell voltages. Criteria for the end of discharging are usually defined by the manufacturer. The implementation of such cell-specific voltage values for the end of discharge requires special measuring devices for batteries with a high nominal voltage and high discharge currents, as the measuring time for manual individual voltage measurement is too long.
  However, infrequent and short discharges beyond the voltage limit are insignificant in terms of the effects on the service life of batteries that can withstand cyclic loads.

12.11.2 Monitoring Requirements During Discharging

For service life reasons, it is necessary to limit the depth of discharge. The easiest way to do this is via a voltage-dependent discharge limit. The use of a state-of-charge-dependent end-of-discharge criterion is particularly useful for applications with opportunity charging or cycling in a partial state of charge (*pSoC*).[27]

The overdischarge reactions present on both electrodes prevent safety problems during discharging if there are cells with low-capacity or defective cells in a string. Battery management systems or individual cell voltage measurements are therefore not necessary. In many applications, however, there is an additional voltage measuring point after half of the cells. The differences between the voltages of both halves of the string indicate aging effects, a 2 V jump indicates the overdischarge reaction of an electrode in a cell.

12.11.3 Power Output and Internal Resistance

The ability of the battery to absorb or release a high-power levels over a short period of time with an acceptable voltage change is important for many applications. The internal resistance must be defined as an incremental resistance dU/dI.

Of all commercially available electrochemical battery systems, only certain alkaline nickel batteries have an internal resistance as low or lower than lead–acid batteries at the same capacity and low temperatures. The authors are convinced that the good low-temperature properties at high currents will ensure that lead–acid batteries will continue to be used for starting combustion engines for a long time to come.

From the dominance of the BV equation for the discharging and charging behavior of the battery, it can be directly seen that the internal resistance depends on the state of charge, the current amplitude, and the current direction. A fully charged battery will have a low internal resistance in the discharging direction and a high internal resistance in the charging direction. The reverse is true for discharged batteries, which have a low internal resistance in the charging direction but a high one in the discharging direction, see Figure 12.22. The temperature is also very important. Battery power data must therefore be recorded as a function of the state of charge, temperature, and current direction.

It is also clear that internal resistance decreases with increasing capacity because the total surface area of the active mass, macroscopic surface area of the electrodes, and their number increase, while the electrolyte resistance decreases.

Opposing effects must be taken into account when considering the dependence of internal resistance on current amplitude. The BV resistance decreases with increasing current amplitude, but the current inhomogeneity increases so that the effective lengths and areas of the ohmic resistors become smaller. Overall, however, internal resistance decreases as current amplitude increases. The dependence of the internal resistance on the current amplitude also has an effect

27 Opportunity charging refers to operating conditions, where the battery is recharged whenever this can be done, and partial state of charge operating conditions, where a full charge is not possible for long periods of time.

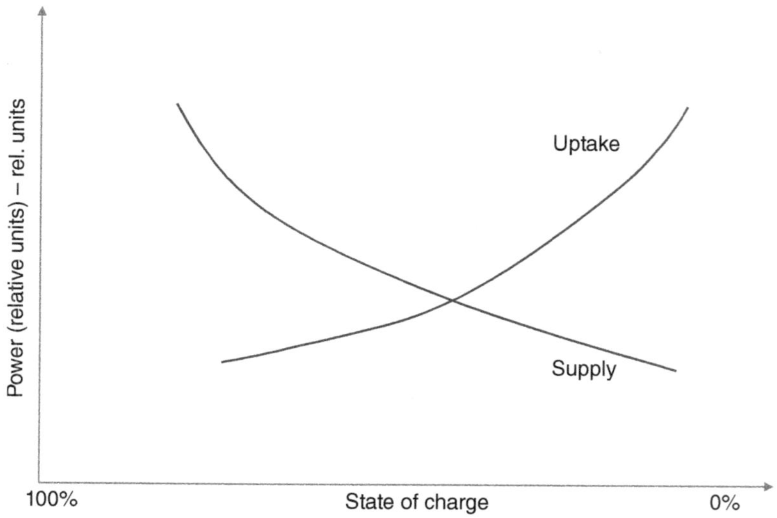

Figure 12.22 *SoC*-dependent power of batteries for mild hybrid applications.

on the measurement results with different measurement methods. Impedance spectroscopic measurements lead to different internal resistances than methods using high current amplitudes.

The capacity of a battery and its internal resistance are different properties, but in the case of lead–acid batteries they are usually particularly closely linked. Determining the capacity of batteries by measuring the internal resistance of a fully charged battery is, for example, a standard procedure for evaluating batteries for emergency power systems.

12.12 Charging and Charging Characteristics

12.12.1 Basic Information on Charging Lead–Acid Batteries

Analogous to Figure 12.18, the voltage curve during charging can also be represented by the changes in the voltage drop across the components in the current path of a cell.

Voltage values for controlling the charging characteristic are specified in the following. It is important to note that these values refer to the voltage between the terminals of the battery. In the case of long charging cables with a limited cross-section, the voltage drop along the cables is not negligible and can lead to considerable technical problems. Therefore, either a four-point measurement must be carried out at the battery terminals or the voltage drop across the cables must be compensated for in the charger by taking into account the cable properties (length, cross-section, material). Some standards (e.g. Ref. [32]) contain corresponding requirements. The consequences of the voltage drop across the cables are of course also present when discharging a battery, but they only lead to a reduction in capacity utilization.

Special aspects of charging are the existence of side reactions, the solubility of lead sulfate as a preliminary stage of the actual charging process, the elimination of acid stratification, and a short, but sometimes significant increase in voltage after a deep and long discharge.

- Competition between main and side reactions

When charging, the competition between the main reactions and side reactions must always be taken into account. At the beginning of charging a discharged battery, the side reaction current is still largely negligible due to the low BV resistance of the main reaction. With a fully charged battery in trickle charge mode – no discharged active material in the negative or positive electrode – the battery current can only flow via the side reactions. In between, it is unclear and cannot be measured[28] along which reaction paths the charging current flows. Estimates of the side reaction currents are important for calculating the state of charge and are dealt with there.

At the end of charging, the behavior of the battery is essentially determined by the side reactions, the higher the charging voltage the more. Aging-related increases in the exchange current densities of the side reactions are relevant for charging.

- Solubility of lead sulfate

As a precursor to the conversion of lead sulfate into charged active material for the positive or negative electrode, lead sulfate, which is present in crystalline form, must first go into solution. The availability of lead sulfate for charging therefore depends not so much on the amount of lead sulfate present as on the total surface area, which decreases during long phases in a partially or fully discharged state, on the surface condition of the lead sulfate crystals and on the temperature, as a low temperature reduces the solubility of lead sulfate. At the end of charging, the low solubility of lead sulfate can slow down the charging process or increase the cell voltage faster than usual at a constant current.

- Removal of acid stratification (flooded batteries only)

At the end of charging, acid stratification started during discharging must be completely removed. If no further measures are taken, the gassing reaction is used for this purpose.

- Voltage peak at the start of charging

In the case of deep discharge and long phases without a full charge, a voltage peak occurs at the start of charging, which can be attributed to the high active mass resistance. In extreme cases, this can impact the charging process (desulfatation charging).

---

28 In principle, a calculation is possible based on the ratio and the amount of oxygen and hydrogen produced. However, the internal oxygen cycle cannot be neglected. The technical complexity of the measurement makes this unrealistic.

12.12.2 IU Charging Characteristic

The characteristic curve for charging electrochemical systems with a constant current I followed by a constant voltage U (for lead–acid batteries, the term "IU characteristic curve" is often used) shown in chapter 3 is very similar and widely used for all battery technologies. The general curve can be explained using a simple model, namely the linear decrease in the discharged material as a function of the amount of current charged and its effect on the BV equation. The notes on more precise models of the voltage curve mentioned for discharging also apply here, but are not pursued further.

Differences between sealed and flooded batteries relate to the effects of the higher electrolyte density and the oxygen consumption in sealed batteries. This leads to differences in the end-of-charge voltage, but has no other relevant effects, so that, apart from the problem of acid stratification, this will not be discussed further in the following.

The usual parameters for an IU charge are as follows, depending on use (standby parallel operation, cyclical load and their mixed forms) in accordance with [33]:

- Charging current assignment:

10–50 A/100 Ah, some battery manufacturers do not specify a current limit for some types.[29]

- End-of-charge voltage:

2.23–2.48 V/cell (flooded batteries) and 2.27–2.48 V/cell (sealed batteries).

It should be noted that the parameters specified in standards cannot always be used because they are intended to fulfill other purposes. For fully charging a lead starter battery, for example, DIN EN 50342 suggests charging with a 16 V voltage limit for 24 h to ensure that the battery is fully charged chemically (no more lead sulfate) and to remove acid stratification – obviously damaging parameters that should not be used during operation.

The higher the charging current, the shorter the charging time, but the sooner the end of charging voltage is reached and the charger is switched off. The time gain when charging with very high currents is therefore limited. The charging speed is ultimately limited by the solubility of lead sulfate, which can only be influenced externally by increasing the temperature, and the elimination of acid stratification by gassing. A shortening of the constant current charging phase is not achieved by high initial charging currents alone. Due to the sharp rise in costs for chargers with a high rated current, there is therefore an economic incentive to use low charging currents if the available charging time is not a restriction. The influence of the charging current on the service life and capacity of the battery is discussed in the following.

The lower the end-of-charge voltage is set, the longer a full charge takes, and the higher the voltage, the greater the water losses (flooded battery) or oxygen transport rates of the oxygen cycle (closed battery) as well as corrosion processes.

---

29 There are no studies that show negative effects of high initial charging currents on the service life of batteries. However, there are indications of positive effects of high initial charging currents.

Considerations based on the agglomerate of spheres model also show that the microstructure of the active masses forms larger contact zones at low end-of-charge voltages, which lead to higher conductivity and better mechanical stability.

The initial charging current and final charging voltage are therefore always a technical and economic compromise, which also depends on the cycle depth and available charging time.

In addition to these two points, however, it must also be noted that lead–acid batteries should be continuously charged in so-called trickle charge mode after being fully charged to compensate for self-discharge, minimize corrosion and keep the microstructure in the optimum range. If trickle or float charging is not possible, then the charging process must be started again after some time.

In trickle charging mode, the charging current is dependent on the temperature in addition to the voltage. To improve the service life, many battery manufacturers therefore require temperature compensation of the charging voltage in the range of approx. 0.005 mV/K/cell in the temperature range from 0 to 40 °C. This requires a temperature measurement of the cell or block, ideally on the housing.

12.12.3 IUoU Charging Characteristic

Charging curves that switch to trickle charging mode after an IU charge that has more or less fully charged the battery are referred to as IUoU curves. This characteristic curve is the standard for many applications except in the traction sector. The point at which the charging voltage of the IUoU characteristic curve switches to the lower voltage of the float charge mode and the extent to which the battery is already fully charged is application-specific. The main advantage of the IUoU characteristic curve is the decoupling of rapid recharging and further operation in a charging mode designed to maximize service life.

The float charge voltage must be above the equilibrium voltage of the two electrodes to maintain the fully charged state and a sufficient charging current must flow. To guarantee this for batteries with several cells connected in series, in which there may also be individual slightly higher-resistance cells, the difference between the trickle charge voltage and the equilibrium voltage must be sufficiently great. For flooded stationary batteries with an electrolyte density of 1.24 g/cm^3 (approx. 2.08 V equilibrium voltage), the frequently recommended float charge voltage is 2.23 V/cell and the difference to the equilibrium voltage is 0.15 V/cell. For sealed batteries with an electrolyte density of 1.34 g/cm^3 (approx. 2.18 V equilibrium voltage), the recommended float charge voltage is often 2.27 or 2.29 V/cell and the difference is therefore only approx. 0.10 V/cell.

It is known from cyclic operation of sealed batteries that the risk of sulfatation of the negative electrode is higher than for flooded batteries. During the charging process, the charging current for the conversion of lead sulfate in the NAM to lead not only competes with the formation of hydrogen, but also with the reaction of oxygen to water.

Many standards and data sheets (DIN 41772) recommend this charging characteristic with specified voltage values, current values in relation to capacity and times.

The current and voltage values recommended by manufacturers are based on the above considerations and also on usage restrictions:

- Available, in particular plannable[30] charging capacity,
- Maximum charging current,
- Charging time until the start of the next discharging phase,
- Time until full charge is reached,
- Need to remove acid stratification (flooded batteries only).

These usage restrictions can only be met by an IUoU characteristic curve for batteries in trickle charge mode and infrequent discharges. Even then, however, a pure IUoU charge cannot guarantee the permanent full charge of all electrodes of all cells in series connection and the maximum service life expectancy. For this reason, so-called equalization or boost charges are carried out at regular intervals, every few weeks for a few hours. DIN EN 62485-5:2019-02 specifies 2.4 V for flooded batteries and 2.45 V/cell for sealed batteries, but no time specifications concerning duration and frequency are given.

For all IU and IUoU characteristics, the trickle charge voltage is selected to be as low as possible so that corrosion attack and, in the case of flooded batteries, water consumption is low, but all electrodes of all cells remain fully charged. With aging processes, short charging times, which do not always lead to a fully charged state for all cells (small differences in the exchange current density of the side reactions of the individual electrodes and slight temperature differences between cells, etc.), insufficient charges can occur in individual cells in series connections. The more cells are connected in series, the greater the risk. Despite a low charging current, one or both electrodes of the affected cells may discharge and lead sulfate may form.

The advantage of complete recharging and removal of all lead sulfate through equalization or boost charging outweighs the disadvantage of a short-term increase in corrosion. Equalization charging is generally suggested for batteries in standby parallel operation, but is also used for starter batteries in modern vehicles.

12.12.4 Further Charging Characteristics

In most cases, the speed of recharging and low corrosion attack contradict each other, so that many different characteristic curves are used for lead–acid batteries. The most important characteristic curves and some interesting special characteristic curves are described in the following.

1) IUIa characteristic curve,
2) W, Wa, and W0Wa or WUIa characteristic curves,
3) IR-free charging.

30 In autonomous energy systems with photovoltaic and wind power plants, the charging capacity depends on the weather and also, as with vehicles, on the current load.

1) IUIa characteristic curve
   This charging characteristic is used for cyclic operation of flooded lead–acid batteries in industrial trucks. The initial charging current is usually in the range of 16–25 A per 100 Ah capacity. The constant voltage phase thereafter is usually 2.4 V/cell until the charging current has dropped to 4 A/100 Ah. During the subsequent constant current charging phase, the cell voltage rises to values in the range of 2.65–2.75 V/cell. The aim is a short phase of relatively high gas development so that a homogeneous acid density is achieved and the battery is fully charged again for the next shift in eight hours. There must be an automatic cutoff when the fully charged state is recognized, because otherwise long phases of high voltage with high water loss and high corrosion would occur.
   When used in industrial trucks, the battery is often charged again and again at intervals of approx. 1 h with 4 A/100 Ah so that the battery remains fully charged and the remaining acid stratification is removed.
   This charging characteristic reduces service life due to the high voltages,[31] but is required for operational reasons. Numerous modifications of the IUIa charging characteristics are used in the field of industrial trucks.
2) W, Wa, or W0Wa characteristic curves
   With this characteristic curve, the charging current is reduced with increasing counter voltage, that is, corresponding to a higher resistance. Transformers with diode rectifiers as the simplest charging circuit show a W or taper charge characteristic. However, the initial current, which is intended to achieve rapid recharging, and the current at the end of charging, which is intended to ensure a permanent full charge at a low voltage, are technically coupled in very simple technology, so that an automatic termination of the charging process should be provided (Wa characteristic curve). Very inexpensive chargers for the do-it-yourself market often have such a charging characteristic.
   W0Wa charging characteristics are used to decouple the initial charging current and low full charging current or favorable end of charging voltage, in which two taper charge characteristics with different properties are coupled. The replacement of the second W characteristic curve by a UI characteristic curve (WUIa characteristic curve) is also used. Chargers with such characteristics are sometimes very high-quality chargers and charging characteristics to which the criticism of Wa chargers does not apply.
   Another problem with very simple chargers with a W characteristic is the dependence of the charging voltage on the main voltage.
3) IR-free charging
   The aim of this characteristic curve, which is very rarely used for lead–acid batteries, is to compensate for the ohmic voltage drop of all passive components of a battery or cell and to relate the charging voltage to the electrode voltage [34]. This explains the term “IR-free charging” (internal resistance free).
   The voltage drop across all passive elements is estimated by observing the voltage curve during short interruption of the current (see Figure 12.23a). The further

31 Depending on the intensity of use, the service life of lead–acid traction batteries for industrial trucks is still well over four years.

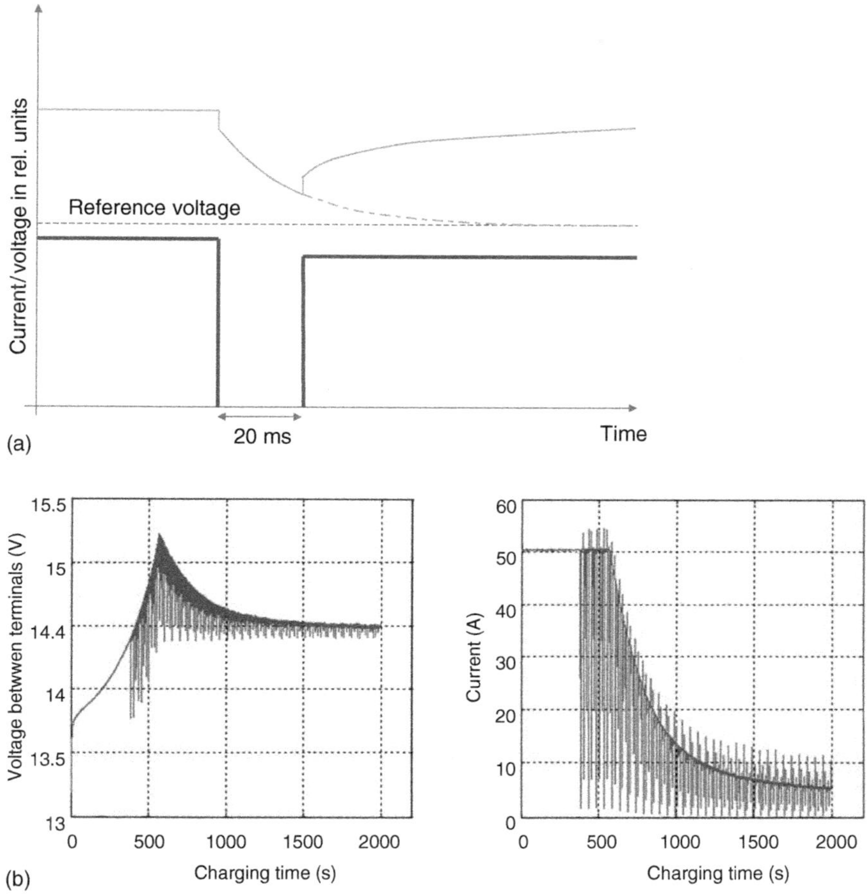

Figure 12.23 IR-free charging: (a) Schematic diagram of the current and voltage curve during a short current interruption with estimation of the electrode potential and (b) charging curve for a lead–acid battery.

course of the voltage can be calculated from the voltage drop and then serves as a reference value for the assumed actual electrode voltage. The terminal voltage can then be significantly higher than the usual end-of-charge voltage during charging without the electrode voltage reaching critical values (see Figure 12.23b). The charging current can then assume higher values and, above all, flow for longer before it is reduced.

With this charging characteristic, charging times in the range of a few minutes were realized, especially for alkaline nickel batteries, but charging times in the range of 1 h were also achieved for lead–acid batteries.

In addition to these charging characteristics, many other characteristics are used in commercial chargers, for example, pulse charging characteristics with different pulse patterns and sometimes with short discharge pulses to "depolarize" the electrodes as well as "pulsers" that are intended to help reduce lead sulfate formation.

However, there is still no clear scientific evidence that the reported improvements in properties can be attributed to these methods. Very-low-capacity batteries that are cycled several times and carefully charged with an IU characteristic often also show a significant improvement in capacity.

12.12.5 Evaluation of Charging Characteristics

The aim of a charge is to completely convert discharged material on both electrodes as gently as possible (low temperature increase, minimization of gassing and corrosion attack) while at the same time removing acid stratification and any passivation layers that may have formed and restoring an optimum microstructure of the active masses for the next discharge.

The voltage curve which is relevant for corrosion and side reactions with the formation of oxygen and hydrogen (both chemically aggressive gases) is different for each charging characteristic. It is therefore expected that the choice of charging characteristic, current amplitude and voltage values should have an effect on the battery's performance, charging efficiency and service life. However, there are few studies on this and only a few indications from experience, because historically the discharge characteristics have always been in the foreground. The following is an overview of some of the results.

12.12.5.1 Good and Bad Charging

Winsel et al. [15] carried out tests with IU charging curves with different charging currents and charge factors and showed that the capacity decreases the lower the charging current and the higher the charge factor. Since the probability of a complete full charge increases with increasing charge factor, the effect cannot be explained by an incomplete conversion of active masses by the charging process. The decrease of capacity is reversible.

In the publication, the capacity loss is explained using the agglomerate of spheres model. Under unfavorable conditions, contact zones with a small diameter are formed between the particles of the active mass. The effect was termed reversible insufficient mass utilization (RIMU).

This explains problems with IUoU charging characteristics, in which a high charging voltage is applied over a fixed period of time even after very short discharge phases to ensure a fast full recharge before switching to a lower charging voltage.

Part of a good charging process is a preceding discharge in which high short current peaks are reduced by technical means. In the course of this investigation, the terms "good charging" and "bad charging" were coined, which also takes into account the discharge, which largely depends on the application (see Table 12.6).

12.12.5.2 Interruption of the Current During Float Charge

Haaben [35] showed that "sawtooth charging," the interruption of the charging voltage to a value close to the equilibrium voltage (which can take several hours), almost doubles the service life of a sealed lead–acid battery under normal operating conditions.

Table 12.6 Good and bad charging according to Ref. [15].

Criterion	Good charging	Bad charging
Current amplitude	High as long as the state of charge is not high	Low
End-of-charge voltage	Low	High
Charge factor	Low	High
End of charge	Termination of charging after reaching full charge and switchover to low float charge voltage, unless used in cyclic operation	No termination of charging after full charge is reached and switchover to low float charge voltage, unless used in cyclic operation

12.12.5.3 Electrolyte Circulation

The elimination of acid stratification by means of high charging voltage phases and gassing rates, as occurs in particular with an IUIa characteristic curve, leads to chemical, thermal, and mechanical stress on the active masses. Chemically aggressive hydrogen and oxygen are produced using charging energy, which heats the cell and forms bubbles that make their way under pressure through the microstructure of the active mass into the electrolyte space between the electrodes. The charge factor until a full charge is reached is approx. 1.15–1.18 for new batteries and significantly more for aged batteries. If air is bubbled in to circulate the electrolyte, the charge factor drops to approx. 1.05 and the charging efficiency increases, and the temperature increase is significantly lower.

Field experience shows that batteries with electrolyte circulation can achieve an unusually high charge throughput, for example, approx. 7000 equivalent full cycles in the battery storage system of Bewag AG [36] and approx. 2300 equivalent full cycles in a municipal bus in Mönchengladbach/Germany, where good thermal management was also used and charging was carried out with high charging currents [29].

12.12.5.4 Constant Current or Constant Voltage Charging During Float Charging

Charging current and charging voltage are coupled via the Nernst voltage, BV equation, and resistances of the passive components. For batteries with several cells, however, there is a difference in whether the total voltage for all cells or the current is specified in trickle charge mode. If the voltage is specified, differences in the internal resistances can result in insufficient charging currents, which cannot occur if the charging current is specified. In Ref. [37], it was shown that the discharge capacity is up to 10% higher with a set current than with a set voltage value. The effect is reversible.

However, the use of current-controlled float charging is technically much more difficult and is therefore rarely used.

12.12.5.5 Charging Current Amplitude

There are indications that the service life slightly increases at high initial charging currents [38], and the active mass structure retains a fine-grained structure of high capacity for longer when high current amplitudes (500 A/100 Ah, [39]) are used, but there is no evidence that it decreases. However, the voltage must be monitored at very high initial charging currents (e.g. >100 A/100 Ah) because cells are destroyed very quickly if they reach the end of charging voltage before the other cells and continue to be charged with high currents.

Starter batteries and UPS batteries are often charged in their system environment with a current allocation of 100 A/100 Ah if they have a low state of charge.

Charging currents cause all components to heat up, so constant cycling with high charging currents can lead to unacceptably high temperatures.

12.12.5.6 Current Ripple

The current supplied by the charger normally has a low current ripple in the nominal current range. However, when the current is reduced during float charging, the current ripple depends on the power electronics used. The battery current ripple often leads to discharge currents, but not necessarily Faraday currents, with a frequency in the range of twice the mains frequency or more.

In one of the first very large battery systems for the electricity industry [36], it was found that equalizing charges, whose low average current was generated by very short, high current pulses, led to the shedding of active materials. The use of a special charger with a low ripple current prevented further shedding.

Many battery manufacturers require a current ripple of less than 5 A_{eff}/100 Ah during float charging or maximum heating of the battery of 3 K due to the current ripple of the charger.

12.12.6 Full Charge Criteria

Until now, a full charge has been understood to mean "full chemical charge," that is, the complete conversion of lead sulfate into charged active material on both electrodes without any discharged material left. Due to the reconstitution mechanism, the full charge can be clearly defined in contrast to lithium-ion batteries and is also technically achievable due to the side reactions, regardless of aging processes and production differences.

A reliable criterion for reaching full charge is the measurement of the rate of change of current and voltage in trickle charge mode. If there is no change over time in the current at a fixed voltage or the voltage at a fixed current, then electrochemically constant conditions have been achieved at both electrodes. This can only be the case if the total current only flows via the side reaction paths. This criterion is used in IEC 60245. This does not necessarily guarantee the elimination of acid stratification.

However, other criteria are also used to determine full charge, for example:

- Time:
  As soon as a certain charging voltage is reached or the charging current has fallen below a fixed value (the main criterion for lithium-ion batteries), charging continues for several hours.
- Determination of the charge factor:
  The amount of charge removed during discharging is measured or estimated and then the amount of charge to be recharged is determined using a fixed charge factor.

Both full charge criteria only lead to useful results for new batteries with deep discharges, but tend to overcharging in case of shallow cycles. With old and warm batteries, however, these full charge criteria do not necessarily lead to a full chemical charge.

When troubleshooting and dealing with battery problems, it is therefore always necessary to ask what state of charge the battery actually has when it is "fully charged" and whether the acid stratification has been removed from flooded batteries.

12.13 Aging Effects

12.13.1 Overview of Aging Effects

Aging includes all effects that irreversibly change the performance of the battery and other required functions (self-discharge, maintenance requirements, charging efficiency, etc.) and contribute to a reduction in service life. Acid stratification, on the other hand, is not understood here as an aging effect because acid stratification is an effect that can be expected with every discharge and is in principle completely reversible. However, some of the aging effects described in the following are accelerated by long-term exposure to acid stratification which leads to irreversible damage.

The aging effects discussed in this section are:

- Reduction of the effective surface area of active materials,
- Sulfatation,
- Passivating layers – PCL,
- Shedding or softening of active masses,
- Corrosion of the separator,
- Drying out of the electrolyte (sealed batteries),
- Dendrite formation, and
- Under pressure in sealed cells.

Corrosion of the positive and negative grid and the top lead[32] are electrochemically driven processes and are dealt with separately in section 12.14 due to their importance.

Table 12.7 summarizes the changes in the components of a battery during aging.

32 Head lead is the lead required for terminal bridges and terminals in cells.

Table 12.7 Overview of the components of a lead–acid battery and changes in their properties during aging.

Component	Change due to aging
Cell connector	Corrosion may occur in the event of electrolyte leakage and damage to insulation, an increase in contact resistance if the connection is poorly made
Terminals/group bars, intercell connectors	Positive electrode: Corrosion of the terminals and group bars due to stress corrosion cracking with risk of breaking, flaking of corrosion layers from terminals, corrosion of the terminal lugs
	Negative electrode: comparatively low corrosion attack with flooded batteries, in sealed batteries also stress corrosion cracking with a risk of breaking
Current collectors	Different corrosion rates near and at a distance from the terminal lugs; lugs can become separated from the group bar by corrosion
Passivation layers	Between the active mass and electrolyte, premature capacity loss
	Conditions for the formation of high-resistance layers are complex and can be avoided with the correct choice of alloy components and additives
Active materials	Loss of material due to shedding or softening as a result of irreversible loss of contact zones between crystallites; large sulfate crystals, which can no longer be charged in normal operation and thus reduce capacity, mechanically stress the material; reduction in surface area and an increase in average particle sizes
Electrode/electrolyte interface (charge transfer resistance)	Main reaction: Dependence on the effective surface area and changes in the diffusion properties of the active masses
	Side reaction: Increase in reaction rate due to catalytic properties of alloy components from the materials of the positive electrodes and deterioration of additives of the negative electrode to reduce hydrogen evolution
Separators	Damage may be caused by dendrite growth and corrosion due to the oxidizing properties of lead dioxide
Electrolyte	Possible contamination may occur due to insufficiently purified water

12.13.2 Reduction of the Surface Area of the Active Masses

During the use of the battery, the large effective surface area of the active masses on both electrodes decreases, especially on the negative electrode. This process changes the porous lead sponge to a more compact structure and is slowed down by the addition of organic additives (expanders). The slow degradation of expanders in aqueous sulfuric acid reduces their effect over time. High temperatures accelerate the degradation process of the expanders and thus causes a reduction in the active surfaces.

12.13.3 Sulfatation

Lead sulfate, the active mass discharged, is always produced during discharging and can be completely removed during charging. During deep discharges with high currents and the resulting local deep discharge on the surface of the electrodes (see Figure 12.15) as well as during the formation of acid stratification, lead sulfate accumulates in a specific pattern: on the outside of the surface and at the bottom, where the distance to the terminals is greatest.

Table 12.8 shows measurements of the local content of lead sulfate in various electrode areas of a battery from an off-grid photovoltaic system, where a full charge is rarely, if ever, achieved and long phases of acid stratification must be expected.

Lead sulfate crystallites are in a constant chemical, non-electrochemical dissolution and deposition process in the electrolyte. This leads to large crystallites growing at the expense of smaller crystallites (Ostwald ripening). If batteries are left in a partially or fully discharged state for a long time, this leads to more and more large and fewer lead sulfate crystals overall. The charging process becomes increasingly difficult due to the decrease in the surface area of all crystallites, especially at the end of the charge and at low temperatures. Furthermore, lead sulfate crystals that have already been present for some time are no longer so easily accessible for charging (the so-called hard sulfatation).

Acid stratification, which leads to equalizing currents between electrode areas with high and low acid density and causes a slow, continuous discharge in the area of high acid density, increases the formation of large lead sulfate crystals that can no longer be charged later. Operating strategies that lead to recharging immediately after the end of use are therefore necessary. Increased temperatures at the end of charging can have a positive effect on the complete conversion of lead sulfate.[33]

Residual lead sulfate, which is still present after a "full charge," reduces the capacity.

Table 12.8 Measurements of the lead sulfate content in various electrode areas of an aged lead–acid battery from an off-grid photovoltaic system.

Negative electrode	Top	Middle	Bottom
Cell 2	36.6	40.9	75.3
Cell 3	37.5	41.7	74.0
Positive electrode cell 3	Above	Center	Below
Inside electrode	13.3	17.6	37.3
Outside electrode	55.4	56.0	65.5

33 Sauer [23] uses this example as a warning against viewing increased temperatures as a general acceleration factor for aging tests. According to the view expressed in the article, an increased temperature during charging can also have a life-extending effect in certain operating modes because it reduces the risk of sulfatation.

When charging heavily sulfated batteries, it can happen that the final charging voltage is reached within a very short time immediately after the start of charging and the charger reduces the charging current. The battery is no longer charged. With so-called desulfatation chargers, there is therefore the option of operating the battery at the end-of-charge voltage, possibly even with a slightly higher voltage value, for a very long time until the voltage drops and the charging process can be continued normally. Batteries treated in this way must be classified as damaged, even if their capacity returns to acceptable levels.

Figure 3.7 shows that the voltage of the negative electrode of sealed batteries is lower than that of the reference electrode compared to flooded batteries. An age-related increase in internal resistance and other aging effects, in particular an increase in hydrogen evolution at the negative electrode, reduces the electrode voltage of the negative electrode even further until the electrode voltage is below the $Pb/PbSO_4$ equilibrium voltage. As soon as this occurs, lead sulfate is formed, which can no longer be removed without increasing the charging voltage (equalization charge/boost charge), and the negative electrode sulfates despite trickle charging.

Under good operating conditions, the formation of sulfatation can be avoided.

12.13.4 Passivating Layers – Premature Capacity Loss (PLC)

The term "PCL" refers to the formation of a passivation layer consisting of lead oxide (PbO) on the positive electrode. The capacity of the battery collapses after just a few cycles. Additives to the positive grid, particularly tin, have made it possible to eliminate this effect so that it no longer plays a role in modern batteries.

12.13.5 Shedding and Softening of Active Mass

The porous structure of the active mass is mechanically stressed by the formation of lead sulfate crystals, especially large crystals during deep discharge, long periods in a partially discharged state, and gassing. In addition, depending on the voltage level, the contact zones between particles of the active material may deteriorate. In a post-mortem analysis, the active material is noticeably softer than when new. Particles from the mechanically damaged electrode can fall to the bottom of the cell container (shedding). If positive and negative electrodes touch the dislocated material at the bottom, a (possibly high impedance) short circuit occurs between electrodes within a cell. Design countermeasures include the use of gauntlets, separators designed as pockets or fleece for mechanical stabilization of the surface of the electrode, a firm assembly of the plates using a mechanically stable cell container and, for industrial batteries, a free electrolyte space below the electrodes (sludge space).

A high energy throughput, which leads to the formation of a lot of lead sulfate and frequent charging processes with high gas formation, causes shedding and mass softening to be particularly severe. Sealed batteries are not affected by this aging process to the same extent.

12.13.6 Corrosion of the Separator

Lead dioxide is a very strong oxidizing agent. Separator materials and other plastics can be damaged by this, forming CO_2.

$$1.5PbO_2 + \text{"}CH_2\text{"(from separator)} \rightarrow 1.5Pb + CO_2 + H_2O + \text{damaged separator.} \quad (12.16)$$

Contact between the separator and the positive electrode cannot be avoided by design, because a tight assembly is necessary for many other reasons (minimization of electrolyte resistance, equal distance between the electrodes over the entire surface, mechanical pressure on the electrodes to reduce shedding).

However, separator corrosion normally no longer contributes to the end of life of modern batteries with high-quality separator materials.

12.13.7 Drying Out of the Electrolyte (Sealed Batteries)

Flooded batteries lose water during charging. A low electrolyte level is therefore not an aging effect, but a maintenance problem.

Sealed batteries lose small amounts of water during operation because the internal oxygen cycle cannot completely suppress water loss. At high temperatures and charging voltages, so much water can be lost that the gel or AGM material can dry out. This can lead to contact problems between the electrode and the electrolyte. Parts of the electrode are effectively cut off from the discharging and charging process. Above all, however, the resistance of the electrolyte increases.

Toward the end of the expected service life, drying out of the gel or AGM accelerates the reduction of the internal resistance of the battery.

12.13.8 Dendrite Formation

Dendrites are metallic threads that can form on negative electrodes during charging. This effect is known in many battery systems. In lead–acid batteries, the formation of dendrites only leads to high-resistance short circuits, which discharge the cell or increase the float charge current at a constant float charge voltage. The materials for separators are selected appropriately to avoid dendrite formation.

Dendrite formation can become noticeable through very short, small voltage dips during float charging.

12.13.9 Oxygen Consumption and Formation of Under Pressure in Sealed Batteries

Sealed batteries have a resealable valve that opens in the event of overpressure, whereas flooded batteries have gas vents allowing always an exchange with the ambient air.

Depending on the voltage, oxygen reacts with lead to form lead oxide and lead dioxide and is consumed by the oxygen consumption reaction. If the oxygen required for this is not supplied from the oxygen development of the positive electrode, under pressure is created in sealed batteries. Occasionally, therefore, a corresponding deformation of the battery housing can be observed in sealed batteries.

12.14 Corrosion of the Positive Grid and Connector Lead, Negative Terminals and Intercell Connectors

12.14.1 Corrosion of the Positive Grid

The reason for the corrosion is the coupling of two reactions on the surface of the current collector of the positive electrode in contact with aqueous sulfuric acid:

$$Pb + HSO_4^- \rightarrow PbSO_4 + H^+ + 2e^-. \quad (12.17)$$

This reaction is the main reaction of the negative electrode (see Eq. (12.2)) and has a concentration-dependent equilibrium voltage of approx. −0.4 V.

The coupled reaction leading to corrosion is the discharge reaction of the positive electrode:

$$PbO_2 + HSO_4^- + 3H^+ + 2e^- \rightarrow PbSO_4 + 2H_2O. \quad (12.18)$$

This reaction with a concentration-dependent equilibrium voltage of approx. 1.7 V consumes the charge provided by the anodic reaction (12.1) and thus allows the following continuous reaction:

$$Pb + PbO_2 + 2HSO_4^- + 2H^+ \rightarrow 2PbSO_4 + 2H_2O. \quad (12.19)$$

This overall reaction at the lead grid of the positive electrode is driven by the very high voltage difference of approx. 2.1 V. The anodic reaction (12.16) takes place immediately when lead and sulfuric acid come into contact, but can only continue if the charge carriers produced (protons and electrons) can be transferred by the cathodic parallel reaction (12.17) are consumed. An external current flow is not necessary and the reaction also takes place at rest.

As no external current flows, this is a reversible self-discharge of the PAM, which can be reversed by a charging current, but destroys the current collector, because only the active material PbO_2 can be formed by charging. The increase in active mass leads to an increase in capacity, which is often observed in lead–acid batteries during use.

Due to the coupling of the anodic and cathodic reaction, the lead of the current collector is not thermodynamically stable and there is no way to prevent the process of converting the lead of the positive electrode to active mass. There are no known alternatives to lead as a current collector for the positive electrode. Despite the high voltage difference, the conversion rate is very low due to the kinetic inhibition of the reaction. The reasons for the kinetic inhibition are:

- The current collector has a smooth surface and thus a very low effective surface area, which is a proportionality factor in the BV equation.

- The surface of the current collector is also surrounded by the active mass of the positive electrode so that the availability of sulfuric acid on the surface of the lead grid is reduced.
- The reaction product lead sulfate partially covers the surface and reduces the surface even further. During charging, this forms a protective layer of PbO_2 that tightly covers the current collector. The further reaction can then only continue to grow through diffusion effects of sulfuric acid through the PbO_2 layer and cracks in the protective layer. Cracks are caused, for example, by different expansion coefficients of Pb and PbO_2, resp.

In addition, the local concentration and pH conditions at the interface are different from those of normal battery conditions [40].

The charging voltage required during charging increases the voltage difference between the coupled reactions and thus the corrosion rate. The lead sulfate formed is already converted back into active material in the positive electrode during charging, so the corrosion process is a parasitic side reaction of the charging current that only needs to be taken into account in detailed investigations. In sealed batteries, the voltage of the positive electrode is higher than that of the electrolyte due to oxygen consumption at the negative electrode (Figure 12.5). For this reason, sealed batteries can show higher grid corrosion of the positive electrode than flooded batteries.

The corrosion of lead of the positive electrode in aqueous sulfuric acid without a layer of active mass was investigated by Lander [41, 42]. Figure 12.24 shows his results. There is a very narrow window of low corrosion slightly above the equilibrium voltage of the positive electrode and a strong increase in corrosion at both lower and higher voltages. The reactions in the higher voltage range are shown above. The corrosion attack at lower voltages is much higher, but only occurs at low voltages caused by acid stratification and in a discharged state.

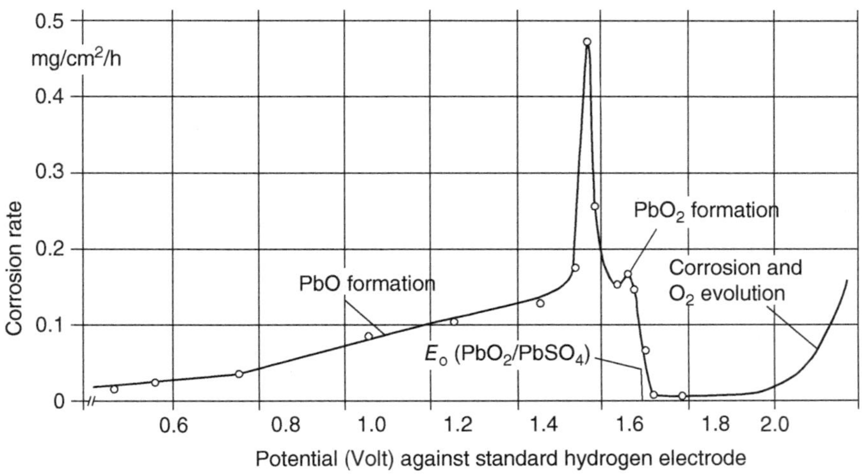

Figure 12.24 Lander curve, voltage-dependent corrosion rate of lead at the positive electrode.

For an explanation of the process below the equilibrium voltage of the $PbO_2/PbSO_4$ reaction [e.g. formation of metallic nonconductive lead oxide (PbO), conversion of PbO to lead dioxide (PbO_2), influence of alloy components, etc.]. With regard to the transferability of Lander's results to grids with coverage of the current collector lead by the PAM, reference is made to further literature (e.g. Ref. [3]).

12.14.2 Effects of Grid Corrosion

As a result of the grid corrosion according to Eq. (12.19), an increase in the capacity of a lead–acid battery can be observed at the beginning of the service life until the electrode loses capacity again, also due to the destroyed grid. The increase in capacity can be up to 20–25%. At the end of the service life, however, the grid of the positive electrode is damaged to such an extent that it can no longer fulfill its function of mechanically stabilizing the active masses and improving the conductivity of the electrode. The high current capability decreases. The examination of corroded grids shows:

- The protective layer of PbO_2 formed by the corrosion process and low float charge currents only grows slowly, but is subject to increasing mechanical stress the thicker it becomes due to different expansion coefficients. Cracks in the protective layer allow more aqueous sulfuric acid to reach the metallic grid and the corrosion process continues, particularly at grain boundaries. The lead sulfate that forms has twice the volume of lead or lead dioxide and builds up mechanical pressure. There is a mechanical deformation of group bars and terminals because the grid expands. Aged batteries with high corrosion show growth of the positive terminals by several millimeters. This leads to deformed lids of batteries if the terminal is firmly connected to the lid, or a clear protrusion of the terminal through the lid if the terminal is installed movably via a seal.
- The frame of the grid at the top of the electrode is broken and bent, in the worst case the broken parts can touch the negative group bar and thus cause a possible low-resistance short circuit in the cell. If the contact occurs above the electrolyte level, there is a high risk of explosion. The same is observed with expanded metal grids, which bend even more due to the growth pressure of the active mass because of their low mechanical strength.
- After removing the active mass from the grid, a large-scale destruction of the grid is observed; the lead of the grid is largely converted into active mass. Sometimes the lattices are so mechanically unstable that they disintegrate even under low mechanical stress.
- Acid stratification leads to significantly higher electrolyte densities in the lower electrode area than in the upper electrode area. If the acid stratification is not removed quickly, this can lead to much more serious and faster destruction of the grid in the lower electrode area than in the upper electrode area.
- Acid stratification leads to a low acid concentration above the electrodes particularly where the metal is not covered by active material, that is, in the area of the

pole lugs. According to the Lander curve, a particularly high corrosion attack is therefore expected in the area of the terminal lugs. The terminals and group bars of heavily corroded batteries can often be easily removed from the electrodes.

Corrosion is also dependent on the alloys used, temperature, and production technology. However, the thermodynamically driven process ultimately limits the service life when all other aging problems are solved. Batteries with a very long service life are therefore built with very thick and massive current collectors. For example, a service life expectancy of 20–25 years can be expected for GroE batteries with their thick lead plates as current collectors.

A side effect of corrosion is the leaching of alloy components from the positive current collector, which are then deposited on the negative electrode in a galvanic process. For this reason, copper cannot be used to improve conductivity on the positive electrode because copper massively accelerates the development of hydrogen on the negative electrode. Antimony used to produce hard grids is leached and leads to higher hydrogen evolution (antimony poisoning). More detailed data can be found in Ref. [5], for example.

12.14.3 Corrosion of the Positive Terminals and Group Bars

The reaction mechanisms described earlier (Eqs. 12.17 and 12.18) lead to particular effects in the area of the top lead (lead of terminals and group bars) of the positive electrodes above the electrolyte level. Due to different local acid densities, the corrosion attack is based on deviations of the local equilibrium voltage in the area of the top lead from the equilibrium voltage of the current collectors determined by the active mass.

It must always be assumed that all surfaces above the electrolyte level are wetted with aqueous sulfuric acid (aerosols during charging due to gassing, the surface tension of aqueous sulfuric acid, consequences of the production process, etc.). The thin film of aqueous sulfuric acid is electrically connected to the potential of the electrolyte between the electrodes. Figure 12.25 shows the reactions shown earlier coupled via an electrolyte film.

The electrolyte film has a very high resistance (e.g. 10 μm thickness, 1 cm width, and 5 cm length correspond to approx. 1 kΩ). The local concentration of the electrolyte is lower in the area of the top lead because of the consumption of sulfuric acid and the formation of water as a result of the reactions taking place in the thin electrolyte film. Therefore, the local Nernst voltage drops. The low diffusion coefficient of sulfuric acid in water does not allow a noticeable or rapid equalization of concentration. Kirchhoff's laws require that the sum of all voltages in a closed circuit is zero. A corrosion current must therefore flow so that the potentials are balanced. However, the current, which discharges the area with the higher Nernst voltage and charges the area with the lower Nernst voltage, is very small due to the high resistance of the electrolyte film.

During charging and trickle charging, lead sulfate is converted into active material which increases the local sulfuric acid concentration. When batteries are opened, it

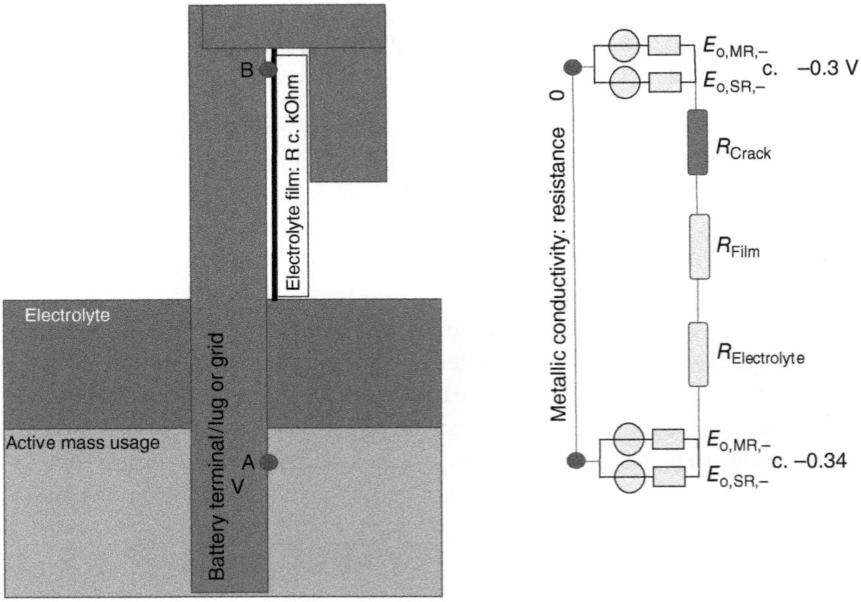

Figure 12.25 Schematic diagram of the local voltage conditions at the group bar compared with the areas below the electrolyte level. Due to the high conductivity of the metallic components, no resistor for the electron current in the lead is shown in this equivalent circuit diagram.

is thus observed that the positive top lead is coated with a layer of corrosion products, in contrast to the top lead of the negative electrodes, where the terminals and group bars have shiny metallic surfaces.

Technical problems are not observed due to this low corrosion attack on the surface compared to the areas below the electrolyte level. However, in batteries with a very long service life for stationary applications, a flexible tube made of plastic often surrounds the positive terminal posts to prevent corrosion layers from flaking off. Flaking of the material can thus no longer fall onto the plates and electrically connect the positive and negative electrodes (high-impedance short circuit).

The corrosion process is problematic if it takes place in cracks in the material, for example, at the point where terminal lugs of the grids are welded to the group bar, at casting defects or grain boundaries. In the first step, the electrolyte film in cracks leads to the formation of lead sulfate. The locally low electrolyte concentration increases the solubility of lead sulfate so that the surface is free to form new lead sulfate. In addition, the increase in volume during the conversion of lead to lead sulfate (approx. a factor of 2) leads to mechanical stress in the crack, which expands the crack and, in conjunction with the sulfuric acid film creeping into the crack, opens up new areas for corrosion attack. This process is known as stress corrosion cracking and is also observed in many areas of mechanical engineering.

Stress corrosion cracking can cause major damage, for example, individual electrodes can be electrically separated from the group bar and then no longer participate

Figure 12.26 Section through the positive group bar of a starter battery after a standard corrosion test. For some electrodes the contact to the group bar is completely intact, for others a corrosion attack along the terminal lugs and further corrosive destruction of the group bar can be seen. Source: Courtesy of E. Cattaneo.

in discharging and charging processes. Figure 12.26 shows a section through a group bar of a starter battery with remnants of terminal lugs from the individual electrodes.

Stress cracking processes can also take place below the electrolyte level, but less quickly due to the higher electrolyte density compared to an electrolyte films.

To reduce stress corrosion cracking, terminal lugs are therefore often immersed in tin before welding – albeit an expensive process. In the prevailing voltage range, tin is stable against aqueous sulfuric acid and no further corrosion occurs.

12.14.4 Corrosion of the Negative Grids, Terminals and Group Bars

When charging lead–acid batteries in trickle charge mode, the potential of the negative electrode is above the $Pb/PbSO_4$ equilibrium voltage. Lead sulfate, generated during discharge and in short phases below the equilibrium voltage, can be completely converted back into lead during charging. In flooded lead–acid batteries, no corrosion of the negative grids is observed, although there are reports of corrosion of the negative terminal lugs in operating profiles similar to start-stop operation. However, the assessment of the relevance of this failure mechanism is controversial [43, 44].

In sealed lead–acid batteries, on the other hand, there are corrosion problems due to the reduction in the voltage of the negative electrode caused by the oxygen consumption reaction (Figure 12.4).

- If the float charge voltage of a cell falls below the $Pb/PbSO_4$ equilibrium voltage despite careful setting of the charging voltage for the battery string, sulfatation will occur. This effect can be reversed by regular boost charging. Corrosion of the negative grid therefore does not occur even with sealed batteries.
- Analogous to the corrosion of the positive group bars and terminals, there is a reduction in the local voltage level at the terminal lugs, group bars, and terminals in sealed batteries above the electrolyte immobilized in gel or AGM, which leads to corrosion attack. The equivalent circuit diagram shown in Figure 12.25 applies accordingly. Wetting of the terminal lugs, group bars, and terminals of the

negative electrode is unavoidable because surface tension forms an electrolyte film, which then allows the formation of water through the oxygen consumption reaction. Due to the low acid concentration in the surface film, the equilibrium voltage is significantly reduced compared to the current collector covered with active mass. It is not always possible to keep this low local potential constantly above the local $Pb/PbSO_4$ equilibrium voltage during trickle charging because the resistance of the electrolyte film is too high. Corrosion attacks occur, particularly in gaps and cracks, which lead to stress corrosion cracking, just as with positive group bars and terminals.

In flooded batteries, this type of corrosion attack should only be expected if there is a large loss of water and a low electrolyte level.

One way to reduce this corrosion attack is to reduce the resistance of the electrolyte film in the areas where the grid is not covered with active mass, for example, on the lugs. Corrosion can be reduced by covering the parts at risk with materials soaked in aqueous electrolyte with the nominal density of the electrolyte and by tinning the terminal lugs.

12.14.5 Explosion Risk

If the corrosion attack is very strong, both the terminal lugs of the negative and positive electrodes can detach from the group bar. This takes place above the electrolyte level, because only there is a correspondingly large corrosion attack. The voltage of an electrode that is no longer connected to the group bar assumes the equilibrium potential and differs slightly by approx. 0.1–0.2 V depending on the current amplitude. This voltage difference is not sufficient to trigger a spark. However, the capacitive and/or inductive energy stored in the detaching electrode may be so high that it is sufficient as an ignition source for the hydrogen/oxygen gas mixture in this area, especially if the current amplitude is very high at the same time. It should also be noted that high current amplitudes, for example, during a starting process, also exert mechanical forces on all current-carrying parts and promote breakage. It has actually been observed that the risk of a battery explosion is particularly high during inrush currents and starting processes.

Countermeasures are, for example, a sufficiently high electrolyte level in flooded batteries, which covers the critical areas in all operating states and also during aging.

12.15 Corrosion of the Intercell Connectors

Intercell connectors are subject to massive corrosion attacks if a sulfuric acid film connects the electrolyte volumes of neighboring cells and wets the intercell connector. Figure 12.27 shows this schematically.

The diagram shows that the positive electrode of one cell is connected to the negative electrode of the neighboring cell with low resistance via the metallically conductive intercell connector and that both electrodes have the same potential. The electrolyte film along the intercell connector electrically connects the electrolyte volumes of the neighboring cells, which have a voltage difference of

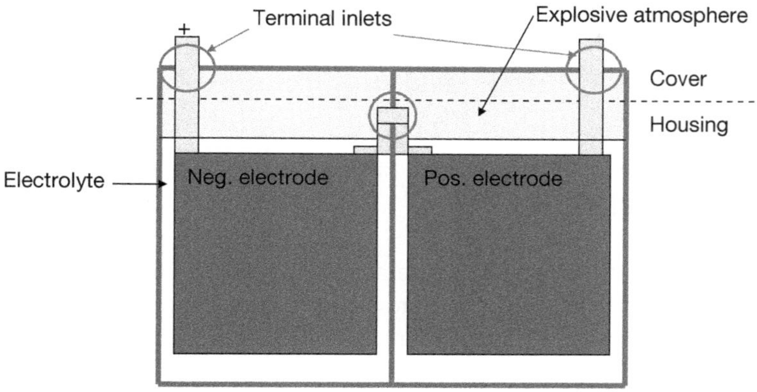

Figure 12.27 Schematic diagram of a two-cell block battery with intercell connector (top) and section of the intercell connector with equivalent circuit diagram (bottom), in which voltage sources with BV resistance are shown at four points to illustrate the electrochemical reactions taking place at the contact surface.

approx. 2.1 V at rest. The short circuit between the cells has a very high impedance due to the high specific resistance and the thickness of the electrolyte film. A significant self-discharge is therefore not expected.

The electrolyte film in contact with the lead of the intercell connector enables electrochemical reactions. The potential between the intercell connector and the electrolyte film spans the entire voltage difference of the electrolyte volumes in the neighboring cells. The local potential along the intercell connector, local reactions, and corrosion rates correspond to the values shown in the Lander diagram (Figure 12.24).

The effects of corrosion are a reduction in the cross-section and therefore an increase in the resistance of the intercell connector, for very advanced corrosion a significant voltage drop across the intercell connectors with high heating power in the case of high currents. If the intercell connector only has one high-resistance contact point and breaks at high current amplitudes or current pulses,[34] an arc may occur in the worst-case scenario, causing an explosion. The observation that starter and UPS batteries explode particularly frequently when switched on at high currents can therefore be explained not only by terminal lug corrosion but also by the intercell connector breaking.

Corrosion of the intercell connector cannot be prevented once an electrolyte film has formed parallel to the intercell connector. In addition to monitoring the production processes, some manufacturers use additional sealing rings during production and others place the terminal of each cell in an area separated from the electrolyte, where a conventional connection is made to avoid this corrosion attack and its effects. By arranging the intercell connector in such a way that the electrolyte level is permanently above the intercell connector, an explosion can be avoided if an intercell connector breaks.

34 Large current changes lead to significant mechanical forces.

It is difficult to detect this fault during maintenance work and an increase in internal resistance of a cell cannot be clearly assigned to intercell connector corrosion.

12.16 Operating Strategies and Design Implications for Lead–Acid Batteries

From the earlier considerations, general recommendations for operating strategies and design measures can be derived to ensure a long service life and reliability. The energy or charge throughput of a battery is essentially determined by the requirements of the loads and the design of chargers (maximum power, availability of charging energy in energy systems without grid connection). The effects on the service life and security of operation depend on the choice of battery (type and capacity) and the operating strategy. To achieve a long service life, the following aspects must be observed for all applications, which sometimes have to be weighed up against each other and affect the ratio of operating and investment costs.

- Maintenance
  Maintenance requirements must be planned carefully to ensure serviceability.

Some battery types are marketed as maintenance-free, but this only ever refers to the fact that they do not need to be topped up with water. It must be possible to carry out work required for the diagnosis and inspection of batteries during operation without any special equipment, the need for dismantling all or part of the battery, and without endangering occupational safety.

- High temperatures must be avoided in all operating phases.
  Design measures for a uniform cell temperature are important.
- Discharge current
  Loads and connected power electronics should not cause high and short current pulses.
- Depth of discharge
  The number of cycles until the end of service life and the total amount of charge that can be drawn from a battery decreases with increasing depth of discharge. Choosing cells with a higher capacity therefore increases the service life, although weight, costs, and space requirements increase accordingly. Cells with a higher capacity require more charging power to achieve a full charge and to remove acid stratification, which means that more powerful chargers are required and the energy generation in autonomous energy systems must be dimensioned accordingly.
- End of discharge
  For discharges with current, which fluctuates little, and only short interruption during the discharge, it is sufficient to define falling below a voltage limit as end-of-discharge criterion. In the case of discharges with strongly fluctuating currents, frequent and possibly long interruptions, and possibly even opportunity charging, a fixed voltage limit can end the discharge prematurely (voltage drops

shortly below the limit at short high current pulses), but can also lead to very deep discharges (low discharge current possibly with interruptions when the state of charge is already low). In these cases, it is advisable to use the state of charge of the battery as the end-of-discharge criterion.

- Full charge
  After discharging, lead–acid batteries must be quickly returned to a fully charged state with complete conversion of lead sulfate and, in the case of flooded batteries, removal of acid stratification. However, this does not mean that batteries that undergo short cycles with a low depth of discharge several times a day must be charged with long phases at high charging voltage above the float charge voltage, as this leads to high charge factors, long times at high charge voltages and high temperatures.
  The authors are not aware of any studies on the most favorable time intervals for achieving full charge that also takes application restrictions into account.
- Partially discharged state
  The partially discharged state, in particular cycling in the partially discharged state, must be avoided if at all possible due to the formation of large lead sulfate crystals, the development of large differences in acid concentration, and for reasons of corrosion. Particularly in the case of batteries under cyclic operation, where charging is terminated when the final charging voltage is reached or the constant voltage or constant current charging phases are too short, a very strong acid stratification forms, which is then very difficult to remove. The aim must be a full charge and not a very high state of charge.
- Acid stratification (flooded batteries)
  The formation of acid stratification cannot be avoided. Some auxiliary equipment makes it possible to reduce the concentration gradient in the battery and the acid stratification can be removed gently.
  - Electrolyte circulation for batteries in cyclic operation and also for unavoidable cycling in a partially discharged state.
  - Mammoth pump principle for batteries that are moved during use (e.g. starter batteries).
  - Installation direction of the battery in vehicles.
  - Appropriate design of the battery (Task 12.6).

For applications where the removal of acid stratification is difficult due to application-specific restrictions, the use of sealed batteries should be considered as an alternative.

- Trickle or float charge mode
  Batteries in trickle charge mode must be operated continuously at voltages above the equilibrium voltage to convert the lead sulfate produced by corrosion back into active material and to minimize the corrosion attack according to the Lander curve. The float charge voltage must be set as low as possible, preferably with temperature compensation. To compensate for differences between cells, a boost charge should be carried out at regular intervals. Short phases of high charging voltages are less damaging than trickle charge voltages without sufficient charging current.

- Type of charger
  The current assignment of the charger (maximum current in relation to the nominal capacity) must not be too low (at least approx. 10 A/100 Ah). In complete systems, a battery often supplies additional consumers so that the charging current of the battery is only partially available. It then becomes necessary to calculate the energy balance to determine the required charging current.
  Some standards suggest values in the range of approx. $4 \cdot I_n$, corresponding to 20 A/100 Ah for a 20-hour nominal capacity. This value should not be understood as an upper limit but as a guide value. Many manufacturers also do not limit the current amplitude for initial charging. UPS batteries and starter batteries are charged with current amplitudes of up to approx. 100 A/100 Ah.
  Chargers that provide low float charge current as an average by pulses with the mains frequency lead to higher heating and stress on the active masses (shedding) compared to chargers with highly smoothed charging currents.
- Charging characteristics
  Only standard characteristic curves should be used and nonstandard characteristics should be avoided without prior study of the effects on the function and service life of batteries.
- Resistance of charging cables
  The voltage drop along the charging cables is relevant when high currents and long cables are used. Chargers therefore often allow a mathematical correction of the voltage by entering the length and cross-section of the cables. A four-point connection for direct measurement of the voltage at the terminals is recommended in some standards.
- Storage time
  Lead–acid batteries should always be in trickle charge mode. Where this is not possible, a full charge should be carried out at intervals of several months so that self-discharge effects can be compensated for. However, the storage time, regardless of the temperature, leads to irreversible aging processes.

How and whether these requirements can be implemented in the various applications is addressed again in chapters 16–20.

12.17 Determination of Battery States

12.17.1 State of Charge

Determining the state of charge of lead–acid batteries is more complex than with other battery systems. This is because the open-circuit voltage is only linked to the state of charge in rare operating cases and after long waiting times. In addition, side reactions, especially at the end of the charge, distort the measurement of the charge throughput, along with other measurement problems. Pure charge balancing ("Coulomb counting") is therefore not possible. Figure 12.28 schematically shows

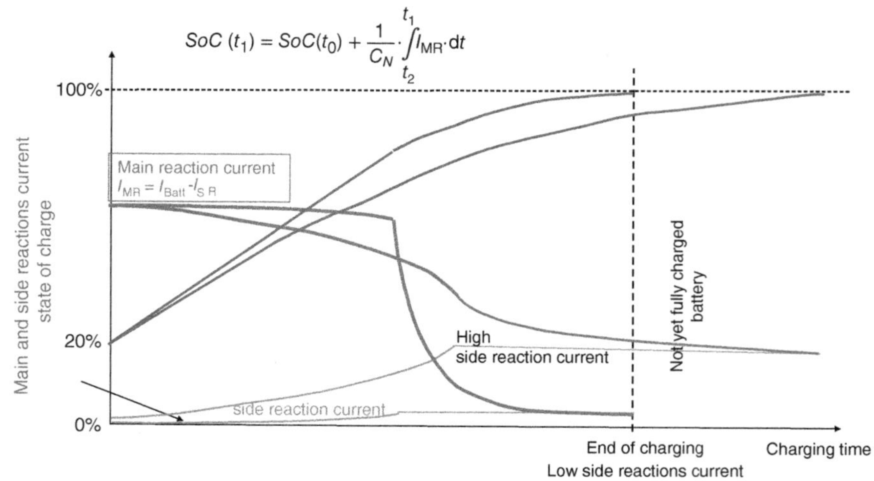

Figure 12.28 State of charge and current curve when charging two batteries with different side reaction currents. The longer time to reach full charge with the same battery current can be clearly seen.

the current and voltage curve of a battery with an IU charging characteristic with very different side reaction rates. In one battery, the side reaction rate is approximately five times higher than in the other, which corresponds to the difference between a new and an aged battery. Due to the faster decrease of the main reaction current in the aged battery, it takes longer to reach the end-of-charge voltage. Consequently, the time to full charge, as well as the amount of charge and total energy required for charging, is also higher.

Various methods are used to determine the state of charge.

- Voltage measurement
  In the case of cyclic operation, in which full charge is reached frequently, sometimes even daily, the state of charge is often only of interest during discharging, because charging "always" leads to a full charge in the application.
  Very simple discharge indicators only use the battery voltage during discharging and average the voltage values over a certain period of time. These very inexpensive devices are surprisingly efficient under cyclic operation with a daily full charge.
- Voltage measurement with estimation of side reactions
  If a good state of charge determination is also necessary during charging, the proportion of the side reaction current in the battery current is estimated using the electrochemical characteristic values of the gassing reactions. It can be calculated from the battery voltage and the current during float charging. Changes as a result of aging processes are implicitly taken into account.
- Use of the OCV characteristic curve
  After long rest periods (approx. 12–24 h) and after removal of the acid stratification, the open-circuit voltage of the battery corresponds to the state of charge.

Due to the difficulties in meeting all requirements of the method during operation, algorithms are used with which the open-circuit voltage can be deduced from the voltage curve over time during rest[35] periods. It is doubtful whether this method can really deliver good results.

- Comparison with discharge curves
  Discharge curves of the battery from laboratory measurements, which show the voltage expected at a certain current as a function of the amount of charge that has been drawn respectively the state of charge, make it possible to estimate the state of charge when measuring current and voltage in phases without a very high dynamic load.
- Algorithms using expert rules
  In these methods, expert rules are used to correct the state of charge, which was estimated by voltage measurement and current integration, possibly coupled with a correction for side reaction currents [45]. An example of such rules is the knowledge that a battery must have, depending on the charging current, a state of charge of at least 85%, but no more than 95% when the end-of-charge voltage has been reached. These methods are very powerful and require neither an initial state of charge nor information about the battery capacity as input.

In every case, the calculated state of charge value must be reset to 100% when the full charge status is detected.

12.17.2 Capacity or State of Health

Knowing the capacity of a fully charged state is of great importance for all applications. Under cyclic operation, it is often very easy to determine the battery capacity. The decrease in utilization time is sometimes clearly noticeable, for example, in industrial trucks that perform similar tasks every day. A simple analysis, possibly measuring the amount of charge removed up to the end of discharge, is possible without major errors. By comparing the amount of charge removed since the last full charge, the discharge current, and the voltage with stored discharge characteristics, it is possible to estimate not only the remaining capacity but also the capacity of the fully charged battery.

It is difficult to assess the capacity of a battery in trickle charge mode without prior discharge. Possible diagnostic methods for this, using internal resistance measurement are presented in chapter 19.

For batteries where long operating phases without a full charge are unavoidable, especially batteries in renewable energy systems, the capacity must be estimated based on models using expert rules (e.g. Ref. [45]).

12.18 Safety

In addition to the general electrical hazards (see Appendix), there are three particular hazards associated with lead–acid batteries:

35 That is without control or monitoring units, etc. that are supplied by the battery, etc.

- Gas evolution
  Risk of explosion due to the formation of oxyhydrogen gas, both in flooded and sealed batteries.
  Under very rare circumstances it has been reported that hydrogen evolution can also lead to the formation of H_2S and other components (H_3As, H_3Sb). Little is known under which circumstances these substances which are all very toxic, may develop. Good ventilation which is required as protection against explosions is likely to dilute the substances below dangerous levels.
- Burns caused by aqueous sulfuric acid.
- Handling lead as a toxic heavy metal.

In addition to the marking of batteries and battery systems as electrical hazards, all other hazard symbols that must be used are based on these three hazards.

12.18.1 Risk of Explosion Due to Oxyhydrogen Gas

The formation of oxyhydrogen gas cannot be prevented, even by using sealed batteries with an internal oxygen cycle. It can be assumed that an explosive gas mixture always exists inside cell housings above the electrolyte level, and that explosive gas is emitted or could be emitted from all cell openings.

There are no reliable and up-to-date figures on the number of batteries that explode. In a 2002 press release, the Federal Institute for Risk Assessment in Germany spoke of 175 incidents involving lead–acid starter batteries in Germany in 2001, most of them described as "gas deflagrations" (a slow speed, mild detonation) or explosions. Batteries for emergency power systems also explode time and again. Damage is caused by sharp-edged polypropylene parts, which can be produced when the cell housing is destroyed and ejected at high speed, along with escaping acid and the pressure wave, which can lead to massive damage to buildings.

The high risk of damage requires compliance with very high safety standards with regard to ventilation of the installation room, avoidance of ignition sources and maintaining a minimum distance to openings in cell or block housings.

Hydrogen can no longer explode at a concentration of more than 75% hydrogen in air or less than 4%. To avoid the risk of explosion, dilution to at least 4% with a safety factor of 4 is therefore required.

Sealed batteries emit less hydrogen than flooded batteries with the same capacity because the formation of hydrogen is largely suppressed by the oxygen cycle during normal operation. The formulas for calculating the ventilation requirements explained in the Appendix are therefore adjusted by a factor. However, it is advisable to size ventilation systems for flooded batteries so that the gassing rate of a sealed battery which is comparable to that of a flooded battery under some fault conditions is also taken into account. No safety problems will then arise when replacing a sealed battery with a flooded battery.

DIN EN 62485-5:2019-02 summarizes the requirements for ventilation systems. These are described in detail in the annex when dealing with the safe and environmentally compatible handling of batteries.

Design measures to reduce the risk of explosion are:

- Design of the cell plugs to minimize electrostatic discharge into the cell interior.
- Preventing sources of ignition that may cause an explosion inside the battery.
  An external spark in the immediate vicinity of a cell opening can cause the oxyhydrogen gas located there to explode. Under good ventilation conditions, the amount of gas outside the cell vessel is generally so small that hardly any damage needs to be expected. If the inside and outside of the cell are not separated by a frit with backfire-inhibiting properties,[36] an explosion can be caused inside the cell.
- Areas of high corrosion risks under the electrolyte level.
  An ignition spark within the electrolyte cannot cause an explosion because the concentration of hydrogen and oxygen dissolved in the electrolyte is too low. Therefore, if the electrolyte level is permanently so high that all intercell connectors and group bars are below the electrolyte level, they cannot cause an explosion.

12.18.2 Aqueous Sulfuric Acid

Required protective measures exist in many occupational health regulations and are summarized, for example, in DIN EN 62485-5:2019-02. No distinction is made between sealed and flooded batteries, as sealed batteries can also cause chemical burns if the battery housing is damaged. The use of protective gloves and protective aprons is mandatory. A source of clean water in the form of a tap or a sterile water container must be provided in the immediate vicinity of the battery, for example, in the entrance area of the battery room, or a portable disposable eye shower must be carried in the toolbox during maintenance work.

- Aqueous sulfuric acid irritates the skin, but can be easily rinsed off with plenty of water and rarely causes health problems in the concentration range used in batteries. However, a doctor must be consulted if skin irritation persists.
- Aqueous sulfuric acid in contact with the eyes is a serious problem. If aqueous sulfuric acid gets into the eyes despite wearing safety goggles, rinse as quickly as possible with plenty of clean water over a long period of time and consult a doctor.
- Aqueous sulfuric acid is corrosive and attacks most materials. Protective clothing made of polypropylene is advisable; normal textiles can only be protected from destruction if they are washed off very quickly with diluted caustic soda.
  Very large quantities of aqueous sulfuric acid can flow onto the floor if cells are overfilled or if a container breaks (material defect, explosion, etc.). Without protective measures, the floor and the concrete underneath can be destroyed and lead to costly construction work. Many battery rooms therefore have tiled floors or special coatings or the battery rack is placed in a drip tray. Regardless of the design of the floor, however, the requirements for the permissible resistance range must always be complied with.

---

36 Inserts made of highly porous material that increases the free path length of gas molecules.

- Aqueous sulfuric acid that is sprayed into switch cabinets, a car, or similar products due to a battery explosion can lead to total destruction because corrosion attacks in the entire cable and wiring system must be expected.

12.18.3 Handling Lead

Lead is a toxic heavy metal. The risks when handling the final product are low compared to the risks and precautions during production. It is sufficient to refrain from smoking during work (which is completely prohibited due to the risk of explosion), avoid drinking or eating, and wash up after work. Nevertheless, in European lead–acid battery factories, it is common practice for employees to have their blood lead levels checked regularly and pregnant women are prohibited from working on the factory floor.

Damp cloths used for cleaning batteries and removing acid splashes are commercial waste and do not need to be disposed of separately. However, as soon as lead parts are removed, for example, by removing corrosion products from terminals, the cleaning agents must be disposed of as hazardous waste.

After a battery accident (fire, massive overcharging, explosion, etc.), the entire battery, including all materials used for absorbing acid, must be treated as hazardous waste.

12.19 Battery Problems

Lead–acid batteries are technically mature and robust products with high and consistent production quality. However, occasional production defects, such as leaking vent plugs, blowholes in casting processes (group bars and terminals), and welding problems when connecting terminal lugs and group bars, are unavoidable given the high production volumes. These aspects are not discussed further in the following.

In many systems, batteries are the components that react most sensitively to problems with the design of the overall system and faults in other components or their operation. In failure statistics from automotive associations (AA, ADAC, etc.), batteries are at the top of the list for vehicle breakdowns. However, a closer analysis shows that alternator faults, usage errors and special application conditions have led to battery failure. Faults specific to batteries, on the other hand, are very rare.

Many cases of battery failure and accidents are caused by insufficient charging (charging processes with too low voltage and too short a time, sometimes followed by too low trickle charge voltage), overcharging (too high voltages, long charging phases after short, shallow discharges) or infrequent full charges (no complete conversion of all lead sulfate into charged active material and removal of acid stratification). The aim of minimizing the risk of corrosion through "gentle" charging comes at the cost of insufficient charging and sulfatation.

Furthermore, determining the condition of the battery during maintenance is borderline in many cases. It is difficult to understand why batteries in multi-redundant power supply systems for the protection of systems and high property values as well as the protection of human life receive only a minimum of monitoring and maintenance, and why proposals from standards bodies for the testing of batteries, for example, IEC TR 62060:2001, are ignored. In contrast, most other battery systems are monitored extensively and are fitted with battery management systems as a matter of course.

Monitoring and maintenance aspects are discussed further in the description of the applications.

The problem of insufficient knowledge has hopefully been addressed in this book. One of the authors has (April 2024) been made aware that batteries in a professional environment are discharged at times to "get rid of the memory effect" and FSB batteries, designed for start-stop applications in vehicles, are likely to be charged no more than 4000 hours during their whole lifetime are used as standby batteries without topping up of water and now have failed after three years of use with some even have exploded.

References

1 Sauer, D.-U. (2000). *Optimization of the use of lead-acid accumulators in photovoltaic hybrid systems with special consideration of battery ageing (Original German title: Optimierung des Einsatzes von Blei-Säure-Akkumulatoren in photovoltaik-Hybrid-Systemen unter spezieller Berücksichtigung der Batteriealterung)*. Dissertation. University of Ulm.

2 Meissner, E. (1999). How to understand the reversible capacity decay of the lead dioxide electrode. *Journal of Power Sources* 78: 99–114.

3 Pavlov, D. (2011). *Lead-Acid Batteries Science and Technology*, 1e, 260–266. Elsevier.

4 Meissner, E. (1997). Phosphoric acid as an electrolyte additive for lead acid batteries in electric-vehicle applications. *Journal of Power Sources* 7: 135–150.

5 Berndt, D. (1993). *Maintenance Free Batteries*. New York: Wiley.

6 Fischer, D., Lohner, I., and Mauracher, P. (1996). *Battery Management – Increasing the Reliability of UPS (Original German title: Batteriemanagement - Erhöhung der Zuverlässigkeit von USV). etz 4*. Berlin: VDE-Verlag GmbH.

7 Schmidt, H. and Siedle, C. (1994). *Which ChargeEQualizer for Which Application? (Original German title: Welcher ChargeEQualizer für welche Anwendung?)* VDI Report No. 1168. Düsseldorf: VDI-Verlag.

8 Siedle, C. (1997, 1998). *Comparative investigation of charge equalization devices to improve the long-term performance of multicell battery banks (Original German title: Vergleichende Untersuchung von Ladungsausgleichseinrichtungen zur Verbesserung des Langzeitverhaltens vielzelliger Batteriebänke)*. Dissertation. University of Karlsruhe. Düsseldorf: VDI-Verlag.

9 Hostetler, E., Duchateau, K., and Waterhouse, R. (1993). *Measurement of Oxygen Transport Through Separators in Flooded Batteries*, 51. The Electrochemical Society, Meet. Abstr. MA2020-01.

10 Meissner, E. (2017). Oxygen cycle with flooded-electrolyte lead-acid batteries: quantification of oxygen transport rate in case of starter-type battery designs. Workshop “High-Temperature Durability Tests for Advanced Lead-Acid 12-V Batteries”. Eberbach Monastery.

11 Döring, H., Jossen, A., Köstner, D., and Garche, J. (1995). PV operation and failure mode of different storage batteries and recommendations for operation management (Original German title: PV-Betrieb und Fehlermode verschiedener Speicherbatterien und Empfehlungen für die Betriebsführung). In: *10th Symp. Photovoltaic Solar Energy*, 549–553. Bad Staffelstein.

12 IEC TR 62060:2001 (2001). *Secondary Cells and Batteries – Monitoring of Lead Acid Stationary Batteries – User Guide.*

13 Bashtavelova, E. and Winsel, A. (1996). The PbO_2 agglomerate-of-spheres: investigation of four grid alloys. 5th European Lead Conference, Barcelona.

14 Geuer, W. (1992). *Investigations on the aging behavior of lead accumulators (Original Germna title: Untersuchungen über das Alterungsverhalten von Blei-Akkumulatoren).* Dissertation. RWTH Aachen.

15 Winsel, A., Voss, E., and Hullmeine, U. (1989). *The Agglomerate of Spheres Electrode (Original German title: Die Kugelhaufenelektrode).* Bd. 121, Frankfurt am Main: Dechema Monographie.

16 Edwards, B.E. and Schmitz, C. (2000). Theoretical calculations for using positive electrode compression to increase lead-acid battery life. *Journal of Power Sources* 85: 63–71.

17 Stauffer, D. and Aharony, A. (2017). *Introduction to Percolation Theory*, 2e. London: Taylor and Francis.

18 Karden, E. (2001). *Using low-frequency impedance spectroscopy for characterization, monitoring and modelling of industrial batteries.* Dissertation: RWTH Aachen University.

19 ZVEI Leaflet 03 (2011). *Requirements for Electrolyte and Refill Water for Lead-Acid Batteries.*

20 Euler, J. and Horn, L. (1965). Current distribution over the length of accumulator plates made of tubular elements (Original German title: Stromverteilung über die Länge von Akkumulatorenplatten aus rohrförmigen Elementen). *Archiv für Elektrotechnik* 50: 85–90.

21 Warlimont, H., Hofmann, T. and Gelbke, M. (2004). *Grid for an Electrode of a Lead Accumulator.* European patent EP1738426B1.

22 Lülsdorf, A. and Mathes, S. (2008). Motive power cell connectors – comparison in terms of performance, cost and safety. Proceedings of the 8th European Lead Acid Battery Conference, Warsaw.

23 Sauer, D.U. and Wenzl, H. (2008). Comparison of different approaches for lifetime prediction of electrochemical systems – using lead-acid batteries as example. *Journal of Power Sources* 1176: 534–546.

24 Varta Batterie, A.G. (1986). *Lead-Acid Accumulators (Original German title: Bleisäureakkumulatoren).* Düsseldorf: VDI-Verlag.

25 Kappus, W. (1983). Homogeneous nucleation, growth and recrystallization of discharge products on electrodes. *Electrochimica Acta* 28: 1529–1537.
26 Schulte, D., Sauer, U., Ebner, E. et al. (2014). "Stratifiability index" – a quantitative assessment of acid stratification in flooded lead acid batteries. *Journal of Power Sources* 269: 704–715.
27 Ebner, E., Wark, M., and Börger, A. (2011). Passive mixing elements for electrolyte convection in lead-acid wet batteries (Original German title: Passive Mischelemente zur Eletrkolytkonvektion in Blei-Säure-Nassbatterien). *Chemie Ingenieur Technik* 83: 1–9.
28 Ebner, E., Metge, I., Keil, S.N., and Börger, A. (2013). Electrolyte convection in lead-acid wet batteries – testing of passive mixing elements (Original German title: Elektrolytkonvektion in Blei-Säure-Nassbatterien - Erprobung von passiven Mischelementen) in vehicles. *Chemie Ingenieur Technik* 85: 1294–1301.
29 Klein, F. (1989). The lead-acid battery: improving economic efficiency through peripheral measures (Original German title: Die Bleibatterie: Verbesserung der Wirtschaftlichkeit durch periphere Maßnahmen). In: *Elektrostraßenfahrzeuge – Technik, Entwicklungsstand und Einsatzbedingungen*, vol. 255 (ed. D. Naunin and C. Bader). Kontakt & Studium. Böblingen: Expert-Verlag.
30 Desmettre, F., Matteria, F., Malbranche, P., and Métais, S. (1997). *Qualibat – Publichable Final Report*. JOR 3-CT97-0161. GENEC.
31 Symons, P.C. (1996). Optimizing the performance and life of batteries in continuous power systems. Proceedings of the Power Quality solutions/Alternative Energy Conference, Las Vegas, 169–177.
32 DIN 14679:2008-03 (2008). *Firefighting – Chargers for Trickle Charging of Starter Batteries and Auxiliary Batteries for Special Applications – Requirements and Testing.*
33 ZVEI Information Sheet 32 (2016). *Charging Lead Starter Batteries.*
34 Kordesch, K.V. (1997). Charging methods for batteries using the resistance-free voltage as endpoint indication. *Journal of the Electrochemical Society* 119: 1053–1055.
35 Haaben, K. (1989). Charging methods in the military sector (Original German title: Lademethoden im militärischen Bereich). 6th International Technical Symposium on Emergency Power Supply with Batteries, Munich.
36 Krämer, K. (1996). Battery energy storage Bewag (Original German title: Batterie-Energiespeicher BEWAG). In: *Batterie in Energiespeicher in der Elektrizitätsversorgung* (ed. H.J. Haubrich). Verlag Mainz: Wissenschaftsverlag Aachen.
37 Hullmeine, U., Winsel, A., and Voss, E. (1989). Effect of previous charging/discharging history on the capacity of the $PbO_2/PbSO_4$ electrode: the hysteresis or memory effect. *Journal of Power Sources* 25: 27–47.
38 Vutetakis, D.G. and Wu, H. (1992). The effect of charge rate and the depth of discharge on the cycle life of sealed lead-acid aircraft batteries. Proceedings of the 35th Power Sources Symposium, Cherry Hill, New Jersey, June 21–25, 103–105.

39 Valeriote, E., Nor, J. and Ettel, V. (1991). Very fast charging of lead-acid batteries. 5th ILZRO International Lead-Acid Battery Seminar, Vienna, Virginia.

40 Hu, H. (1993). Experimental study of the alkalization model for the $PbSO_4$ anodic layer on lead/acid positive-plate grids. *Journal of Power Sources* 46: 263–267.

41 Lander, J.J. (1951). Anodic corrosion of lead in H_2SO_4 solutions. *Journal of the Electrochemical Society* 98: 213–219.

42 Lander, J.J. (1951). Some preliminary studies of positive grid corrosion in the lead acid cell. *Journal of the Electrochemical Society* 98: 220–224.

43 Börger, A. and Keil, S.N. (2018). On the field relevance of negative current arrester corrosion due to micro-hybrid operation in vehicles (Original German title: Über die Feldrelevanz von negativer Stromableiterkorrosion durch Mikro-Hybrid-Betrieb im Fahrzeug). *Chemie Ingenieur Technik* 90: 272–276.

44 Takeuchi, T., Sawai, K., Tsuboi, Y. et al. (2009). The partial state-of-charge cycle performance of lead-acid batteries. *Journal of Power Sources* 189: 1190–1198.

45 Kaiser, R.J. (2005). *Optimization of a battery management system for use in photovoltaic power supply systems (Original German title: Optimierung eines Batteriemanagementsystems zum Einsatz in photovoltaischen Stromversorgungssystemen).* Dissertation. University of Ulm.

Further Reading

Bode, H. (1977). *Lead Acid Batteries*. New York: Wiley.

DIN 41772: 1979-02 (1979). *Power Converters; Semiconductor Rectifiers, Shapes and Abbreviations of Characteristic Curves.*

Euler, J. (1971). On the determination of the effective electrical resistance of the grids in lead accumulators (Original German title: Zur Ermittlung des effektiven elektrisdhen Widerstands der Gitter in Bleiakkumulatoren). *Archiv für Elektrotechnik* 54: 122–126.

Pavlov, D., Monakhov, A., Kirchev, D., and Valkovsky, D. (2006). Thermal runaway in VRLAB – phenomina, reaction mechanisms and monitoring. *Journal of Power Sources* 158: 689–704.

Tasks

Task 12.1

Sizing of active materials and specific energy

When designing a battery, it is important to know how much active material and aqueous sulfuric acid has to be provided so that, at the end of discharge, the electrodes are still sufficiently electron conducting and the electrolyte still

(Continued)

Task 12.1 (Continued)

sufficiently ion conducting. Use a density of 1.28 g/cm^3 for the fully charged battery and 1.15 g/cm^3 at the end of discharge. Assume a mass utilization of 40%. Data for molar weights:

- Lead: 216 g/mol
- Lead dioxide: 240 g/mol
- Aqueous sulfuric acid
  1.28 g/cm^3: 4.876 mol/l
  1.15 g/cm^3: 2.507 mol/l
- Concentrated sulfuric acid H_2SO_4: 108 g/mol

Questions

1) How much active material and aqueous sulfuric acid has to be used?
2) How much does the specific energy of the system differ from the theoretical specific energy?
3) How will the specific energy change when making the calculation for a sealed battery?
4) What changes will occur when taking passive components into account?

Task 12.2

Calculate the resistance of an electrode made of tubular plates with and without active material. Conventional tubular plate electrodes have a diameter of approx. 8 mm, consisting of a cast lead rod in the middle with a diameter of approx. 1.5 mm surrounded by lead dioxide.

What effects does it have if the lead core is broken at one point, and only the active material ensures electrical conductivity?

Task 12.3

The terminal lugs of electrodes are arranged differently, sometimes far apart, sometimes as close as possible. Discuss the effects on the internal resistance at different currents and states of charge, and the shape of the discharge curve that results in each case.

Task 12.4

Ventilation calculation at other temperatures and voltages

Sealed batteries are often housed in enclosures that are subject to special requirements, for example, splash water protection. The batteries become warm during operation and are often operated at higher voltages due to frequent short discharges.

Calculate how much larger the ventilation openings need to be, assuming an operating temperature of 45 °C and a charging voltage of 2.35 V/cell.

How do you combine ventilation requirements and splash protection?

Task 12.5

Voltage drop across corrosion areas in batteries with series connection

Some emergency power systems use batteries with a nominal voltage of up to 800 V, corresponding to 400 cells connected in series. In trickle charging mode at 2.23 V/cell, the charging voltage is then 984 V, the current is in the range of 20 mA/100 Ah capacity for new cells and the superimposed alternating current of the charger is 1 A.

For a defective cell with an internal contact resistance of 10 Ω, determine the thermal output at the defective point and the voltage drop across the defective point. What effects does this have on the good cells without such faulty contact resistances?

Task 12.6

Design of a battery that develops a low acid stratification

Develop concepts for the design of a battery in which acid stratification development is as low as possible.

13

Lithium-Ion Batteries

The term "lithium-ion batteries" refers to a family of closely related battery technologies with mostly complex metal oxides as cathodes, graphite as the anode, and electrolytes made from organic solvents and a lithium-based salt (usually $LiPF_6$). The choice of cathode and anode materials leads to significantly different voltages, specific energy, and performance data with regard to high-current and low-temperature properties and expected service life. It is therefore often necessary to specify the cathode and anode materials.

During discharging and charging, neither the cathode nor the anode material forms reaction products with fundamentally different chemical and physical properties. Instead, both the cathode and anode material serve as a host grid into which lithium ions are deposited (intercalation). The electrolyte does not take part in the charging and discharging processes. The formation of a complex protective layer on the graphite anode, the so-called SEI (solid–electrolyte interface), is essential for the way the battery works. Lithium-ion cells do not have powerful overcharge or overdischarge reactions, so safety measures must be taken for series and parallel connections. Extensive monitoring is essential for all but the smallest battery systems, for example, through individual cell voltage measurements of all cells in a battery system. Prismatic, pouch, and round cells are used as cell formats and are connected in series and parallel to form battery systems with a complex battery management system (BMS).

For some time now, particular attention has been paid to thermal runaway (TR) processes of a cell and its propagation to neighboring cells. However, experience shows that safety problems as a result of a TR can be minimized using a good BMS and other measures.

Despite billions of cells being used already, the technology is still undergoing rapid development as regards cost, performance, and safety. The recycling rate for lithium-ion batteries is still poor, as fundamental technological and economic challenges remain.

Battery Technology: Fundamentals of Battery Electrochemistry, Systems and Applications,
First Edition. Alexander Börger and Heinz Wenzl.

Companion website: www.wiley.com/go/Börger/Battery_Technology

13.1 Introduction and Economic Significance

As early as 2018, lithium-ion cells with an energy content of well over 150 MWh were produced, according to www.avicenne.org. This is a rapid increase for a battery technology that has only been commercialized since the 1980s.

This progress is due to the wide range of applications, which extend to:

- All portable consumer applications: cell phones, laptops, digital cameras, etc.;
- Power tools;
- E-scooters, e-bikes;
- Military and aviation;
- Traction batteries for pure and hybrid electric vehicles;
- Starter batteries for motorcycles (for standard motor vehicles: so far still only a niche product);
- Stationary battery systems (though currently still a small market share compared to lead–acid batteries);

Table 13.1 Advantages and disadvantages of lithium-ion batteries.

Feature	Advantage (+) or disadvantage (−)
Highest specific energy and energy density of all rechargeable batteries commercialized to date	+
Good high-current properties, except at very low temperatures	+
Very good cycle stability	+
Fast charging possible	+
Easy to store (low self-discharge)	+
Maintenance-free	+
Very good energy and Coulombic efficiencies	+
No memory effect	+
Many possible chemistries allow use of customized materials	+
Comparatively high costs	−
Elaborate production	−
Various safety-critical scenarios: severe overcharging, charging at low temperatures, mechanical damage, internal short circuit	−
Battery monitoring is essential due to safety issues	−
Supply of all required raw materials may be problematic in the long term	−
Recycling concepts only in the trial phase and not yet implemented on a large scale	−
Significant performance losses at low temperatures	−
Interconnection of several cells to form a group requires complex system monitoring	−

- Other mobile and stationary applications, for example, in the medical sector; and
- Second-life concepts, that is, the subsequent use of aged traction batteries, for example, as storage for electrical energy.

The various advantages and disadvantages of lithium-ion batteries are shown in Table 13.1 which illustrates their wide range of applications.

13.2 Electrochemistry

13.2.1 Basic Principle

The fundamental difference between lithium-ion batteries and other battery systems, such as lead–acid batteries, is that intercalation processes (intercalation and deintercalation) take place in the anode and cathode materials, also referred to as the host lattice. In contrast to other battery systems, there is no reaction of the anode or cathode material with ions to form new materials with completely different electrical and thermal properties and crystal structures. The ions that are intercalated or deintercalated during charging or discharging are not converted into a lithium atom but occupy areas with a favorable energy level. As no chemical bond is formed between molecules of the host lattice and a lithium ion, the electrochemical process[1] is written in the following form, see Eqs. (13.1)–(13.3), as already introduced in chapter 2. The ions constantly "rock" back and forth between the two electrodes (see Figure 13.1), which is why the process is also referred to as the "rocking chair" principle.

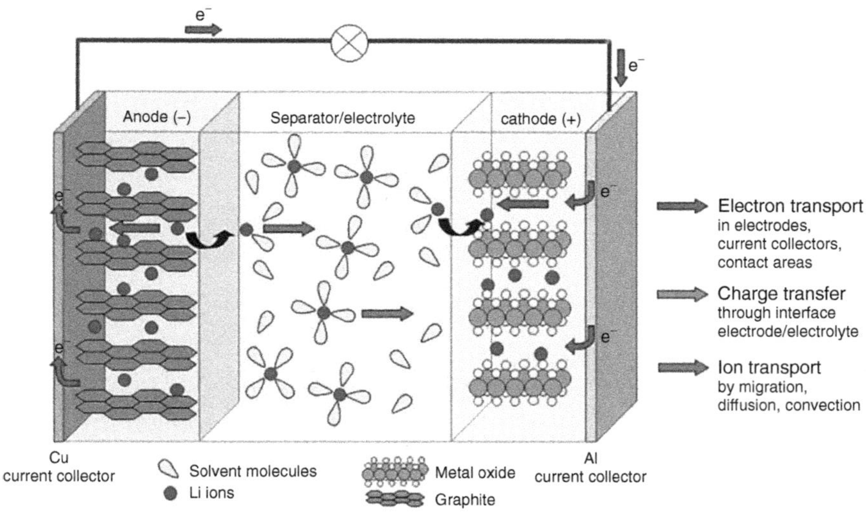

Figure 13.1 Rocking chair cell principle.

1 The term "electrochemical reaction" can be used alternatively and is also used in this book, but it should always be remembered that there is no chemical bond between a lithium ion and a molecule of the host lattice.

When it comes to the specific choice of intercalation materials for the anode and cathode, different classes of materials are in principle possible and are used in different applications. The same applies to the electrolyte. In contrast to lead–acid batteries, in which the electrochemical principles are the same for all battery types and are known very precisely, the term "lithium-ion batteries" refers to a whole family of very different battery systems with different choices of anode, cathode, and electrolyte. Depending on the application, it is therefore possible to tailor properties (costs, service life, performance data, safety, etc.) by selecting the most suitable or most appropriate chemistry for the specific application. However, the other cell components (electrolyte additives, separator, current collector, housing, etc.) must also be modified for the specific combination of electrodes. This makes the development of lithium-ion batteries much more complicated than that of lead–acid batteries, which explains why lead–acid battery factories can be found all over the world, including emerging and developing countries, while lithium-ion battery factories are concentrated in a few leading countries (currently mainly East Asia, with a massive increase in efforts in North America and Europe in recent years).

13.2.2 Overview of Active Materials

Some relevant combinations of anode and cathode materials are shown in Figure 13.2. It becomes clear that cathode materials are mostly metal oxides

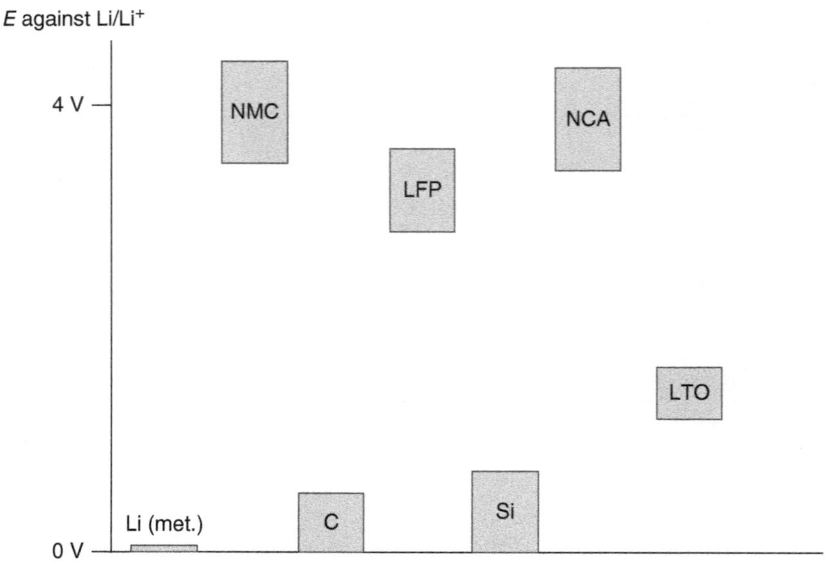

Figure 13.2 Relative voltage of cathode and anode materials for lithium-ion batteries, compared to the reference voltage of Li/Li$^+$. Combining different materials results in different cell voltage windows. Due to different degrees of intercalation processes and locally different lattice properties, only energy ranges ("bands") are shown and not just lines. Since not all energy levels are accessible, there are also the so-called forbidden voltage ranges.

Table 13.2 Common material combinations and voltage levels for lithium-ion batteries. The data are empirical, as correct thermodynamic data are difficult to obtain.

Material combination	Equilibrium voltage in fully charged state [V]	Nominal voltage [V]
LMO, NMC, NCA/graphite	Approx. 4.0–4.2	3.6
LMO, NMC, NCA/titanate	Approx. 2.7–2.9	2.5
LFP/graphite	Approx. 3.6	3.2
LFP/titanate	Approx. 2.5	2.1

or phosphates. The basic choice of anode materials is broader, although graphite and similar carbon materials dominate by far at present and will continue to do so in the near future, with some applications for lithium titanate (LTO) and many research activities on silicon and similar materials.

The combination of different cathode and anode materials changes the equilibrium voltages of cells. Table 13.2 provides an overview.

The range of equilibrium voltages in the fully charged state is based on the differences between the cathode materials, but also on the criteria defined by the manufacturers for a full charge. Data for cells with LMO, NMC, and NCA as cathode materials and graphite as anode material are approx. 4.0–4.2 V. Precise thermodynamic evaluation is very difficult and would presumably result in somewhat different values.

Differences in the specific capacity of the materials and their voltage levels lead to significant differences in the specific energy. The safety and service life advantages of the LFP/titanate material combination must therefore be bought at the price of comparatively low specific energy.

13.2.3 Overview of the Most Important Reactions at the Positive and Negative Electrodes

In analogy to chapters 2 and 12, a distinction is made here between the main voltage-forming reaction and side reactions.

Due to the rocking chair principle, the electrolyte in lithium-ion batteries is not involved in the reactions. However, it plays a decisive role in the structure of the so-called SEI (see section 13.5).

Both the anode and cathode undergo intercalation processes, which can be described using the following reaction equations, shown here for discharging in the forward direction and charging in the reverse direction using a metal oxide $LiMO_2$ (with M for metal) as the cathode and graphite as the anode materials:

$$\text{Cathode: } Li_{1-x}MO_2 + xLi^+ + xe^- \rightleftharpoons LiMO_2. \tag{13.1}$$

$$\text{Anode: } Li_{y+x}C \rightleftharpoons Li_yC + xLi^+ + xe^-. \tag{13.2}$$

$$\text{total: } Li_{1-x}MO_2 + Li_{y+x}C \rightleftharpoons LiMO_2 + Li_yC. \tag{13.3}$$

These equations can be applied analogously to all commercial lithium-ion cells with any cathode and anode materials.

Under operating conditions, the intercalation limits, that is, all sites for lithium ions in the host lattice are either empty or full – x or y = 1 0 – are never reached. Although the (unsolvated) lithium ion is the smallest ion after H^+, it still leads to a change in volume during intercalation or deintercalation. Complete delithiation of an electrode material hinders the intercalation process. This leads to problems at the start of charging or discharging, with correspondingly large losses, and affects the mechanical stability of the electrode particles and the service life that can be achieved. The electrode materials must therefore already contain a minimum amount of lithium during the production process and before they are interconnected to form a cell. During operation, it is not possible to completely delithiate electrode materials if the voltage limits of a cell are adhered to. A degree of intercalation of approx. 0.1 or 0.9 is often specified for a complete discharge or full charge, although the actual limits are material and manufacturer-specific. The fully charged state is therefore not defined by the complete conversion of the discharged material of the anode and cathode (chemically complete charge), as is the case with lead–acid batteries, but is determined by a charging process defined by the manufacturer.

Keeping the crystal structures of the two electrode materials intact during charging and discharging has several advantages.

a) It enables significantly faster reactions and therefore higher charging and discharging currents, because the intercalation and deintercalation processes have no reaction constants, and transport problems for the converted substances, which have a rate-inhibiting effect, do not exist.
b) The formation and dissolution of a crystal structure are always associated with activation barriers and nucleation processes, and crystal growth may lower the overall surface area, etc. All these problems are avoided, which explains the significantly higher cycle stability of lithium-ion batteries compared to batteries with reconstitution processes, such as in lead–acid batteries.

However, there are also some disadvantages:

a) The change in volume of the electrode materials during intercalation and deintercalation is a process relevant to the service life (see also sections 13.4, 13.15, and 13.16).
b) The atoms of the crystal lattice are actually electrochemically passive masses in which the ions are stored. This limits the specific energy of the whole system.
c) The materials used only have low electrical conductivity (for electrons), especially in the largely delithiated state, so additives must be used for good electrical contact between particles and/or to increase bulk conductivity.
d) Finally, although the electrolyte does not take part in the actual electrochemical reaction, it must be stable, that is, (electro)chemically inert, in relation to the potentials of the electrode materials used. The latter excludes the use of water as an electrolyte, for example, because water decomposes as an electrolyte above

a cell voltage of 1.23 V (there is no significant kinetic inhibition [low exchange current density] for hydrogen development at the electrodes, compared to lead electrodes). Therefore, organic solvents and conductive salt additives must be used. However, the electrical conductivity of the electrolyte is orders of magnitude lower than that of aqueous electrolytes.

In contrast to lead–acid batteries, it is not possible to calculate the equilibrium voltage from basic thermodynamic data based on the material composition and crystal structures for either the anode or the cathode reaction. The difficulties of applying the concept of the Nernst voltage and the Butler-Volmer equation to lithium-ion batteries (partly due to the use of nonaqueous solvents) have already been pointed out in chapter 2. The basic concepts and laws such as equilibrium voltage, open-circuit voltage, Butler-Volmer equation, and charge transfer resistance can, however, be used to describe the system. Specialist literature must be consulted for details.

13.2.4 Side Reactions

Lithium-ion cells, like all electrochemical systems, have side reactions that run parallel to the main reactions. This can be seen, for example, in the fact that the charging efficiency of lithium-ion cells is slightly below one, and it has been observed that gas at increased pressure (up to 2 bar has been measured in some cases) is released when aged cells are opened, whereby part of the gas development is certainly also due to non-electrochemical decomposition reactions. In contrast to lead–acid batteries, side reactions are much more complex and less understood. In particular, there are no system-inherent side reactions that provide a high-ampacity current path during overdischarge or overcharge. As a result, there is no protection for cells in series connections, and it is not possible to reach a fully charged state of both electrodes of a cell in the case of differences in the properties of electrodes. However, due to additives and aging processes or impurities, there are reactions at both electrodes that are known as redox shuttles which can carry a very low charging current without intercalation processes of the main reactions (see section 13.2.5).

The importance of side reactions with high current-carrying capability in normal operation was described in detail in chapter 3. The absence of such side reactions in all lithium-ion cells, regardless of the material combinations, has serious consequences for operation in series and parallel connections of cells, the effects of which must be limited by material selection (e.g. additives in the electrolyte), design (fuses and similar safety elements), individual cell voltage monitoring, and a BMS. The operation of many lithium-ion cells in series connection without any protective measures is not permissible, in contrast to lead–acid batteries, where comparable measures do not necessarily have to be taken, even for systems with several hundred volts nominal voltage and energy content in the range of over 100 kWh.

As side reactions in lithium-ion batteries are generally associated with aging processes, they are discussed in sections 13.15 and 13.16.

13.2.5 Overcharging and Overdischarging Reactions

Overcharging lithium-ion batteries can produce gases, in particular oxygen and its subsequent reaction products, because the fully oxidized metal oxides can release oxygen during further charging, for example, according to:

$$Li_{1-x}MO_2 \rightarrow Li_{1-x}MO_{2-\delta} + 0,5\delta O_2. \tag{13.4}$$

This reaction normally leads to the destruction of the oxide lattice and is therefore irreversible.

When lithium-ion batteries are overdischarged, on the other hand, the copper current collector of the anode is often dissolved, that is,

$$Cu \rightarrow Cu^{2+} + 2e^-. \tag{13.5}$$

In subsequent charging reactions or in subsequent reactions of the copper ions with other cell components, the copper ions can be reduced so that copper is formed again. This can then form dendrites.

13.3 Active Materials

13.3.1 Cathode Materials

For the cathode, three main classes of materials are used, all of which were developed in the 1970s in the working group of John Goodenough, who was awarded the Nobel Prize in Chemistry in 2019 for his work. In all cases, these are compounds of transition metals into which lithium ions can be deposited – intercalated, the distinguishing principle being the crystal structure.

a) Compounds with a spinel structure such as lithium manganese oxide (LMO),[2]
b) Compounds with superstructures of the common salt structure such as lithium cobaltate (LCO) or lithium nickel manganese cobalt oxide (NMC) or lithium nickel cobalt aluminum oxide (NCA), and
c) Compounds with an olivine structure such as lithium iron phosphate (LFP).

The corresponding crystal structures are shown in Figure 13.3; in all cases, these are a type of layered structure or channels into which lithium can be easily incorporated.

In the field of high-performance materials, materials with superstructures of the common salt structure are currently the most interesting for electromobility, for example. The elements most frequently used there are manganese, cobalt, nickel ("NMC") and aluminum (in combination with cobalt and nickel: "NCA"). The exclusive use of one metal, as was common in the early days of technology development, is now rarely used. The specific material is selected based on a combination of safety, cost, and performance, as shown in Figure 13.4.

---

2 Please note that these abbreviations do not use the element symbols, but one-letter abbreviations of the elements, that is, N instead of Ni for Nickel.

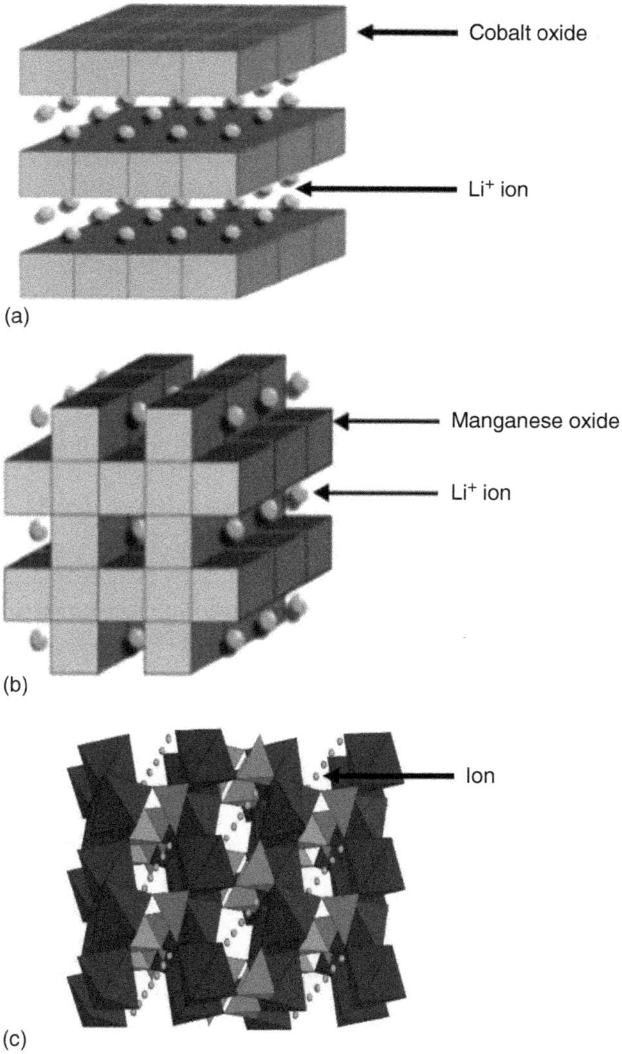

Figure 13.3 Overview of cathode materials for lithium-ion batteries. (a) Superstructure of the common salt structure using LCO as an example, (b) spinel structure using LMO as an example, (c) olivine structure with green phosphate units and black FeO_6 octahedra. Source: Reproduced with permission from Ref. [1]/Elie Riviere.

Further development is going in various directions. On the one hand, material compositions are being optimized, for example, away from $LiNi_{1/3}Mn_{1/3}Co_{1/3}O_2$ (also abbreviated to "111" or "333") to $LiNi_{0.6}Mn_{0.2}Co_{0.2}O_2$ (the so-called "622") or $LiNi_{0.8}Mn_{0.1}Co_{0.1}O_2$ (the so-called "811") to lower costs and improve certain properties. On the other hand, other (but fundamentally related) crystal structures are being investigated, or the extent to which the so-called interstitial sites can still be occupied with lithium ions (the so-called overlithiation) is being investigated,

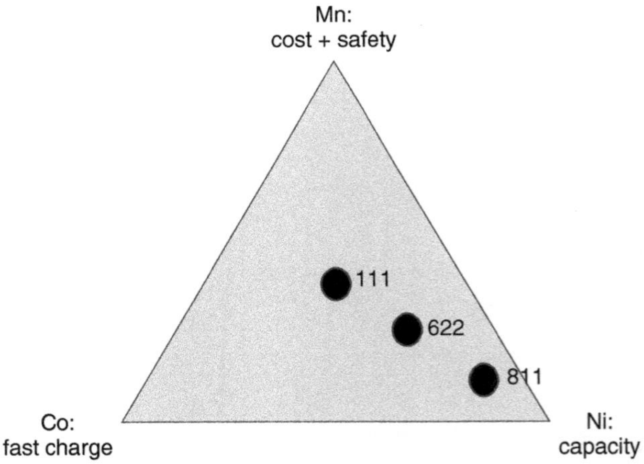

Figure 13.4 Triangle of property optimization in the NMC system.

Table 13.3 Comparison of some selected cathode materials.

Material	LMO	NMC ("333")	LFP	NCA ("333")
Voltage range with graphite anode	2.0–4.0 V	2.5–4.2 V	2.0–3.6 V	2.5–4.2 V
Specific energy	High	High	Medium	High
Safety	Medium	Low	High	Low
Cycle stability	Medium	Medium	High	Medium
Costs	Low	High	Low	High

and finally, attempts are being made to use iron in some compounds to save costs. A comparison of the advantages and disadvantages of different cathode materials is shown in Table 13.3.

Unfortunately, all materials used have poor electrical conductivity. Therefore, additives are required to increase conductivity, generally graphite or conductive carbon black, and other additives that ensure sufficient cohesion of the cathode mass, the so-called binders. These are usually fluorine-containing polymers, such as polytetrafluoroethylene (PTFE) or polyvinylidene fluoride (PVdF) (see Figure 13.5). This is because lithium intercalation in the cathode materials results in volume changes that must be compensated for.

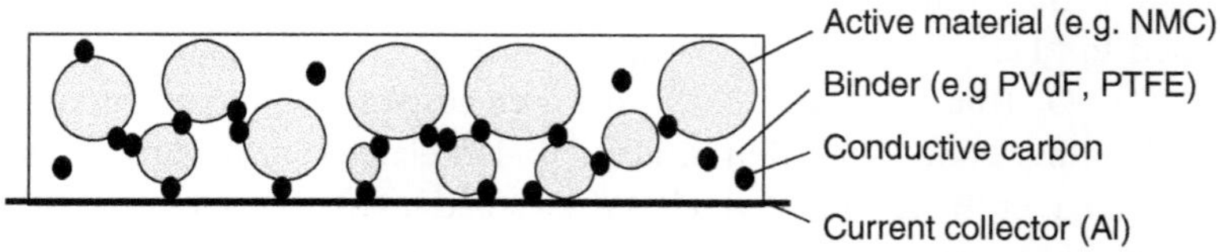

Figure 13.5 Structure of the cathode with current collector, active material, and binder.

In contrast to the active materials of lead–acid batteries, which exhibit a pronounced dependence of conductivity on the state of charge, the electron conductivity of the cathodic active masses is considerably less dependent on the state of charge. The specific electron conductivity of the active masses is much lower than that of the active masses of lead–acid batteries, despite the addition of conductive carbon blacks and other additives. In lead–acid batteries, additives to improve conductivity are used only occasionally and usually only in the negative active mass.

It is beyond the scope of this textbook to describe all current developments in active materials for cathodes in greater detail and discuss the range of properties that may be achieved.

13.3.2 Anode Materials

In principle, lithium metal could be used on the anode, as is common with non-rechargeable primary cells. However, the dissolution of metallic lithium to $Li^+ + e^-$ is a relatively slow process, so only cells with low discharge currents can be built, and the redeposition during recharging often leads to dendrite formation and other structures that are unsuitable for rechargeable batteries. Anodes made of lithium metal have therefore only been used for lithium primary cells to date. Lithium metal anodes have been the subject of research for many years (and will remain so), but have not yet achieved sufficient stability in practice, as accidents have occurred due to the deposition of lithium in dendrite form and the resulting formation of short circuits.[3]

The deposition of metallic lithium and the formation of lithium dendrites is prevented in lithium-ion batteries by the choice of anode materials that have a voltage level above that of Li/Li^+ (see Table 13.4). This applies to carbon or graphite with a voltage level of approx. 0.2 V in the fully charged state and for LTO with a voltage

Table 13.4 Comparison of some selected anode materials.

Material	Carbon/ graphite	Lithium titanate	Silicon	Lithium metal
Potential compared to Li^+/Li [V]	0.1	1.55	0.1	0
Specific capacity [Ah/kg]	360	160	3580	3860
Safety	Controllable	Very high	Low	Low
Cycle stability	Good	Excellent	Bad	Bad
Costs	Low	High	Low	High

Source: Adapted from Ref. [2]/McGraw Hill.

3 This is why the term "lithium batteries" sometimes used for lithium-ion batteries is incorrect. Lithium-ion batteries do not normally contain metallic lithium. However, battery systems that actually work with metallic lithium (such as lithium-sulfur, lithium-air, and in some cases lithium-solid-state batteries, see Chapter 14) can be referred to as lithium batteries.

level of approx. 1.3 V in the fully charged state. With graphite anodes, the voltage difference to the voltage enabling the deposition of lithium is only small, so under certain conditions (aging, very high currents, low temperatures), an anode voltage below 0 V compared to Li/Li^+ becomes possible and thus the formation of metallic lithium (lithium plating) and dendrites. It should always be noted that the voltage level of an electrode is always dependent on the properties of the counter electrode (see chapter 3).

Graphite or similar carbon materials are by far the most common materials used as anodes. Silicon, tin, and various alloys have been intensively investigated in the last decade – however, due to large volume changes during the charging/discharging processes, these materials have so far only been used as additives in anode mixtures and not as pure anode materials. An alternative material to graphite that is also used commercially is LTO. Among other things, this material has the advantage of high stability but the disadvantage of a significantly higher voltage level compared to graphite and thus a lower cell voltage.

Graphite and similar carbon materials dominate by far as anode materials in lithium-ion cells and it is worth taking a closer look at the various materials and their differences. All of these materials are made up of layers of graphene sheets, that is, layers of condensed unsaturated carbon C_6 rings. Along these layers, which are held together by covalent bonds, metallic conductivity exists due to the delocalization of the π electrons. However, only van der Waals-like bonding forces exist between the individual layers, and there is very low electrical conductivity in the direction perpendicular to the layers, which is why the materials are also referred to as two-dimensional metals. The individual layers can be arranged in the sequence A–B–A–B–... (the so-called hexagonal graphite, normal form), in the form A–B–C–A–B–C–... (the so-called rhombohedral graphite), in complex mixed forms of both, or in a disordered stacking sequence (the so-called turbostratic arrangement) (see Figure 13.6). In noncrystalline material, the individual graphene layers no longer lie on top of each other but are tilted against each other, resulting in an amorphous solid. However, amorphous carbon has a lower intercalation

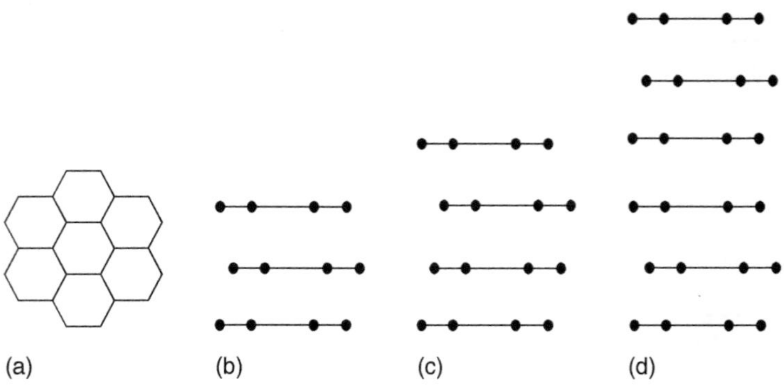

Figure 13.6 (a) Graphene layer, (b) hexagonal graphite with stacking sequence A–B, (c) rhombohedral graphite with stacking sequence A–B–C, and (d) turbostratic arrangement of graphene layers.

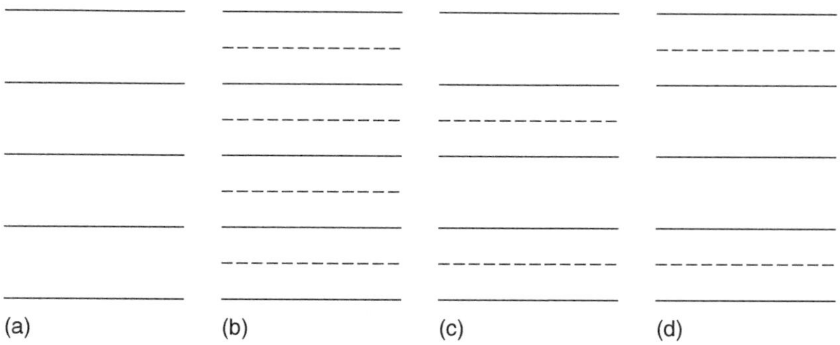

Figure 13.7 Stepwise lithium-ion intercalation in graphene: Layers (a) pure graphite, (b) graphite with lithium intercalation between each layer, (c) graphite with lithium intercalation between every second layer, and (d) graphite with lithium intercalation between every third layer.

capacity than crystalline graphite, as no intercalation is possible between graphene layers that are disordered in relation to each other.

During the intercalation process, lithium ions are intercalated between the individual graphene layers. This does not occur in a completely random manner but according to the principle of minimum energy or activation energy. Several intercalation stages can therefore generally be clearly distinguished from each other and also identified from electrical measurements (see chapter 10, ICA [incremental capacity analysis] and DVA [differential voltage analysis] methods, and Figure 13.7). Ions are first removed from storage locations with the highest potential, and the storage locations with the lowest potential are occupied first during charging (storage of ions). With careful measurements of discharge curves, such voltage levels can be observed, and it has also been shown in aging studies [3] that passing from one intercalation level to the next, after all potential sites have been filled, causes degradation.

The degree of order of the graphene layers can be described quantitatively by the so-called degree of disorder P (proportion of neighboring disordered graphene layers in relation to the total number of layers), which is also experimentally accessible from X-ray diffraction data. The lower the degree of disorder, the more crystalline the product and the higher the intercalation capacity; however, highly ordered crystalline graphite is also correspondingly expensive to produce.

With graphite anodes, an intermediate layer permeable to lithium ions, the so-called SEI, is formed on the anode (usually during the first charge). This is discussed in more detail in a separate section 13.5.

LTO ($Li_{4/3}Ti_{5/3}O_4$) is one of the few fully commercialized alternatives to graphite. It is a compound with a spinel structure. It can be intercalated up to the composition $Li_{7/3}Ti_{5/3}O_4$. Remarkably, intercalation does not change the space requirement, as non-intercalated and fully intercalated substances have the same lattice constant. This is also referred to as a "zero-strain material," and the lack of volume work leads to an extremely high cycle stability on the anode (similar materials for the cathode are not known). Since $Li_{4/3}Ti_{5/3}O_4$ and $Li_{7/3}Ti_{5/3}O_4$ also coexist as a two-phase mixture, which results in a very flat characteristic voltage curve over a very wide state of charge range. Due to the higher voltage level compared to the electrolyte, LTO does not form an SEI and is not necessary, as LTO is stable in contact with the solvents commonly used for lithium-ion systems. This makes LTO a very safe system. The disadvantage, however, is the significantly higher voltage level compared to Li^+/Li than with graphite, which leads to significantly lower energy densities and specific energy, making such systems particularly interesting for stationary applications or as starter batteries due to their very good low-temperature properties compared with all other cell types. Finally (like cathode materials), LTO particles have a very low electron conductivity, so electron-conducting particles and binders have to be added.

In recent years, the focus has increasingly shifted to alloys made of silicon, tin, or complex intermetallic alloys of these and other elements. The background to this is the observation that lithium is even more densely packed in $Li_{22}Si_5$ than in metallic lithium. Similar to graphite, the intercalation here also takes place in several, well-defined stages, for example, in the case of silicon:

- First stage: Li_{12} Si_7 (approx. 600 mV vs. Li^+/Li),
- Second stage: Li_7 Si_3 (approx. 540 mV vs. Li^+/Li),
- Third stage: Li_{13} Si_4 (approx. 450 mV vs. Li^+/Li),
- Fourth stage: Li_{22} Si_5 (approx. 200 mV vs. Li^+/Li).

However, the major and as yet unsolved problem with this class of material is the extraordinarily large volume expansion, which can be up to 400%, and the resulting inadequate mechanical stability against intercalation and deintercalation processes. Therefore, such anode materials cannot withstand cycling. Use in primary cells, on the other hand, is conceivable.

For this reason, it has been decided not to use these systems as the sole anode material but only in combination with graphite anodes, to which up to 10% silicon is added, for example, to increase the capacity.

Similar to the cathode materials, binders are added to the anode material to increase mechanical stability, but also conductivity additives so that the negative active mass becomes macroscopically electron-conductive. The high two-dimensional electron conductivity of the individual graphite crystallites is not sufficient because no stable electron-conducting contact zones are formed between them.

13.3.3 Ionic Conductivity of the Active Masses

The voltage-forming reaction, that is, the transfer of a lithium ion from the electrolyte (which contains the anion of the conducting salt as a countercharge) to the active material with an electron as a countercharge, takes place at the cathode/electrolyte or SEI/electrolyte interface. The lithium ion must then be transported away from the reaction site into the particles of the active material or transported from inside the particles to the reaction site. Both the cathode and anode materials therefore are and must also be ion-conductive and not only electron-conductive.

When there is no current and after all equalization and diffusion processes have subsided, lithium ions in the active masses are evenly distributed in the particles and occupy the locations with the lowest potentials, whereby their thermal energy enables constant movement between neighboring locations. During charging and discharging, however, the local concentration of lithium ions in the particles changes considerably. If the local current density exceeds the transport capacity of the active material, mass transport phenomena, as described in chapter 4, become noticeable. Technically relevant are local current densities during charging that exceed the transport capacity of graphite. Metallic lithium can then form (lithium plating). The lithium metal formed in this way is then no longer accessible for subsequent cycles. There are various approaches to increase the transportability of lithium ions into the particles, for example, partial coating with copper.

Models that focus on ion transport within an active mass particle are referred to as "single-particle models."

13.4 Electrolyte

13.4.1 Fundamentals

In principle, five different types of electrolytes can be used for lithium-ion batteries:

1) Liquid organic electrolytes, which are the standard,
2) Organic electrolytes that are fixed in a polymer (the so-called gel electrolytes), which are also widely commercialized – such batteries are sometimes differentiated from type a) by the term "lithium polymer battery,"
3) Solid polymer electrolytes without solvents (SPE), which are used in niche applications,
4) Ionic liquids,[4] which are still at the research stage,
5) Solid electrolytes: The small-scale commercial use of which had already been announced by, for example, Toyota for 2020 (which was not realized) and promise high potential for the future, but whose large-scale use still requires enormous research efforts. Different classes of materials are being discussed: oxide or silicate glasses as well as complex phosphosulfides, both in crystalline and amorphous (i.e. glassy) state (see also chapter 14).

The following generally applies to materials of all classes: The electrolyte must ensure a high mass transport capacity for lithium ions over the entire temperature range and, in contrast to the electrolyte in lead–acid batteries, is an electrochemically passive component. The basis of conductivity in liquid systems is conductive salts, which provide freely movable lithium ions in suitable solvents. In contrast to the transport of protons in aqueous electrolytes with an enormous surplus of protons, lithium ions must actually be transported through the electrolyte during discharging and charging.

---

4 Ionic liquids are low-melting organic salts. Due to the strong electrostatic forces of attraction, these systems naturally have very high dielectric constants and a nonmeasurable vapor pressure, which would bring a number of safety advantages in handling. The disadvantage, however, is the very high price of this class of substances.

13.4.2 Organic Solvents

As the vast majority of lithium-ion batteries still use liquid organic solvents, these will be discussed further. Cyclic esters of carbonic acid (the so-called organic carbonates) such as ethylene carbonate ("EC") belong to the most important class of compounds used as components of solvents. EC itself has many advantages but a melting point that is too high (above room temperature, see Table 13.5), so it can only be used in a mixture with other cyclic compounds (e.g. propylene carbonate, "PC") or acyclic compounds (e.g. dimethyl carbonate, "DMC"). Furthermore, ethers, cyclic esters (lactones), or completely different compounds such as acetonitrile are also used as the main components of the solvent mixture to produce the desired viscosity, temperature, and solubility properties. In recent years, there has also been an increase in the use of fluorinated compounds (e.g. fluorinated EC), which, however, further increases the amount of the element fluorine in the lithium-ion battery, which is, in principle, problematic.

All of the solvents mentioned here are flammable, so the fire load of lithium-ion batteries represents an increased safety risk compared to many other battery systems for this reason alone.

13.4.3 Other Components

In addition to the main organic components, conventional electrolytes for lithium-ion batteries also contain a range of other components that are essential for the functioning of the batteries and are therefore listed below. In particular, these are:

- Conducting salts,
- Additives for overcharging protection,

Table 13.5 Properties of some important solvent components of lithium-ion batteries. The table does not claim to be exhaustive; in particular, fluorine-containing solvent components are missing.

Name	Ethylene carbonate	Propylene carbonate	Dimethyl carbonate	1,2-Dimethyl ether	Tetra-hydrofuran	γ-Butyrol-actone
Abbreviation	EC	PC	DMC	DME	THF	GBL
Boiling point [°C]	248	242	90	84	66	206
Melting point [°C]	39	−48	4	−46	−108	−43
Dielectric constant	89.6[a)]	64.4	3.1	7.2	7.8	39.0
Sum formula	$C_3H_4O_3$	$C_4H_6O_3$	$C_3H_6O_3$	$C_4H_{10}O_2$	C_4H_8O	$C_4H_6O_2$

a) At 40 °C.
Source: Adapted from Ref. [2]/McGraw Hill.

- Additives to trap traces of water,
- Additives to trap traces of hydrogen fluoride,
- Additives to support the build-up of the so-called SEI (see section 13.5), and
- Flame retardant additives.

The exact composition and selection of additives (which can also be supplemented by additional special additives for specific purposes) are trade secrets of the respective manufacturing companies.

13.4.3.1 Conducting Salt

The organic electrolytes used have no inherent conductivity, unlike sulfuric acid or even pure water, as they do not break down into ions. To achieve a certain electrical conductivity, without which no conductive connection between the electrodes would be possible, the addition of a salt is essential. Due to its good solubility and conductivity properties, lithium hexafluorophosphate ($LiPF_6$) is the dominant salt in commercial applications.[5] Its conductivity maximum is at concentrations of approx. 1 mol/l; however, the conductivity of even such solutions is still at least 1 to 2 orders of magnitude lower than that of aqueous electrolytes. The comparatively high specific resistance caused by this contributes little to the overall internal resistance of a cell due to the design (see chapter 4) with a very large surface area and a very small distance between the electrodes.

The disadvantages of lithium hexafluorophosphate are

a) The comparatively high price,
b) The strong reactivity with even traces of water (the so-called hygroscopicity),
c) The fact that hydrogen fluoride or hydrofluoric acid is released on reaction with water (hydrolysis) during fires,[6] and
d) In the event of a fire, volatile phosphorus compounds with a very high toxic effect can also be produced.

Fluorine-free alternatives for conductive salts have been researched for a long time, but promising candidates such as lithium bisoxalatoborate ("LiBOB") have not yet been able to establish themselves, as the performance data of the corresponding batteries do not come close to those with $LiPF_6$ electrolytes.

Other fluorine-containing conductive salts with special applications, but often also serious disadvantages, are $LiBF_4$, $LiAsF_6$ and $LiN(CF_3SO_2)_3$. More recently, lithium difluorophosphate ($LiPO_2F_2$) and lithium bisulfonylimide ($LiN(FSO_2)_2$) have become more widespread due to their life-extending effect.

13.4.3.2 Overcharge Protection

Since lithium-ion batteries do not have an inherent overcharging reaction, such as water decomposition or the oxygen cycle in lead–acid batteries, the addition of an

---

5 In solution, it decomposes into the lithium ion (Li^+) and the hexafluorophosphate anion (PF_6^-).

6 Gaseous hydrogen fluoride (HF) or hydrofluoric acid (HF dissolved in water) are chemically extremely aggressive substances. Even small traces of HF on the skin can cause massive health risks, as HF is able to diffuse through the skin and the underlying tissue and then damage the bones. Working with HF therefore requires special protective clothing. HF produced in the event of fire poses a further serious hazard.

overcharge protection additive is very useful. Otherwise, even slight overcharging could lead to serious damage to the battery due to the very high cell voltages that occur when connected in series. A typical representative of overcharge protection additives is biphenyl ($C_{12}H_{10}$), which undergoes oligomerization[7] and polymerization reactions in the event of overcharging and also leads to a reduction in the overcharge current due to increases in viscosity and other mass transport restrictions.

In addition, additives are used that enable ion transport, for example, Fe^{2+}/Fe^{3+} compounds, which can lead to low overcharge currents, but also reduce charging efficiency (the so-called redox shuttle). However, the use of such additives does not allow the state of charge of both electrodes to be adjusted, as the discharging current is the same for both electrodes.

13.4.3.3 Water Scavenger

As even traces of water can massively impair the performance of lithium-ion batteries with organic electrolytes, production must be carried out in an atmosphere with very low humidity. It is also necessary to effectively remove water from the electrolyte. This can be achieved, for example, with water scavengers such as hexamethylsilazane $HN[Si(CH_3)_3]_2$, which hydrolyze to form compounds with Si—O bonds that are generally insoluble in the electrolyte and thus permanently remove water molecules. The sensitivity of the electrolyte to water is one of the reasons why the housing of lithium-ion batteries must be tightly sealed against the ambient atmosphere.

13.4.3.4 Removal of Hydrogen Fluoride

Due to the strength of the Si—F bond, compounds similar to water scavengers can also be used to trap HF, which can form in various side reactions of the conducting salt or certain solvent components. Suitable compounds are, for example, silicon dioxide (SiO_2) or aluminum oxide (Al_2O_3). In these cases, too, insoluble products are formed, which lead to the permanent removal of HF molecules.

13.4.3.5 Additives for the Formation of the SEI

The formation of the SEI is a particularly important step for all lithium-ion batteries with a graphite anode. Its formation can be supported by various substances, for example, the aforementioned LiBOB or vinylidene carbonate ("VC"), an unsaturated cyclic carbonate. The aim is to create a dense but good lithium-ion conductive layer within a very short time, which then creates high (electro-)chemical stability between the anode material and the electrolyte.

13.4.3.6 Flame Retardants

Organic phosphates, for example, have proven their effectiveness as flame retardants in many applications. As the safety requirements for lithium-ion batteries constantly increase, not least due to legislation, it can be assumed that further developments will show in the near future.

---

7 In oligomerization reactions, only a few molecules react with each other, for example, in the form of dimerization, trimerization, etc.

The use of organic phosphate compounds may lead to the formation of very toxic, herbicide- or insecticide-like volatile organic phosphorus compounds in the event of a fire.

13.5 Solid–Electrolyte Interface and Its Significance for Lithium-Ion Batteries

With graphite anodes, a lithium-ion permeable intermediate layer (usually only a few nanometers thick) forms on the anode, called the SEI. The exact structure and formation of the SEI are still the subject of research, but there is a consensus that it consists of different decomposition products of all components of the electrode and electrolyte. Some of these are purely inorganic, for example, lithium fluoride (LiF), lithium hydroxide (LiOH), or lithium carbonate (Li_2CO_3), some are purely organic in nature, for example, nonconductive oligo- or polymeric materials like hydrocarbons, organic ethers, or organic esters; and some are of a mixed inorganic–organic nature, for example, lithium salts of organic acids (see Figure 13.8).

The part of the SEI in which lithium ions react with electrode components and electrons of the electrode can be described as a type of side reaction or self-discharge reaction. The part of the SEI that is formed by the reaction of lithium ions with counterions of the electrolyte, on the other hand, is not a side reaction but a decomposition reaction of the electrolyte, in which, among other things, gaseous by-products can be formed.

One of the main reasons the structure of the SEI is still not fully understood is the difficulty of analysis (see chapter 10). While many surface methods work in a vacuum and therefore cannot be used under in-situ conditions in a battery, the usual

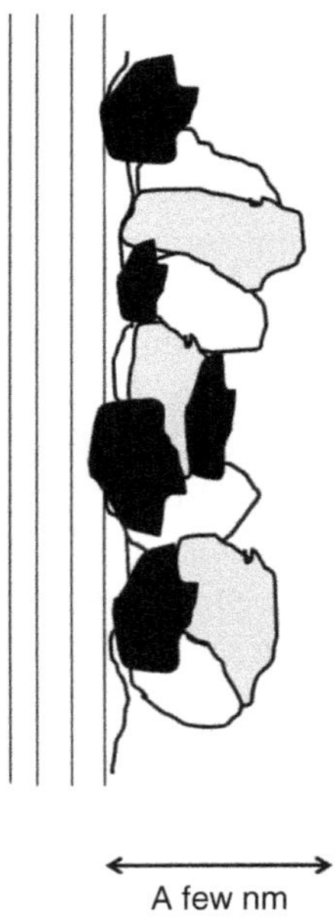

Figure 13.8 Components of the SEI. White: Inorganic lithium compounds such as LiOH, LiF; gray: nonconductive organic polymer (decomposition product of the electrolyte); black: lithium salts of organic acids and other mixed organic-inorganic compounds; hatched: Graphite anode.

spectroscopic or diffraction techniques require larger quantities of material to obtain results that can be interpreted well.

The formation of a stable SEI is essential for lithium-ion batteries with graphite anodes, as this is the only way to protect the anode materials from damaging reactions with electrolyte components and other components of the lithium-ion batteries. If recurring damage to the SEI occurs during the life of the battery, the graphite anode reacts with the electrolyte whenever the electrolyte is in direct contact. This leads to a reduction in capacity due to permanently bound lithium that is no longer available for the electrochemical processes and an increase in the internal resistance of the cell. In the discussion of aging processes, the processes that take place over time will be discussed in more detail.

On the other hand, the SEI must not become too thick and must remain sufficiently conductive, that is, permeable to lithium ions, otherwise the cell reaction will be inhibited. It is clear from all these remarks that the SEI plays a key role in lithium-ion batteries.

Similar boundary layers on the cathode have been described in the literature under the name SPI (solid permeable interface), but generally do not play the same role as the SEI on the anode. The main reason for this is that the cathode is basically stable in the electrochemical voltage window normally used. Nevertheless, space charge effects on the electrolyte-electrode interface and similar causes also lead to surface and near-surface structural changes at the cathode, as well as changes in the chemical composition, which must be considered on a case-by-case basis.

13.6 Current Collectors

Current collectors are usually made of copper on the anode and aluminum on the cathode. The use of copper on the anode is not possible, as this leads to electrochemical side reactions and/or chemical corrosion processes with lithium, which irreversibly damages the current collector (see Figure 13.9). In contrast to other battery systems, the use of perforated electrodes or similar design solutions is not common. A compact, that is, metal sheet, collector is considered necessary to minimize the electrical resistance of the current collectors, especially under high-current loads. Copper and aluminum are the metals with the lowest specific resistances in relation to volume and weight. The disadvantage is, of course, the high cost of both metals.

Depending on the application and in particular requirements for high current capability, the thickness of the foils is approx.10 – 30 μm, whereby the thickness ratio of aluminum to copper is often approximated to the ratio of their specific electrical conductivities (see section 13.4.2 regarding the necessity of this design). Normally, the current collectors are coated on both sides with active mass, which is mechanically compacted during production.

Although alternative current collector designs (e.g. metallic collectors welded into electronically conductive polymer film) are known from the literature and are also commercially available, they are not widely used despite certain safety advantages.

(a) (b)

Figure 13.9 Current collector for (a) the anode (Cu) and (b) the cathode (Al). Source: Courtesy of Clarios.

13.7 Production of Electrodes

The production of electrodes, consisting of a current collector coated on both sides with active materials, is carried out in the following steps, where differences between the various materials are not considered here:

1) *Preparation of the active masses*
   The properties of the active masses relevant to their function as a battery are closely linked to their structure (structure–property relationship). The raw materials are processed into fine-grained particles using conventional mechanical process engineering methods (grinding, grading, mixing, etc.) and then mixed with the required additives (binders and conductivity additives).
2) *Production of the current collectors*
   Aluminum and copper of the appropriate purity are rolled to the desired thickness.
3) *Application and compaction of the active masses on the current collectors*
   The active masses are applied to both sides of the current collectors and compacted to the desired thickness and porosity, both using conventional mechanical process engineering methods. Usual capacities per unit area are in the range of $0,5-2\,\text{mAh/cm}^2$. The thicker and more compacted the active mass layer is, the more suitable it is for cells with a high energy density and lower current-carrying capability. The electrodes produced in this way are manufactured continuously when used for wound cells (including wound prismatic ones); for pouch cells, the active mass layer is also limited to the electrode area used. An important process step here is the detection of large mass particles that could damage the separator

during assembly into a cell and possibly even puncture it due to mechanical stress during further processing or during operation. If a short circuit between electrodes in cells were to develop, this could lead to a TR with all its safety risks.

When manufacturing the anodes, care must be taken to ensure that the copper is completely covered in the edge area. Copper that is not completely covered with active mass can assume a slightly lower voltage than the active mass during charging and increases the risk of lithium deposition.

13.8 Separators

The separator is often the most expensive single component of a lithium-ion cell and fulfills a very important function because without it there would be a spontaneous short circuit with the risk of a serious thermal event. At the same time, the separator must have sufficient ion permeability for lithium ions, as otherwise, the current-carrying capacity of the battery is negatively affected. To minimize the overall resistance, separators should be as thin as possible (typical thicknesses are in the range of 20–30 μm with a trend to even lower thickness). In addition, it must have high chemical stability, as direct contact with the oxygen-containing and, in principle, oxygen-releasing cathode can otherwise lead to oxidation phenomena.

Separators made of polymer materials have become established, especially polyethylene and polypropylene. Pure ceramic separators have not been able to establish themselves despite their non-combustibility or reduction of the fire load of a cell; however, many polymer separators with some ceramic content are used today. Typical for pure polymer separators is the use of so-called shutdown separators, which consist of a three-layer composite of polypropylene, polyethylene, and polypropylene again (see Figure 13.10). At temperatures above around 120 °C, the polyethylene begins to melt, closing the micropores so that the current flow is interrupted locally. However, the lithium-ion cell does not lose its mechanical stability, as the polypropylene parts only melt at higher temperatures (approx. 170 °C). This therefore represents an effective protective mechanism against excessively high temperatures (and the associated excessively high currents). It is known that

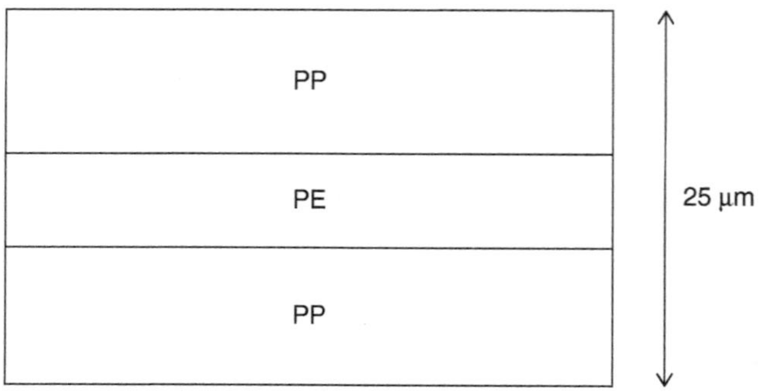

Figure 13.10 Principle of the shutdown separator.

the current flow through a cell, which is heated evenly and slowly so that the temperature distribution in the cell is largely homogeneous, is interrupted within a very narrow temperature range. Further heating due to current flow via the terminals is then no longer possible. Only in the event of an internal short circuit due to a hole in a damaged separator can current continue to flow and heat the cell further. In real applications, however, a homogeneous temperature distribution in a cell is not to be expected in the event of a fault, so the cell can continue to conduct external current and continue to heat the cell, even if there are already very high local temperatures.

An important control step in production is checking the separators for small holes (the so-called pinholes), which can lead to contact between the active masses of different electrodes and thus to an internal short circuit.

13.9 Safety Measures

This chapter deals with safety measures at the cell level. What is meant by safety and the term “state of safety” (*SoS*), and what must be done on a module and system level are dealt with in sections 13.19–13.23.

Some safety measures have already been discussed in the description of the individual cell components, see the discussion of additives or the shutdown separator. Nevertheless, these measures are by no means sufficient for complete protection against hazards, and commercially available cells contain further design measures.

In the case of cells with a prismatic or round solid housing, a rupture valve[8] in particular is important. The valve opens if the internal pressure in the cell becomes too high, as can occur due to gas-forming reactions, especially when a cell undergoes TR, and releases the gases formed (this is known as venting). By (irreversibly) opening the cell at an internal pressure of a few bars (the usual specifications for opening the rupture valves are in the range of 5–7 bar), the current flow can be permanently interrupted if the cell is appropriately designed. For example, the current is conducted via a metal foil under the rupture valve so that the flow is interrupted when the rupture valve opens. Another possibility is to use the cell housing as a current-conducting component: when the cell housing has a predetermined weak point, it breaks in two so that the current flow is interrupted. To protect against very hot particles ejected from inside the cell and capable of causing a fire, a gas-permeable film is now also used to cover the vent.

There is surprisingly little research concerning the optimum pressure for venting. Recent research has shown that lowering the opening pressure has significant safety advantages. The cell changes from a closed system (apart from heat loss) to an open cell, which loses heat much more easily, for example, by means of the electrolyte’s enthalpy of evaporation. Experiments have shown that all safety-relevant parameters (heat energy required to cause a TR and time required until a TR occurs) increase, while the maximum cell temperature and maximum gas temperature decrease [4].

8 The valve opens only once.

Cells with a flexible housing, that is, pouch cells, usually do not have a rupture valve, as the seam of the foils ruptures at a sufficiently high internal pressure and thus assumes the function of the rupture valve. However, this has the disadvantage that the location of the cell opening cannot be precisely controlled.

There are additional components in the cell that limit or interrupt the current when it is too high (current interrupt device [CID] or current limiting device [CLD]). CLDs are usually made of a material with a greatly increasing electrical resistance when the temperature rises (PTC elements, PTC being positive temperature coefficient). An alternative is the use of bimetals, whose thermal expansion leads to the (reversible) interruption of the current. Fuses are also used, which irreversibly interrupt the current flow and then have to be replaced if outside the cell, or the cell is damaged beyond repair. Typical data sheet values for fuses to cut the current flow are around 10 C after 200 milliseconds. For low-resistance short circuits, the current limitation or current interruption is triggered after a very short time. For short circuits with higher resistance, which cause currents in the range of c. 1 A (400 W at 400 V), no effect can be expected or only after a much longer time. The design problem for fuses, and in particular the handling of fault currents well below the tripping limit of fuses, is a fundamental electrical engineering problem that also applies to lithium-ion batteries.

A pressure element can also be installed. This is, for example, a prestressed metal sheet curved into the interior of the cell, which changes its radius at high internal pressure and then short-circuits the cell at the terminals or via the cell housing inside the cell (see Ref. [5]). The short-circuit current that then flows is so high that the cell's CLD is activated, and the cell can no longer carry any external current. The further supply of heat to the cell by an external current is thus interrupted, possibly before the rupture valve opens.

Figure 13.11 shows an example of the interaction of various safety devices as an equivalent circuit diagram. This type of interaction between different safety features is common for cells in the automotive sector.

Finally, the BMS, as a monitoring and control unit, should also be seen as a safety measure. In contrast to lead–acid batteries, where, until a few years ago, even larger

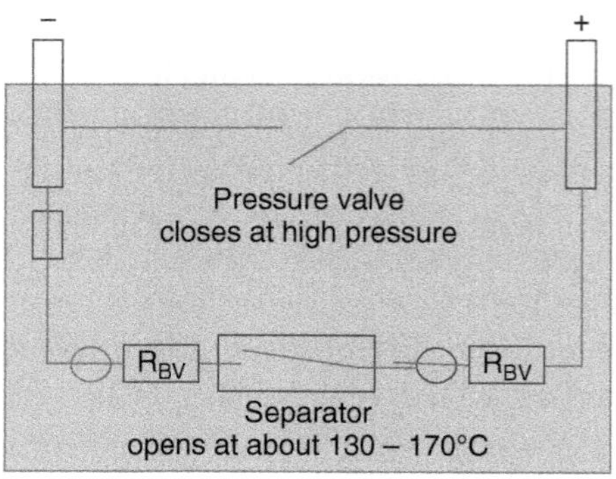

Figure 13.11 Equivalent circuit diagram of various safety components installed in an automotive cell.

starter batteries were usually operated without a BMS and this is still often the case for industrial batteries with several strings despite high voltages and capacities, at least a rudimentary BMS for small lithium-ion batteries in the consumer electronics sector is considered state of the art. The design details – that is, which parameters are recorded with which frequency, wherein the battery-appropriate measuring points should be, and how to analyze data – must be adapted to the specific system requirements. It should be pointed out that there are still no appropriate standards and the way a BMS works is considered a commercial secret.

However, the range of existing safety measures should not lead to the conclusion that lithium-ion batteries can no longer reach safety-critical states. Particularly in cases of misuse (forced overcharging, excessive heating, mechanical damage to the battery by nail-like objects, etc.), safety-critical scenarios involving fire and the release of toxic substances may very well occur. In addition, there is no inherent protection mechanism against an internal short circuit, such as that caused by large active material particles that accidentally get into the cell during the manufacturing process. For further details, see sections 13.19–13.23.

13.10 Types of Lithium-Ion Batteries

13.10.1 Cell Designs

There are three common types of lithium-ion batteries, namely cells with a cylindrical housing, the so-called round cells (a typical representative is the so-called 18650 cell with a length of 6.5 cm and a diameter of 18 mm, hence the name); cells with a so-called prismatic (cuboidal) housing (a typical representative in the automotive sector is the so-called PHEV2 cell with capacities in the range of 20–40 Ah); and cells with a flexible housing, the so-called pouch cells (see Figure 13.12). Round cells dominate the consumer electronics sector but are also used in the automotive sector, particularly by Tesla. Prismatic cells are mostly used in electric vehicles (e.g. in the BMW i3), but pouch cells are also used there (e.g. in the Renault Zoe). There are hardly any standardized dimensions in the automotive industry, but efforts are underway at the company level to limit the number of different cell dimensions for their electric vehicle fleet. Currently, the largest commercially readily available cells have a capacity of approx. 250 Ah (e.g. Kokom), which leads to the provision of very high amounts of energy through the parallel connection of many strings. The largest cell known to us for military applications has a capacity of 500 Ah – a standard industrial size for lead–acid batteries.

However, the basic design of the various formats is fairly similar and can be described using Figure 13.13. According to this, the anode material, that is, graphite with additives – is coated onto a copper foil as a current collector. A separator prevents the anode and cathode from touching under any circumstances but at the same time minimizes the distance between the electrodes and ensures that the distances are as equal as possible at the end of production. The cathode materials are coated on an aluminum foil, which serves as a current collector. For round cells

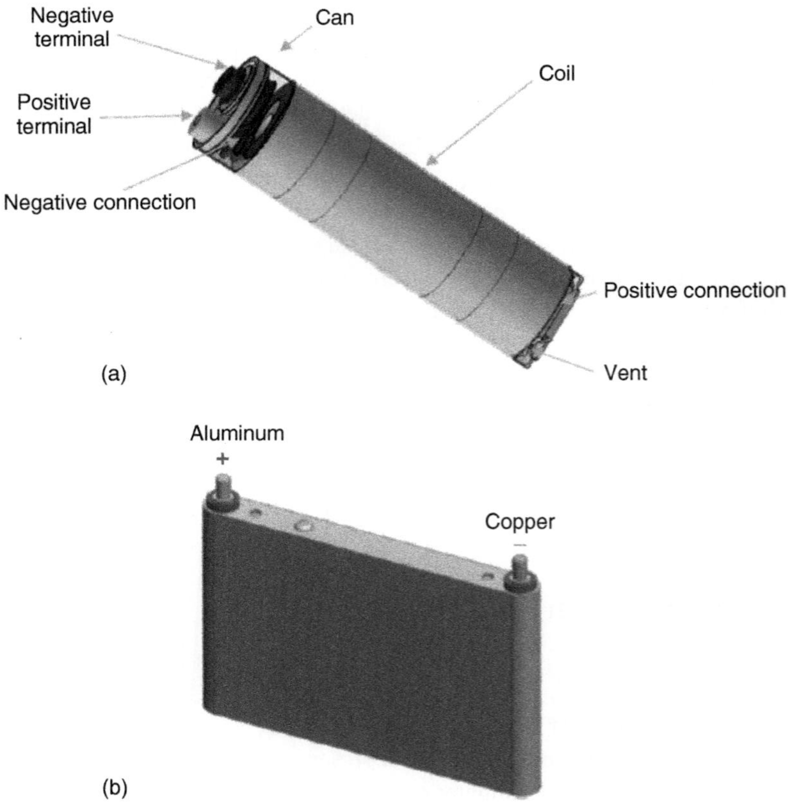

Figure 13.12 (a) Round cell, (b) prismatic cell. Source: Courtesy of Clarios.

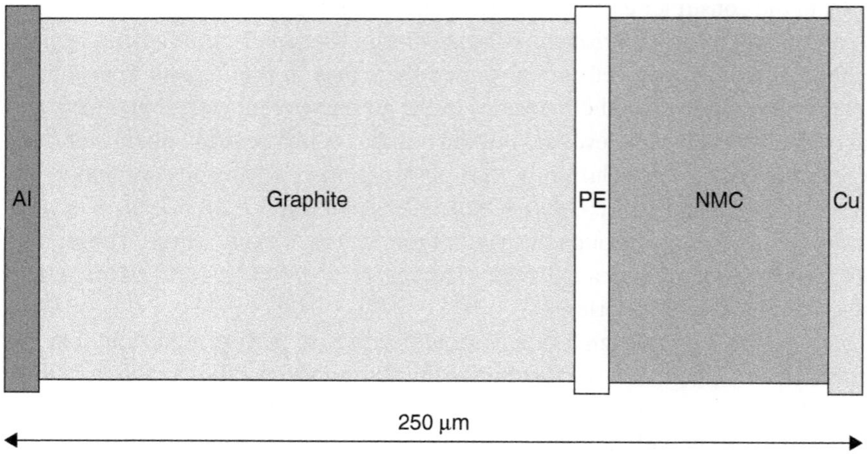

Figure 13.13 Internal structure of a lithium-ion cell.

and wound prismatic cells, all of the components are wound together, whereas pouch cells often have electrode stacks consisting of individual electrodes – the same design.

In contrast to lead–acid batteries, it is not common for lithium-ion batteries to contain several cells in a single housing. Typically, each cell has its own housing that is hermetically sealed against the environment and, in particular, humidity. The cells are connected to each other externally to form cell assemblies (modules) or a complete system (battery system).

There are very different advantages and disadvantages between the three formats:

a) The use of a large number of small round cells leads in principle to increased susceptibility to faults in the overall system, as the failure of even a single cell leads to a fault, and the failure rates of larger and smaller cells seem to be not significantly different. In Tesla cars, the overall battery system consists of several thousand cells with a correspondingly large number of connection points to cell connectors or the next cell. However, the amount of energy that can be released by small cells during a fault is significantly lower than that of larger cells, making it much easier to locally contain a fault (e.g. by switching off a string of modules).
b) Round cells cannot be assembled without space between cells and have fundamental disadvantages in terms of the space required compared to pouch cells or prismatic cells. On the other hand, they provide the space required for additional cooling with a geometry optimized for such a purpose.

Both round cells and wound cells require an internal core, which leads to an unusable volume inside the cell. The volume or the radius of this core can only be reduced as far as the mechanical stability of the active mass coating on the current collectors allows. For energy cells with a thick active mass coating and high compacting, the core must therefore be several millimeters thick. Pouch cells, for this reason, have the highest energy density (Wh/l) of all cells.

In addition, pouch cells have very thin foils as cell housings which weigh little and require little space. Nevertheless, the housing has to ensure both secure separation against solvent vapors from the inside to the outside and, in particular, for water vapor from the outside to the inside. The mechanical protection of pouch cells against damage from the outside is lower.

c) In general, smaller cells have more homogeneous current and temperature distributions than larger ones because larger temperature differences can build up more easily, which then leads to local hotspots. On the other hand, individual cell voltage and temperature monitoring are more complex with smaller cells because considerably more sensors have to be used when the energy content is similar.
d) As round cells in particular dominate the consumer sector, there is the advantage of economies of scale and more learning from experience gained in areas that tend to be less safety-critical. On the other hand, there is less willingness to create cells for vehicle-specific requirements, whereas prismatic cells, for example, are usually developed to meet a specific requirement.

The advantages and disadvantages of the individual designs, especially in applications where many cells are needed, make clear that there will continue to be competition between different formats for the foreseeable future.

It is obvious that the production processes for round cells are the easiest to master. Round cells are also the most mature due to the very high quantities required for the consumer sector and the similarity in terms of the structure of alkaline round cells. With wound prismatic cells (see chapter 4), the uniformity of the electrode spacing is difficult to control. For stacked cells, the handling of very thin electrodes and separators, which can be in the range of $200 \times 200\,\text{mm}^2$ for large cells, can lead to problems. In addition, much more care must be taken with single electrodes than with wound cells to ensure that edges as a result of cutting electrodes to size do not damage the separator. With otherwise identical conditions (electrode and separator thickness), the cell types have slight differences in terms of energy density (Wh/l) because the winding cores have to be taken into account. In the production of modules, prismatic cells have a significantly higher energy density than round cells, and pouch cells have a higher energy density than prismatic cells.

The three cell types also differ in terms of the contact between the electrodes and the terminals. Ideally, the current flow through the cell is always in one direction so that a homogeneous current density distribution can be reached (see chapter 4). In small round cells, the terminals or connection points of the cell are on opposite sides of the cylindrical housing. In the case of larger round cells and wound prismatic cells, on the other hand, the terminals are often on the same side. However, the current flow inside the cell is still only in one direction; the terminals are simply located on the same side to allow easier handling. The housing is often also used as a current collector. This must be observed when assembling cells. Touching cell housings in series or parallel connection leads to short circuits and the discharge of all cells in between, possibly with very high currents.

In the case of stacked electrodes, as used in pouch cells, on the other hand, the electrodes are connected to the terminals, often on the same side of the electrode, similar to what is shown in chapter 4 for lead–acid batteries. The local temperatures and states of charge are correspondingly uneven, although the differences are much smaller compared to lead–acid batteries and have less serious effects.

13.10.2 Buildup of Modules and Batteries

For most applications, cells must be connected in series (string) to obtain the required voltage, and strings are often also connected in parallel to increase the battery capacity or energy content. The usual designation is then *nsmp*, where *n* stands for the number of cells connected in series and *m* for the number of parallel strings.

In the case of large batteries and batteries with a very high nominal voltage, for example, 800 V for emergency power systems, modules are manufactured for practical reasons, which are then connected in series and possibly also in parallel. Which sensors a module has and whether the BMS of the entire battery system has a decentralized structure depends on the application and manufacturer.

When designing modules and batteries, the following aspects must always be taken into account:

a) The capacity of all cells and their state of charge must be as similar as possible. In chapter 3, it was stressed that the properties of both cathodes and anodes of a cell, as well as cells connected to each other in series, in particular their capacity, had to be carefully balanced. An example of how much attention is paid to this aspect is the assembly of a lithium-ion starter battery (LTO/graphite) by one manufacturer. The time between the final inspection of the cells after manufacturing and the beginning of the assembly of six cells for a starter battery can vary by a few weeks depending on production schedules and shipping time. The cell manufacturer provides the open circuit voltage (OCV) voltage of each individual cell, and when the cells are placed into the battery box for interconnection with other cells, the OCV voltage is measured again. If the difference from the end-of-production OCV voltage exceeds a threshold value, the cell is discarded, and it is assumed that there is an unacceptable high discharge rate due to a production problem. Six individual cells with very similar OCV voltages are then selected to be interconnected.
b) Connectors must enable stable electrical and mechanical connections, and mechanical stress must not be induced on the cells.
c) Sufficient mechanical protection from the environment, especially in-vehicle applications.
d) Protection of the environment from hazards caused by the cells (release of gases during venting etc.) so that no critical exposure can occur for vehicle occupants or operating personnel, in particular from CO and HF.
e) Fire protection from the outside, but also from fires within the battery.
f) Cooling or heating.
Technical solutions are the use of air (fans, natural convection) or other media (either distilled water or water-glycerol mixtures – flammable oils are rarely used due to the additional fire load). Passive elements such as heat pipes can dissipate large amounts of heat well through phase transitions (usually evaporation). Heaters are predominantly designed as electrical resistance heaters, although heating of batteries is still rare for lithium-ion batteries and most likely to be used in the commercial vehicle sector or for special applications in cold climate zones. Efforts must always be made to ensure that all cells have very similar temperatures. In electric vehicles, air conditioning and heating systems are usually used for temperature control of the battery as well.
All cooling and heating devices have to be controlled appropriately by the BMS.
g) Fuses, switches, etc., and
h) Current, voltage, and temperature sensors.

The time and cost of meeting these requirements, additional weight, as well as the additional volume, are high, so for batteries that are not yet manufactured in large quantities, the price of the battery system is many times the price of the cells alone. When comparing battery systems, these aspects (see also chapter 1) must be considered very carefully.

13.11 Dimensioning of Cells and Design Differences for Different Applications

13.11.1 Dimensioning of Cells

Chapters 3 and 4 showed that attention must be paid to adjusting the electrode capacities and states of charge during the construction of the cell. This is particularly important for lithium-ion batteries because the lack of side reactions makes it impossible to subsequently change the state of charge of an electrode in relation to its counterpart. It should also be noted that graphite anodes age faster than the known cathode materials. However, charging a cell with an aged electrode and thus reduced capacity means that the cell voltage can affect both electrodes in such a way that the graphite electrode reaches the potential of the Li/Li^+ reference electrode and may fall below it. The resulting deposits of metallic lithium reduce the capacity, increase the internal resistance, and lead to safety problems. As a result, graphite anodes are normally designed with a slightly higher capacity than cathodes.

For the layout and dimensioning of a lithium-ion cell, it is important to keep the intended applications and operating ranges in mind from the outset, as cells for high-performance and high-energy applications differ significantly. However, safety aspects and cost reasons can also be decisive for the choice of a particular cell chemistry or design. For example, the (at least supposedly) "safer"[9] chemistries such as LTO or LFP are more likely to be found in stationary batteries, while high-end automotive and consumer applications tend to use different variations of NMC or NCA chemistry (using graphite anodes, possibly with Si and Sn alloy additives). Other design criteria include the quantity of active masses in view of the lithiation limits chosen for lifetime reasons, the thickness of the current collectors (because these affect the maximum current-carrying capability), the quantities of conductivity additives used, etc.

A fundamental distinction in cell design is that of high-performance or high-energy applications versus intermediate applications. This distinction is very important in the automotive sector, for example, if the cells are to be used in a hybrid vehicle, a pure electric vehicle, or a plug-in hybrid vehicle. The following applies based on comparable size and weight:

1) High-energy cells are primarily optimized to have the highest possible capacity. This leads to comparatively thick layers of anode and cathode materials, whereby only relatively low maximum currents can be achieved due to the associated large distances between the current collectors and small specific surfaces.
2) Conversely, high-performance cells are able to guarantee a very high current capability due to the very high specific surface area, which may be further increased by suitable additives as part of the SEI formation, by using as many layers of anode and cathode as possible. However, cell areas that are not or

9 To the knowledge of the authors there are no general technological differences concerning safety features between "safer" cells and others, as the potential hazards are in both cases great. Individual cell voltage measurement systems and other safety features do not seem to differ much.

Table 13.6 Performance data of some commercial lithium-ion batteries.

Cell designation	**INR 18650MJ1**	**INR 18650HG2**	**US 18650VC7**	**n.a.**[a]
Type	Energy cell	Power cell	Energy cell	Power cell
Manufacturer	LG Chem	LG Chem	Sony	n.a.
Chemistry	NMC/C	NMC/C	NCA/C	n.a./LTO (nominal voltage 2.4 V)
Format	Round cell	Round cell	Round cell	Round cell
Capacity [Ah]	3,4	2,9	3,5	10
Specific energy [Wh/kg]	266	228	269	84
Max. current [A]	10	n.a.	n.a.	60
Internal resistance [mΩ]	33 (DC)	n.a.	31 (DC)	n.a.

a) https://www.evlithium.com/Lithium-Titanate-Battery-LTO/10ah-lto-battery.html.
n.a.: not available.

not completely filled with active material (e.g. by not compacting the oxides used to the maximum extent possible) lead to a reduction in the available capacity. Since cells with a high current-carrying capability will also generate higher amounts of heat, additional cooling measures or concepts for equalizing the temperature rise may have to be implemented, which can further reduce the available capacity and increase the volume or total weight of the entire battery system.

3) To be able to allocate cells to as many different areas of application as possible (especially as vehicle manufacturers in the field of plug-in hybrid vehicles do not necessarily know from the outset whether hybrid or purely electric operation will be used), designs that represent a good compromise between the two extremes are often the means of choice.

Some typical performance data of commercial cells are summarized in Table 13.6.

13.11.2 Electric Traction Batteries

The advance of electromobility is largely responsible for the boom in lithium-ion technology from around 2010. Surprisingly, the competition between the various formats is still open and even within one car manufacturer, the decision for a specific cell type may not be decided from the outset for a particular vehicle project. Pouch cells have the best space utilization, but are the most demanding in terms of safety. In the case of prismatic cells, the challenge is most likely to be ensuring current and temperature homogeneity and therefore reducing aging effects, as local inhomogeneities in the individual cell (see chapters 4 and 5) can have a very negative effect on overall performance and safety behavior. In principle, round cells provide economies of scale, as they are already available for many consumer applications; however, the special requirements of electromobility normally require certain adjustments to the cell chemistry and/or other cell components. There is currently no end in sight to the technological competition. Although many small cells require an enormous number of cell connectors and individual cell voltage measurements, they also offer advantages in terms of safety, as a defect in one cell does not necessarily have to propagate to other cells due to the comparatively small maximum amount of energy that can be released. The larger space requirement for round cells is considerable for batteries with a high energy content, but it is accepted by manufacturers such as Tesla.

13.11.3 Starter Batteries

In contrast to traction batteries for road vehicles, where lithium-ion batteries are currently by far the dominant battery technology, lithium-ion batteries in the starter battery sector are in strong competition with traditional lead–acid batteries and have so far only been able to gain a foothold in isolated cases. The specific requirements for starter batteries demand completely different solutions than those in the electric traction sector.

This starts with the chemistry: due to the voltage range, lithium iron phosphate (LFP) batteries have clear advantages in this area, as four LFP cells can reproduce the voltage range of a lead–acid battery well. This is important as lithium-ion starter batteries usually have to be able to replace lead–acid batteries one-to-one or are connected parallel to them, and the vehicle electrical system (with all its consumers etc.) cannot tolerate a wider voltage range than that of lead–acid batteries. NMC batteries, in contrast, do not match the voltage window of a lead–acid battery well with either three or four cells. The voltage window of six cells with LTO as the anode material also fits the existing voltage range of lead–acid batteries well, but at a higher price and lower specific energy than other lithium-ion cell chemistries. Table 13.7 shows the corresponding voltage ranges. The ranges shown there cannot be specified more precisely due to manufacturer-specific requirements.

LFP cells allow a wider temperature range during operation. Lead–acid batteries are often installed in the engine compartment, which may be prohibitive for lithium-ion starter batteries without very complex and therefore expensive cooling measures.

LTO cells have better properties than all other lithium-ion cells in terms of high-current load, even at low temperatures.

Compared to lead–acid batteries, lithium-ion batteries of similar capacity have a lower internal resistance at room temperature and only a very small dependence on the discharge current. The nominal capacity of a lithium-ion battery could therefore be lower than that of a lead–acid battery in terms of starting requirements as long as the temperature does not approach freezing point. However, under today's requirements for the safe starting of internal combustion engines at temperatures around −20 °C (depending on the country and applicable standards), the high rate, low-temperature capacity of a starter battery is the overriding factor

Table 13.7 Voltage comparison of batteries for vehicle electrical systems.

Battery type	Maximum operating voltage	Minimum voltage outside the starting process	Limitations
Lead–acid batteries – six cells	Approx. 15.5 V during recuperation	Approx. 9 V	Wiring system components are designed for this voltage range
NMC, NCA/graphite – four cells,	Approx. 16.4–16.8 V	Approx. 11.0–10.0 V	No full charge possible, capacity cannot be fully utilized
NMC, NCA/graphite – three cells	Approx. 12.3–12.6 V	Approx. 8.25–7.5 V	Capacity cannot be fully utilized when discharging
LFP/graphite – four cells	Approx. 14.4 V	Approx. 8.8 V	Good overlap
NMC, NCA/LTO – six cells	Approx. 16.8 V	Approx. 10.8 V	No full charge possible, capacity cannot be fully utilized
NMC, NCA/LTO – five cells	Approx. 14.0 V	Approx. 9.0 V	Capacity cannot be fully utilized when discharging

and gives lead–acid batteries an advantage. At present, it is very difficult and expensive to replace lead–acid batteries if the vehicle must be able to start an engine at temperatures that may exist in Arctic or continental climates (−20 °C or lower).

13.11.4 Batteries for Stationary Applications

Lithium-ion batteries are in fierce competition with lead–acid batteries in stationary applications. Lithium-ion batteries only have advantages if installation space or weight problems[10] as well as the perceived longer service life and higher reliability are important and low costs are not necessarily the main criterion.

The requirements that lead to the selection of a particular technology or chemistry depend very much on the specifics of the application and the properties needed (see chapter 19). General trends are difficult to predict. It is worth mentioning that lithium-ion batteries can have a much smaller capacity than lead–acid batteries for short bridging times, which may be a reason for their use. However, service life considerations should not play a role, as high-quality lead–acid batteries in this area have a proven service life of 12 years if properly serviced. At present, lead–acid batteries still dominate this application mainly for cost reasons.

It should be mentioned at this point that there are many discussions about using discarded electric traction batteries (which often still have between 70% and 80% of their original capacity when they are replaced) for further use in the stationary sector; this is known as "Second Life." So far, however, the framework conditions for

10 Large stationary lead-acid battery systems weigh several tons. An exchange with a larger battery capacity and thus higher weight is then often no longer permissible.

when a lithium-ion battery can actually be used for a second time, again for several years, are still unclear, and complicated questions regarding product liability, for example, still need to be clarified. Large-scale use as Second-Life battery systems is therefore not yet an option – especially as the large quantities of discarded electric traction batteries with suitable properties are not expected for several years.

13.11.5 Consumer Batteries

Portable devices such as cell phones and laptops were the original motivation for the large-scale introduction of lithium-ion batteries, which have now completely replaced the alkaline nickel systems that were initially common in this area. As the batteries in these applications often only consist of only a few cells, sometimes just a single cell, standardized round cell formats are often the preferred technology, with pouch cells also being used, but prismatic cells rather infrequently. As mobile devices today are generally used for a significantly shorter period compared to electric vehicles, requirements concerning the number of cycles and calendar service life tend to be lower, while no compromises can be made in terms of capacity, performance, and safety. Similar to the automotive sector, fast-charging capability still plays a special role, so specific charging characteristics (e.g. with pulses) are often used for the first time in this sector.

The cells used in this area of application generally have a much smaller capacity than those required in other applications, and the nominal voltages of the battery are low. BMS can therefore be rudimentary or even completely dispensed with if the internal safety measures of the cell (CID, PTC, etc.) are designed to be sufficiently safe.

13.12 Properties

13.12.1 Discharge Behavior and Capacity

Figure 13.14 shows typical discharge curves of a prismatic lithium-ion cell with NMC graphite chemistry.

The discharge curve shown in the diagram is similar in shape to the discharge curves of lead–acid batteries, as there is a relatively clear and easily recognizable relationship between voltage and state of charge. Compared to lead–acid batteries, however, the much higher current-carrying capability is striking; even at very high currents such as 10 C, the immediate voltage drop is still comparatively moderate.

However, the comparatively good relationship between state of charge and voltage measurement is by no means a given for all lithium-ion systems. Cells with LFP chemistry, for example, often have very flat discharge curves, which poses major challenges for battery diagnostics, as direct determination of the *SoC* is only possible with a particularly precise voltage measurement or a combination of voltage measurement and ampere-hour integration.

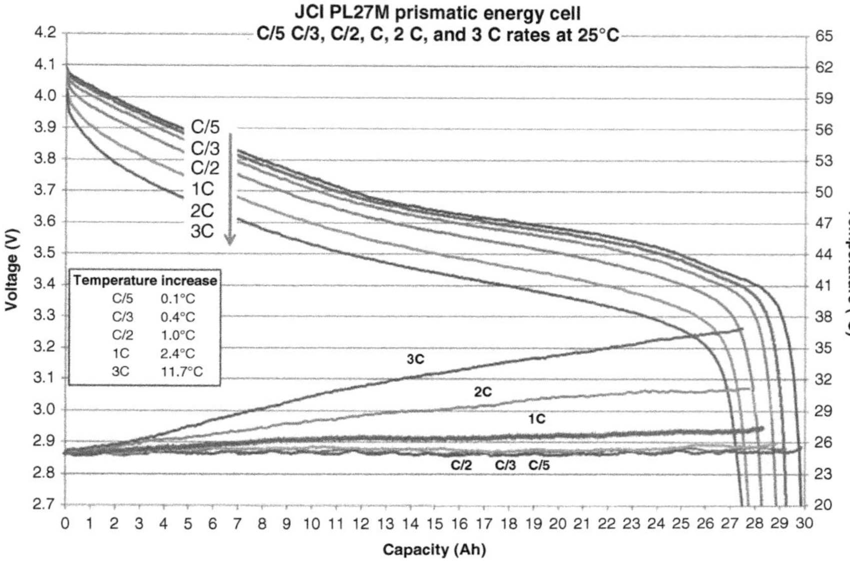

Figure 13.14 Discharge curves of a prismatic cell from Clarios with discharge currents of C/5, C/3, C/2, C, 2 C, and 3 C at 25 °C.

The curves at the lowest voltage are those of cells with LTO anodes, which leads to the lowest specific energy.[11] Such cells are therefore not usually used in the field of electric traction with its driving range issues, but rather in stationary applications, although the excellent cycle stability properties hardly play a role in this area of application.

The ratio of maximum charging voltage and minimum voltage during discharge is of great technical importance in many energy systems because a high voltage range for the permanently connected loads and chargers requires additional design considerations. Compared to lead–acid batteries, the voltage range, that is, the difference between the final charge and discharge voltage, is higher for many commercial lithium-ion batteries.

13.12.2 Capacity Specification and Capacity Measurement

The capacity specification for lithium-ion batteries usually refers to a one-hour discharge current (1 C) until a fixed end-of-discharge voltage, which is the same for all currents, is reached and a temperature in the range of 20–25 °C. Various standards, for example, the ISO 6469 series, stipulate that such a discharge process should be repeated several times (typically three to five times) and an average value should be calculated, provided that the individual measurement results do not deviate too greatly from one another. Between the individual capacity measurements, a certain waiting time must be observed in each case. However, tests by various

11 A cell with an LFP cathode and LTO anode with a nominal voltage of 2.1 V was produced for research purposes. The service life and cycle stability are exceptionally high, but at the expense of a significantly lower specific energy.

research institutes have shown that the capacity of new cells is usually very similar and corresponds to a Gaussian distribution with small variations. There is no preselection of cells necessary before connecting them in series besides their state of charge. Recharging between discharges is usually carried out using the constant current/constant voltage (CC/CV) method until the charging current falls below C/20 or C/100, depending on the manufacturer's specifications.

Lithium-ion batteries usually have Peukert coefficients close to one, which results in a comparatively low current dependence of the capacity. Only at very high currents does the capacity of lithium-ion batteries decrease significantly. The temperature dependence of the capacity, on the other hand, is usually much greater, and at temperatures below 0 °C, significantly lower capacities must be expected. In practice, at higher currents and very low temperatures, there is often a strong heating effect, which has a positive effect on capacity. In discharge tests at very low temperatures, the heating may even result in an increase in voltage during a constant current discharge. Comparisons to the so-called voltage sag in lead–acid batteries, which has a different cause, are not appropriate.

13.12.3 Monitoring Requirements

To ensure good performance in the long term, it is important that the specified operating parameters are adhered to and that the individual cells do not have voltage values that deviate too greatly from one another. For safety reasons, it is also necessary to detect deviations in individual cells at an early stage and to be able to initiate countermeasures if threshold values are approached or even exceeded. The voltage is usually measured for each cell and the current for each string, and the temperature and possibly other parameters (pressure, concentrations of selected gases, etc.) are also measured at selected points. The evaluation of the measurements and the resulting countermeasures, and the role of a BMS, are discussed in detail (see sections 13.17–13.19). Operating lithium-ion batteries without a BMS compromises safety.

13.13 Internal Resistance Measurement

For power-oriented applications, the internal resistance is usually determined by a current interrupt method, as described in chapter 9, or (in research environments, rarely in the industrial sector) by methods of electrochemical impedance measurement. Data usually refer to a state of charge of 50% and in the direction of discharge. Fully charged batteries have a slightly higher internal resistance, and discharged batteries a significantly higher one, both in the direction of discharging. As with all electrochemical systems, the internal resistance depends on the direction of the current and the current amplitude (see chapter 3).

The dependence of the internal resistance on the temperature is of great practical importance. With the exception of cells with a titanate anode, the internal resistance at low temperatures is so high that lithium-ion batteries cannot be used, for example, for starting processes at low temperatures (the usual temperature requirement is −18 °C). However, compared to lead–acid batteries, cells with a titanate anode still have significantly poorer low-temperature properties.

13.14 Charging and Charging Characteristics

13.14.1 Charging Characteristics

A number of different charging processes exist for lithium-ion batteries, which are still being continuously developed and optimized. In the laboratory, the CC/CV method is mostly used, in which charging is first carried out at a constant current until the end-of-charging voltage is reached, and then at a decreasing current at a constant charging voltage until the current has decreased. Voltage thresholds depend on the cell chemistry and the manufacturer's specification. Full charge is usually reached when the current falls below C/20 or C/100 (or other values depending on the manufacturer's specifications).

An example is that special pulse charging characteristics have been developed to shorten the charging time by using higher currents but incorporating rest periods and sometimes even discharge phases into the complex pulse sequences to avoid battery damage caused by excessively high voltages, temperatures, or currents. In some cases, even advanced battery analysis methods such as impedance spectroscopy are used. One example is the method shown in Figure 13.15, which also uses temperature detection to control the charging process [6].

13.14.2 Full Charge

In contrast to lead–acid batteries, where there is a precisely defined and practically achievable state of full charge (no discharged active material in either electrode), the state of full charge of a lithium-ion cell, that is, the degree of intercalation, and the end-of-charge criterion are specified by the battery manufacturer. At full charge, there are still places in the graphite where lithium ions could be stored, and there still are lithium ions available in the host lattice of the cathodes.

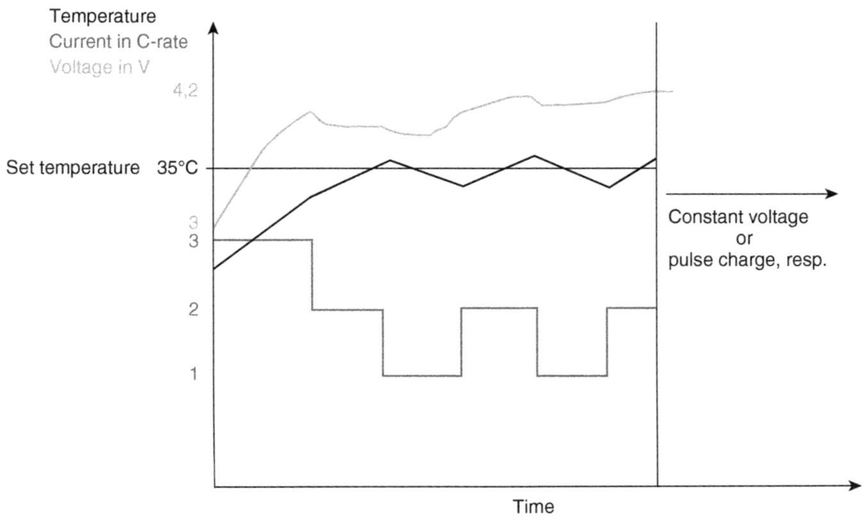

Figure 13.15 Temperature-controlled charging process. Source: Reproduced with permission from Ref [6]/with permission of Marcel Thiele.

As lithium-ion batteries age significantly over time, especially when fully charged, a full charge is not normally desired, and the maximum degree of intercalation specified by the manufacturer is relevant for the service life; a "practical" fully charged battery may therefore have an *SoC* of, for example, only 95% to achieve a longer service life – at the cost of the specific energy and driving range.

Please remember (see chapter 3) that the absence of appropriate side reactions has two consequences:

- It is not possible to fully charge both electrodes if their capacity begins to differ as a result of aging, and
- Cells in a string can have at the end of the charging process, only a very similar state of charge if the BMS of the battery provides electronic circuits for balancing.

13.14.3 Solid-State Diffusion During Discharging and Charging

Lithium ions stored in the anodic and cathodic host lattices have a uniform distribution respectively the lowest energy state in the host lattice, after all diffusion processes have stopped. The charging and discharging process disturbs the equilibrium and causes concentration gradients of lithium ions in each particle of the active mass. During charging, each particle of the anode contains more lithium ions at the interface with the electrolyte or SEI layer than in the center. During discharging, the concentration gradient is reversed. Similar concentration gradients exist also in each particle of the cathode. It should also be noted that intercalation in graphite takes place in an ordered form, that is, the lithium ions first intercalate in every third layer, then in every second layer, then in every layer. During this process, lithium ions are rearranged, which is controlled by solid-state diffusion and can cause additional aging effects.

13.14.4 Charging at Low Temperatures

The limited diffusion rate of lithium ions in the active mass particles can lead to the deposition of metallic lithium on the outside of the SEI on the anode, so-called lithium plating, particularly at high currents and low temperatures. The charging current of lithium-ion batteries at low temperatures and/or high currents must therefore be limited by the BMS. Once formed, metallic lithium reduces the capacity and increases the internal resistance because the effective total surface area of the anode and the total amount of freely available lithium ions for the discharge process are reduced. In addition, metallic lithium reduces the safety of the cell, even without current flow, because the chemical reactions of lithium metal with components of the cell do not stop just because there is no current and accelerate considerably, for example, when the cell is subsequently warmed up to room temperature.

In applications where very low temperatures must be expected (starting of engines and power supply systems in Arctic regions), the use of lithium-ion batteries is only possible with good thermal protection and heating.

13.14.5 Fast Charging

Fast charging is particularly important for electric vehicles. It is not clearly defined which current amplitude or charging time the term “fast charging” refers to and whether it refers to reaching the full charge criterion or a practically useful high state of charge, for example, reaching the end-of-charge voltage. Usually, at charging currents above 2 C,[12] that is, a discharged battery is charged to its final charging voltage in just under 30 minutes, corresponding to approx. 90% state of charge, it is referred to as fast charging. It is clear that the cells are thermally stressed, and the mechanical stress on the active masses also increases due to the limitations of the transport mechanism within the active mass particles. The lower the temperature of the cells, the more problematic high charging currents are.

Frequent fast charging is expected to reduce battery life. The extent of this is the subject of ongoing research activities.

13.15 Aging Effects

13.15.1 General Aging Effects

For all electrochemical systems, a distinction must be made between calendar aging which also takes place without current flow, and cyclic aging, that is, aging caused by successive charging and discharging processes. The fundamental aging mechanisms have been known for some time and have been described in various review articles, see for example Ref. [7]. New phenomena are constantly being reported due to the rapid development of technology and known aging processes are understood better. In principle, almost all components in lithium-ion batteries are affected by aging processes and have an impact on battery properties.

The aging mechanisms in lithium-ion batteries are generally highly dependent on the voltage both at high voltages during charging (a high state of charge) and at low voltages at the end of discharging. A good overview is provided by Figure 13.16. In contrast to lead–acid batteries, the lowest rate of aging of lithium-ion batteries does not occur in a fully charged state, but in partially discharged states at a low *SoC* range, that is, between 20% and 40% (depending on the technology or chemistry). The usual state of charge of lithium-ion batteries during transport is therefore in this range. However, this comparatively low state of charge also means that the voltage level or state of charge of lithium-ion batteries must be checked regularly so that they do not fall into the range of impermissibly low voltage values due to self-discharge reactions.

In addition, all aging processes depend on temperature, and exceeding temperature threshold values is both a safety and a lifetime risk.

12 A 2 C current for a 100 kWh battery, including losses in the battery and power electronics, exceeds 200 kW. For the low voltage electricity grid this will usually mean strengthening the grid and making an additional connection to the medium-voltage distribution grid.

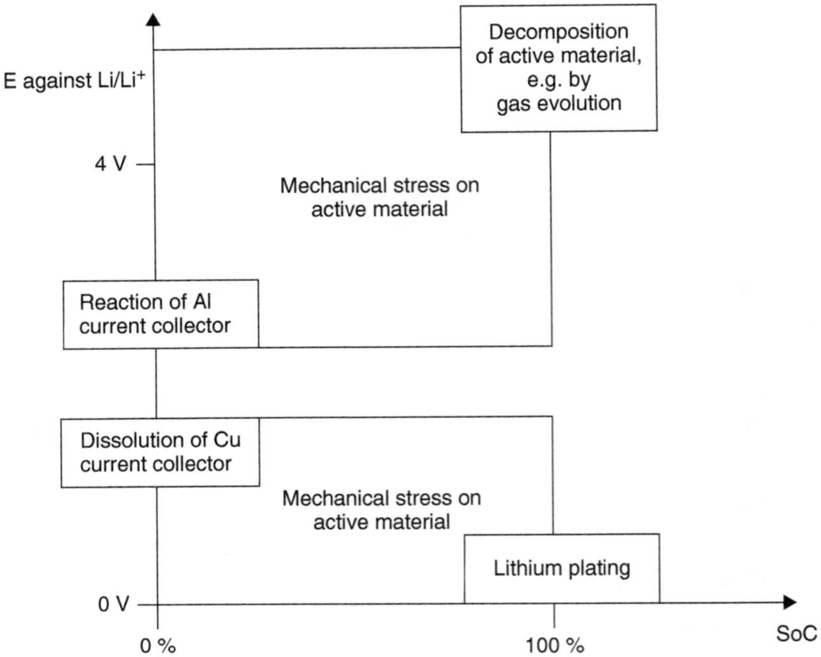

Figure 13.16 Voltage dependence of the most important aging processes in lithium-ion batteries.

13.15.2 Aging of the Cathode

The possible effects of aging of the cathode are shown in Figure 13.17. Structural changes of the cathode materials are most frequent, resulting in the loss of electrical contact between electrochemically active particles, binder materials, and current collectors due to, for example, decomposition of binder material or oxidation of conductive additives. This occurs particularly at high voltages.

Structural changes within particles of the cathode material or on the surface are the result of internal redistribution processes, which occur particularly at higher temperatures. They often lead to uniform distributions, but sometimes also cluster formations of similar ions. At comparatively low voltage levels, on the other hand, certain ions, usually Mn^{2+}, can be released from the crystal structure. They then lead to structural instabilities on the one hand, but also act as redox shuttles on the other (for details, see section 13.15.4) and thus reduce the current yield and ultimately the overall efficiency of the battery. Microcracks in the cathode material as an aging process, on the other hand, are more likely to be the result of mechanical stresses caused by frequent cycling, high currents, and the resulting fluctuations of concentration gradients in particles and temperature. In contrast to lead–acid batteries, however, direct electrical contact between active mass particles only plays a minor role. At high voltages, gaseous substances (especially oxygen) can also be released, which can then cause a build-up of pressure or harmful subsequent reactions (e.g. on the separator) and contribute to the formation of fire and oxygen supply to it.

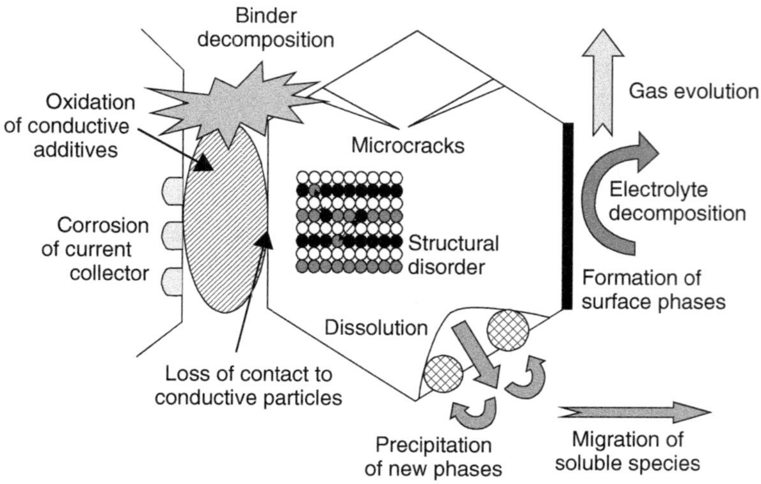

Figure 13.17 Points of attack for aging mechanisms on the cathode.

Cell voltage, current amplitude, and temperature have the effects shown in Table 13.8 and result in changes to the properties of the cathode. Exceeding threshold values must be reliably prevented by a BMS.

In many cases, the changes in properties caused by the various aging processes of the cathodes are less significant than those of the anode material.

Table 13.8 Summary of cathode aging.

Component	Aging phenomenon	Impact
Active material	Structural disorder due to lattice defects and/or phase transitions or overlithiation of the host lattice	Capacity reduction
	Dissolution of metals resulting in the formation of dissolved metal ions, deposition of new phases or formation of barrier layers on surfaces	Decrease in capacity and – with the formation of barrier layers – increase in resistance
	Release of oxygen from the cathode material, which has an oxidative effect on inactive components of the cathode and the electrolyte	Increased resistance, irreversible increase in pressure likely
Inactive components (binders, current collector, etc.)	Decomposition of the binder	Increased resistance due to loss of contact
	Destruction (e.g. through oxidation) of the conductive carbon	
	Corrosion of the current collector	

13.15.3 Aging of the Anode

There are many different aging processes of the anode. They can be divided into aging processes of the SEI and aging mechanisms that affect the bulk of the anode material (see Figure 13.18). Due to volume changes during cycling, both the particles in the cathode and in the anode are mechanically stressed, but more significantly in the anode. Cracks form, which can lead to the disintegration of the particles. An interesting approach to modeling this process is the application of fracture mechanics concepts from fatigue strength theory, which even provides a quantitative approach to the loss of capacity (see chapter 7).

There are four aging mechanisms in particular:

a) Growth of the SEI due to progressive electrolyte decomposition and incorporation of other battery materials in the SEI: The irreversible loss of lithium contained in the SEI reduces the capacity, the electrical resistance of the overall system increases and heat generation increases at high currents, which ultimately leads to reduced performance data and early failure of the battery. Here, too, higher temperatures are particularly damaging.

The increase in thickness x of the SEI is a diffusion-controlled process over time t, whose general form is well known and is described by a square root law:

$$x = kt^{0.5}. \tag{13.6}$$

According to Arrhenius law, the proportionality constant k is temperature dependent. The diffusion induced growth of the SEI increases the risk of SEI

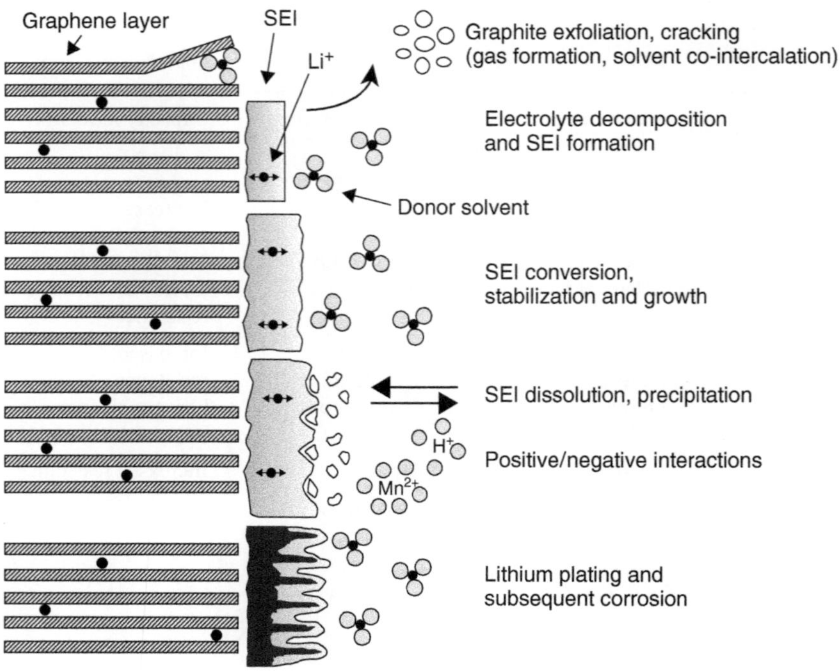

Figure 13.18 Aging of the anode and the SEI.

cracking and not remaining compact. Cracking of surface layers is a phenomenon that occurs wherever the mechanical properties of the surface layer differ from those of the bulk material. This occurs, for example, due to temperature changes or due to volume changes in the anode material under high cyclic operation. Cracking of the SEI exposes unprotected anode surfaces, which are immediately covered by a fresh SEI layer.

b) Slow exfoliation of individual graphite layers because larger particles (molecules from the electrolyte, solvated ions, etc.) are deposited in the graphite close to the surface or SEI, thereby generating mechanical stresses that ultimately cause individual graphite layers to peel or flake off. This mechanism plays a role in particular when the SEI layer cracks.
c) When lithium-ion batteries exceed their permitted temperature window, the first exothermic reaction which may ultimately lead to TR is the decomposition of the SEI and its disappearance. As soon as the electrolyte comes into contact with the unprotected surface of the anode, a new SEI is formed in an exothermic process which brings a cell closer to a TR. The newly formed SEI has different properties, among others, a better thermal stability has been reported [8]. At temperatures above c. 120–130 °C, the heat generated by these exothermal processes can contribute to initiating a TR.
d) Lithium plating, that is, the deposition of metallic lithium on the anode surface and areas of the current collector not covered by anode material, plays a role, especially at high currents and low temperatures. This mechanism is – presumably due to transport inhibition – more relevant the lower the temperatures, but it can also occur at higher temperatures. This process is critical because it is not only relevant during active battery use, but also during rest periods. Subsequent chemical reactions can lead to the formation of gaseous components, and the subsequent pressure buildup can cause the rupture valve to open (venting) or even the cell housing to burst. However, lithium plating can be largely suppressed of an appropriate BMS and is then no longer a relevant aging mechanism.

Finally, a special aging mechanism of the anode is the dissolution of the copper collector, which can occur in particular at low electrode voltages, for example, low states of charge during deep discharge. As a result of redox processes, the copper ions that dissolve can form dendrites, just like lithium ions. Depending on the surface energy of lithium, metallic lithium can also form dendrites. Dendrites cause short circuits between the anode and cathode after penetrating the separator. Due to the small cross-section of dendrites, however, these are high-impedance short circuits that cause self-discharge effects rather than thermal problems. In addition, a dissolution of a lithium dendrite is to be expected because the entire cell voltage drops across it and lithium is not stable above the equilibrium voltage of Li/Li^+.

The graphite anode ages faster than any cathode materials used. As soon as this loss of the anode capacity becomes too great, the cell voltage which is divided between the two electrodes leads to the anode voltage falling into the range or below the Li/Li^+ voltage where metallic lithium is rapidly deposited on the anode.

Oversizing the anode and reducing the maximum cell voltage for aged cells prevents the voltage at the anode from becoming too low.

Other anode materials such as silicon and similar intermetallic alloys age primarily due to the strong volume changes during intercalation. This results in mechanical stresses that can lead to the breaking of silicon particles or loss of electrical contact. So far, it has not been possible to reduce these aging processes in pure materials to a technically acceptable level, so these materials are currently only used in commercial cells as additives for graphite anodes.

As a zero-strain material (no volume change during charging and discharging and therefore no mechanical stresses), LTO shows few aging effects, and lithium plating is unlikely due to its voltage level compared to Li/Li^+. An SEI is not formed. In general, it is not the anode that limits the service life of cells with an LTO anode, but the cathode or electrolyte. However, increases in resistance that are attributed to the anode may be due to microcracks in the LTO, and aggressive decomposition products from the electrolyte can also damage the current collector of LTO anodes.

13.15.4 Aging of the Electrolyte

In addition to the reactions between the electrolyte with anode and cathode materials, there may also be changes in the electrolyte itself. There are additives to prevent the accumulation of damaging substances such as water, HF etc. However, once these are used up, gas-generating reactions can occur.

As the electrolyte in lithium-ion batteries is not an electrochemically active component, its quantity can be limited to that required for transport from the pores of one electrode via the separator to the pores of the other electrode. The amount of electrolyte that may be damaged before the function of the battery is impaired is therefore relevant to the service life. In general, the extent to which aging processes affect the performance of a battery or influences its service life depends on its design, see section 13.10.2.

Metal ions released from the cathode, such as Mn^{2+}, can also take over part of the current through the electrolyte by changing the oxidation state, which then leads to a drop in the coulombic efficiency (redox shuttle, see above). In addition to Mn^{2+}, other ions also relevant – copper has already been mentioned, but nickel, iron and other elements that are present as additives or impurities can also go into solution and cause undesired reactions. A possible reaction equation for the phenomenon using the example of dissolving manganese ions is as follows:

$$\mathrm{Li}_{1-x}(\mathrm{Ni}_{1/3}\mathrm{Mn}_{1/3}\mathrm{Co}_{1/3})\mathrm{O}_2 \rightleftarrows \mathrm{Li}_{1-x}(\mathrm{Ni}_{1/3}\mathrm{Mn}_{1/3-y}\mathrm{Co}_{1/3})\mathrm{O}_2 + y\mathrm{Mn}^{2+} + 2y\mathrm{e}^- \tag{13.7}$$

$$\mathrm{Mn}^{2+} \rightleftarrows \mathrm{Mn}^{3+} + \mathrm{e}^-. \tag{13.8}$$

13.15.5 Corrosion of and Damage to the Separator

The separator can be oxidized by contact with oxidizing agents (e.g. metal ions with a high oxidation state), which ultimately leads to the formation of holes and thus possible to (micro) short circuits and releases gases, mainly CO_2. This can result in increased self-discharge rates, among other things. The formation of dendrites (from lithium or copper deposits) can lead to the piercing of the separator and low-resistance short circuits between the electrodes, which may initiate TR. Finally, particles that become detached from either the anode or cathode due to mechanical stress of the materials can clog the microscopic holes in the separator, resulting in increased internal resistance. Larger particles, which may be present in the anode and cathode as a result of quality problems in production, may dislodge during use – this may take many months – and may cause a hole in the separator and thus a low-resistance short circuits. This effect is considered to be a major cause of TR and one that cannot be prevented by a BMS.

13.15.6 Other Aging Processes

The formation of gaseous components leads to increased internal pressure inside cells. When aged cells are opened for further investigation, it is essential that protective clothing be worn, as there is a risk of contamination with HF and its reaction products. A pressure of 2 or 3 bar can be expected in cells at the end of their lifetime. In prismatic and round cells, the rupture valves have an opening pressure in the range of 5–7 bar, so aged cells do not vent if no abuse occurs. Pouch cells, on the other hand, bulge visibly, as the mechanical strength of the foil making up the cell housing is limited. The pressure can even lead to the cell bursting if there is no external protection or stabilization. In any case, pouch cells are mechanically problematic. Swelling of the cell increases in particular if the aluminum barrier applied to the foils to prevent the diffusion of water vapor into the cell is damaged, either by mechanical or (electro)chemical influences.

13.16 Influence of Calendar and Cyclic Aging and Modeling

13.16.1 Aging and the Need for Modeling

Each individual aging process has its own temperature, current, state of charge, changes in state of charge, voltage, and time dependencies, so predicting the effects of all aging processes together on cell and battery properties is a very complex process. In addition, the operating conditions of batteries are application-specific, and the various aging processes are subject to complex interactions, see the explanations in chapter 6. The effects of aging processes on cell properties and thus service life can be modified by material selection and design adaptations. Any statements concerning service life can therefore only be made for specific products. In the following, experimental results are first summarized, and then modeling approaches are presented. Depending on the use of a battery, the load profiles are very different, and either calendar or cyclic aging processes or a combination of these must be taken into account.

To separate the different aging effects from one another, the use of modeling is essential. Modeling can start at different levels, at the level of the cell subcomponents, at the level of the entire cell, and finally at the level of the entire system. Modeling for larger systems with many cells must take into account differences in cell properties and temperature differences. Battery models that assume identical cell properties are only meaningful for temperature models to determine the temperature differences between cells, see chapter 6.

13.16.2 Modeling and Simulation of Aging

Modeling and simulation are frequently used tools in the battery sector, especially in the automotive sector, both in product development and for characterizing system performance. For larger systems in particular, they are used for:

- Evaluation of design requirements and their limits,
- Cost-effective investigation of a large number of different possible scenarios,
- Identification of "worst-case" conditions as part of safeguarding system performance.

Particular attention is paid to safety behavior and especially the topic of TR and thermal propagation. In addition to the information given above, the focus in the following is on the simulation of battery packs and systems. Simulations are essential in the context of a complete safety assessment of lithium-ion batteries because

- The individual experiments at module, system, or even vehicle level are far too expensive and complex for experimental data alone to be sufficient,
- Regulatory requirements already include safety verifications in accordance with current standards, which also require simulation or an interaction of simulation and experiment,
- Test capacity for all conceivable tests is not available, and
- Sometimes simulations help to verify hypotheses from test results or steer development in new directions.

On the one hand, three different levels or length scales, and, on the other hand, two general types, can be distinguished in battery modeling (see Table 13.9). Since modeling in the lithium-ion sector is more important compared to other battery technologies due to the large number of possible aging and safety-relevant mechanisms, it is explained in more detail in this chapter.

The two types of models at the system level are

- Empirical models, that is, those without physicochemical information or those that can only be traced back to physicochemical information through interpretation, such as equivalent circuit diagrams, and in the field of impedance spectroscopy where physical data can be assigned to the model elements, and
- Analytical models based on (electro)chemical, thermal, and/or other physical principles.

A further subdivision of the model classes is not made here.

Table 13.9 Modeling for battery systems at different levels.

Level	Material level	Cell level	System level
Length scale	Microscopic	Mesoscopic	Macroscopic
Processes considered	Basic processes such as charge transfer, lithium diffusion through active materials and electrolyte or particle deformation	Thermodynamic, thermal, mechanical, kinetic, and mass transport phenomena, as far as necessary for modeling cell behavior	Phenomena that arise from the integration of several cells into a module; necessity to allow variations of cell parameters
Restrictions	Too computationally costly and therefore time-consuming for larger systems; the effort required to determine the material properties with sufficient accuracy is very high	For reasons of efficiency, microscopic physicochemical models can only be used in a reduced form on this scale	Differentiations between individual cells are generally neglected, which limits its value

At the system level, a model basis is currently used relatively frequently, which was originally developed as part of the so-called CAEBAT project (the acronym stands for "Computer-Aided Engineering for Electric Drive Vehicle Batteries"), starting in 2010 under the leadership of the US Department of Energy. The project is led by the National Renewable Energy Laboratory (NREL) in the United States and involves various US laboratories, several industrial partners, and other research institutes.

The aim is to develop a particularly powerful battery simulation tool. The program develops a flexible model for predicting battery behavior on a macroscopic level with a wide variation of system performance and abuse conditions. The program's model basis is currently used in several commercial software solutions. Historically, the development of CAEBAT took place in three stages. First – from 2011 – a multiscale and multidomain model (MSMD model) with a modular multiphysical framework was developed to serve as a software tool for cell and battery design. CAEBAT-II, which was available from 2016, then focused on safety aspects and coupled MSMD-like electrochemical, thermal, and chemical models with a dynamic-mechanical crash package. CAEBAT-III, which is currently undergoing further development, focuses in particular on the effect of the microstructure to understand the influence of material composition and production control.

The general advantages of such MSMD models result from their applicability to large-format prismatic cells and battery systems, whereby the interaction of physical phenomena on different scales is addressed, for example, in the case of simultaneous inhomogeneity of the electrical potentials in individual cells in modules, such as in the case of inhomogeneity of the temperature in the cell. Such models therefore have a hierarchical structure in which solution variables in a lower hierarchy have a greater spatial resolution than variables in a higher hierarchy level. Independent coordinate systems are used at each level to spatially discretize the variables in that domain. This decouples the geometry but couples the physics. By selecting suitable submodel options, this results in a high level of physical and computational granularity, whereby the choice for each domain can be made independently of the choice for other domains.

Another approach was developed by Spotnitz et al. [9], who presented a new modeling approach for estimating the tolerance to thermal abuse for lithium-ion battery systems in 2007. The exothermic cell behavior is described on the basis of accelerated rate calorimetry measurements (ARC measurements), and the energy balance is then solved at the system level.

The most important and most widely used commercial and industrial software solutions are currently:

- COMSOL (for multiphysical battery and fuel cell models),
- Fluent from Ansys,
- Battery Design Studio & Star-CCM+ from Siemens,
- GTSuite in combination with Autolion1D, AutoLionGT, AutoLion3D (all from Gamma Technologies) or AutoLionST (the latter from EC Power), and
- TAITherm and Battery Module from ThermoAnalytics Inc.

However, the existing models can still not be regarded as meaningful enough to allow a comprehensive prediction of the aging process of existing or future battery technologies.

13.16.3 Quantitative Model Approaches for the Description of Aging Effects on Battery Properties

A large number of models for quantitatively describing the effects of aging[13] of lithium-ion cells on service life have been described in the literature. Below, the approach using exponential functions with the parameters current amplitude I and the number of equivalent full cycles is described in more detail. In Refs. [9–11], using exponential functions, (Z) was proposed as a mathematical description for the capacity $C(I, Z)$ as a function of current amplitudes I and the number of equivalent full cycles after a certain operating time:

$$C(I, Z) = C_0 \cdot e^{-(a_1 \cdot I + a_2) \cdot Z}. \tag{13.9}$$

Provided that the cells are charged under constant, usual nominal current and temperature conditions, $C(I, Z)$ in Ah predicts the capacity of the cell after full charging. The parameters a_1 and a_2 are current-dependent. C_0 is the capacity when the cell is new. The parameters $a_1 a_2$ can be determined using the measured capacities and optimization algorithms (such as the least squares method) with standard programs such as MATLAB, etc.

13 Please note that none of the equations in this chapter describe aging processes but only the cumulative effects on battery properties irrespective of the underlying aging processes, such as SEI growth, degradation of particles of electrodes, etc.

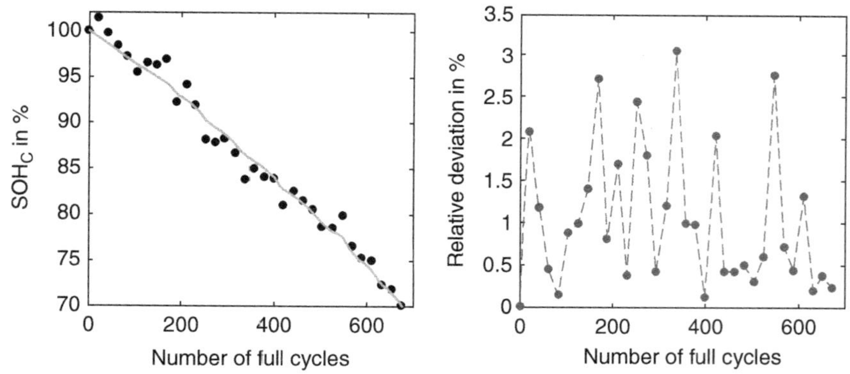

Figure 13.19 Capacity of an 18650 NMC cell, cycled at 30 °C with 100% *DoD* and a constant current amplitude of 1 C. The dots represent the measured capacities of the cell after 20 cycles, and the solid line corresponds to the calculated function $SoH_C(I, Z)$ determined. Source: Reprouced with permission from Ref [12]/Eric Tchoupou Lando, 2020.

The SoH_C can be calculated as a function of current amplitude and number of equivalent full cycles according to Eq. (2.6).

$$SoH_C(I,Z) = \frac{C(I,Z)}{C_0} \cdot 100 \tag{13.10}$$

resp.

$$SoH_C(I,Z) = 100 \cdot e^{-(a_1 \cdot I + a_2) \cdot Z}. \tag{13.11}$$

Figure 13.19, which was taken from Lando [12], shows the capacity curve of an 18650 NMC cell cycled at 30 °C with 100% *DoD* and a constant current amplitude of 1 *C*, and the curve determined according to these equations. The parameters $a_1 a_2$ can be determined using the measured capacities and optimization algorithms (such as the least squares method). The curve of the simulated $SoH_C(I, Z)$, which is marked by the solid line, is not smooth because the current values used for the simulation are based on the real, measured values; in these experiments, the current amplitudes were adjusted to the capacity of the cell which was determined after every 20 equivalent full cycles.

13.17 Battery Management Systems and Battery Operating Strategies

13.17.1 General

According to Gabbar et al. [13], the general task of a BMS is to monitor, control, and optimize the behavior of an individual cell and cells interconnected to form a battery. It may be considered safe to operate a single lithium-ion cell with low capacity without a BMS, relying only on internal safety devices, such as fuses, CIDs, or CLDs. As soon as cells are connected in series and/or cells with a higher energy content are used, a BMS is essential. In the automotive sector, this includes in particular energy management and safety during driving, charging and rest periods.[14] The BMS may consist of several levels:

a) (Rudimentary) monitoring at the cell level,
b) Monitoring (low complexity) and control system at the module level (in case the system uses modules), and
c) A complex control system at the level of the overall battery system.

Not all of these levels are always present.

The main tasks of a BMS for lithium-ion batteries are to ensure safe operation and a long service life. To achieve this, *SoC*, *SoH*, and any other important battery state variables (*SoF*, *SoE*, *SoS*, …) must be calculated as accurately as possible. In addition, a BMS should be able to control or modify the charging process, control the power to auxiliary functions and switch off the battery. In contrast to a simple measuring device, the battery history is usually also recorded to at least a limited extent, for example, by recording the ampere-hour throughput, calculating average values (*SoC*, operating temperature, etc.), storing times outside permissible or critical operating ranges, etc.

Each battery system requires its own BMS that has been optimized and tested for its intended use.

Hybrid energy systems require an energy management system for controlling the energy flow through the battery, possibly additional energy storage systems, loads, and energy generators to optimize the energy efficiency and the lifetime of the various components of the system. Such energy management systems may include functions of a BMS. For the purpose of this book, the discussion here is restricted to those functions which are required for the safe operation of batteries and achieving a high service life. The focus is on multicell battery systems with an energy content of a starter battery at least.

The following discussion therefore applies only to a limited degree to small batteries for consumer products, where the low energy content of the few cells connected to each other poses only little hazards.

Although there are general requirements for batteries compiled in various standards, these have so far not been broken down into general requirements for a BMS. There are not yet standards that describe minimum requirements for a BMS nor publicly accessible documents[15] from manufacturers. BMSs are considered by most companies as trade secrets, both as far as the hardware (sensors and sensor locations, wiring) is concerned as well as the algorithms used. It is therefore up to the manufacturer's know-how to which level a BMS that claims to ensure high safety levels contains all features which are known to provide early warning for malfunctioning and minimize the risk of TR.

---

14 The development of a thermal runaway and propagation to neighboring cells is also possible when batteries are not in operation. Deactivation of the BMS and its monitoring functions, which are important for the safety of the battery system, is therefore not permitted. Deactivation of a BMS corresponds conceptually to the deactivation of fire alarm systems when no one is in a building.
15 One rare exception: OpenSource BMS hardware and software project FoxBMS https://foxbms.org.

Future standardization is likely because of the following requirements: cybersecurity (it must not be possible to manipulate a BMS from the outside), requirements for battery replacement, easier battery reuse (the so-called second life of discarded electric car batteries in stationary applications, etc.), and the requirements of battery recycling. These requirements make it necessary to improve and facilitate the exchangeability of the data measured by the BMS and the results of the data evaluation.

Further requirements for data processing by a BMS arise, particularly in the automotive sector, from the increasing networking of vehicles with each other and with infrastructure such as charging stations. Monitoring points at governmental (eCall,[16] etc.) or nongovernmental (conceivable, e.g. TÜV or manufacturer-specific warning systems) levels can be expected, through which data relevant to safety and performance (e.g. remaining range) must be transmitted at an early stage, but also stored long-term and in a way that cannot be manipulated. To this end, data formats must be compatible with each other, so that data can be exchanged and evaluated in time.

As a BMS is the key safety unit for a battery system, it could be expected that standards and specifications will also include issues of redundancy and fault tolerance.

- Detection of measurement errors or sensor failure.
- Analysis of sensor values for realistic values: Temperature, current, and voltage values must be within the expected ranges. Voltage values can be checked by two independent measurements, for example, by comparing the module or string voltage and the sum of the individual cell voltages. Temperature values cannot jump. Current and voltage values must be in line with the power drawn by loads or supplied by power generators and chargers.
- Function of switches to disconnect the battery from other components and communication devices to be checked at regular intervals.
- Application of safety principles (fault propagation, fault mitigation analysis).

So far, all of this is only in development. However, it can be assumed that manufacturers of high-quality battery systems will follow and implement principles of safety engineering and contribute to standardization activities in their own interest.

13.17.2 General Requirements for a Battery Management System for Lithium-Ion Batteries

A BMS must have the following features.

1) Measurement of individual cell voltages,
2) Limiting current through each string of a battery,
3) Disconnecting the battery in case of safety problems, and
4) Provide advanced warning if a cell in the battery or the battery as a whole can no longer be considered safe.

16 eCall is an automatic emergency call system that is mandatory for new vehicles in the European Union.

There are different ways to fulfill these requirements. The following is therefore neither a complete list of requirements nor are all technical solutions for their implementation listed, for example, measuring the evolution of hydrogen as a first indication of a major thermal event.

13.17.3 Technical Implementation

Once the design of a battery has been finalized, the aim of a suitable BMS will be to constantly check the assumptions made about the expected operating conditions during the design phase. If the operating conditions differ too greatly and lead to a prediction of excessive aging, the BMS should document this concerning warranty issues and, as far as possible, restrict or change operating limits to such an extent that premature aging can be avoided. This is an important conceptual difference from lead–acid batteries, which generally have far less elaborate BMSs, if any, and where the inherent robustness and safety of lead–acid batteries mean that no special assumptions about battery operating conditions need to be incorporated into the design of a BMS. A BMS for lithium-ion batteries, on the other hand, has the following degrees of freedom, among others:

- Avoiding high state of charge by reducing the permissible *SoC* from 100% to, for example, 95% or 90% and limiting the minimum acceptable *SoC*,
- Earlier activation of cooling and heating depending on current and expected use,
- Reduction of the permissible charging current depending on temperature, charging throughput, the number of fast charging processes already carried out, and any other aging-relevant parameters,
- Controlling and monitoring of balancing algorithms to equalize cell voltages and/or *SoC* with each other,
- Changing the power requirements of loads and, in hybrid systems, altering the energy flow between the battery and other power-generating units.

Obviously, most of these options have an impact on the effective specific energy of the battery, bridging times, and cost structure, particularly in hybrid systems.

1) Voltage measurement
   Measuring the cell voltage of each individual cell in a lithium-ion battery is common practice and state of the art in terms of safety. Any deviation from this in commercial BMS is unlikely to be acceptable from a product liability point of view. Chapter 3 explained why cell voltages in series connections with slightly different cell properties differ, and in Task 3.3, it was shown that a high-resistance cell or a cell with a significantly lower capacity due to selective aging can have a cell voltage of, for example, 20 V at the end of the charge when 100 cells are connected in series.
   When measuring the voltage, there are a number of aspects to consider:
   - Either the voltage is measured between the terminals of each cell or, alternatively, between the terminals of the same polarity of adjacent cells. The latter will simplify the wiring but lead to uncertainties in case of an increased resistance of connectors.

- Ideally, all cell voltages have very similar values. Deviations must be detected and classified as either a measurement issue (e.g. disconnected sensing wire) or a cell problem by using appropriate software. The input range of the voltage measurement must be analyzed and voltage values above or below threshold values must lead to actions by the BMS. This will also detect sensor failures.
- For a fail-safe BMS, a redundant voltage measurement, for example, measuring the string voltage and the individual cell voltages, provides added safety.
- The voltage values of cells cannot be fed directly to an AD converter input port of a microcontroller, even when just a few cells are connected in series. Instead, galvanic isolation must be provided between cells and microcontrollers.
- Batteries with a high nominal voltage and several parallel strings can have several thousand cells installed. An error-free and fault-tolerant design and operation with single-cell voltage measurement are no longer trivial. An attempt is then made to limit voltage measurements to a few cells and use special evaluation algorithms to determine whether any cell has an unacceptable voltage deviation.
- The voltage measurement is also used for cell balancing and must be able to detect which cell has a voltage level that deviates too much from the other cells, for example, more than 30 mV. Measuring such voltages in a battery with very high current change rates and therefore high exposure to electromagnetic fields is no longer simple.

Abnormal conditions, such as mechanical damage to the cell, dendrites, or internal short circuits, all of which may possibly trigger a TR, result in very short (a few milliseconds up to c. 1 second) voltage spikes. Their detection requires either very fast sampling of the voltage, for example, 1 kHz which is unrealistic, or special electronics that record such events. It may be considered sufficient to detect such a fault in a string and not necessary to detect which cell in the string caused the voltage dips.

It seems appropriate to use three different threshold values, one that leads to the storage of the event (which cell, when, how long), one that must initiate further checks by skilled personnel and one that indicates imminent danger.

2) Temperature

Temperature measurements are usually carried out using thermocouples. Temperature measurement via impedance spectroscopy is possible but not widely used in practice. Radiation measurements, for example, by means of thermography, are also normally only used in the laboratory. The number of thermal sensors and their location is more important than the type of temperature measurement. It is possible that a single cell can have very different temperatures at different points. Close to TR conditions, temperature differences of more than 100 K at points a few centimeters apart have been observed [14]. It is therefore not easy to detect critical temperature hot spots. In chapter 5, data were shown to illustrate temperature differences in a module. Single-cell temperature measurement is therefore unusual in larger battery systems. Often, there are temperature sensors only at two or three points: the negative terminal of a cell or the module, a second point relatively far away (to detect temperature deviations in the module), and a third point for verification (e.g. to check that the entire module is within the temperature window calculated from the measurements at points 1 and 2). The module temperature or the temperatures of a cell in the module can be calculated from these measurements using more or less complex algorithms.

3) Current
The battery current must be measured for each parallel string. It is difficult to measure both very small (mA) and very large (e.g. 100 A) currents accurately, so the choice of sensors is not trivial.
4) Limiting current through each string of a battery
At the end of discharging, it must be possible to disconnect the battery from the loads, reduce power, or switch off loads that are not important or critical at this point in time, for example, heating or cooling a car. At the end of charging, the BMS must be able to stop the charging process or reduce the charging current in such a way that no cell in a string exceeds the permitted voltage threshold.
It is standard for the charging process to be enabled by the BMS to ensure that there is no mismatch in terms of voltage or current between the battery and charger. Communication between the charger and BMS must be maintained at all times during the charging process.
5) Disconnect the battery in case of serious problems
The BMS must be able to disconnect the battery from all loads if a serious fault has been detected, for example, short-circuit currents below the threshold at which fuses are triggered. In battery systems with parallel strings this requirement may lead to a number of switches to be installed, possibly one per string to avoid high internal currents – discharging one string to feed a short circuit in a parallel string.
6) Provision of advanced warning if a cell in the battery or the battery as a whole can no longer be considered safe.
The BMS as the control unit needs to be able to evaluate the data it has received. Comparing individual cell voltages for all cells and drawing conclusions, that is, differences are the result of aging, a particular cell or one or more cells in a string need to be checked further, or immediate actions have to be taken to mitigate risks, for example, of a TR. Such algorithms may include:
    - Identifying cells which require balancing more or less frequently than other cells, and
    - Comparing measured voltage, current and temperature values of the battery and individual cells with those calculated from models.

13.17.4 Balancing

For long battery life, it is essential that the cells are used as evenly as possible, as the weakest cell always limits the overall system both during discharging and charging. This is of particular important for lithium-ion batteries as there are no side reactions. Many BMSs provide the function of equalizing cell voltages, known as balancing.

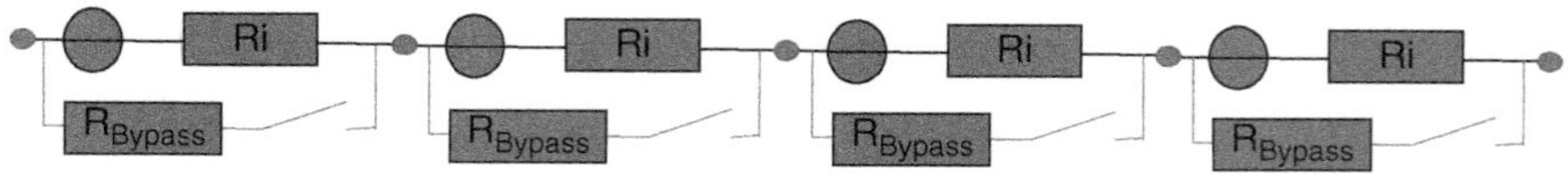

Figure 13.20 Schematic diagram of four cells in series with a passive balancing system.

There are two different types of balancing. In both cases, electronic circuits have to be installed across each cell which are used to ensure that all cells have very similar voltages at all times. Figure 13.20 shows schematically the equivalent circuit diagram of four cells in series with passive balancing.

a) Passive balancing
The balancing circuits are activated either during charging or discharging or at rest.Each cell is fitted with a bridging circuit consisting of a switching element that can be controlled by the BMS or another monitoring unit and a resistor. The resistance must be selected high enough so that the cell is discharged via the resistor with sufficiently low currents when the switch is closed, effectively short-circuiting the cell. It must be possible to dissipate the heat generated without any problems. The size of the resistor depends on when the balancing system is activated and the battery current that can be expected during balancing.
Balancing during charging: Balancing during charging has the advantage that all cells in series can be charged to their maximum capacity. The resistor must be able to conduct the charging current without heating up the cell. The BMS must therefore reduce the charging current as soon as the first cell exceeds its voltage threshold to a level that can be thermally managed. For a battery system with a maximum charging current of 1 A during charging, a resistance of 4.2 Ω will lead, according to Ohm's law, to a discharge current in the cell of 1 A when the switch of the balancing system is closed at a cell voltage of 4.2 V. There will be no net charging current in the cell and the charging process can continue for all other cells unchanged. Once the constant voltage charging phase starts, the cell will be discharged and the switch for the bypass current will be opened. The charging current will then lead to a fast voltage increase of the cell and the switch will close again to take up the current. This will ensure that the cell will remain within a narrow voltage band.
If slightly higher resistance values are chosen, the bypass will be activated earlier, for example, at a cell voltage of 4.1 V, and there will be a very small net charging current. In all cases the switch is opened again when the cell voltage has fallen to a lower value, for example, the average value of all other cells.
Balancing during rest periods: Alternatively, the cell balancing system is activated at times when there is no current, Cells with a higher voltage compared to the other cells are then discharged with a very low current via a much greater resistance so that all cell voltages are as similar as possible. This option is particularly suitable for battery systems that remain in a medium state of charge at all times, for example, batteries for hybrid vehicles or lithium-ion starter batteries if their maximum charging voltage is above the maximum voltage level of a 12 V onboard system.

Balancing during discharging: In addition, or as an alternative to activating the balancing system at the end of charging, the balancing system can also be used during discharging. The bypass circuit is activated once a cell falls below a threshold voltage limit and reduces the discharging current through the cell and the capacity of all other cells can still be used. Obviously, the resistances have to be chosen such that the cell is not discharged via the bypass. For safety reasons, such a balancing system cannot replace load management functions, which limit or stop discharging if one cell falls below its safety-critical voltage limit.

b) Active balancing
Active balancing involves selectively recharging cells with a lower voltage without removing them from a string. This can be done by an external charging circuit or by discharging cells with comparatively high voltage and using their energy to recharge cells with lower voltage. This is usually carried out at rest periods or when only a small current flows for residual loads.
When redistributing energy between cells, an intermediate storage device, usually a capacitor, is required. The capacitor is connected alternatively to a cell with a high voltage where it is charged and then to a cell with a low voltage. The advantage of not having to convert energy into heat (which in this case cannot be used any further) comes at the cost of a complex circuit. As a result, active balancing is used very rarely.

The technical implementation of balancing is complex, even with a passive system and used for one string only, and the battery system must be prepared for it from the outset. Additional hardware and algorithms for balancing parallel strings are required to prevent currents in parallel strings from being divided according to the total internal resistance of each string and, if necessary, to completely remove individual modules or strings from the circuit. Of course, the more complex the balancing circuit is, the more error-prone it is. And a balancing attempt with a faulty circuit can cause much more damage than good under certain circumstances.

Not to be underestimated is the additional acceleration of aging during active balancing due to equalizing currents during rest phases. These internal currents must also be carefully recorded, as they increase the total charge and total energy throughput of the cells in question, making corrections and cell-specific aging predictions necessary.

Due to the many difficulties, the use of balancing algorithms in real lithium-ion battery systems, even in electric vehicles, is therefore by no means a matter of course, but the result of a careful cost-benefit analysis in each specific case. The higher the product quality in terms of cell properties and the smaller changes in properties during aging, the more likely it is that cell balancing can be dispensed with.

Information from the balancing system, for example, for which cells the balancing system has been activated when and how much provides information about the general condition of the battery. For a new battery with high-quality cells with very small variations in internal resistance and capacity, the balancing system is unlikely to be activated. It is a sign of aging processes when more and more cells need to be balanced and a cause for alarm when the same cell always needs to be balanced.

13.17.5 Data Analysis and Fault Detection

Data analysis is of particular importance in a BMS. Not only must incoming data be permanently processed and converted into outgoing signals, but the incoming and outgoing signals must also be checked for plausibility, and warning signals concerning possible faults must be sent if data is missing or the analysis shows that the battery or an individual cell is approaching a critical state. The sampling rate of all measurements is usually in the range of 0.1–1 Hz. Detection of internal short circuits, which cause very short voltage dips, may then no longer be possible. The criteria and algorithms used to classify cells as being outside the expected values (outlying cells) are not trivial.

If a critical state is reached, a disconnecting switch may have to be triggered to disconnect the battery from all loads and the charging unit. If necessary, individual modules can be switched off respectively a bypass activated, if available.

In battery systems that work with active cooling or heating, the BMS is also responsible for temperature control, monitoring, and, if necessary, taking countermeasures if the temperature becomes too low or too high. In addition, the BMS must also have sensors and signals that detect when a cooling or heating device is unable to perform the specified tasks.

Concerning the extent of redundancy required for battery sensor technology to enable correct data analysis, please refer to the literature [15].

13.18 Determination of Battery States and Parameters

Since operation outside the specifications must be avoided at all costs when using lithium-ion batteries, it is very important to accurately determine the battery condition, particularly with regard to the state of charge and state of health (both in terms of capacity and internal resistance), as well as the safety of the overall system.

13.18.1 State of Charge

For the definition of the state of charge *SoC* and the fundamental difficulties concerning the concept, please refer to chapter 7. With lithium-ion batteries, the state of charge is usually determined by direct cell voltage measurements, because equalization and diffusion processes lead to the rapid development of a virtually constant voltage after current flow has stopped. The voltage is often more or less constant after approx. 15 minutes, usually at the latest 30 minutes.[17] However, there are some points to bear in mind:

a) Very flat discharge characteristics such as those of LFP cells require either very precise voltage measurements (0.1 mV or even less) or a combination of voltage measurements with other methods, see chapter 7 for an overview.

17 With lead–acid batteries, on the other hand, a waiting time of 24 hours or more is required in the worst case, and even then it is not possible to draw direct conclusions about the state of charge from the open-circuit voltage due to the phenomenon of acid stratification.

In high-voltage systems, the requirements for the accuracy of the voltage measurement are reduced because the average voltage of cells and thus the state of charge of the batteries is easier to measure.

b) The cell voltage characteristics depend on various other parameters, for example, the aging condition, the temperature (although generally not very pronounced) and the operating conditions, for example, whether the OCV is measured after discharging or charging. This generally requires the determination of areas of usage, which are used to parameterize different lithium-ion battery technologies and products. Determining these maps can be very time-consuming and may well take several months.
c) It is common practice to measure the voltage of each individual cell separately from a cell capacity of a few Ah upwards. However, this does not apply to other parameters, for example, temperature. As a result, *SoC* differences between individual cells in larger cell groups may initially remain undetected. To avoid these problems, validation of the battery state of charge determination derived from the cell voltages using additional methods can therefore be useful or even unavoidable.

Other methods already used widely to determine the state of charge are ampere-hour integration of charging and discharging currents and, in a laboratory environment, impedance spectroscopy. Methods such as those based on ultrasound or optical measurements are currently being developed on a laboratory scale.

13.18.2 Capacity, Internal Resistance or State of Health

Determining the capacity of a battery is not trivial under operating conditions. A discharge test can normally not be carried out because the necessary loads are not available and/or the residual capacity cannot be fully used. In most cases, the use of the battery is restricted or not possible. In addition, standard conditions with regard to temperature, pretreatment, etc., can usually only be used when the battery has been taken out of service and requires specialized test equipment.

In general, a different approach is therefore taken, and the capacity or state of health (*SoH*) is estimated using more or less complex algorithms using various existing measurements (voltage, current, temperature, charge throughput, etc.). The accuracy of these algorithms must be validated beforehand by laboratory tests in which real measured values and those calculated by the BMS are compared. These algorithms are manufacturer- and product-specific and represent core know-how for the respective product.[18]

It is easiest to determine the capacity of batteries that are frequently fully charged under controlled charging conditions and whose voltage, current, and temperature

---

18 The accuracies quoted by manufacturers are high, although there appears to be no published data on the accuracy that has been achieved. The effort required for this (measurement of the capacity that can still be removed in a particularly interesting state of charge range at nominal temperature, full charge, and capacity test) is in the range of approx. 4–5 hours for lithium-ion batteries.

curves as well as ampere-hour throughput can be measured during operation, for example, electric road vehicles with daily full charging. In such cases, the current during operation tends to be high and very long (many days) periods without or with only a very low current are seldom or can be neglected for the evaluation.

- Determination of full charge after completion of the charge by the charger and thus, by resetting the *SoC* value, correction of all errors that have accumulated since the last full charge.
- Evaluation of the voltage curve as a function of the temperature, current amplitude, and the accumulated ampere-hour throughput, and comparison of the measurement results with previously experimentally determined characteristic values.

On the other hand, it is difficult to determine the capacity and state of charge of batteries that have to be kept at a constant medium state of charge for operational reasons, for example, batteries for hybrid vehicles in which batteries have to be able to both absorb and release considerable amounts of energy. Difficulties also exist for batteries in stationary applications that are not discharged for very long periods of time.

13.19 Safety

13.19.1 General Safety Aspects

The safety debate has accompanied lithium-ion batteries since their market launch. A first generation of batteries with lithium metal as anode material had to be withdrawn because recharging due to dendrite formation was problematic. Although increasing use in various areas has led to a high level of trust in the technology, incidents such as those that have occurred with starter batteries in aircraft or smartphones have repeatedly shown how critical safety issues are for lithium-ion batteries. This is all the more true in the automotive sector because the amount of energy stored and the rate at which it can be released are much greater and could endanger a considerable number of lives in the event of an underground garage fire, for example. However, the fact that lithium-ion cells have been used billions, if not trillions, for many years shows that they are a technology that is suitable for everyday use and, if handled correctly, safe with manageable risks. Basically, all types of battery technology have their own safety problems. Of the systems currently being commercialized on a large scale, lead–acid batteries are affected the least by safety concerns and lithium-ion batteries the most. Nevertheless, safety problems such as TR are not limited to lithium-ion batteries, but are also exist in lead–acid batteries, for example, but with much lower danger.

The hazards of lithium-ion batteries can be classified into different categories. In particular, the following hazards exist:

a) Electrical hazards, see appendix,
b) Venting of gases and reaction products of decomposition processes such as toxic electrolyte vapors or their hydrolysis products (e.g. hydrofluoric acid)

or asphyxiating gases (e.g. CO, CO_2) or phosphorus compounds of very high toxicity similar to insecticides and pesticides, and

c) Spontaneous release of large amounts of energy, whether in the form of an explosion or a fire.

The appendix on safety provides a more detailed overview.

These risks can be minimized and mitigated by suitable protective measures and, furthermore, by taking precautions to reduce their effects should they occur despite all countermeasures and monitoring. Particular attention is currently being paid to the phenomena of TR and thermal propagation, that is, the spreading of a TR to neighboring cells, because the amount of energy and the rate of release seem impossible to contain.

Despite various attempts, the quantification[19] of safety aspects is a task that has not yet been fully solved. A first and widely used attempt is the so-called EUCAR hazard level classification, in which the impact of different outcomes of a TR is grouped into classes.

13.19.2 Definition of Safety

Safety is the absence of unacceptable risks for causing damage and health hazards outside the physical boundaries of a system (see ISO/IEC Guide 78:2012 for further details). Safety requirements therefore go beyond product reliability issues as failed products, too, must not cause unacceptable risks and a failed product should not reduce required safety levels. Safety must therefore be distinguished from reliability, but both are only possible for cells designed, manufactured, integrated into a complete battery system, monitored, and maintained at the highest quality level. Safety can only be ensured in cooperation between manufacturers of cells and batteries, system design engineers, and users, and requires the commitment to protect the system from operating conditions outside permissible limits and to ensure protection against unavoidable events. For further details, see the appendix and Ref. [16].

Environmental protection aspects (see ISO/IEC Guide 64:2008 [17]) go further and require minimizing the impact on soil, air, water, and biodiversity.

13.19.3 Hazard and Safety Levels

The classification into hazard levels is of particular importance for the discussion of safety. The levels shown in Table 13.10 come from EUCAR (European Council for Automotive R&D), but have been used in a similar form for years. It should be noted that the safety levels refer to the cell level and that additions or changes may be necessary when considering modules and overall systems.

For functional safety, a term from ISO 26262:2018 "Road Vehicles – Functional Safety", Ashtiani has differentiated between three categories for batteries in vehicles:

1) Impact severity with seven levels from 0 (no impact) to 6 (explosions, damage to the environment due to thermal and kinetic forces, and release of toxic substances);

19 Quantification of safety, that is, defining a numerical state of safety, is a difficult task.

Table 13.10 Various hazard levels according to EUCAR for use in electric vehicles at the cell level.

Hazard level	Description	Classification of risks and effects
0	No effect	No effect, no loss of functionality
1	Activation of passive protection	No defect, no leakage or gas escape, no bursting of the cell, no explosion, no exothermic reaction or thermal runaway; cell reversibly damaged, repair of protective device required
2	Defect/damage	No leakage or gas leakage, no bursting, no explosion, no exothermic reaction or thermal runaway; cell irreversibly damaged, repair (e.g. replacement of cell) required
3	Leakage, mass loss <50%	No gas leakage, no fire or flames, no bursting, no explosion: loss of weight of the electrolyte (electrolyte and conductive salt) <50%
4	Gas leakage (venting), mass loss >50%	No fire or flames, no bursting, no explosion: weight loss of the electrolyte (electrolyte and conducting salt) >50%
5	Fire or open flame	No bursting, no explosion (i.e. no flying parts)
6	Bursting	No explosion, but flying parts of the active masses
7	Explosion	Explosion (i.e. destruction of the cell)

The classification used by Ashtiani essentially corresponds to the classification used by EUCAR above. The reversible loss of function (level 1) and the irreversible loss of function (level 2), both without effects on the environment, are classified as hazard levels although these are more a question of reliability than safety.

2) Probability of occurrence (likelihood) with nine levels from 1 (very low probability, less than 0.01%) to 10 (extremely high probability, ≥10%).
3) Controllability (hazard control) with 10 levels from complete avoidability to low possibility of reducing the impact.

It is obvious that this classification of risks requires careful analysis of all possible failures and their effects, for example, using the procedures according to ISO 26262:2018 "Road Vehicles – Functional Safety," and requires an appropriate assessment using experimental investigations.

The three categories proposed by Ashtiani are also used in ISO 26262:2018,[20] but are only divided into three or four well-described categories. Depending on the classification of a possible event, four overall assessments are made, which are referred to as the Automotive Safety Integrity Level (ASIL). Table 13.11 shows

20 The analogous use of ISO 26262:2018 for batteries instead of road vehicles is possible without restrictions and is also mentioned in other sources.

Table 13.11 ASIL determination according to ISO 26262.

		Controllability class		
Severity class	**Exposure class**	**C1**	**C2**	**C3**
S1	E1	QM	QM	QM
	E2	QM	QM	QM
	E3	QM	QM	A
	E4	QM	A	B
S2	E1	QM	QM	QM
	E2	QM	QM	A
	E3	QM	A	B
	E4	A	B	C
S3	E1	QM	QM	A
	E2	QM	A	B
	E3	A	B	C
	E4	B	C	D

ASIL assessments derived for severity levels S1–S4, probability-of-exposure levels E1–E4 and controllability levels C1–C3. The cases designated as QM in the table are classified as quality management problems because they do not represent a relevant risk due to appropriate product design, manufacturing and monitoring. The highest level D (major impact, high risk of exposure) requires extensive countermeasures for the early detection and mitigation of potential.

13.19.4 Safety Limits

The discussion of hazard and safety levels does not answer the question of whether technical solutions are possible so that there will be no risk, that is, where the classification of events will always lead to ASIL assessment QM or hazard levels 0–2, or possibly also 3, according to Table 13.10 of EUCAR.

In the chemical industry, safety concepts have been developed to avoid high, uncontrollable temperatures, the application of which leads to the so-called safety zones. These are operating parameters of the overall system that, if adhered to, cannot, as a matter of principle, lead to hazards (Semënov theory, Frank-Kamenetzki theory, Thomas theory, see Ref. [18]), and design solutions, including, for example, cooling units, whose implementation prevents leaving the safety zone. Essentially, it is required that, at any temperature above the initial temperature T_1, both the heat generation rate and its temperature gradient are lower than the maximum possible heat dissipation rate or its temperature gradient:

$$\left(\frac{\mathrm{d}\dot{Q}_{\mathrm{gen}}}{\mathrm{d}T}\right)_{T_1} > \left(\frac{\mathrm{d}\dot{Q}_{\mathrm{rem,max}}}{\mathrm{d}T}\right)_{T_1}. \tag{13.12}$$

Börger et al. [19] and Klink et al. [20] have applied this concept to batteries (see also section 13.22). Heat dissipation mechanisms and additives to materials can be used to ensure that a cell cannot undergo a TR.[21] However, this does not mean that it remains undestroyed and that all other risks when handling cells or batteries are eliminated, for example, electric arcs. In any case, the existence of a safety zone may depend on the fact that safety components outside a cell must function as part of a battery system and must not fail.

13.19.5 State of Safety

Due to the great importance of battery safety, which is particularly relevant for lithium-ion batteries, the following attempts to summarize the current discussion in the technical literature on the term "state of safety" (*SoS*). All battery systems contain safety risks of various kinds; the higher the maximum electrical power and the specific energy (Wh/kg) or power (W/kg), the greater the safety risks. It is therefore not surprising that the term has gained importance primarily in connection with lithium-ion batteries.

There is no generally accepted definition with the possibility of calculating a value from measurement data to quantitatively describe a safety condition or at least to qualitatively describe when a safety condition is reduced.

It is not difficult to select operating conditions for batteries that lead to the end of their service life in a very short time, and it is even quicker to treat batteries in such a way that explosions and fires with massive destruction of the environment are triggered. The term "safety" cannot therefore mean the impossibility of such catastrophic developments with effects on the environment, but only their avoidability as far as possible. It is therefore important to differentiate safety from reliability, which refers to loss of function.

The following approaches to defining the term "state of safety" should be considered:

a) Ramadass et al. [21] define the term "state of safety" under the premise that there is a mutual agreement between manufacturers, system integrators, and users of cells and batteries that cells or batteries are only operated within the permissible operating limits and that cells or batteries are safe within these limits. Additionally, abuse conditions that are unavoidable or difficult to avoid[22] do not compromise safety, that is, do not cause fires or explosions.
   This consideration implies that safety can only be considered in conjunction with a BMS and the other safety components (fuses, circuit breakers, valves, cooling systems, etc.). Events that cannot occur when using a BMS with a fail-safe structure (e.g. overcharging of a single cell in a battery string) or faults that can only

21 For lead–acid batteries, a comparable example of the creation of a safety zone is the ventilation requirement for battery rooms in DIN EN 62485-2:2019-04. Compliance with this requirement means that the air in a battery room cannot be explosive. However, the risk of explosion within a cell is not eliminated by the creation of this safety zone.

22 An immediate loss of performance after abusive conditions or the need to remove cells or batteries from service after abusive conditions is not a safety issue, but a lifetime issue.

occur by manipulation (e.g. use of the wrong charger, provided that appropriate precautions – matching plugs, communication between BMS and charger – have been taken) are not a relevant object of investigation for safety considerations. Ramadass does not make any further considerations regarding a quantitative or qualitative parameter for a safety status.

b) Cabrera-Castillo et al. [22] have proposed calculating a state of safety from the operating data of voltage, current, and temperature. If the measured values lie within a certain range, the system is described as safe with an *SoS* of one. The further the measured values lie outside this range, which does not necessarily have to correspond to the permissible operating conditions, the more the value for *SoS* is reduced. The term used thus describes the proximity of the battery to operating conditions and states in which catastrophic destruction may be initiated or occur. As soon as the voltage, temperature, and current are back within the defined limits, the *SoS* has the value one again. The potential reduction in the safety of a cell after such problematic operating conditions is therefore taken into account.
c) Lamb et al. [23] define the state of safety, also referred to in parallel as the state of stability, as a measure that can be used to estimate the probability of catastrophic failure of the battery. A distinction is made between diagnosis, which relates to the current state, and prognosis, which describes the longer-term assessment. Impedance spectroscopy methods (see chapter 10 or Appendix) are used to detect the development of a TR as early as possible. The lead time compared to evaluations of voltage and temperature values is short. It is not discussed whether there is sufficient time to take measures to prevent catastrophic damage or to protect people. If an advance warning could be issued with a sufficiently long lead time, then a measured value would be available that would indicate a probability value for catastrophic destruction or a period of time until its probable occurrence.

A proposal for evaluating the complete history of a battery's operation together with currently measured values and the calculation of a scalar value has not yet been presented in the scientific literature. Such a value could and should enable a statement on whether and to what extent a battery system, which has been used for some time has a higher probability of catastrophic destruction than a new battery. The underlying assumption is that the future operating conditions correspond to those originally specified by the manufacturer and monitored by the BMS.

13.19.6 Attempts to Define of State of Safety

A possible definition of the term "state of safety" for lithium-ion batteries, therefore might be:

The state of safety describes the probability that an unsafe behavior of a cell or battery can occur when all safety-relevant components are taken into account, or that the current operating conditions of the battery lead to operating conditions where the probability of unsafe and dangerous behavior increases.

Such a definition calls for methods to detect cells that unusual behavior, for example, cells requiring frequent balancing. In a recent dissertation [24] cells were stressed thermally to the point where parts of the separators melted and it was observed that the relaxation process after a discharge was markedly different. In another dissertation [14] it was shown that micro short circuits which may later lead to TR can be detected with appropriate voltage measurement systems.

Obviously, a lot of work still needs to be done both for identifying damage in cells and defining categories of increased risks and/or a quantitative value.

13.20 Causes and Test Conditions for Thermal Runaway and Thermal Propagation

13.20.1 Problems and Field Situation

A TR is the uncontrollable increase in cell temperature caused by exothermic reactions in the cell. Such a phenomenon has long been known from chemical reaction technology and is therefore by no means specific to lithium-ion batteries, but it has become much more important due to the increase in electromobility. While it might still be possible to handle a TR of one cell, the propagation of such a TR to neighboring cells and possible so all cells of the entire battery can no longer be handled due to the amount of energy released and the high temperatures of gases. Since modern electric vehicles contain many cells, this problem is inextricably linked to the progress of electromobility. The concern here is particularly for internal short circuits, either caused by undetected large particles of active materials (a clear production fault), other production defects, or processes that could lead to short circuits under unfavorable operating conditions. What processes and operating conditions this might be remains unclear for the time being. Only operating conditions within the permissible limits controlled by a safe BMS are relevant topics of investigation. A safe BMS is able, for example, to prevent impermissibly high charging voltages and overcharging of individual cells and entire modules or battery systems with a very high degree of safety. Tests to investigate a TR which are carried out by overcharging can therefore only be transferred to real battery systems to a very limited extent. Experiments with local heating of cells and the simulation of short circuits, on the other hand, are more realistic.

Despite the correspondingly high level of attention that the topic has now received, empirical evidence shows that it is an extremely rare phenomenon, as several independent studies have shown, particularly with regard to electric vehicles. It has been shown:

a) 28 electric vehicle fires in Europe [25] were reported in publicly available sources in the period 2010–2018, and not a single one could be clearly linked to thermal propagation. In 24 cases, other causes such as accidents, clear abuse, or an external heat source were identified, and in only 4 cases did the published information not rule out the possibility of an internal TR.
b) In the United States, 39 electric vehicle fires were reported in the media in a similar period [26], 20 of which were related to accidents. Again, not a single case could be clearly linked to an internal TR and subsequent thermal propagation, leaving only three possible suspected cases due to unknown causes.

No independently verifiable information is available from other countries.

However, this rather positive field data is no reason to ignore the issue because

a) The requirements for future electric vehicles in terms of greater range, faster charging, etc. will tend to lead to larger and more powerful batteries with greater energy.
b) The development of battery and electromobility is a very dynamic field with high pressure to innovate and many new, that is, less experienced in safety engineering in development, companies such as newcomers in the field, etc.
c) Media coverage has focused on fires involving electric vehicles.
d) In many global markets, there is a high level of political support for a very rapid and comprehensive expansion of electromobility.
e) The phenomena of TR and thermal propagation are not yet sufficiently understood.

Research into understanding the general phenomena in this area therefore has to continue, as will the development of tests and prevention, detection, and mitigation measures. In the meantime, there are further efforts to establish legal requirements in this area as part of the so-called GTR 20 (GTR: global technical regulation) process. There is currently no standardized test requirement that is internationally recognized, but individual countries, especially China, have already taken the lead. So far, the only binding international agreement (i.e. GTR 20) is that electric vehicles must in future be equipped with a warning signal that prompts the driver to leave the vehicle five minutes before the occurrence of a thermal propagation event and that compliance with this requirement must be documented. It must be demonstrated that the vehicle design is sufficiently safe against a TR and thermal propagation, for example, using the ISO 26262 standard for functional safety.

13.20.2 Causes of a Thermal Runaway

There are a number of events that may lead to a TR. Those used for laboratory tests are listed below in section 13.20.3. This chapter only deals with those root causes that are unavoidable but not causes which a redundant fail-safe BMS would prevent, for example, overcharging or deep discharging of an individual cell in a string or of the whole string. Despite its importance, the requirements for a safe, fault-tolerant, and sufficiently redundant BMS have so far been not been standardized.

All causes lead to one or a number of cells will experience excessively high temperatures, either localized or homogeneously, which then initiates a TR and possibly thermal propagation.

<u>External short circuits and very high currents</u>

Most cells, and certainly all cells above a certain capacity, have internal devices that either limit the maximum current or interrupt the current (section 13.9). In addition, modules and battery systems contain fuses and switches that will interrupt the current. However, if the short-circuit current is not high enough and only affects a part of a string due to an insulation fault (cell terminals touch the housing), the current may not be high enough to trigger fuses inside the cell. Additional fuses and switches offer only protection if they are sized for fairly low currents, which is not possible for all applications, and if they happen to be located where the short-circuit current exists. The maximum heat generation of a short circuit occurs if the resistance of the short circuit is equal to the internal resistance of the cell, leading to the cell voltage dropping to 50% of its value. At such high currents, the fuses within the cell will cut the current. In the worst-case scenario, the heat generated inside the cell and by the short-circuit current outside the cell add up to cause local heating and problematically high temperatures. Protection against external short-circuiting requires careful application of electrical engineering principles.

Internal short circuit

An internal short circuit may be the result of external damage to a cell, dendrites which are most likely caused by operating conditions (see section 13.15, aging) or production problems (excessively large mass particles, holes in the separator, for example). Cells with damaged containers or modules that have suffered mechanical stress during production (e.g. dropping onto the production floor) therefore must not be used further.

Internal short circuits are considered the main cause of a TR and it is consensus that, despite intensive monitoring during production, they cannot be prevented with certainty for years to come. However, it is known from real applications that such short circuits do not necessarily lead to major damage to the cell and/or a TR. The reasons for this include the following:

a) When contacting a dendrite with the cathode, the short-circuit current can be calculated using Ohm's law. Although the cross section of dendrites is only a few square micrometers, copper dendrites with a length of approx. 50 μm (this corresponds to the thickness of the active layer and the separator) have a resistance in the range of approx. 0.1 Ω and the current then flowing is some 40 A. At this current, very high amounts of heat (160 W) are released locally, but these can be dissipated quickly. In addition, the entire electrochemical potential drops along the dendrite in the electrolyte, which leads to rapid dissolution – neither copper nor lithium is electrochemically stable at the potential of the cathode. Although a brief voltage dip is therefore to be expected due to the short-circuit current, this is not necessarily the start of a TR.
b) The same applies to excessively large particles that can push through the separator and touch the active masses of the counter electrode.
c) Pinholes in separators do not necessarily lead to electrodes of opposite polarity touching each other.

External fire

External fires can heat up cells in a module beyond their critical value. To minimize this risk, modules in electric vehicles are tested and need to withstand a fire of a certain temperature for a given time.

Mechanical damage

Mechanical damage to cells and batteries is unacceptable during production and installation. However, in the case of an accident, cells in a module can be pierced, cut open or crushed. The battery system in electric vehicles is therefore built to withstand great mechanical stress and has to be tested according to corresponding manufacturer's specifications.

13.20.3 Abuse Tests

Since the behavior of lithium-ion batteries is particularly important and interesting in borderline situations respectively when the permissible operating range recommended or required by the manufacturer is exceeded, various tests have been developed that correspond to different abuse scenarios. These can be found in norms and standards (e.g. in the ISO 6469 series), but in some cases, even in legal regulations such as vehicle approval regulations like ECE R100, as well as transport regulations for hazardous goods known as UN 38.3. Such tests must be carried out in highly specialized laboratories, as each individual test can result in the release of highly toxic substances, fires, and explosions.

Abuse tests are carried out both to study cell properties under critical conditions and to initiate a TR and thermal propagation. They can be carried out with cells as well as modules and complete batteries and can be divided into different groups:

a) Electrical abuse tests that involve overcharging (often up to 200% of the nominal capacity), unacceptable high voltage, deep discharge (i.e. discharge beyond the intended voltage limits but without reversing the polarity), overdischarge (i.e. discharge until the polarity of a cell is completely reversed), and very high current (i.e. a current above the maximum permissible current as a result of an external short circuit).
   Cells are sometimes modified, that is, deactivating safety features of a cell for these tests.
b) Thermal abuse tests in which lithium-ion batteries are brought to such high temperatures, that the intended investigation can be carried out (testing the function of shutdown separators, destroying the separator, TR). Depending on the design of the experiment, cells can be heated up very uniformly (thermal insulation, very low-temperature rate) or locally. The methods to cause high temperatures differ greatly.
   Sometimes, cells are brought to such low temperatures that the electrolyte freezes. Finally, there are also temperature cycling tests in which a cell or battery is subjected to repeated strong temperature fluctuations and thus thermal shocks, which have significantly higher temperature differences and gradients compared to field conditions.
c) Mechanical abuse tests in which cells or batteries are dropped from a certain height (e.g. 1 m), in which they are shaken according to certain (often noise-like) profiles, that is, subjected to vibration loads in one or more directions, or in which they are crumpled, crushed, subjected to a crash test or deliberately penetrated with a pointed (metallic or ceramic) object.
d) Environmental abuse tests in which cells or batteries are exposed to changing environmental conditions. This can be low air pressure (as in the UN 38.3 tests) or permanently high humidity, possibly with simultaneous high temperatures. Immersion tests in water (often even salt water due to the increased electrical conductivity) or watt tests for electric vehicles also fall into this category.

Such abuse tests are constantly being developed further to be able to react to new requirements. For example, it is already foreseeable that targeted cyberattacks on the BMS will also have to be tested in future, and as inductive charging technologies become more widespread, it is conceivable that tests in magnetic fields of varying strength will also be part of the standard repertoire of such tests. There is also a trend toward not only determining the hazard level of a cell or battery in such tests, but documenting the entire test, for example, by video, and recording additional information, for example, by measuring any harmful gases released during the tests.

13.20.4 Abuse Tests to Trigger a Thermal Runaway and Thermal Propagation

Causing a TR in the laboratory is referred to as triggering a TR, in thermal propagation tests where there are many cells, the cell, in which the TR is initiated is referred to as the trigger cell.

There are two issues concerning the interpretation of data when using trigger methods:

- Most trigger methods transfer energy into the cell.
- For experimental reasons, in particular for investigation thermal propagation, trigger methods are sometimes used that cannot occur in real systems with an appropriate BMS, for example, overcharging a cell.
- Preparing a module with a trigger cell and fitting it out with sensors will change the integrity of the module.

When interpreting results, these issues should be born in mind.

In general, all trigger-based tests have the disadvantages of being error-prone, time-consuming, expensive, not resistant to manipulation and not resistant to cheat devices (cheat devices are devices that only help to pass a test but do not generate a real safety increase) and not universal (i.e. require considerable adaptation for each new battery system or in some cases cannot be used at all). For this reason, proposals have also been made for non-trigger-based procedures that focus more on existing safety functionalities and their effects.

13.21 Thermal Runaway

A theoretical treatment of the TR problem is possible with the help of the classical Semënov theory. In this theory, an energy balance is required. Energy is generated in the system through exothermic reactions and dissipated via heat transfer processes (convection, radiation, conduction) and/or endothermic reactions such as the melting of separators. As long as the amount of heat generated Q_p is equal to the amount of heat dissipated

$$Q_\mathrm{p} > Q_\mathrm{r} \tag{13.13}$$

and as long as this also applies at any higher temperature, that is,

$$\mathrm{d}Q_\mathrm{p}/\mathrm{d}T > \mathrm{d}Q_\mathrm{r}/\mathrm{d}T, \tag{13.14}$$

a TR is impossible. The two equations can be expressed in the formula

$$\mathrm{d}(\ln Q_\mathrm{p})/\mathrm{d}T > \mathrm{d}(\ln Q_\mathrm{r})/\mathrm{d}T \tag{13.15}$$

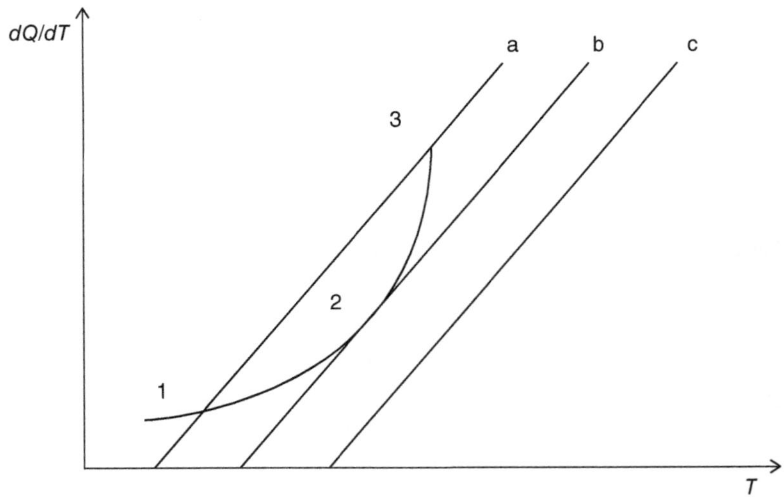

Figure 13.21 Semënov diagram for a thermal runaway. In the temperature range between 1 and 3, a so-called safe zone exists, in which no thermal runaway can occur because the heat dissipation is greater than the heat generation.

Exothermic reactions usually follow Arrhenius behavior, that is, the reaction rate and therefore the amount of heat released increases exponentially with increasing temperature. The heat dissipation in most systems takes place primarily through convection, which leads to a heat loss proportional to the temperature difference (Fourier relationship), that is, the following applies to the heat flow of the cooling power $\mathrm{d}Q_\mathrm{c}/\mathrm{d}t$

$$\mathrm{d}Q_\mathrm{c}/\mathrm{d}T \sim \Delta t. \tag{13.16}$$

This results in the so-called Semënov graphs as shown in Figure 13.21 (see also Ref. [27]).

It remains to be seen whether the requirements of Semënov's theory regarding the creation of a safe zone can be implemented without significant restrictions on specific energy and energy density. However, it is clear that, for example, a single-layer cell (only one pair of electrodes) with a low capacity per unit area and good cooling via the outer walls could not undergo a TR because the local heat generation would always be below the heat dissipation.

Due to the difficulties of test development, many activities continue to take place in which both experiments and simulations are carried out to characterize a TR and thermal propagation. It is essential to understand the processes of a TR before the propagation of a single cell undergoing a TR to neighboring cells can be investigated.

Investigations into a TR have shown that the following sequence can be expected with the currently prevailing commercial battery chemistries (i.e. NMC or NCA cathode, organic electrolyte, graphite anode) (see Figures 13.22 and 13.23):

Step 1: Slow, but still manageable temperature increase.

Step 2: Rapid temperature increase as a result of exothermic reactions which accelerates more and more. Once a certain temperature (referred to as T_onset) and

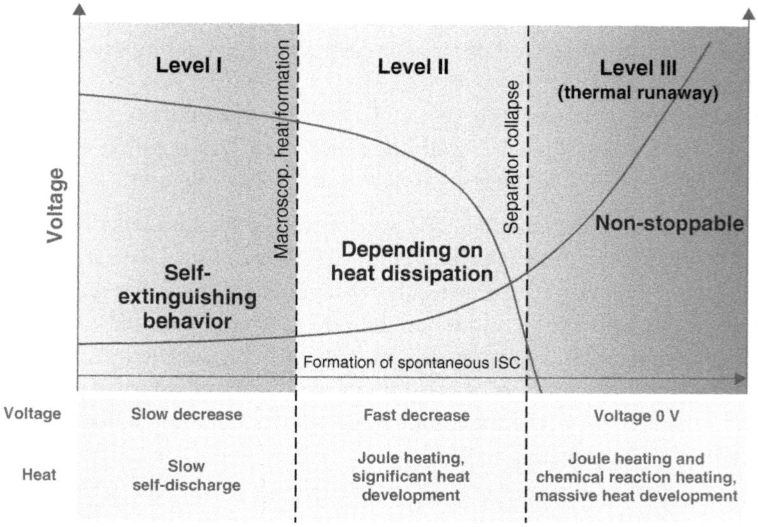

Figure 13.22 Phases of a thermal runaway. Source: Reproduced with permission from Ref. [28]/Elsevier.

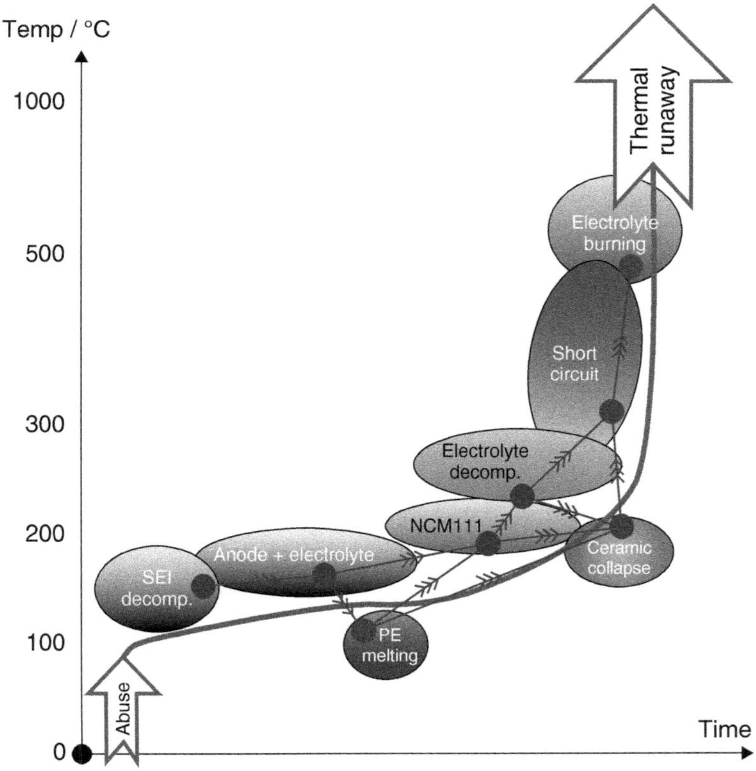

Figure 13.23 Chemical reactions during a thermal runaway. Source: Reproduced with permission from Ref. [28]/Elsevier.

a temperature gradient (often given as 1 K/s) are reached, stopping the further temperature increase is considered impossible (depending on cooling equipment and cooling power).

Step 3: Very fast increase in temperature (e.g. 20 K/s) with a maximum temperature in the range of a few hundred °C (cell container, gas) and possibly exceeding 1250 °C inside the cell. The process cannot be controlled any more.

The most important processes during a TR are considered well understood, and there is a wealth of papers in the scientific literature. It is even possible to simulate fairly accurately the temperature rise using thermodynamic data of the reactions and the Arrhenius law and estimate the cooling power that is required to avoid catastrophic high temperatures. As soon as any process (very high current load, external or internal short circuit, external heat) increases the temperature of the cell or a local area of the cell above the maximum permissible operating temperature, the following processes can take place:

- In cells with a carbon or graphite anode, the SEI layer begins to decompose exothermically. The decomposition of the SEI, which was formed initially during production and then increases as a function of time and temperature lays bare the anode material. A new SEI layer with slightly different properties is then formed immediately in an exothermal reaction, which then also decomposes when the temperature continues to increase. Interestingly, it has been shown that the SEI which has newly been formed at such high temperatures has a greater thermal stability than the original SEI formed during initial cycling [8].
- Some of the first gaseous decomposition products that can be measured are hydrogen and carbon monoxide.
- The pressure in the cell and the temperature rise.
- At about 130 °C, the separator begins to melt. In the case of a multilayer PE/PP/PE separator, the PE melts first and prevents ion transport (shutdown separator). The melting of PE is an endothermic process.
  As soon as the separator is thermally damaged, the cell must be replaced, even if a thermal runaway can still be stopped at this point in time, irrespective of cell properties [24].
- The cathode material, in most cases metal oxides, begins to decompose, releasing oxygen.
- The temperature and pressure curve varies depending on the cell materials and the cause of the heating. The cell voltage begins to fall with fast fluctuating values.
- At about 160–170 °C, the PP portion of the separator starts to melt as well. The pressure inside the cell reaches a pressure in the range of c. 5–7 bar and the rupture valve opens (venting).
- The pressure relief in the cell leads to a short cooling of the cell which can be measured on the cell wall, probably due to evaporation of the solvents used as the electrolyte.
- At some later stage, the separator collapses and leads ultimately to a short circuit, a high current discharge of the cell. At this point, the cell voltage becomes zero. It is no longer possible to stop the processes inside the cell.

- The cathode material, in most cases metal oxides, begins to decompose, releasing oxygen.
- The temperature gradient becomes very high, for example, c. 20 K/s and the pressure increases. As the rupture valve is already open, a jet of gas and hot particles ejected from the cell can be observed.
- Depending on cell size and chemistry, the maximum temperature of the cell housing reaches a temperature in the range of c. 600–700 °C and the temperature of the gas jet c. 800–950 °C. The position of the sensors and their type as well as the details of the experimental setup lead to significant differences. Based on the analysis of the emitted gases and other considerations, it can be assumed that the local maximum temperature inside a cell can exceed 1250 °C. It is unclear how high the maximum pressure can become because lithium-ion cells always have a rupture valve, which leads to the cell being open by this time.
- The temperature and pressure curve varies depending on the cell materials and the cause of the heating.
- The emitted gas is sometimes compared to a high-temperature gas jet, whose flow rate drops sharply after a few seconds because the pressure built up in the cell can relax. With regard to the burning and ignition properties of the gases, a comparison is often made with propane.
- The hot gas can either be ignited by the oxygen present in the ambient air if the threshold of temperature and oxygen concentration are exceeded. Additionally, the gases can be ignited by hot, small particles that are ejected from the cell.
- If the gas can fill a larger volume before ignition, the effect of self-ignition is an explosion.
- The cell continues to burn for a few minutes until the combustible material in the cell has been consumed. The total mass loss in EUCAR hazard levels 4–7 may exceed 50%.
- The TR of neighboring cells can be initiated before the TR of the first cell can be detected.

Explosions are particularly problematic. It can be assumed that the explosion does not take place in the cell, but due to gas accumulations outside the cell. Good ventilation of the cell environment is therefore advisable, but trying to keep the escaping gas in a largely closed container is not.

The amount and concentration of gas escaping from a cell depends not only on the cell chemistry but also on the triggering mechanism. Cells with a titanate anode, which do not form an SEI, therefore show a different temperature and pressure development than cells with a graphite anode, because there is no heat input from the exothermic reaction of the SEI. The same applies to cathode materials with higher thermal stability and lower oxygen release, such as iron phosphate, compared to NMC and NCA cathode materials.

From the data in Table 13.12 it is clear that, depending on the opening pressure and gas temperature, spontaneous ignition may or may not occur on contact with the environment. In the fire protection sector, oxygen-reduced atmospheres are used in the area of document security and storage and it is known that with approx. 15%

Table 13.12 Typical values for the composition of the gas released in a thermal runaway and data on ignition temperature, flame temperature and explosion limit (n.a. = not applicable).

Gas	Concentration overload/nail test [Vol-%]	Ignition temperature [°C]	Flame temperature in air [°C]	Explosion limits in air [%]
Hydrogen	37/36	585	2130	4.0–77.0
Carbon monoxide	37/27	605	2121	10.9–76.0
Carbon dioxide	16/28	n.a.	n.a.	n.a.
Methane	7.4/4.7	595	1970	1.7–10.8
Ethane	0.4/0.3	515	1960	3.0–12,5
Ethylene	0.9/1.2	425		3.0–36,0
Propylene	0.7/1.6	490	1925	4.4–16,5
Rest	0.4/0.2	n.a.	n.a.	n.a.

oxygen in the air, a flame can form using the oxygen formed in the cell, but that the flame will not spread to other areas with a particularly destructive effect. As a rule of thumb, the total energy released in a TR doubles if the gases ignite.

Concepts to minimize the effects of a TR are under investigation. They focus on lowering the pressure at which venting occurs, managing the gas flow, containing the effect to the module housing, etc. Their discussion goes beyond the scope of this book.

13.22 Thermal Propagation

There seems to exist general agreement that, for the foreseeable future, it will not be possible to prevent lithium-ion cells from developing a TR throughout their lifetime. Preventing thermal propagation to other cells, however, may be possible and is also much more important as the energy released is so great that limiting damage seems impossible. The following issues in connection with thermal propagation are currently the subject of intensive research, but must still be regarded as unresolved:

a) How can a TR be recognized in good time (preferably before it even happens) so that thermal propagation can be ruled out under all circumstances?
b) What measures can be taken to mitigate the effects of a TR that has occurred and prevent thermal propagation?
c) Which sensors and actuators, or which combination of suitable sensors and actuators, can be used to achieve this goal?
d) What level of redundancy in the BMS and other components is required?
e) Are there technological solutions that per se prevent the occurrence of a TR or thermal propagation, that is, in analogy to the Semënov theory of safe zones?

Minimizing or preventing thermal propagation is a major regulatory issue. So far, however, discussions are ongoing about what this means. Is venting of neighboring cells permissible as long as they do not undergo a TR respectively what EUCAR level they must not exceed after the TR of a neighboring cell?

It has proven to be very difficult to develop a generally valid test with pass criteria for thermal propagation scenarios. The main reasons are:

- The heat generated, its rate, and the maximum temperature of the cell housing and the gas emitted depend on cell capacity and materials used. Cathodes with NMC111 and NMC811 have different maximum temperatures.
- The occurrence of thermal propagation depends on the details of the module design. It has been observed, for instance, that the probability of a fire differs between tests with individual cells and cells in a module.
- The results depend on the method used to trigger the TR in a cell.
- The results depend on the location in the module.
- The results depend on a very accurate and detailed description of the test procedure.

A large mass particle as a trigger, for example, cannot be easily inserted into a cell at a later date, and even if it is, there is no guarantee that a short circuit will actually be generated, which would lead to a TR with subsequent propagation. Although various test proposals have been made (see Figure 13.24), these have not yet been widely accepted for reasons of practicality and reproducibility. In the tests, a trigger cell in the module is brought to a TR and then the further behavior of the overall battery system, sometimes the entire vehicle, is examined. Even with an appropriate selection of the trigger cell, boundary conditions such as ambient temperature, switching an existing cooling systems or other safety mechanisms on or off, and, of course, the execution of the experiment are important.

The following proposals are currently being discussed as trigger methods, without a final solution having been found or being practicable for all different battery and cell types:

a) Replacement of a cell in the system with an appropriately prepared or specially manufactured cell that contains an internal short circuit or a device that can produce such a short circuit. Various devices have been developed for this purpose (see Figure 13.25).
b) Heating a cell using a local heating element that is as small as possible. The so-called TRIM heater, which was developed and patented in Canada [30],[23] see Figure 13.26, is now of particular importance.
c) Fast or sometimes deliberately slow[24] penetration of a trigger cell with a (ceramic or metal) nail.

---

23 Work is also being carried out on inductive heating methods in which only approx. 1% of the surface is heated with less than 10% energy input in relation to the total energy content of the cell.

24 With a comparatively slow penetration, the voltage drop can be measured precisely and the moment of the internal short circuit can thus be determined well. However, the subsequent

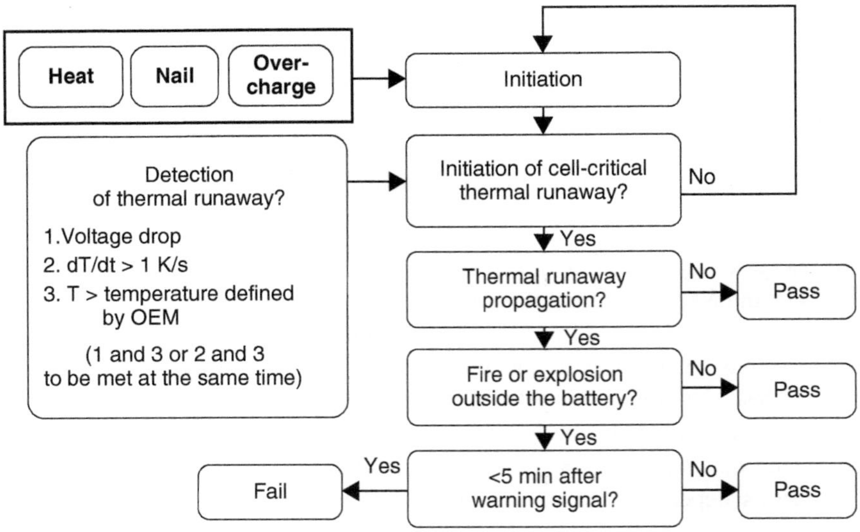

Figure 13.24 Possible, but rejected propagation test.

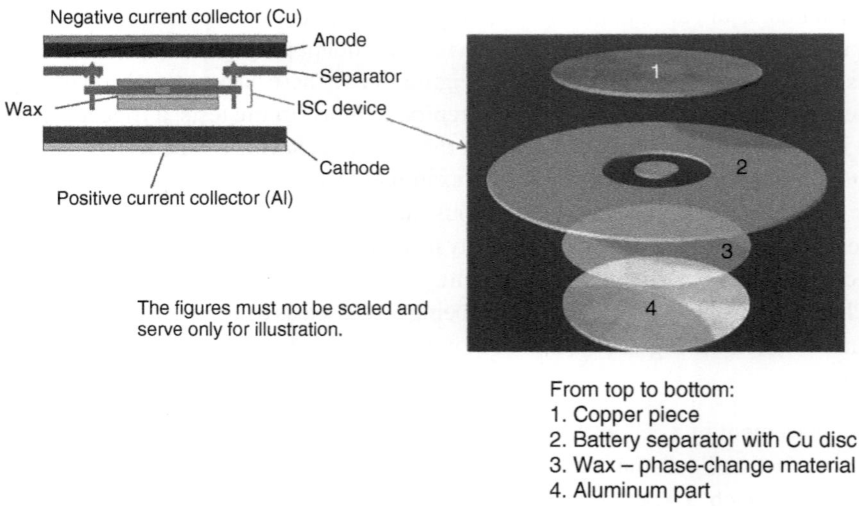

Figure 13.25 Device for generating an internal short circuit, patented by NREL (the United States). Source: NREL [29]/U.S.Department of Energy/Public Domain.

There are differences between the speed at which the nail enters (in the laboratory, very slow speeds in the range of millimeters per minute are usually selected), the cross section, and the conductivity (metallic or nonconductive ceramic) of the "nail."

It is also known that not every penetration of a cell with a nail leads to a TR despite an internal short circuit with very high local temperatures.

---

process does not correspond to what is known about internal short circuits caused by large particles, which are generally characterized by a much quicker temperature rise. The realism and thus the significance of such a trigger method is therefore doubtful.

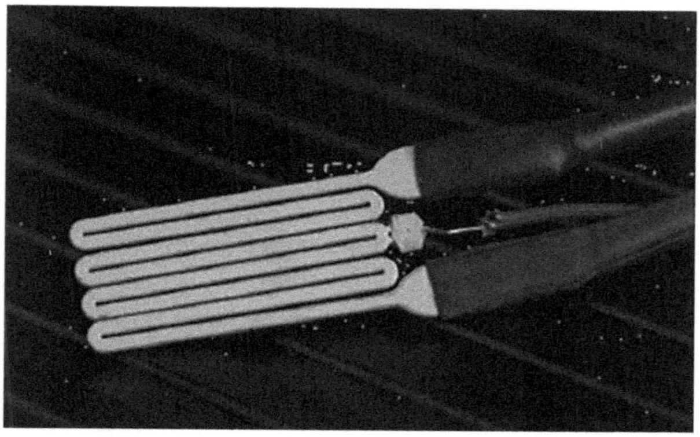

Figure 13.26 Local heating element for triggering a cell by external heating, patented in Canada. Source: Reproduced with permission from Ref. [30]/National Research Council of Canada.

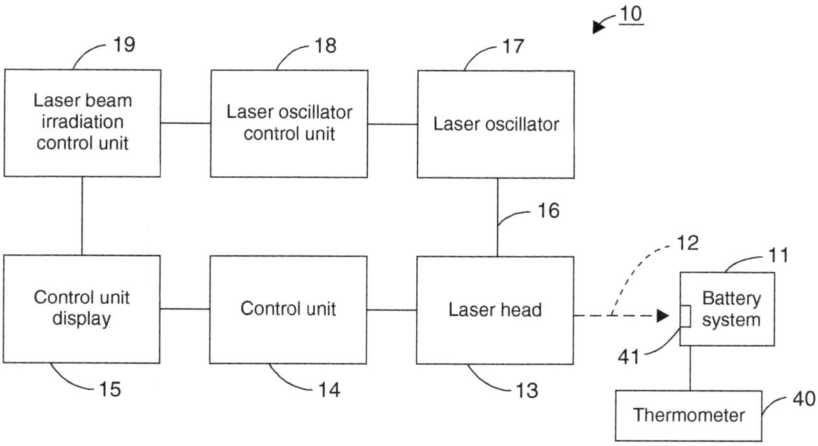

Figure 13.27 Sequence of a laser-triggered propagation test. Source: Reproduced with permission from Ref. [30]/Public Domain.

d) Local heating of a cell using a laser, see Figure 13.27. Although this method is patented, it has been used seldom in practice due to the high experimental cost and difficulties involved, and it is doubtful that this will change in the foreseeable future.
e) Removal of a cell and replacement by a system that can release the same amount of energy in the form of chemical energy (e.g. through the thermite reaction).
f) Forced external short circuit of a cell, where the self-heating of the cell and the heat of the external short circuit together provide the necessary heat input. The disadvantage of this principle is that the short-circuited cell becomes inherently safer due to its lower state of charge and the energy stored, so a worst-case scenario cannot be investigated.

None of the trigger methods mentioned (and possibly others) are free of significant disadvantages, so the discussions will continue, see Table 13.13.

Table 13.13 Trigger methods and commentary on their supposed field correlation.

Trigger method	Comment concerning field experience
Overcharging (i.e. too high voltages)	With reliable monitoring and system design using a redundant and fail-safe BMS, it is not possible for a cell to be overcharged alone or in any series or parallel connection.
External short circuit (i.e. currents too high)	In many cells, either a fuse or a current limiter via a PTC element is installed. At currents that are only slightly above the permissible current load, these elements are not necessarily activated; but here too, a redundantly and safely designed BMS leads to the current being switched off. As the cell is discharged, no worst-case scenarios can be investigated normally.
Internal short circuits	Internal short circuits are usually carried out with specially prepared cells in which the contact area for creating the short circuit is in the range of 0.5 cm^2. In real applications, the short-circuit area is in the range of less than 1 mm^2 or even less as a result of material defects in current collectors and active masses; however, there are currently no ideas for simulating the course of such a small-area short circuit (except by special nail penetration tests with needles with a cross section considerably below 1 mm^2), so that the question of transferability also arises here.
Thermal triggering	The heating of a cell used for triggering, which leads to a homogeneous temperature distribution, is unthinkable in real applications and the same applies to local heating, which cannot occur in normal operation with careful design[a)]; However, in the case of an external fire that heats up the battery, the process is transferable.
Mechanical damage	The current penetration tests correspond to severe accident scenarios – however, the probability that only a single cell has been damaged is quite small and the question arises as to whether such mechanical damage to the cell, which will only occur in very serious accidents, must be simulated in a kind of standard test procedure, because there are already extensive crash test requirements, especially for vehicles (and their traction batteries).

a) Design and layout errors, such as placing a battery directly over the exhaust manifold of an internal combustion engine, need not be taken into account with regard to the risks to the battery.

A comprehensive comparison in laboratories to find a universally applicable method has not yet taken place and is also very costly, if only in terms of the number of samples to be examined and the available laboratory capacities. So far, there are only a few specialized laboratories in the world that are even capable of carrying out the corresponding tests. The number of possible and experienced test laboratories is further reduced if additional boundary conditions such as thermography, gas detection, filming, etc., are imposed.

Under no circumstances should the selection of the trigger cell be underestimated. A cell in the middle is not always the most critical, but due to special thermal conditions, this can also be a cell at the edge in individual cases. Since the tests are also to be carried out on commercial battery systems, a certain modification of the original battery system is usually necessary to be able to use the desired trigger type at all and, if necessary, to use additional measuring accessories. Since the aim must be to make as few changes as possible, this can also lead to the trigger cell of choice being at the edge rather than at the center of the battery.

13.23 Safety Engineering

The questions of when and why cells change and may become unsafe will probably continue to occupy research and development for a long time to come, and it is to be expected that it will not be possible to completely avoid remaining risks. Good engineering practices, that is, failure mitigation analysis and documentation of the results, must therefore be followed. Suitable design solutions are important, for which there are already a number of approaches (without claiming to be exhaustive), for example:

- Reduction of the fire load in the immediate vicinity of the battery.
- Aligning the rupture valve of a cell so that the hot gas jet is not directed against the wall of adjacent cells, but as far as possible into free spaces between cells or the housing of the module.
- Positioning of fuses in the middle of a string so that the likelihood of extinguishing a short-circuit current that affects only a part of the string increases (contact of a cell to housing of the module, insulation faults).
- Use of protective foils over the rupture valve, through which hot gases can flow, but hot particles, which can cause flames together with the oxygen in the air, are held back.

Reduction of the pressure at which a cell vents, that is, at the point when the cell is already irreversibly damaged and unsafe to be used further, to maximize heat loss from the cell (transition from a closed to an open system) [4].

- Controlled discharge of the emitted gases from the rupture valve to the ambient air,
- Circuit breakers that disconnect the electrical connection in the event of impermissible currents and voltages that cannot technically be caused by loads and chargers.

It is conceivable that in the further development of lithium-ion batteries, safety aspects will be of equal or even greater importance than energy density improvements and the like in the future. For example, a ceramic solid electrolyte in a lithium solid-state battery would lead to a considerable reduction in the fire load and the potentially escaping toxic substances. However, it is to be expected that the alternative materials in new battery technologies will bring with them additional safety challenges that may not even be known yet, so there will certainly be a considerable need for research and development in this area for many years to come.

13.24 Further Battery Problems

Commercial lithium-ion cells now have very consistent production quality and a long calendar and cycle life. It is questionable whether it makes technical and economic sense to invest development capacity and possibly additional costs for production to further improve 10 years of calendar aging and 3000 or more full cycles. For an electric vehicle with a range of 300 km, 3000 full cycles correspond to a driving distance of 900,000 km – a requirement that even technically mature combustion engines do not have to meet.

Technical battery problems exist in particular in the area of inadequate system integration (excessively high continuous charging voltages, missing or too simple BMS) and a lack of knowledge about how batteries are actually used (see chapter 15 and following). A traction battery for an electric road vehicle for a high-performance vehicle for hot climates is used completely differently from a battery for delivery vehicles in cold climates with professional maintenance and vehicle deployment management systems.

The major challenges facing lithium-ion technology currently are

a) The still high price of batteries,
b) The scarcity of the raw materials used,
c) The development of sustainability and a viable recycling concept in terms of sufficient sustainability, and
d) The origin of the electricity used to charge the batteries.

The comparatively high price of lithium-ion batteries (2021 approx. 100 €/kWh per cell, and for comparison, 50 €/kWh for complete lead–acid batteries – that is, not just the individual cell, see, and approx. 130 €/Wh for a battery system that is produced in large quantities[25]) is the result of many factors. On the one hand, the raw materials used are comparatively rare and expensive (especially lithium and cobalt) and costly to extract and process. Second, the manufacturing process for lithium-ion batteries is significantly more complex than that for lead–acid batteries, for example, as many operations must be carried out in the absence of atmospheric oxygen and traces of moisture. Impurities in materials must be excluded, which places very high demands on process monitoring and automation. Thirdly (due to the complex manufacturing processes, which require considerable know-how), the competitive situation has been characterized for years by a strong concentration of supply in a few manufacturers in a few countries, so localization efforts and thus the avoidance of transport costs often come to nothing, and security of supply remains an important issue.

The availability of lithium and other metals used in lithium-ion batteries has been investigated many times. In the opinion of the authors, there is no reason to doubt the sufficient availability of lithium in the coming decades, even with a far-reaching switch to electromobility. As there are sufficient alternatives available for the other

---

25 For new applications, the production of a battery system requires a high level of design and development effort, so significantly higher prices are demanded for individual batteries or batteries for niche applications. Please note also that exact prices are only known to the purchasing department of companies buying cells and batteries.

metals, lithium-ion technology is not expected to come to a halt even if a single element (e.g. cobalt) becomes scarce – this will then occur in those areas where it appears to be comparatively easy to switch, that is, those areas of applications where comparable solutions are easiest and cheapest to achieve change first. However, the mining conditions in many regions of the world are currently incompatible with ethical standards, compliance with which is rightly expected by modern industrial companies and demanded socially and politically. It is therefore to be expected that pressure will arise here, which will hopefully quickly lead to a significant improvement in the living and working conditions of the people affected. The ecological consequences of raw material extraction, for example, in South America, will certainly continue to be the subject of public debate in the future, and the resulting consequences will possibly slow down a further drop in the price of lithium-ion batteries.[26]

A potential shortage of raw materials could be mitigated or prevented by comprehensive recycling. It can be expected that recycling will initially be carried out at a large scale in industrialized countries and then have both an economic and ecological impact concerning the extraction of raw materials in developing countries. However, the recycling rate for lithium-ion batteries is currently comparatively low, and a large proportion of the materials used (e.g. lithium or graphite) are not recovered. Although several companies, such as Umicore and Duesenberg, have developed recycling concepts, they are not (yet) sufficiently cost-effective despite the issue being addressed by both legislators and industry.[27] Further developments can be expected here in the future. An example is the future European battery legislation, which not only requires recycling concepts but also demands that users of battery-powered devices must be able to replace batteries easily. A preliminary stage to recycling could be the so-called second-life use, that is, the reuse of aged batteries that are no longer suitable for use in vehicles but could still be used in other applications such as UPS or photovoltaic systems. The Closed Substance Cycle Waste Management Act generally requires the continued use of products before recycling. Despite many discussions and studies, this second-life use is fraught with considerable technical and legal problems: on the one hand, there is the question of quality criteria that permit the use in second-life applications; on the other hand, there are questions of liability in the event of early failures, safety problems, or low performance in the new area of application. Many of the aforementioned challenges could be solved more easily through greater standardization of lithium-ion batteries, but at the possible cost of inhibiting technological progress and making the available solutions for electromobility, for example, more subject to compromises that could affect price, range, and ultimately customer acceptance.

---

26 At this point, it should be noted that comparable discussions are surprisingly rare in relation to oil or gas production. To ensure fair competition between electromobility and internal combustion engines, it is important not to burden electromobility with requirements that are not imposed in other areas. This statement is not an appeal not to subject the extraction conditions for battery raw materials to critical analysis, but on the contrary, an appeal to take a closer look at oil and gas extraction.

27 Market logic suggests that sophisticated recycling systems will emerge for scarce materials even without social and political pressure. Given the general scarcity of raw materials, billions of consumer batteries per year cannot be thrown away in the long term without recycling.

Batteries only represent a truly environmentally friendly alternative in traction applications (both industrial transportation and electromobility) if the electricity used to charge them is produced from environmentally friendly, that is, from renewable energy sources. While some nations, such as Norway, have already come a long way in this direction (helped by a comparatively low population density in combination with the country's particular topography), other parts of the world still have a long way to go. In particular, suitable storage technologies need to be developed to ensure sufficient electricity generation from existing wind farms and photovoltaic systems at times when wind or solar energy is not available or not available in sufficient quantities (incidentally, this can sometimes even apply to hydro power, which is otherwise capable of providing base load). At present, these storage facilities hardly exist, as coal, gas, or nuclear power plants take over the base load function. In principle, a (albeit comparatively small) part of this storage function can also be covered by batteries, but a general solution to this challenge of no wind no sun phases is currently only emerging through the use of other storage systems, such as hydrogen technology in particular, supplemented by concepts such as power-to-heat, better load management [32], and fast-response power generation units based on hydrogen or synthetic methane. However, many technical issues are still completely unresolved, and there is an urgent need for research.

References

1 Riviere, E. (2016). *Détermination in-situ de l'état de santé de batteries lithium-ion pour un véhicule électrique.* Thèse de doctorat. Université Grenoble Alpes (ComUE).

2 Beard, K.W. (2019). *Linden's Handbook of Batteries*, 5e. New York: McGraw Hill.

3 Laresgoiti, A., Käbitz, S., Ecker, M., and Sauer, D.U. (2015). Modeling mechanical degradation in lithium ion batteries during cycling: solid electrolyte interphase fracture. *Journal of Power Sources* 300: 112.

4 Orazov, N., Benger, R., Wenzl, H., and Koch, I. (26–28 April 2023). Increasing the safety of lithium-ion cells through pressure-controlled gas removal. In: *Proceedings of Kraftwerk Batterien, Aachen.*

5 Patent application EP 2 779 271 A2.

6 Thiele, M. (2013). *Fast Charging Process for Lithium Batteries.* Master thesis. Nordhausen University of Applied Sciences.

7 Vetter, J., Novák, P., Wagner, M. et al. (2005). Ageing mechanisms in lithium-ion batteries. *Journal of Power Sources* 147: 269.

8 Feng, X., Ren, D., Zhang, S. et al. (2019). Influence of ageing paths on the thermal runaway features of lithium-ion batteries in accelerating rate calorimetry tests. *International Journal of Electrochemical Science* 14: 44–58.

9 Spotnitz, R.M., Weaver, J., Yeduvaka, G. et al. (2007). Simulation of abuse tolerance of lithium-ion battery packs. *Journal of Power Sources* 163: 1080–1086.

10 Cordoba-Arenas, A., Onori, S., Guezennec, Y., and Rizzoni, G. (2015). Capacity and power fade cycle-life model for plug-in hybrid electric vehicle lithium-ion

battery cells containing blended spinel and layered-oxide positive electrodes. *Journal of Power Sources* 278: 473.

11 Haubrock, A. (2011). *Degradation Studies of Lithium-Ion Batteries When Used in Electric and Hybrid Vehicles*. Dissertation. TU Clausthal. Göttingen: Cuvillier.

12 Lando, E.T. (2021). *Event-Based Lifetime Model for Lithium Ion Cells Using Linear Damage Accumulation*. Dissertation. TU Clausthal.

13 Gabbar, H.A., Othman, A.M., and Abdussami, M.R. (2021). Review of Battery Management Systems (BMS) development and industrial standards. *Technologies* 9: 28.

14 Grabow, J., Klink, J., Orazov, N. et al. (2023). Triggering and characterisation of realistic internal short circuits in lithium-ion pouch cells – a new approach using precise needle penetration. *Batteries* https://www.mdpi.com/2313-0105/9/10/496.

15 Mertens, J. and Börger, A. (2020). Generic hazard and risk assessment for warning and intervention systems for thermal propagation. *ATZ Electronics Worldwide* https://doi.org/10.1007/s38314-019-0157-7.

16 Wenzl, H., Benger, R., and Hauer, I. (2024). Safety and state of safety. In: *Encyclopedia of Electrochemical Power Sources*, 2ee.

17 http://avicenne.org, last accessed on 23.10.2021.

18 Stoessel, F. (2008). *Thermal Safety of Chemical Processes: Risk Assessment and Process Design*. Weinheim, New York: Wiley.

19 Börger, A., Mertens, J., and Wenzl, H. (2019). Thermal runway and thermal runaway propagation in batteries—what do we talk about. *Journal of Energy Storage* 24: 100649.

20 Klink, J., Grabow, J., Orazov, M. et al. (2021). Thermal fault detection by changes in electrical behavior in lithium-ion cells. *Journal of Power Sources* 490: 229572.

21 Santhanagopalan, S., Ramadass, P., and Zhang, J. (2009). Analysis of internal short-circuit in a lithium cell. *Journal of Power Sources* 194: 550–557.

22 Cabrera-Castillo, E., Niedermeier, F., and Jossen, A. (2016). Calculation of the State of Safety (SOS) for lithium-ion batteries. *Journal of Power Sources* 342: 509.

23 Lamb, J. and Jeeravajan, J.A. (2021). New developments in battery safety for large-scale systems. *MRS Bulletin* 46 (10): 1–7.

24 Orazov, N.: *Methode zur Zustandsbewertung thermisch beschädigter Lithium-Ionen-Zellen*, PhD thesis, TU Clausthal, Germany 2024

25 GTR (2019). Retrieved on 23.08.2021 from https://wiki.unece.org/download/attachments/72024290/EVS17-E1TP-0200%20%5BOICA%5DACEA%20study%20European%20field-%20data.pdf?api=v2.

26 GTR (2019). abgerufen am 23.08.2021 unter https://wiki.unece.org/download/attachments/72024290/EVS17-E1TP-0100%20%5BOICA%5DUnited%20States%20EV%20Fire%20-Incident%20Field%20Data%20Review.pdf?api=v2.

27 Wang, Q., Ping, P., Zhao, X. et al. (2012). Thermal runaway caused fire and explosion of lithium ion battery. *Journal of Power Sources* 208: 210.

28 Feng, X. et al. (2018). Thermal runaway mechanism of lithium ion battery for electric vehicles: a review. *Energy Storage Materials* 10: 246–267.

29 NREL (2015). retrieved on 24.08.2021 from https://www.nrel.gov/docs/fy17osti/66958.pdf.

30 NRC (2021). retrieved on 24.08.2021 from https://nrc-publications.canada.ca/eng/view/ft/?id=c32c4684-53e8-422e-9cf5-2845bbe92f33.

31 Honda, K., Tsuruga, H., Kodama, M., Senoo, T., Nishioka, K., and Ichimaru, H. (2017). *Method for a Propagation Test on a Battery System*. DE 10 2016 216 075 A1.

32 ESPEN (2021). last accessed on 23.08.2021 at https://www.ei.tum.de/en/ees/research-projects/completed-projects/espen.

Further Reading

Ashtiani, C. (2008). Analysis of battery safety and hazards risk mitigation. *IDC Transactions* 11 (19): 1.

DIN EN IEC 62485-2:2019-04 (2019). *Safety Requirements for Secondary Batteries and Battery Installations. Part 2: Stationary Batteries.*

EUCAR (2019). *Battery Requirements for Future Automotive Applications (EG BEV&FCEV).*

GTR (2018). Retrieved on December 31, 2020 from https://wiki.unece.org/download/attachments/78742159/EVS18-E1TP-0400%20%5BOICA%5DJourney%20to%20a%20New%20Regulatory%20Option.pdf?api=v2.

ISO 26262:2018 (2018). *Road Vehicles – Functional Safety* (12 volumes).

ISO 6469 (2019). *Electrically propelled road vehicles – Safety specifications* (3 volumes from 2018 and 2019).

Kohlberger, M. (2014). Safety element for battery cell. European patent application EP 2 779 271 A2.

Kriston, A. et al. (2020). Initiation of thermal runaway in lithium-ion cells by inductive heating. *Journal of Power Sources* 454: 227914.

Lamb, J., Torres-Castro, L., Orendorff, C.J. et al. (2017). *Determination of Battery Stability with Advanced Diagnostics*. Sandia Report SANR20176959. Sandia National Laboratories.

OICA (2017). Retrieved on 24.08.2021 from https://wiki.unece.org/download/attachments/54428762/EVS1536-613%20%5BOICA%5D%20Thermal%20Propagation%20Topics.pdf?api=v2.

Ramadass, P., Fang, W., and Zhang, Z. (2014). Study of internal short in a Li-ion cell 1. Test method development using infra-red imaging technique. *Journal of Power Sources* 248: 769–776.

Schmalstieg, J., Käbitz, S., Ecker, M., and Sauer, D.U. (2013). From accelerated aging tests to a lifetime prediction model: analyzing lithium-ion batteries. In: *2013 World Electric Vehicle Symposium and Exhibition*. 2014, 6914753.

Statista (2020), last accessed on 13.06.2020 from https://de.statista.com/statistik/daten/studie/534429/umfrage/weltweite-preise-fuer-lithium-ionen-akkus.

Tasks

Task 13.1

Show the discharge curves for the following lithium-ion battery systems schematically in a diagram, estimating the voltage curve, and explain the curve:

a) NMC – Graphite,
b) LFP – Graphite,
c) LFP – LTO, and
d) $LiNiMnO_2$ – Graphite.

Task 13.2

What types of electrolytes are available for lithium batteries? Name at least one advantage and two disadvantages of each.

Task 13.3

What is an SEI, in which lithium-ion batteries does it occur, and what function does it fulfill?

Task 13.4

Name four different additives in electrolytes of NMC graphite lithium-ion batteries, each of which fulfilling a different function (which?)!

Task 13.5

Name five different safety tests that are usually required for lithium-ion batteries.

Task 13.6

Explain the function of at least two different safety devices that are used in the design of many lithium-ion batteries.

Task 13.7

What is meant by calendar and cyclic aging? Name three aging mechanisms in lithium-ion cells with conventional NMC graphite chemistry.

Task 13.8

Using your own notebook and phone, estimate the number of equivalent full cycles the battery will have per year and the distribution of discharge depths before a full charge occurs.

Task 13.9

Figure 13.28 shows the data of an electric vehicle with a lithium-ion battery.

a) Describe the voltage curve during rest periods in which only consumers with very low energy requirements, such as clocks or some control units, draw current from the vehicle battery. Can a relaxation behavior be seen?
b) What conclusions can be drawn about the validity of an equivalent circuit diagram consisting of a voltage source and a resistor?
c) How do you rate the use of a simple parameter set, as is often used in the form of the so-called look-up tables?

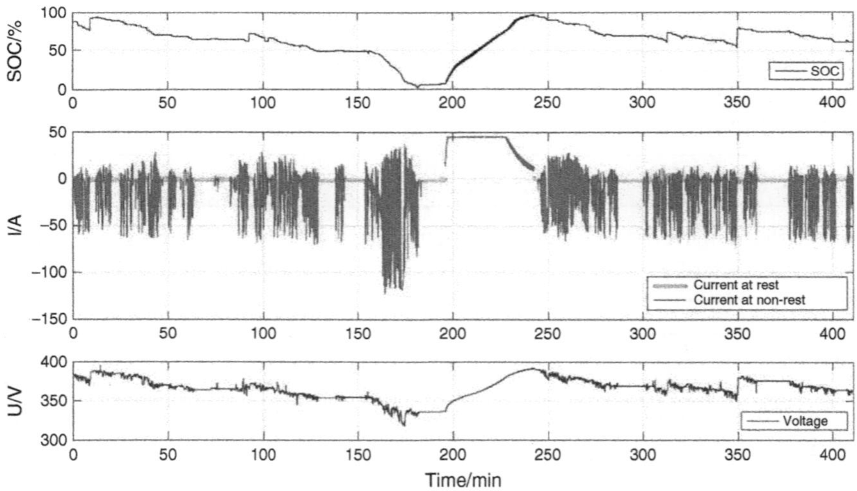

Figure 13.28 Data of an electric vehicle with lithium-ion battery. Source: Adapted from Ref. [12].

Task 13.10

Describe the safety risks that exist when a BMS is switched off while the battery is not in use, that is, when no loads are connected respectively the vehicle or machine that uses the battery is switched off. What additional features that the BMS must have in such a case?

14

Other Battery Technologies

In chapter 11, it has already been explained why, in addition to lead–acid and lithium-ion batteries as the technically and economically most important commercial battery systems, a number of additional battery systems need to be discussed briefly in this book. These battery systems:

1) Either have properties that make them indispensable for special applications, for example, silver–zinc batteries for some military applications or magnesium–air batteries for emergency service applications, both of which are non-rechargeable,
2) may be able to compete with lead–acid and lithium–air batteries in the future, or
3) are still widely used despite the superior properties of lithium-ion batteries.

Further details on these and many other battery systems with a key role in specialized niche markets can be found in the Handbook of Batteries and the Encyclopedia of Electrochemical Power Sources [1, 2].

The first four sections of this chapter describe commercially available battery systems of different technological maturity and market penetration. Alkaline nickel batteries, whose range of applications has been greatly reduced by lithium-ion batteries but are still used in existing systems and may continue to be used in niche areas. The other three battery systems offer properties that are of interest to a number of applications that are outside the range of either lead–acid or lithium-ion batteries. If and when they will disappear completely from the market, however, is unclear.

- Alkaline nickel batteries (section 14.1),
- Zinc–air batteries (primary batteries, non-rechargeable) (section 14.2),
- Redox-flow batteries (section 14.3), and
- High-temperature batteries (section 14.4).

Battery Technology: Fundamentals of Battery Electrochemistry, Systems and Applications,
First Edition. Alexander Börger and Heinz Wenzl.

Companion website: www.wiley.com/go/Börger/Battery_Technology

The former five sections of this chapter discuss some battery technologies that could be long-term successors to existing lithium-ion technology. These battery technologies are often referred to as post-lithium battery systems. They are:

- Lithium solid-state batteries (section 14.5),
- Lithium–sulfur batteries (section 14.6),
- Lithium–air batteries (section 14.7),
- Sodium–air batteries (section 14.8), and
- Sodium-ion batteries (section 14.9).

It will take a long time before these technologies can compete technically and economically with current lithium-ion batteries, and the prospects of success are uncertain.

The last section, ultracapacitors and hybrid battery concepts – section 14.10, contains, for the sake of completeness, additional comments on concepts that are or could be alternatives to conventional batteries in certain areas.

The extent to which these electrochemical systems are treated in this book not only reflects their past or possibly future importance but also aspects that set them apart from lead–acid and lithium-ion batteries. More information about the battery systems discussed here can be found in Refs. [1, 2].

14.1 Alkaline Nickel Batteries

14.1.1 General Information

Alkaline nickel batteries are the battery technology currently most widespread among the systems competing with lithium-ion and lead–acid batteries, including in the automotive sector for hybrid vehicles. The oldest representative of this group of batteries, the nickel–iron battery, was quickly supplemented by batteries with other anode materials. The systems shown here are:

a) Nickel–metal hydride batteries,
b) Nickel–iron batteries,
c) Nickel–zinc batteries.

Nickel–iron and nickel–zinc batteries have not been widely used to date. One system that is now obsolete for environmental and health reasons is the nickel–cadmium battery, which has been banned in the EU for some time due to major environmental and health risks[1] if not properly disposed of but may still be found elsewhere. Nickel–hydrogen is a special battery that has played a role in space travel. Overall, nickel–metal hydride batteries dominate the alkaline nickel battery market, which is still a growth market, even though numerous applications in the household and automotive sectors have been replaced by lithium-ion batteries. However, new markets have also opened up that were previously reserved for non-rechargeable, that is, primary batteries.

1 Cadmium is, in contrast to lead and some other heavy metals, water soluble. Once cadmium enters the environment, it causes major, unacceptable health risks.

14.1.2 Physical-Chemical Principles

In all alkaline nickel batteries, the cathode consists of a nickel compound (generally a nickel (oxide) hydroxide), the electrolyte is a potassium hydroxide solution, and only the anode consists of different materials, that is, a metal hydride, iron, or zinc. Cathode and anode materials are stable in potassium hydroxide, so no corrosion occurs and a very long calendar life can be achieved.

The main reaction of the cathode in all cells leads to the conversion of NiOOH zu $Ni(OH)_2$

$$2NiOOH + 2H_2O + 2e^- \rightarrow 2Ni(OH)_2 + 2OH^-. \tag{14.1}$$

The corresponding anode reactions for iron (Fe), cadmium (Cd), and zinc (Zn) are written here only for zinc (in the case of iron, the reaction can continue to a trivalent iron species):

$$Zn + 2OH^- \rightarrow Zn(OH)_2 + 2e^-. \tag{14.2}$$

For hydrogen from a metal hydride reservoir, where M stands for the metal hydride material, the anodic reaction equation is as follows:

$$xOH^- + MH_x \rightarrow xe^- + xH_2O + M. \tag{14.3}$$

The hydrogen is intercalated in the metal hydride, so the notation already introduced for lithium-ion batteries is used here, which represents intercalation processes.

The combined formula for nickel-iron and nickel-zinc batteries is only given for zinc:

$$2NiOOH + Zn + H_2O \rightarrow 2Ni(OH)_2 + Zn(OH)_2. \tag{14.4}$$

As in lead–acid batteries, the electrolyte takes part in the reaction, but the water consumption during discharge, in relation to the amount of water available in the electrolyte, is negligible.

In nickel–metal hydride batteries, however, the combined reaction is somewhat different:

$$xNiOOH + MH_x \rightarrow xNi(OH)_2 + M. \tag{14.5}$$

Here, no water is consumed during the electrochemical reaction, so the electrolyte volume can be minimized. It is also important to note that the use of an aqueous electrolyte provides inherent overcharge protection via the electrolysis reactions for all alkaline nickel cells, and overdischarge protection can be achieved using additives in the negative active materials. Alkaline batteries, therefore, do not require special monitoring systems for overcharge or overdischarge protection. The formation of hydrogen or oxygen can be completely suppressed as long as suitable voltage limits are maintained during charging. In contrast to lead–acid batteries, cells can therefore be designed as gas-tight and only require a burst valve as protection against misuse, which in many cases is designed as a resealable valve. For safety reasons, venting requirements for battery housings must be fulfilled in case of charger problems or defective cells.

Table 14.1 Data for different alkaline nickel battery systems.

	NiZn	NiFe	NiMH
Nominal voltage U^o in V	1.65	1.2	1.2
Specific energy (Wh/kg)	60–90	30–55	42–110
Specific power (W/kg)	290	25–110	200–1300
Cycle stability	Rather bad	Very good	Good
Costs	Low	Low	High

Source: McGraw-Hill Education.

Important data for alkaline nickel batteries are summarized in Table 14.1.

An important difference from other battery systems is that alkaline nickel batteries exhibit significant temperature effects during charging and discharging due to the reversible heat effect $T\Delta_r S$. These effects can overcompensate for the ohmic heat production and heat generation from the charge transfer voltage, so cells cool down at technically relevant current amplitudes. This effect can be used as a full charge criterion. For example, in NiMH batteries, the endothermic entropic effect at room temperature is approx. −40 kJ/mol. Additionally, for NiMH batteries, the heat of absorption must be included in the overall thermal balance, which can lead to complete compensation of the cooling effect during charging. Heat and temperature considerations are therefore much more complex for alkaline nickel batteries than for other battery technologies.

14.1.3 Cell Structure

14.1.3.1 Cathode Material

The cathode consists of nickel oxide hydroxide ("NiOOH"), which has a low electrical conductivity and an additional conducting structure, either thin nickel metal particles respectively foils or a nickel-coated fleece. In addition, cobalt is usually added to increase conductivity, and a small amount of zinc is added to suppress the formation of γ-NiOOH, which impairs the discharge behavior.

"NiOOH" is not a clearly stoichiometrically defined compound and is therefore often written as $NiO_2 \cdot NiOOH \cdot H_2O$. Due to the unclear stoichiometry and composition, the equilibrium voltage cannot be precisely established.

14.1.3.2 Anode Materials

In NiFe and NiZn batteries, porous metal sponges made of iron or zinc are used, with certain additives (NiFe: iron sulfide, FeS, to extend battery life; NiZn: mainly Ca to reduce the dissolution of $Zn(OH)_2$ as zinc cations).

Hydrogen-storing alloys are used as the anode material in NiMH batteries. There are three types, each with different properties and structures (see Table 14.2).

14.1.3.3 Electrolyte

The electrolyte consists of an aqueous potassium hydroxide solution (approx. 6–7 M, corresponding to a density of 1.24–1.31 g/cm^3, as the maximum conductivity of

Table 14.2 Anode alloys of NiMH batteries.

Type	Example	Properties
AB_5	(La, Ce, Nd, Pr)$_1$Ni$_{3.2}$Co$_{1.0}$ Mn$_{0.6}$Al$_{0.11}$Mo$_{0.09}$	This (still most common) class of materials has made NiMH batteries with a low self-discharge rate possible; Co reduces the volume expansion during hydrogen storage, but is expensive and does not store hydrogen; Al reduces the operating pressure of the battery; Mn increases the hydrogen storage capacity and, with Al, improves the high-current capability
AB_2	$V_{18}Ti_{15}Zr_{18}Ni_{29}Cr_5Co_7Mn_8$	These materials theoretically have higher storage densities for hydrogen than the AB_5 materials, which could lead to advantages for battery systems; however, they are more expensive in terms of individual material prices, often have higher self-discharge rates, and higher hydrogen pressure, which is why further work is being done on optimization
A_2B_7	$Nd_{18.7}Mg_{2.5}Ni_{74.7}Co_{0.1}Al_{3.6}Zr_{0.2}$	

potassium hydroxide solution is in this range). In NiZn and NiFe batteries, water takes part in the electrode reaction, which leads to a slight change in the electrolyte during discharge so its quantity cannot be reduced at will.

Additives in the electrolyte are LiOH or NaOH, which are used, for example, to prevent swelling of the positive electrode.

If CO_2 from the air gets into the electrolyte, carbonate compounds form, which, if the entry of air into the cell is not prevented, will require a change of electrolyte after several years. However, this is normally prevented by the sealed structure of the batteries.

14.1.3.4 Separators

In the early days of alkaline nickel batteries, polyamide was often used as a separator material, but this can be oxidized to nitrate, which then acts as a redox shuttle.[2] In NiMH batteries today, separators are mostly made of polyolefins (polyethylene/polypropylene composites) with special surface treatments (e.g. sulfonation). In NiZn batteries, the separator often consists of a multilayer structure to minimize soluble zinc cations $[Zn(OH)_4]^{2-}$ to reach the counter electrode.

14.1.3.5 Current Collectors

As hydrogen must not be released, nickel-plated steels, which minimize hydrogen evolution, are usually used as electrode materials. Composite materials are also possible, as are sometimes used in the automotive sector. With the correct choice of materials for current collectors, housing, and terminals, there are no corrosion processes in alkaline nickel systems, so a very long service life can be achieved.

2 See chapter 13 for explanations of this term.

14.1.3.6 Cell Assembly

Alkaline nickel systems are either cylindrical or prismatic cells, as described in chapter 4. Usually, all alkaline nickel batteries with low capacity (up to approx. 10 Ah) are sealed. A valve is only necessary in case of some abuse conditions and charger malfunction or wrong sizing as the potential of the negative electrode prevents hydrogen evolution most of the time in contrast to lead–acid batteries. The oxygen cycle described for AGM batteries also plays a role. In the case of high-capacity cells, where they are still built and used, flooded designs are still used, which, similar to lead–acid batteries, must be regularly topped up with purified water or occasionally require a complete electrolyte change to ensure a very long service life. Access of air into the cell must be minimized.

14.1.4 Battery Properties

Table 14.1 shows the at least potentially (depending on the design) significantly higher power density and specific energy compared to lead–acid batteries. Due to the overcharging and overdischarging side reactions, operation of cells in series is possible for batteries with a high nominal voltage[3] without any problems. A complex battery management system, as with lithium-ion batteries, is therefore not absolutely necessary purely to prevent overcharging, but it is often advisable for other reasons.

The charge/discharge characteristics of alkaline nickel batteries often show strong hysteresis effects. That is, the open-circuit voltage is 0.1 V or more higher in the charging direction than in the discharging direction for the same state of charge. This makes determining the *SoC* of alkaline nickel–metal hydride batteries more challenging, as simple voltage measurements provide unreliable information.

14.1.5 Aging and Related Processes and Battery Properties

14.1.5.1 General Aging Characteristics

Nickel–metal hydride batteries can be used in a variety of applications due to the absence of significant corrosion processes, the process of storing and removing hydrogen from an alloy, and the fact that the electrolyte is not an electrochemically active element. This means:

a) No consumption of electrolyte during charging or discharging.
b) For metal hydride anodes: the active materials do not undergo a solution and deposition process and there are no phase changes, which means no dendrite formation or recrystallization processes duging rest periods.
c) For the anode: no crystallization overvoltages and, thus, the discharging processes are largely independent of the previous discharge history.

In principle, this leads to a very long service life. However, alkaline nickel batteries also have other life-limiting aging mechanisms, and NiMH batteries

3 In an energy storage facility in Fairbanks, Alaska, the nominal voltage is 4600 V, that is, 4000 cells in series. For such a large number of high-capacity cells, extensive monitoring is necessary.

in particular generally have poor low-temperature behavior due to the strongly temperature-dependent intercalation process of hydrogen.

14.1.5.2 Self-Discharge

Self-discharge occurs in alkaline nickel batteries mainly due to:

d) Decomposition of higher-valent nickel hydroxides, making the self-discharge rate highly dependent on the state of charge.
e) Redox shuttles (e.g. nitrate/ammonium shuttle), which make the self-discharge rate highly dependent on the separator materials used or impurities, for example.
f) Formation of subtle short circuits, for example, Co/Al/Mn particles (detached from the alloys) in NiMH batteries, which make the self-discharge rate highly dependent on the degree of aging.

In comparison to lithium-ion and lead–acid batteries, the self-discharge rate is high even in new batteries. However, this does not lead to rapid damage, that is, the cells do not need to be recharged during storage.

14.1.5.3 Memory Effect

Alkaline nickel batteries are the best-known examples of the so-called memory effects, that is, batteries that are not fully discharged exhibit, after some time, poorer properties than batteries that regularly experience full cycles [3]. There are two different effects:

a) The so-called classic memory effect. This only occurs in NiCd batteries due to cadmium, which can form an intermetallic compound Ni_5Cd_{21} if the discharge is incomplete, causing a reduction in the discharge voltage of up to 150 mV. This effect can be suppressed by appropriate manufacturing processes of the nickel electrode; nowadays it is only of historical significance (as is the entire NiCd technology in the EU).
b) The so-called memory or lazy battery effect. This can occur in all alkaline nickel batteries because it is caused by the nickel electrode. The physical cause here is the formation of γ-NiOOH, which causes a reduction in the discharge voltage of up to 40 mV.

Both memory effects can be reversed by completely discharging and then recharging. However, this requires that a voltage-based controller does not terminate the discharge prematurely.

14.1.5.4 Aging of the Cathode Due to the Formation of γ-NiOOH

The formation of γ-NiOOH in the nickel cathode, which also triggers the lazy battery effect, is caused by overcharging. It is not only harmful due to the restricted voltage window but also due to the fixation of electrolyte, which increases the internal battery resistance, and leads to a strong volume expansion of up to 44%. The stress for the active material is then further increased by the mechanical stress during discharging to α-$Ni(OH)_2$ with a further 39% increase in volume. The drop in the voltage level not only leads to a reduction in performance but also to a shift of potentials between the positive and negative electrodes. However, excessive discharges are also harmful, as the conductive structure of the nickel cathode formed by the cobalt additives (more precisely: CoOOH) is destroyed – alkaline nickel batteries should therefore not be stored for long periods in a discharged state.

14.1.5.5 Aging of the Metal Hydride Anode

Metal hydride anodes age mainly through the following mechanisms:

a) Corrosion according to a reaction equation such as the following:

$$(1 + x)\mathrm{M} + x\mathrm{H_2O} \rightarrow \mathrm{M(OH)}_x + x\mathrm{MH}. \tag{14.6}$$

This is usually associated with the formation of solid layers covering the electrode surface. The increase in volume due to hydrogen intercalation can lead to microcracks and thus further corrosion. Possible countermeasure: The lower the cycle depth during cyclization, the less this corrosion mechanism plays a role, and the longer the service life.

b) Loosening of the active material due to the increase in volume during hydrogen storage. The reduced contact in the active material leads to an increase in internal resistance. The negative electrode is therefore generally oversized in NiMH batteries. If the overcharge reserve is used up, the oxygen cycle will eventually stop working, and hydrogen will be produced, which can lead to the pressure relief valve opening.

14.1.5.6 Aging of the Electrolyte

The most important aging phenomena of the electrolyte are decomposition into hydrogen and oxygen. In alkaline nickel batteries that are not sealed, carbonation also plays a role, that is, the formation (and precipitation) of potassium carbonate through chemical reaction with carbon dioxide in the air.

14.1.6 Safety Aspects

Alkaline nickel batteries are very safe systems. Despite decades of use, only very few serious accidents have been reported. Possible risks include in particular the leakage of caustic potash solution, the release of hydrogen from metal hydride electrodes at high temperatures and states of charge (therefore transport at max. 30% *SoC*), the opening of the safety valve in the event of constant overcharging and acceleration of the oxygen cycle (therefore, a battery container must not be hermetically sealed, and adequate venting of the battery environment must be ensured), and strong heating in the event of a cell short circuit.

14.1.7 Optimum Operation

Alkaline nickel batteries should be stored between 10% and 100% *SoC*; at charge levels of less than 10%, the conductive elements of the positive electrode can be damaged. The storage temperature should always be below 40 °C, preferably even below

20 °C. After long periods of storage, several full cycles with a low charging current and a time limit are advisable.

Cycling should ideally take place in a temperature window of 10–25 °C and with a suitable charging process that avoids overcharging as far as possible. Charging can be carried out with relatively high currents. In cell groups, periodic equalization charges make sense, which are usually implemented using pulse methods for NiMH batteries.

The recycling of NiMH batteries is not only required by law, but is also economical due to the valuable metals (nickel, lanthanum, etc.).

14.1.8 Outlook

NiMH batteries are still used as traction batteries in some hybrid vehicles (tending to be phased out). Earlier applications in cell phones and laptops have almost completely disappeared due to the emergence of lithium-ion batteries, but in some areas, non-rechargeable primary batteries are being replaced by alkaline nickel batteries, which are a problem-free replacement in terms of dimensions and voltage, but only for applications with a bridging time of a few months due to their high self-discharge. As these batteries are easy to store and maintenance-free and are based on an aqueous solvent, they are considered safe. Their poor low-temperature behavior remains a disadvantage. NiMH batteries are therefore an alternative in applications where very robust, safe, and long-lasting batteries are required, for example, in the aircraft sector, as emergency power batteries for safety lighting, and in the military sector.

The other alkaline nickel battery technologies have so far had a rather marginal existence. Nickel–iron has now been proposed as a combination of battery and electrolyzer. During continuous overcharging, competition with PEM electrolyzers appears to be possible, although energy can also be stored and released. Nickel–zinc batteries are sometimes discussed as an alternative to lithium-ion starter batteries (which are considerably more expensive and more complex in terms of safety) in the automotive sector, especially if there might be a legal ban on the marketing of lead-containing battery systems.

14.2 Zinc–Air Batteries

14.2.1 General Information

Of all the metal–air systems, zinc–air batteries are by far the most important system commercially: they have been used on a large scale for a long time and are currently the battery technology with the highest specific energy available. In zinc–air batteries, the anode consists of metallic zinc. Oxygen from the air is converted to a hydroxide ion at the cathode. As the oxygen required for the reaction does not have to be included in the total weight of the cell, the zinc–air battery (like all metal–air batteries) has a high specific energy density. Commercially available zinc–air batteries have a specific energy of over 400 Wh/kg, which is more than twice as high as most lithium-ion batteries. However, the oxygen required for discharging remains in the cell as ZnO or $Zn(OH)_2$, so that the cells increase in weight during discharge. The disadvantage of the zinc–air system is its poor cycle stability, which has led to the idea of "mechanical recharging." This means that when the zinc anode is fully converted, it is simply exchanged and replaced by a fresh anode, while the old one is recycled. This concept corresponds to the solid-state variant of a redox-flow battery (see section 14.3). Previous trials with the concept, which was marketed as "Electric Fuel," have not been successful, however.

14.2.2 Physical-Chemical Principles

Cells consist of an anode made from metallic zinc, a cathode that converts oxygen from the air to hydroxy ions, and a 30% potassium hydroxide solution as the electrolyte.

The cathode reaction takes place on a membrane that allows oxygen to pass through but prevents the electrolyte used as potassium hydroxide solution from escaping. The membrane contains catalysts (e.g. perovskites of the formula (La, Sr)(Mn, Ni)O_3) and an electron-conducting structure that provides a three-phase boundary for the electrochemical reaction. The reaction (which is analogous to all other metal–air systems) then takes place on the catalysts in the membrane:

$$\frac{1}{2}O_2 + H_2O + 2e^- \rightarrow 2OH^-. \tag{14.7}$$

The anodic reaction is the conversion of zinc to the oxide via the (soluble) zinc ion:

$$2Zn + 4OH^- \rightarrow [Zn(OH)_4]^{2-} + 2e^- \rightarrow ZnO + H_2O + 2OH^- + 2e^-, \tag{14.8}$$

resulting in the combined reaction:

$$2Zn + O_2 \rightarrow 2ZnO. \tag{14.9}$$

In principle, recharging can be carried out electrochemically (by reversing the discharge reaction), although the solubility of ZnO crystals is problematic. The total number of cycles achieved to date is so low that it is not yet technically possible or practical to use them as rechargeable batteries.

14.2.3 Cell Structure

Most zinc–air batteries are built as button cells, which are used in hearing aids. Larger prismatic batteries for other applications are now also available. The zinc anode is separated by a cellulose separator from the cathode, a highly porous film that absorbs electrolyte but does not allow electrolyte to leak from the cell. The cathode material contains a mix of carbon and electrocatalysts that are electrically connected to the current collector. The barrier to the outside atmosphere is usually formed by PTFE films that allow oxygen and water vapor to pass through but prevent liquid from escaping. During storage, the cathode is protected against air by a thin foil.

14.2.4 Properties

The OCV of a fully charged zinc–air battery is around 1.4 V. The batteries can generally be discharged well up to approx. 1.0 V, whereby the discharge curve is quite flat, that is, there is only a significant drop in voltage toward the end of the discharge. As 30% caustic potash only freezes at around −40°C, very low operating temperatures are also possible, although the greatly increased internal resistance causes restrictions in use.

Another characteristic of zinc–air batteries is the existence of a limiting current, a threshold value resulting mainly from mass transport restrictions – limits which cannot be reached technically either in lead–acid or lithium-ion batteries or many other battery systems. If an attempt is made to permanently exceed this limit, the voltage drops and the required current decreases. Sufficiently short current peaks as peak loads are not critical.

14.2.5 Aging Behavior

As cycling does not occur (non-rechargeable system), the most important aging phenomena are caused by environmental conditions. One of these is the carbonatation of the electrolyte through the absorption of CO_2 and subsequent reactions to form insoluble carbonates. If air enters the cell in an uncontrolled manner, zinc is chemically oxidized to ZnO, which results in high self-discharge. The access of air before use is prevented by a sealing foil. An important role in aging is the relative humidity of the ambient air. Ideally, this should be 50%; if it is permanently lower, there is a risk of the electrolyte drying out; if it is permanently higher, there is a risk of mechanical damage to the battery due to humidity penetrating the battery and requiring more volume.

As long as entry of air to the cathode is prevented, the storage time is long and unproblematic.

14.2.6 Optimum Operation

Care should always be taken to avoid extreme temperature and humidity conditions with zinc–air batteries. In the case of chemically rechargeable batteries, air treatment is often required to remove CO_2 or catalyst poisons in particular.

As the oxygen in zinc–air batteries is taken from the ambient air, the geodetic altitude plays a role in battery performance, which is irrelevant for most other battery systems (with the exception of other metal–air systems, of course). As the oxygen partial pressure decreases with increasing altitude above sea level, the battery characteristics deteriorate. This must be taken into account for applications under extreme conditions, such as those found in the military sector, but can also affect the functioning of hearing aids in an aircraft.

14.2.7 Safety Features

From a safety point of view, zinc–air batteries are almost an ideal solution. The systems are nonhazardous and use almost exclusively nontoxic components that are also easy to recycle.

14.2.8 Outlook

One of the largest commercial applications for zinc–air batteries is currently hearing aids. However, substitution by lithium-ion batteries has begun. It is expected that zinc–air batteries will continue to be used in applications where an inexpensive battery with very high specific energy is required for a single use.

Whether, when, and how the problem of non-rechargeability can be solved is unclear. The proposal to mechanically recharge by replacing the used zinc electrode with a new one could possibly provide an option for electromobility in certain applications, for example, the use of vehicles in commercial fleets.

A breakthrough in rechargeability would lead to a serious competitive situation for many applications of lithium-ion batteries, as all materials used are inexpensive.

14.3 Redox-Flow Batteries

14.3.1 General and Physical-Chemical Principles

Flow batteries, also known as redox-flow batteries, are characterized by the fact that the actual electrochemically active materials are present in liquid or dissolved form. They are contained in separate tanks and are pumped through the reaction chamber, or the tanks could be emptied and the system "refueled." The basic structure corresponds to that of a fuel cell or a bipolar plate structure (see Figure 14.1). To create three-phase boundaries, the cathodic ("catholyte") and anodic ("anolyte") reaction chambers, which must be separated by a separator, contain a fleece that conducts electrons, comparable to gas diffusion materials for fuel cells. The performance of redox-flow batteries is characterized by the number of cells and their dimensions, while the energy content is determined by the size of the tanks in which the active material is stored. The term "specific energy" no longer has meaning in these systems without knowledge of the size of the storage tanks.

Although the concept appears convincing, it has not yet been possible to find active materials that allow sufficiently competitive energy densities commercially. The most commonly investigated systems to date are based on vanadium solutions (maximum energy density: 25 Wh/l) or the reaction of zinc with bromine (maximum energy density: 65 Wh/l). Pilot and demonstration projects with an energy content of around 100 kWh have been built. In both cases, there are considerable environmental and safety issues. Vanadium redox-flow systems contain very large quantities

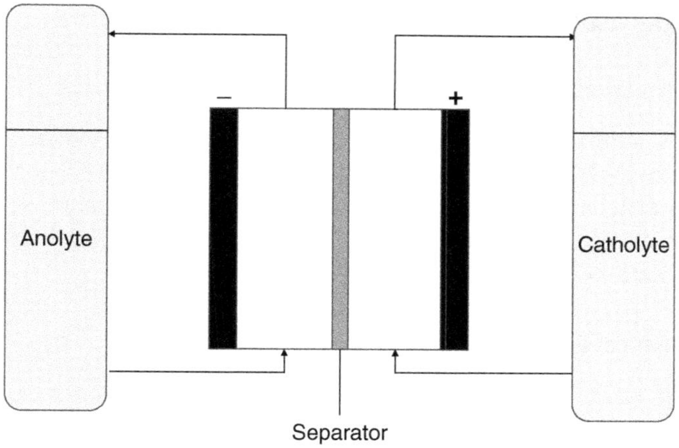

Figure 14.1 Principle of a flow cell.

of aqueous sulfuric acid because of the poor solubility of vanadium oxides in aqueous sulfuric acid, and in zinc-bromine systems, the technically safe handling of the highly toxic and corrosive bromine is difficult. There are no technical or economic advantages in terms of performance data compared to lithium-ion batteries. Organic systems are still only at the research stage. A detailed description is therefore not given here, only the reaction equations of the two systems mentioned here are given for the sake of completeness:

$$VO_2{}^{+} + V^{2+} + 2H^{+} \rightarrow VO^{2+} + V^{3+} + H_2O. \tag{14.10}$$

$$Zn + Br_2 \rightarrow ZnBr_2. \tag{14.11}$$

14.3.2 Outlook

In the long term, it may be conceivable that redox-flow batteries can be used to store very large amounts of energy on an industrial scale, if this application ever becomes relevant (see chapter 19).

The use of vanadium redox-flow batteries with an energy content of several tens of kilowatt-hours as household storage units requires several hundred liters of aqueous sulfuric acid with heavy metals dissolved in it. It is highly doubtful that this should be a desirable development for safety and environmental reasons. In general, the relative excess of attention that has been paid to various redox-flow concepts in the past – and to some extent in the present – in the academic world does not correspond to any foreseeable commercialization prospects and should therefore be viewed critically, despite all the scientific fascination for pure basic research. (For comparison, *The Handbook of Batteries* [1], which contains information on virtually every battery system, devotes just seven pages to flow batteries out of a total of over 1400 pages.)

14.4 High-Temperature Batteries

14.4.1 General Information

The attempt to use sodium ions to transport charge between two electrodes has started significantly earlier than research on lithium-ion batteries. The first commercial battery systems and vehicles were already available in the 1960s. At that time, the specific energy of complete systems, including thermal insulation and housing, was approx. 100 Wh/kg.

14.4.2 Physical-Chemical Principles

The solid electrolyte used in sodium-based high-temperature batteries is β-aluminum oxide which, despite its name, is not pure aluminum oxide but a sodium–aluminum–oxygen compound with a certain conductivity for sodium ions. However, as this sodium-ion conductivity and the associated charge transport only work sufficiently well at higher temperatures, systems with a working temperature of around 300–350 °C have been developed: the sodium–sulfur (NaS) battery and the so-called ZEBRA battery (ZEBRA = zero emission battery research activities) based on nickel.

The anodic reaction is the same in both cases:

$$2Na_{(l)} \rightarrow 2Na^+ + 2e^-. \tag{14.12}$$

The cathodic reaction differs and is complex for NaS batteries:

$$xS_{(l)} + 2e^- \rightarrow S_x^{2-} \tag{14.13}$$

(x can assume values from 5 to 3; any further reduction is usually avoided as it leads to increased losses. The exact reaction equation depends on the respective operating conditions).

or

$$NiCl_2 + 2e^- \rightarrow Ni + 2Cl^-. \tag{14.14}$$

Please note: Sodium and sulfur are liquids at the operating temperatures. Nickel chloride is in contact with liquid $NaAlCl_4$ as the second electrolyte, which acts as a Na^+ ion conductor.

The combined reaction equations are therefore:

$$Na + xS_{(l)} \rightarrow Na_2S_x \tag{14.15}$$

and

$$2Na + NiCl_2 \rightarrow 2NaCl + Ni \tag{14.16}$$

with nominal voltages of up to 2.08 V (Na–S) or 2.58 V (ZEBRA).

14.4.3 Cell and Battery Structure

Individual cells are usually cylindrical cells made of stainless steel with a capacity of a few Ah. The ceramic electrolyte acts as a separator and needs to permanently separate the anodic and cathodic liquids from each other. The batteries are usually built as very large systems with many cells in series and parallel strings, as only in this way are the thermal losses due to the high operating temperatures negligible. Electric vehicles were built, but now the batteries are primarily used for stationary applications where weight and volume restrictions are less severe.

An alternative, which is currently under investigation, is the design of a NaS system as a flow cell design.

14.4.4 Properties

Due to the absence of side reactions and no relevant parasitic reactions, the self-discharge rate is very low if heat losses are not considered. If the energy for maintaining cell temperature has to come from the battery itself, the heating or cooling power is, from a system's point of view equivalent to a very high self-discharge rate.

Cyclic stability is exceptionally good. About 30 years ago, Silent Power presented a sodium–sulfur cell with a service life of 5000 full cycles without noticeable degradation. However, the goal of consistent product quality for the manufacture of multi-cell battery systems with frequent thermal cycles was not achieved at that time.

14.4.5 Aging

A high number of charge/discharge cycles requires effective prevention of deep discharges, as low charge levels greatly accelerate the aging of the materials.

Thermal cycles, which are unavoidable in applications with long downtimes, generate thermal stresses in the ceramic, which can lead to cracking of the ceramic separator/electrolyte and thus to the destruction of the cell due to an internal short circuit.

14.4.6 Safety

The high operating temperatures and use of liquid sodium, which burns immediately in air or on contact with water, and in the case of the NaS battery, the presence of liquid sulfur, which can react to form highly toxic and corrosive gases such as sulfur dioxide SO_2, sulfur trioxide SO_3, or hydrogen sulfide H_2S, make these battery technologies a safety challenge. This is exacerbated by the risk of internal short circuits, which can easily occur with a solid separator or electrolyte destroyed by thermal cycling. NaS and ZEBRA batteries can therefore only be operated safely with consistent battery monitoring, ideally in combination with an effective gas sensor system.

14.4.7 Optimal Operation

Due to the high operating temperatures, use in mobile applications or for electric traction is only possible with difficulty, as longer downtimes can hardly be bridged sufficiently.

In stationary applications, that is, for intermediate energy storage or buffering of peak loads or for use in uninterruptible power supplies, their use is less critical but requires consistent monitoring of temperature and state of charge.

14.4.8 Outlook

The sodium–sulfur battery is potentially one of the most cost-effective battery systems because both sodium and sulfur are available in very large quantities and can be obtained in part from by-products and waste products from industrial processes. With the ZEBRA battery, on the other hand, the use of nickel poses a cost problem, comparable to alkaline nickel batteries and lithium-ion batteries.

Both systems were tested in vehicles in the early days of electromobility (when neither nickel–metal hydride nor lithium-ion batteries were available), but were not able to establish themselves due to difficulties, for example, with the heat-up period and the energy required to compensate for the heat demand when stationary. To the knowledge of the authors, there are no projects to use high-temperature batteries of newer designs for vehicles.

There are currently some applications in large energy supply systems to compensate for fluctuations in production and demand of electricity, particularly in Japan. Whether these large-scale systems with an energy content of several megawatt-hours represent economic options in the longer term and without state subsidies is currently not foreseeable. Thermal cycling and heat losses play no particular role in these applications because the modules can be kept permanently at operating temperature, and the losses caused by constant charging and discharging processes compensate for the low heat losses achievable in large systems. NaS batteries are therefore a conceivable alternative to lithium-ion batteries for large battery storage systems in the electricity industry due to the very low cost of the materials.

Research and development activities continue in this area.

14.5 Lithium Solid-State Electrolyte Batteries

14.5.1 General Information

As things stand at present, the lithium solid-state battery may be the most promising candidate for a successor to the present lithium-ion systems, as it reduces size and volume and increases safety. Toyota has announced that it will go into series production in the early 2020s (Asia Nikkei), but this has not really happened; also Chinese battery manufacturers claim to commercialize this battery type soon [4, 5].

14.5.2 Physical-Chemical Principles

The key difference between lithium-ion batteries and lithium solid-state batteries is the use of a solid-state ion conductor, which also acts as a separator. In principle, this can also enable the use of lithium metal anodes, as the risk of dendrite

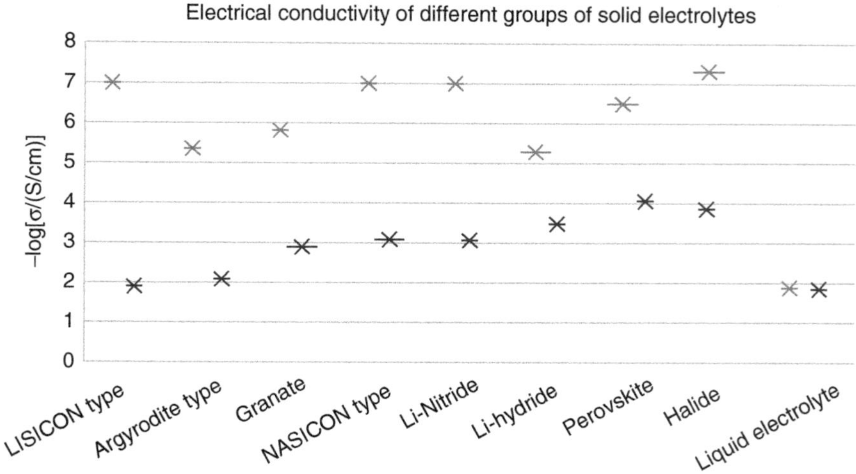

Figure 14.2 Conductivity for lithium ions of promising material classes for lithium solid-state batteries. LISICON-type materials are chalcogenide phosphides, argyrodite-type materials are chalcogenide phosphide halides, garnets are zirconium tantalum niobium oxides, and NASICON-type materials are phosphates. The crosses indicate the start and end points of the conductivity range. In the case of Li halides and liquid electrolytes, both crosses coincide.

growth in the current porous separators is eliminated by using solid electrolytes. However, this is not necessarily required, and systems with graphite anodes are also being further researched. The electrochemistry of both systems is identical, but no SEI is formed with its negative effects on the performance and safety of cells, and the aging processes directly associated with the current electrolytes (degradation of the electrolyte, storage of solvent molecules in the graphite, etc.) are eliminated. If the stability and service life of the solid electrolytes meet the requirements, an even longer service life than that achieved for today's lithium-ion batteries can be expected.

As solid-state separators, the following three basic material classes can be considered:

- inorganic oxidic materials,
- inorganic non-oxidic materials, and
- polymeric organic materials.

Figure 14.2 shows further details of such electrolytes, including their ion conductivity and the materials they contain.

14.5.3 Outlook

Lithium solid-state batteries have two major problems:

1) Ionic conductivity of the materials used, especially at low temperatures.
   The current state of the art is shown in Figure 14.2.

Current electrolytes have an ionic conductivity that is orders of magnitude lower than organic electrolytes in lithium-ion batteries. However, the resistance increases much less due to the significantly lower thickness of the electrolyte layer. In lithium-ion and lithium solid electrolyte cells, the electrolyte does not take part in discharging and charging processes, so the thickness of the electrolyte layer depends only on how well the production processes can be controlled and the safety margin with regard to the mechanical stress on the cell and electrodes during operation. How thick the solid electrolyte must be cannot yet be conclusively determined.

2) Production process for the solid electrolyte.
   Since the solid electrolyte must be very thin and at the same time completely free of holes (to avoid short circuits), high-vacuum processes (sputtering, etc.) are currently mainly used, but these are extremely expensive in terms of large-scale use. In contrast to conventional lithium-ion batteries, direct contact with the electrode material plays a very important role, that is, there is usually a sequence of electrode 1 – solid ion conductor = separator – electrode 2. To achieve good conductivity, it may be necessary to incorporate additional transition layers or to achieve a reduction in contact resistance by means of special measures (e.g. high magnetic fields). Grain boundaries must also be reduced within the active masses. As the chemistries used are very different, it is difficult to make a general statement about the improvement in specific energy and energy density through lithium solid-state batteries. Theoretical energy densities of up to 1000 Wh/l can be achieved when using lithium metal anodes (see Ref. [6]).
3) Cost structure.
   It is not clear whether the complexity of the production processes will allow production costs comparable to conventional lithium-ion batteries and whether the cost of the electrolyte material will be below the presently high costs of electrolytes.
4) Safety.
   It is sometimes argued that the elimination of the organic electrolyte and the gain in safety is the decisive advantage in favor of the lithium solid-state battery. However, the safety of a battery system with such high energy density also depends on production quality. Short circuits may not occur over the whole lifetime for billions of cells in many different applications and operating conditions. Both anode materials, which are being considered, metallic lithium and graphite, ignite spontaneously in air at high temperatures.

14.6 Lithium–Sulfur Batteries

14.6.1 General Information

While sodium–sulfur batteries can only be operated at higher temperatures, lithium–sulfur batteries can also be used at room temperature and lower temperatures. As a very common element that is readily available, sulfur offers potentially significant cost advantages. In addition, the lithium–sulfur system has a theoretical specific energy of 2500 Wh/kg, which is the highest theoretical specific energy of all sealed (i.e. not using air) systems based on liquid electrolytes. However, as the technological difficulties are still considerable, there are no cells produced in reasonable quantities as yet, and a large-scale market launch is only expected on a limited scale in the medium term.

14.6.2 Physical-Chemical Principles

In the lithium–sulfur battery, sulfur serves as the cathode and lithium as the anode. Various intermediate soluble and insoluble polysulfides are formed. Organic liquids such as THF, dioxolane, or tetra(ethylene glycol) dimethyl ether (TEGDME) are usually used as solvents, but solids, gel polymer electrolytes and ionic liquids are also being investigated. In general, the following reactions take place in organic solvents:

$$\begin{aligned}48\mathrm{Li} + 3\mathrm{S}_8 &\rightleftarrows 42\mathrm{Li} + 3\mathrm{Li}_2\mathrm{S}_8 \rightleftarrows 40\mathrm{Li} + 4\mathrm{Li}_2\mathrm{S}_6 \rightleftarrows 36\mathrm{Li} + 6\mathrm{Li}_2\mathrm{S}_4\\ &\rightleftarrows 32\mathrm{Li} + 8\mathrm{Li}_2\mathrm{S}_3 \rightleftarrows 24\mathrm{Li} + 12\mathrm{Li}_2\mathrm{S}_2 \rightleftarrows 24\mathrm{Li}_2\mathrm{S}. \qquad (14.17)\end{aligned}$$

The respective end products sulfur (S_8) and Li_2S_2 or Li_2 S are insoluble, while the middle, longer-chain polysulfides have a certain solubility in the electrolyte. This allows specific system energies of 500 Wh/kg to be achieved. The nominal cell voltage is 2.5 V, but the cell voltage can deviate considerably from this value in some cases due to the complex discharge characteristics.

As a nonconductor, sulfur cannot be used in its pure form but must be mixed with a conductivity additive – usually carbon-based. The cathode materials in lithium–sulfur batteries can be classified as follows:

a) Porous sulfur–carbon nanocomposites that can be considered to be point-like (zero-dimensional),
b) Sulfur nanotube or nanofiber composites (one-dimensional),
c) Graphene–sulfur cathodes (two-dimensional),
d) Structured sulfur-carbon nanocomposites (three-dimensional),
e) Core-shell/core-shell yolk structures,
f) Polymer-sulfur nanocomposites, and
g) Li_2S-cathodes.

For details, see Ref. [7].

$LiNO_3$ is added to the electrolyte to form covering layers on the lithium metal surface and lithium polysulfides to reduce polysulfide dissolution (disadvantage: increased viscosity).

14.6.3 Outlook

The unsolved problems of the lithium–sulfur battery are as follows:

- Poor conductivity of sulfur at room temperature.

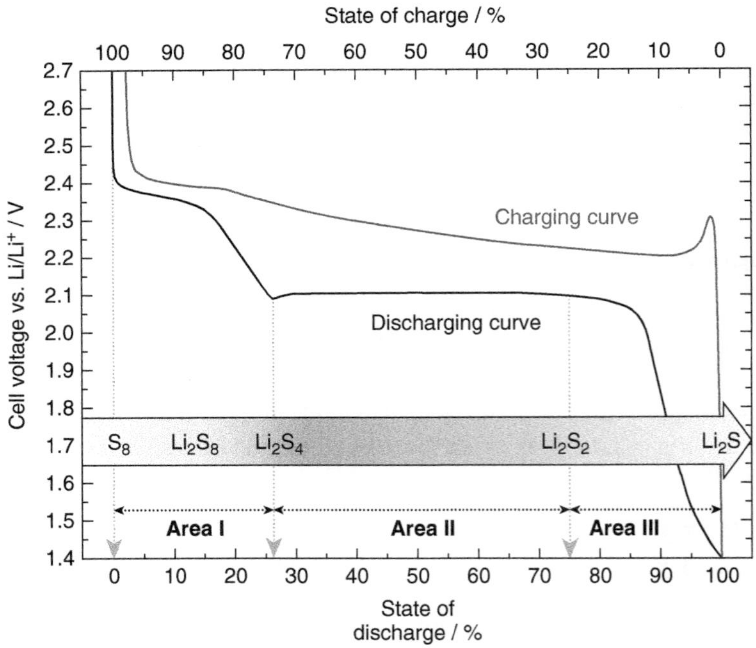

Figure 14.3 Charge and discharge curve of lithium–sulfur batteries. Source: Reproduced with permission from Ref. [8]/Springer Nature.

- Dissolution of polysulfides in the electrolyte (loss of active material, side reactions and efficiency losses due to the so-called polysulfide shuttle, etc.).
- Strong volume expansion (up to 80%) of the sulfur cathode, which causes poor cycle stability.
- An often "pathological" discharge and charge curve, see Figure 14.3, which reflects a sequence of several electrochemical reactions with different thermodynamic data during discharge. Electrically, these curves are easy to understand qualitatively as a parallel connection of many voltage sources, which represent the various reactions with their equilibrium voltages and exchange current densities.

These charging and discharging curves are so complex because there are generally three areas:

Area 1: Start of discharge with dissolution of polysulfides.
Area 2: Formation of insoluble $Li_2S_{(2)}$ phases, resulting in crystallization overpotentials.
Area 3: Area of (mostly incomplete) discharge, which is limited by solid-state diffusion, among other factors.

There is still a great need for research into lithium–sulfur batteries, as even fundamental reactions have not been sufficiently clarified. For more, see, for example, Ref. [7].

14.7 Lithium–Air Batteries

14.7.1 General

Lithium–air batteries are potentially one of the battery systems with the highest specific energy (up to 5000 Wh/kg, if the weight of the oxygen to be taken from the air is disregarded). The cells are widely used as a primary system. Particularly advantageous is the fact that, as with all metal–air batteries, one of the substances involved in the main electrochemical reaction, namely oxygen, does not have to be stored but can be taken from the air. The air cathode, where atmospheric oxygen is reduced, corresponds to that of zinc–air batteries in terms of its structure.

However, it has not yet been able to recharge such systems, and at present, they are only primary battery systems.

14.7.2 Physical-Chemical Principles

The reaction of lithium with oxygen can lead to the oxide (Li_2O) or to the peroxide (Li_2O_2), as shown in the following reaction equations:

a) Reaction to form oxide (standard reaction voltage under equilibrium conditions: 2.91 V):

$$\begin{aligned} &\text{anode reaction}: \ 2\text{Li} + 2\text{Li}^+ + 2\text{e}^- \\ &\text{cathode reaction}: \ 0.5\text{O}_2 + 2\text{e}^- \rightarrow \text{O}^{2-} \\ &\text{entire reaction}: \ 2\text{Li} + 0.5\text{O}_2 + \text{Li}_2\text{O}. \end{aligned} \tag{14.18}$$

b) Reaction to form peroxide (standard reaction voltage under equilibrium conditions: 2.96 V):

$$\begin{aligned} &\text{anode reaction}: \ 2\text{Li} + 2\text{Li}^+ + 2\text{e}^- \\ &\text{cathode reaction}: \ \text{O}_2 + 2\text{e}^- \rightarrow \text{O}^{2-} \\ &\text{entire reaction}: \ 2\text{Li} + \text{O}_2 + \text{Li}_2\text{O}. \end{aligned} \tag{14.19}$$

For thermodynamic reasons, peroxide is usually produced, which is also much easier to convert, that is, rechargeable.

A sufficiently electron-conductive and electrocatalytically active carrier material (usually a combination of mesoporous carbon with a precious metal catalyst such as platinum or palladium) with the largest possible surface area is required for sufficiently fast and complete conversion at the cathode.

14.7.3 Current Status

So far, no sufficiently stable electrolyte for a rechargeable lithium–air battery system has been found; many experiments have been carried out with dimethyl sulfoxide (DMSO) or certain ethers. Organic carbonates, that is, esters of carbonic acid, as used in lithium-ion batteries are, however, decomposed. The high energy inefficiency of the charge reaction of the air cathode is a major obstacle, and the search for suitable catalysts has so far been unsuccessful. Furthermore, in contrast to zinc–air batteries, atmospheric components (CO_2, H_2O, N_2) lead to rapid poisoning of the electrode. However, air pre-treatment would make the overall system much more expensive and significantly reduce the high specific energy at the system level.

For the foreseeable future, lithium–air batteries will therefore remain the exclusive preserve of basic research. For more information, see also, for example, Ref. [9].

14.8 Sodium–Air Batteries

14.8.1 General

Sodium–air batteries have the same structure as zinc–air and lithium–air batteries and the same basic advantage that the active mass material, oxygen, can be taken from the air.

Although sodium has very similar chemical properties to lithium, it differs considerably in its intercalation behavior due to the significantly larger radius of the ion. This favors the formation of defined chemical compounds instead of compounds with varying stoichiometry, whose intercalation locations can be filled successively. There are therefore no direct analogs to LFP, NMC, or similar materials.

It was therefore surprising that a few years ago it was shown that sodium in sodium–air batteries appears to have clear advantages over lithium–air batteries in some respects [10, 11].

14.8.2 Physical-Chemical Principles

The decisive factor in sodium–air systems is the reversible formation of sodium superoxide:

$$Na + O_2 \rightarrow NaO_2. \tag{14.20}$$

In contrast to the situation with lithium–air batteries, atmospheric media such as carbon dioxide and water are not necessarily harmful to sodium–air batteries but are instead helpful under certain circumstances [10]. Stabilized DMSO in particular has proven to be a suitable solvent.

14.8.3 Outlook

In principle, the higher weight of sodium is a disadvantage compared to lithium. However, sodium is a much more common element on Earth and is therefore significantly cheaper. A sodium–air battery would still have clear advantages in terms of performance data compared to existing lithium-ion batteries with organic electrolytes; the conceivable theoretical specific energy is around 1600 Wh/kg at a cell voltage of max. 2.3 V. Further basic research activities are therefore expected in this area, see also Ref. [11]

14.9 Sodium-Ion Batteries

14.9.1 General

Solid-state lithium batteries, lithium–sulfur batteries, lithium–air batteries, and sodium–air batteries share the potential for electric and energy storage properties beyond present-day lithium-ion batteries. This is not the case for sodium-ion batteries: Their decisive advantage is lower cost due to sodium being omnipresent, in contrast to lithium, whose availability is often discussed as a major issue. Although research activities were not very prominent in this field, the market introduction of some electric vehicles in China with sodium-ion batteries in 2023 has stimulated research and technical development in the field [11, 12]. However, the actual market share is still very low.[4]

14.9.2 Physicochemical Principles

Sodium-ion batteries resemble lithium-ion batteries in many aspects. Instead of lithium ions being transported from the anode via the electrolyte to the cathode, sodium ions are transported.

However, the intercalation behavior of lithium and sodium is quite different despite similar materials. The much larger sodium ion tends not to intercalate into compounds with a wide range of possible composition (i.e. degrees of intercalation) but to form definite, stoichiometric chemical compounds.

Anode materials are either graphite, hard carbons alloyed with silicon, tin, or other elements, as in the case of lithium-ion batteries. As for the cathode, layered oxides on Ni, Mn, and Fe bases, and transition metal phosphates or fluorophosphates are used, which is still analogous to lithium-ion batteries. A materials class unique to sodium ions is the use of Prussian Blue analogs based on hexacyanoferrate ions $[Fe(CN)_6]^{2-/3-}$ (some other metals than iron are also possible). With respect to the electrolyte, linear and cyclic esters like propylene carbonate or diethyl carbonate tend to be used (as for lithium-ion batteries), with $NaPF_6$, $NaBF_4$, or sodium trifluoromethanesulfonate (NaOTF = $Na(CF_3SO_3)$) as examples of conductive salts. However, sodium-ion batteries[5] have a significant difference from their lithium counterparts: there is no SEI that forms on a graphite or hard carbon surface. This makes ether-based compounds also interesting as electrolytes since they facilitate intercalation and de-intercalation in the graphite structure. Potential ether solvents are dimethoxyethane (DME, monoglyme), diglyme, also known as bis(2-methoxyethyl) ether, or cyclic ethers like 1,3-dioxolane (DOL). These solvents have high reduction stability and strong solvation capability for Na^+ ions.

Achievable gravimetric energy density ranges between 75 and 200 Wh/kg, based on prototypes and product announcements. This is at least a factor of 2 worse than

---

4 Besides sodium-ion and lithium-ion batteries, both of which are treated in this book, many other metal-ion systems are also under investigation like aluminum-ion or zinc-ion and others. However, the most promising and technologically most advanced technologies are lithium-ion and sodium-ion batteries. See Ref. [2, 12], for more details.

5 Sodium-ion batteries are also abbreviated as NIBs as lithium-ion batteries are referred to as LIBs.

lithium-ion batteries, but cost reduction potentials of more than 60% could make the battery concept economical. The nominal cell voltage is about 3 V, which is significantly lower than in the case for lithium-ion batteries.

14.9.3 Outlook

This system is in active development, and there is interest in large-scale industrialization. More details on the fundamentals can be found in, for example, Ref. [12]. The most attractive market for sodium-ion batteries is most probably low-cost electric vehicles.

14.10 Ultracapacitors and Hybrid Batteries

14.10.1 General

Ultra- or supercapacitors, also known as ultracaps or supercaps, are electrochemical double-layer capacitors (EDLC or DLC) with an exceptionally large capacitance for storing charge. Since the chemistry used is partly similar to that of lithium-ion batteries (e.g. in the use of graphite electrodes, organic electrolytes with lithium-containing conductive salts, and separators), it is appropriate to provide brief information on these energy storage devices at this point. The specific energy of ultracapacitors is, however, very small compared to any battery system.

The main difference from batteries is that electrochemical reactions only take place at the surface of the electrodes, and the charge transfer processes are not affected by diffusion processes. The charge transfer processes are analogous to those in batteries. The electrochemical signature of EDLCs, their voltage curve during charging and discharging, is that of a conventional capacitor and not that of an electrochemical cell due to the dominance of surface properties of the materials instead of their bulk properties. CoO_2, a material used in lithium-ion batteries, although no longer in its pure form, exhibits more and more the electrical properties of a capacitor the smaller the crystallites of CoO_2 become – a linear rise of voltage when charging and a linear decrease of voltage during discharging [13]. This explains the term "electrochemical double-layer capacitor" or "double-layer capacitor," and the storage process is sometimes referred to as pseudocapacitance. The process is fully reversible and virtually free of any losses. If electrochemical reactions only take place on one electrode and the other electrode is that of a conventional supercapacitor, they are also referred to as hybrid capacitors [13].

Ultracapacitors are used in particular for applications with very high power demand or supply for a very short time, for example, in the traction sector, due to their high power density and specific power. Their great advantage is their extremely good cyclability (due to the lack of chemical side reactions, so even longer periods in a fully discharged state can be survived without any problems). The great disadvantage is the much faster self-discharge rate compared to batteries (after a few days, all ultracapacitors are fully discharged) and the low energy content compared to lithium-ion cells.

14.10.2 Physical-Chemical Principles

In conventional capacitors, energy is stored in the electric field between two opposite electrode surfaces, where the dielectric properties of the material insulating the electrodes from each other are an important factor. Voltage differences of more than 1000 V are possible, and the capacitance for the largest capacitors is in the mF range (Farad (F) = As/V), that is, the amount of charge that can be stored is very small. In EDLCs, in contrast, a fully reversible electrochemical reaction or rather a charge transfer from the electron-conducting electrons to molecules and ions adsorbed on the electrode surface stores charge. EDLC are normally constructed with carbon electrodes (activated graphite, etc.) with extremely high surface areas, but some inorganic materials (such as MnO_2) can be used too. Lithiated graphite is also used as a material. The largest possible active surface area is important for all electrode materials, which places particular demands on the manufacturing process.

Acetonitrile is used as the electrolyte in many EDLCs, despite its toxicity, because it is a good solvent for inorganic salts with a wide temperature application window, comparatively low electrical resistance, and low viscosity. Where acetonitrile is not an option, it can be replaced by propylene carbonate, which is also used in lithium-ion batteries. The large surface area allows for storing large amounts of charge; in large EDLC, many thousands of As or a few Ah, however, at only low voltage differences. The largest commercially available EDLCs have a capacitance of 3600 F.

Because of the absence of electrochemical reactions, the energy content of an EDLC can be calculated in the same way as that of conventional capacitors:

$$E = \frac{1}{2}CU^2. \tag{14.21}$$

Technically, a capacitor is usually discharged only to half the voltage of a fully charged capacitor, so that only 75% of the energy content can be used. The voltage change during discharging is significantly greater than that of any battery system and therefore often leads to the use of DC/DC converters between the capacitor and loads.

The usual nominal voltage is 2.5 V. Attempts to increase the voltage by using other, more voltage-stable electrolytes have not been commercially successful to date. Increasing the voltage from 2.5 to 3.5 V would almost double the energy content.

Commercial double-layer capacitors with a capacitance of 3000 F have an energy content of approx. 3.9 Wh at a nominal voltage of 2.5 V. Small commercial lithium-ion round cells (e.g. 18,650 format) with a capacity of 2.25 Ah have around twice the energy content with a significantly smaller size and weight. When comparing double-layer capacitors with lithium-ion cells of similar weight and size, the already cited advantage of the high cycle stability of double-layer capacitors is considerably reduced, because the often given value of 1,000,000 full cycles for double-layer capacitors has to be compared with the number of partial cycles of lithium-ion batteries. However, double-layer capacitors still have major advantages in applications that require very high power but low energy requirements (emergency opening of aircraft doors, etc.). It remains to be seen whether and in which niches they will be able to hold their own against newly developed lithium titanate cells with very high cycle stability and good low-temperature properties.

Double-layer capacitors do not have side reactions for charging. Even small differences in self-discharge rates therefore lead to the need for a cell balancing system. On a system level, the complexity and design of monitoring systems for EDLCs and lithium-ion batteries is very similar. Monitoring systems and cell balancing are also required due to very high self-discharge rates, which lead to a complete discharging of cells in approx. four weeks. After a short time without charging, the cell voltages begin to differ and recharging cells in a string is then only possibly when using cell balancing systems.

The aging phenomena of EDLC are still under investigation.

More information about supercapacitors can be found in Refs. [13–18]

14.10.3 Hybrid Battery Concepts

It has been discussed for some time whether the combination of electrodes with electrochemical energy storage and electrodes with electrical energy storage, as used in double-layer capacitors, could result in technically and economically attractive property profiles and products. No proposed product has yet achieved convincing market penetration.

References

1 Beard, K.W. (ed.) (2019). *Linden's Handbook of Batteries*, 5e. McGraw-Hill.

2 Garche, J. (ed.) (2025). *Encyclopedia of Electrochemical Power Sources*, 2e. Elsevier.

3 Jossen, A. and Weydanz, W. (2006). *Using Modern Accumulators Correctly*. Bonn: Reichardt-Verlag.

4 Asia Nikkei: https://asia.nikkei.com/Spotlight/Most-read-in-2020/Toyota-s-game-changing-solid-state-battery-en-route-for-2021-debut (last accessed 31 December 2020).

5 https://www.idtechex.com/en/research-article/the-shift-to-solid-state-battery-technology-is-next-reports-idtechex/21642 (last accessed 14 November 2021).

6 https://www.idtechex.com/en/research-article/the-shift-to-solid-state-battery-technology-is-next-reports-idtechex/21642 (last accessed 09 June 2022).

7 Chen, L. and Shaw, L. (2014). Recent advances in lithium sulfur batteries. *Journal of Power Sources* 267: 770.

8 Korthauer, R. (ed.) (2013). *Handbook of Lithium-Ion Batteries*. Berlin: Springer.

9 Wang, C., Xie, Z., and Zhou, Z. (2019). Lithium-air batteries: challenges coexist with opportunities editor's pick. *APL Materials* 7: 040701.

10 Hartmann, P., Bender, C., Vracar, M. et al. (2012). A rechargeable room-temperature sodium superoxide(NaO_2) battery. *Nature Materials Letters* 12 (1): 228–232.

11 Pozo-Gonzalo, C. and Ortiz-Vitoriano, N. (2022). Recent progress, advances, and future prospects in Na-O_2 batteries. *Current Opinion in Electrochemistry* 36: 101120.

12 Titirici, M.-M., Adelhelm, P., and Hu, Y.-S. (ed.) (2023). *Sodium-Ion Batteries. Materials, Characterization, and Technology*, vol. 2.

13 Jinag, J. (2019). Liu: definitions of pseudocapacitive materials: a brief review. *Energy & Environmental Materials* 30–37.

14 Beguin, F., Frąckowiak, E. et al. (2013). *Supercapacitors*. Weinheim, New York: Wiley VCH.

15 Bharti, A., Kumar, G., Ahmed, M. et al. (2021). Theories and models of supercapacitors with recent advancements: impact and interpretations. *Nano Express* 2: https://doi.org/10.1088/2632-959X/abf8c2.

16 Wenzl, H., Benger, R., and Hauer, I. (2025). Capacitance. In: *Encyclopedia of Electrochemical Power Sources*, 2e. Elsevier.

17 Conway, B.E., Pell, W., and Liu, T. (1997). Diagnostic analyses for mechanisms of self-discharge of electrochemical capacitors and batteries. *Journal of Power Sources* 65: 53–59.

18 Pean, C., Rotenberg, B., Simon, P., and Salanne, M. (2016). Multi-scale modelling of supercapacitors: from molecular simulations to a transmission line model. *Journal of Power Sources* 326.

Tasks

Task 14.1

Give the main advantages and disadvantages of lithium-ion batteries with traditional NMC graphite chemistry and organic electrolyte in comparison to:

a) Lithium–sulfur batteries,
b) Lithium–air batteries,
c) Lithium solid-state batteries,
d) Sodium-ion batteries with liquid electrolyte and sodium anode, and
e) Alkaline nickel batteries based on nickel–metal hydride.

Differentiate between:

i) Fundamental difficulties, and
ii) Difficulties that one can hope to overcome as a result of scientific progress.

Task 14.2

Fuel cells, zinc–air batteries, and redox-flow systems have some things in common. Explain similarities and differences.

Task 14.3

Draw a schematic diagram and provide the reaction equations for a conceivable sodium–sulfur battery, designed as a redox-flow system. What areas of application do you see for such a system, and what would be its advantages and disadvantages?

15

Overview of Applications

In real applications, batteries are usually subjected to highly dynamic currents that can fluctuate rapidly and with high amplitude between charging and discharging directions and included long periods of very low or no current. Constant current and constant voltage phases rarely occur.

It is crucial to understand the exact operating conditions, especially variations in state of charge and temperature, when selecting the most suitable battery technology, capacity, nominal voltage, chargers, and auxiliary components. For new applications, in particular, generalized information about average depth of discharge, average current amplitude, and temperature may lead to wrong decisions. When analyzing faults after a battery failure in a power supply system, the operating conditions of the battery must be investigated as much as possible.

In this section, the major areas of application for batteries are presented along with specific operating conditions, which can vary significantly within the same area of application. The requirements, technical and economic conditions, influence the choice of battery technology, total capacity to be installed, nominal voltage, and current rating of the charger, leading to the use of additional components for safe and reliable operation. Sometimes there are different battery technologies to choose from, but often one technology is clearly superior to all others considering all necessary properties. When discussing areas of application in the following chapters, sometimes only one battery technology is described in detail.

In all planning processes, it should always be kept in mind that battery systems not only compete with each other, but also with alternatives that require batteries with lower energy content or power capability, or sometimes even manage without batteries.

15.1 General Remarks

For the selection and use of batteries, precise knowledge of the power profile, ambient temperature, chargers, availability of external power to recharge the battery, and

Battery Technology: Fundamentals of Battery Electrochemistry, Systems and Applications,
First Edition. Alexander Börger and Heinz Wenzl.

Companion website: www.wiley.com/go/Börger/Battery_Technology

operating strategy is crucial. This applies not only to planning and energy management but also to diagnostics and troubleshooting. Average cycle depths, states of charge, and the number of cycles in the period under consideration can be used for designs for well-known applications, but lead to major inaccuracies if the application is new and unusual or widely varying operating conditions are expected. Examples of this are batteries for autonomous renewable energy systems and electric vehicles with different application profiles. The performance of electric vehicle batteries when driving infrequently and mainly in the city, for delivery services with a fixed route or for average use mainly to a fixed place of work differ greatly.

The long service life of batteries in most applications and often many hours of operation without an external power supply makes data collection difficult, even if data loggers with very low energy consumption are used. As a result, there are no complete data sets of batteries over their entire service life in the scientific literature that also include the determination of capacity and internal resistance at the end of life. When analyzing the past or future use of batteries, it is therefore always necessary to assume that the available measurement data or simulations of the operating conditions are representative of the use of the battery over its entire service life.

The following remarks are therefore a plea to always critically scrutinize the operating conditions of batteries and to always consider a battery as a component of the overall system. Technical problems in complex power systems often first affect the battery used, but are not caused by defects in the battery or the battery design, but by the inadequate functioning of other components or their monitoring. Examples for questions are:

- What is the exact curve of current and voltage over time? Where and how are voltage values measured? Are battery voltage limits truly observed at the terminals of the battery, or only at the point of connection to measuring devices, loads, or chargers because the connection cables are sometimes very long and do not have a sufficient cross-section? In Ref. [3], for fire department emergency vehicles, for example, it is explicitly stated that voltage values for chargers must refer to the voltage at the battery terminals, if necessary by using a separate voltage measurement cable (four-wire measurement method, as in electrochemical impedance spectroscopy).
- What are the actual maximum and minimum values for temperature and state of charge of the batteries during operation? What does the temperature and state of charge curve of the battery look like?
- What current ripple (frequency and amplitude) is caused by the charger or the loads?

Chapters 16–20 show battery operating data that illustrate this problem. Damage and failure of a battery can sometimes be assigned to individual components or the aging process of the battery using analysis methods (see chapter 10). However, understanding the reasons for their occurrence requires a precise analysis of past operations and the consideration of all other components of the system.

15.2 Use of Battery

In the descriptions provided so far, the behavior of batteries has mostly been explained under constant current or constant voltage conditions. Constant current conditions represent a constant electrochemical reaction rate and are advantageous for laboratory tests and achieving a desired state of charge.

However, in most cases, the operating conditions of batteries can be expected to involve highly dynamic currents and fluctuating temperatures. In some cases, the battery is connected to either a charger or loads (such as portable devices, electric vehicles, and industrial trucks), while in other cases the battery is constantly connected to a charger and loads at the same time (renewable energy systems, vehicles with combustion engines, or hybrid drive trains).

15.2.1 Simultaneous Connection of Batteries with a Charger and Loads

Figure 15.1 schematically shows the structure of an energy supply system. Various power generation components, including grid-connected chargers, vehicle alternators, photovoltaic systems with their controllers, etc. connected in parallel, are possible. The loads can be pure resistance loads, DC loads with or without DC/DC converters, AC or three-phase loads with or without reactive power requirements. The battery in Figure 15.1 is drawn with a temperature management system because attention must always be paid to the temperature environment, heat generation in the battery and heat dissipation.

In most cases, the battery shown in Figure 15.1 is the component with the lowest impedance. The equivalent circuit diagram of a cell shows that due to its very low inductance (usually in the range between 0.01 and 0.1 μH), it can react to requests for power consumption or supply within an extremely short time (nanoseconds). Benger [2] has shown that batteries can react more quickly to power requirements than ultracapacitors because their inductance is often higher than that of batteries.

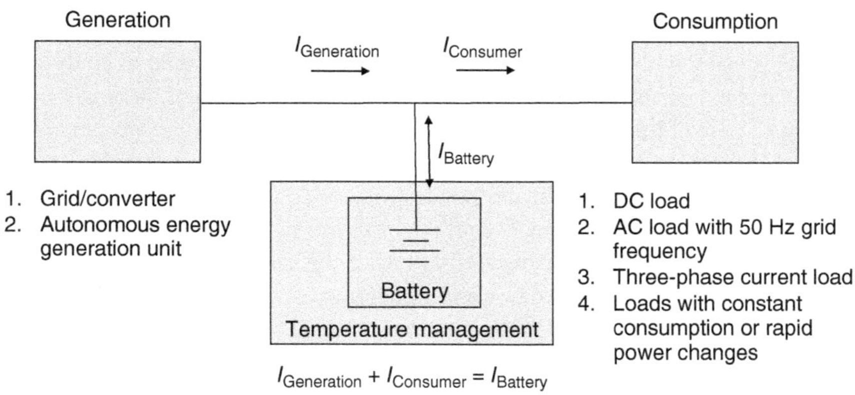

Figure 15.1 Diagram of an energy supply system in which the battery is constantly connected to energy generation units and energy consumption components at the same time.

In technical systems, it is always expected that the response time of batteries to power requirements does not represent a technical restriction.

The slight differences between power generation and power consumption are immediately compensated for by the battery, leading to a highly dynamic current profile with constant changes in the direction of the battery current. High-frequency current ripple is caused by the control characteristics of the power generators and loads as well as the switching frequency of semiconductors, capacitors, and inductances in the system. Task 15.1 illustrates that a battery with an AC load that requires reactive power is subject to rapid changes from charging to discharging currents during most of its operating time.

$$I_{\text{Batt}} = I_{\text{Gen}} - I_{\text{Load}}. \tag{15.1}$$

Figure 15.2 shows an example of the battery current caused by an alternating current load supplied by an inverter. The power fluctuation of an alternating current load is twice the mains frequency, that is, 100 Hz in a 50 Hz grid, so that the current and voltage profile with a period of 10 ms. In most cases, the power supply is unable to compensate for the rapid changes in the power demand of the load, so the battery has to take over this task. It should also be noted here that the voltage falls below the battery's equilibrium voltage every 10 ms, which means that Faraday discharge currents could occur at the electrode.[1]

When analyzed over a longer period of time, the current can also be regarded as a combination of a DC and AC current and can be written as:

$$I_{\text{Batt}} = I_{\text{DC}} + I_{\text{AC}} \tag{15.2}$$

Whether the superimposed alternating current I_{AC} or effective alternating current I_{eff} is significant for the functional safety and service life of the battery may need to be checked. For lead–acid batteries for some applications, it is specified that $I_{\text{eff}} < 5\,\text{A}/100\,\text{Ah}$ or temperature increase of the battery as a result of such an AC heating is less than 5 K [4]. However, it should always be noted that only a measurement with high time resolution and a precise analysis of the battery current and voltage can describe the operating conditions of a battery comprehensively.

Operating phases with constant current or voltage are only possible in charging phases when the available power exceeds the power requirements of the loads and the battery. In case of limited charging power, constant current or voltage can only be expected if the charging and discharging times are different, for example, when recharging a vehicle battery from the power grid. In contrast, when a battery is recharged by a photovoltaic system, even if the loads are switched off, there are only charging phases with constant current or voltage if the photovoltaic power is permanently higher than the required power for charging.

In most operating phases, frequent fluctuations between charging and discharging currents with high current amplitudes and dynamics are expected.

---

1 Whether and to what extent the electrode voltage actually falls below the equilibrium voltage can only be determined by analyzing the capacitive, inductive, and ohmic voltage drops in the battery system.

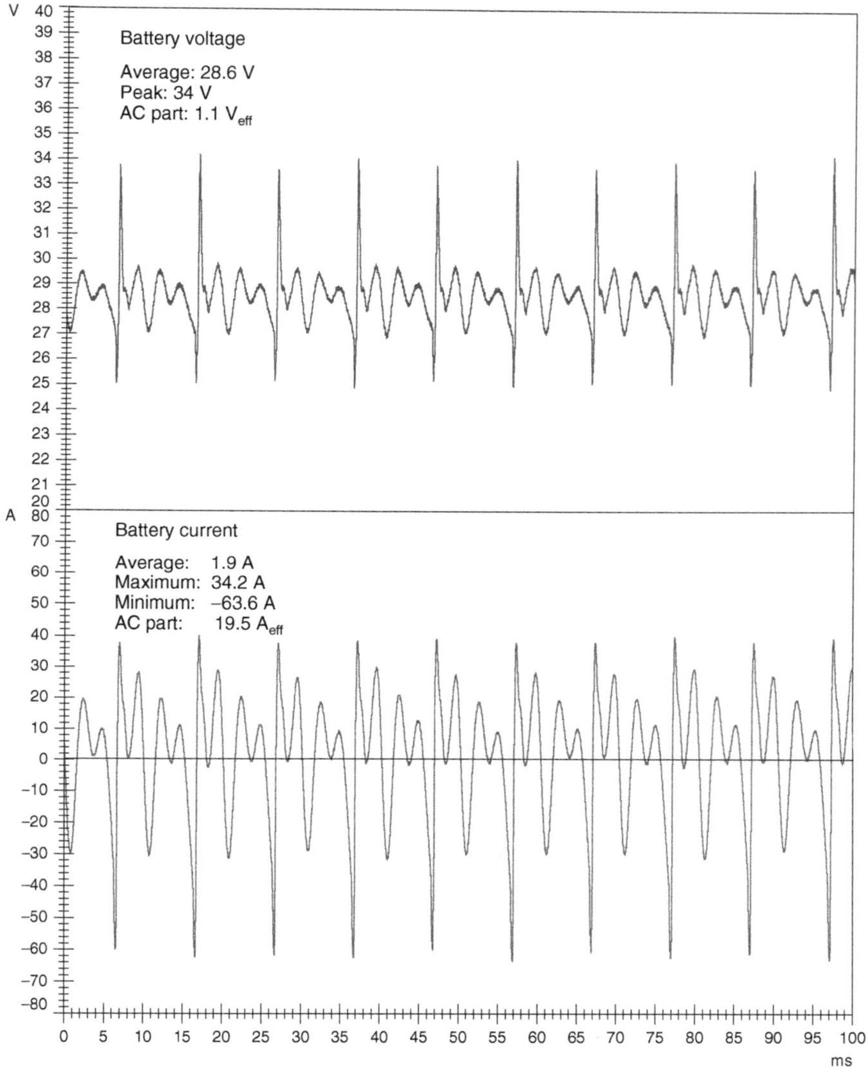

Figure 15.2 Current and voltage curve of a battery in trickle charging mode with simultaneous supply of a 50 Hz alternating current load. Source: Courtesy of DB Systemtechnik GmbH.

15.2.2 Connection of Batteries to Either Charger or Load

Examples of this are portable devices and purely battery-powered vehicles. With most of the charging characteristics used (exceptions are pulsed charging characteristics with currentless pauses or short discharge pulses), charging takes place with a highly smoothed charging current. Depending on the charger technology, it is possible that toward the end of charging and in trickle charging mode or at very low charging currents, a highly dynamic current of low amplitude flows through the battery, alternating between charging and discharging directions at several times the mains frequency.

When discharging, a variable discharge current is often expected, occasionally also short discharge pulses of high power (GSM current profile for mobile phones).

In some applications, however, a highly dynamic change from charging and discharging currents is also possible in some phases of operation (regenerative braking of motors during braking, reactive power requirements of connected loads).

15.3 State of Charge and Remaining Capacity

The state of charge of the battery is of great importance for its use and service life. Knowledge of the state of charge curve is also important for understanding the operating conditions.

Knowledge of the battery current, voltage, and temperature alone does not provide sufficient information to estimate the state of charge. Calculating the *SoC*, the state of health, and the residual capacity of the battery, which is important for understanding the operating conditions and operating energy management systems, is difficult (see chapters 7–9) and the accuracy expected must be carefully checked when using the value provided by measuring devices and algorithms.

15.4 Efficiency

In the chapters so far, the efficiency of batteries has only been discussed occasionally. The reason for this is that ultimately only the efficiency of the overall system and not the efficiency of cells under standard conditions is relevant.

15.4.1 Efficiency Under Cyclic Load

The efficiency, the ratio of energy drawn to energy charged, can only be compared under the same cyclical load, for example, a full cycle. Lithium-ion batteries generally have a significantly higher efficiency than lead–acid batteries:

- Average voltage during discharging and charging
  The internal resistance of lithium-ion batteries is lower at a comparable energy content and therefore the voltage difference between the discharge and charge voltages is also lower.
- Charge throughput
  The amount of charge required for recharging a lead–acid batteries in relation to the amount of charge removed is significantly higher than for lithium-ion batteries due to side reactions and the need to remove acid stratification. To quantify the amount of charge, the terms “charge factor” (amount of charge for recharging/amount of charge discharged) and “Coulomb efficiency” (amount of charge/amount of charge for recharge) are used.

For new sealed lead–acid batteries, the charge factor is approx. 1.04–1.07, for new flooded batteries approx. 1.15–1.17, for flooded batteries with electrolyte circulation approx. 1.05–1.07. For aged batteries, the charge factor is significantly higher. For lithium-ion batteries, on the other hand, the charge factor is more in the range of 1.005–1.01. In contrast to lead–acid batteries, the charge factor or Coulomb efficiency of lithium-ion batteries hardly changes with age.

As a result, efficiency figures of just over 95% are realistic for lithium-ion cells, whereas for lead–acid batteries, the best-case scenario is just under 90%, and often only just over 80% for flooded batteries.

However, the efficiency of chargers is also important when considering overall efficiency. The high efficiency of chargers with modern power electronics must be evaluated with regard to the power curve during charging. At full load during the constant current phase, charger efficiencies of over 95% are state of the art. However, these same chargers have significantly lower efficiency at low currents at the end of charging.

With constant cycling of a battery, the overall efficiency therefore depends on the average depth of discharge, the current amplitudes, and the frequency of constant voltage phases during charging, and in the case of lead–acid batteries also on the temperature due to the increase in side reactions. It is difficult to calculate the efficiency of dynamic power profiles. Our measurements of batteries for industrial trucks have shown that the efficiency of some batteries was only around 10% due to frequent but low energy discharges which were followed by a complete recharge after each use.

15.4.2 Standby Losses

In most applications, batteries are subject to long rest periods; in stationary applications, there may only be a partial discharge once a year for testing, and in some applications, there may be no discharge at all for a whole year. The term "efficiency" is then not very meaningful and may be zero for a whole year. It is more important to consider the standby losses which may only lead to low energy consumption of the battery, but technically and economically relevant costs for cooling.

Lead–acid batteries are usually kept in trickle charge mode to ensure optimum functional reliability. The trickle charge current for new batteries in stationary applications is approx. 10–30 mA per 100 Ah capacity, that is, approx. 22–66 mW and approx. 200–600 Wh in relation to the energy content of a 100 Ah cell of 200 Wh. At the end of the service life, the float charge current increases by a factor of approx. 5–6 with correspondingly higher losses. The energy consumed in the battery for float charging is therefore between 100% and 1800% of the energy content of a new cell. Although an efficiency can be calculated from this for occasional use, it is not particularly relevant.

It is expected that the energy requirements of battery management systems, which are essential for the operation of lithium-ion batteries and should not be switched off for safety reasons, will be lower and that the float charge currents will be lower.

When considering the overall energy efficiency of a battery system, the efficiency of power electronics in standby mode as a function of their power output in relation to the rated power is important. The system topology plays a decisive role. If a separate charger is used, then the efficiency of the charger in trickle charging mode is relevant. In double-conversion systems, that is, a rectifier to produce a DC voltage to which the battery is connected and an inverter to supply loads from the DC voltage, the efficiency of both the rectifier and inverter at their respective power outputs must be considered. For medium-sized UPS systems (e.g. with a rated output of 100 kW), this can amount to 5 kW of losses with a corresponding cooling requirement. The losses of the battery in trickle charging mode can be very low in comparison.

15.4.3 Relevance of the Efficiency of the Battery

The total costs of battery systems and the operating costs are in most cases much higher than the energy costs, so that precise knowledge of efficiency is rarely relevant to costs or decision-making.

15.5 Safety and Environmentally Friendly Handling of Batteries

In the previous chapters, the potential hazards of batteries were already explained. Examples include:

- Thermal runaway of lithium-ion batteries,
- Risk of explosion due to emissions of hydrogen in lead–acid batteries, and
- Emission of toxic and aggressive substances in the event of an accident or malfunction (such as aqueous sulfuric acid in lead–acid batteries, toxic gases and substances in lithium-ion batteries after fires).

Avoiding and dealing with these and other hazards, for example, very high short-circuit currents over a long period of time, must be solved on an application-specific basis.

The handling of hazardous products and substances is regulated by law, for example, in the "Act on the Implementation of Occupational Health and Safety Measures to improve the safety and health protection of employment at work" [1] or comparable national laws. Regulations of the employers' liability insurance associations, technical rules for operational safety, and standards must be complied with or the choice of other solutions must be equivalent. Failure to comply with the recognized state of the art, as described in standards for example, can result in serious legal problems and litigation. The most important standards for lead–acid and lithium-ion batteries are presented in Appendix C.

In contrast to standards, which are based on the cooperation and consensus of people from many different organizations and interests, leaflets from industry associations, for example, ZVEI, and VdS or the Institute of Electrical and Electronics Engineers (IEEE), provide valuable information on the use of products, but cannot claim to provide balanced information that takes all interests equally into account. Nevertheless, it is useful and advisable to check the suggestions made there with regard to their applicability.

Appendix B contains a description of measures that must always be taken to ensure the safe and environmentally sound handling of batteries, which in part corresponds to the content of some standards. In the following application-oriented chapters, additional application-specific aspects for dealing with hazards are described and highlighted where necessary.

15.6 Subdivision into Application Areas

The division into application areas chosen here is based on the requirements placed on batteries in each case. Special applications, for example, for military purposes (submarines, torpedoes) and aerospace, are not covered because these are very small areas, albeit with high technical requirements. An overview about the market share of the different applications is given in Table 15.1.

15.6.1 Starter Batteries for Vehicles (Starting, Lighting, Ignition, SLI)

All starter batteries for vehicles are covered in chapter 16. The term "starting, lighting, ignition" (SLI) indicates that, in addition to starting, the batteries are also responsible for lighting when the vehicle is stationary and for all other consumers in the vehicle electrical system, including the supply to the spark plugs. Batteries for vehicles which have a rudimentary automatic start-stop system (micro-hybrid vehicles) are also assigned to the starter battery application area. As soon as batteries support acceleration or absorb braking energy, they are assigned to the application area of batteries for electromobility.

Starter batteries for emergency power systems, for example, diesel generators, differ significantly in terms of operating conditions and are dealt with in the section on stationary batteries (chapter 19).

Table 15.1 Market share for different battery applications.

	Global market share (revenue, 2022)	Dominating technology
SLI	34.6%	Lead–acid
Electromobility		Lithium-ion
Industrial trucks	35.8%	Lead–acid
Stationary		Lead–acid
Portable	29.6%	Lithium-ion
Total in 2022	$ 104.31 billion	

Source: Data taken from: https://www.grandviewresearch.com/industry-analysis/battery-market Battery Market Size, Share & Trends Analysis Report By Product (Lead Acid, Li-ion, Nickle Metal Hydride, Ni—Cd), By Application, By End-use, By Region, And Segment Forecasts, 2023–2030.

In the application area of starter batteries, it is necessary to distinguish between batteries for on-board power supply used in special vehicles (such as ambulances, motorhomes, special transportation vehicles, and construction machinery) and the actual starter battery. On-board power supply batteries with high cyclic loads, for example, for work platforms etc., are often chosen from the "semi-traction batteries" product group and are dealt with in chapter 18.

15.6.2 Batteries for Electromobility

This area of application includes all vehicle batteries used to power electric vehicles other than industrial trucks for material handling in factories and warehouses, from hybrid vehicles to pure electric vehicles. Included are also batteries that, in addition to starting an internal combustion engine, are used to assist propulsion (acceleration and absorption of braking energy) because the requirements differ from those for conventional starter batteries. requirements.

Of all the application areas covered in this book, this is currently (2024) still the smallest in terms of energy content and the value of batteries produced. However, it is expected to catch up with the market size of other areas in a few years due to the high development speed and legal requirements.

15.6.3 Batteries for Industrial Trucks for Material Handling

This area of application is treated separately from electromobility and includes all battery-powered industrial trucks of any size. In contrast to batteries for electromobility this is a mature market for industrial customers with significantly different requirements compared to electric vehicles. This area of application is currently significantly larger than that for batteries in electromobility.

15.6.4 Stationary Applications

Stationary applications refers to batteries in electrical power supply systems that must fulfill the following functions:

1) Bridging the public power supply with a backup power supply in the event of voltage dips or complete power failures for special power supply systems (secured loads); this also includes batteries in some vehicles, such as electric trains.
2) Batteries that are integrated into the normal operation of a power supply system:
   - Providing high power ancillary services to ensure the stability of the energy supply system (frequency stabilization, secondary reserve power).
   - Batteries for balancing electricity generation from renewable energies and electricity demand.
3) Batteries for autonomous energy supply systems that are not connected to the utility grid

This chapter also covers starter batteries for emergency power systems (such as emergency diesel generators).

15.6.5 Batteries for Portable Devices (Tools, Communication, Etc.)

This area of application differs from others due to the much larger number of cells and modules, but lower capacities and nominal voltages. While cells or blocks with a capacity of c. 6 Ah or less are rarely used in other application areas (such as motorcycle starter batteries or batteries for small UPS applications), most cells and batteries in this area have lower capacities.

Primary cells, which play an important role, are not treated.

References

1 *Occupational Health and Safety Act – ArbSchG.* Cited from https://www.gesetze-im-internet.de/arbschg, (last accessed 03 September 2022) (Arbeitsschutzgesetz, German law; member states of the European Union will have very similar legal requirements. It should be expected that in most countries, in particular industrial countgries, a set of rules and laws exists which requires a high level of protection, e.g. OSHA, the Occupational Safety and Health Administration in the USA).

2 Benger, R. (2012). *Dynamic behavior of inverter-fed energy storage systems (Original German title: Dynamisches Verhalten von umrichtergespeisten Energiepsiechersystemen).* Dissertation. TU Clausthal.

3 DIN 14679:2008-03 (2008). *Firefighting – Chargers for Trickle Charging of Starter Batteries and Auxiliary Batteries for Special Applications – Requirements and Testing.*

4 IEEE 1188 (1995). Recommended Practice for Maintenance, Testing, and Replacement of Valve-Regulated Lead-Acid (VRLA) Batteries for Stationary Applications.

Task

Task 15.1

Alternating current load of batteries

A photovoltaic module with the power P_F supplies an inverter that supplies an AC load (single-phase) with the power P_L. Calculate the battery current for a systems topology as shown in Figure 15.3, that is, a battery that is directly connected to the DC link.

Make the following assumptions:

- There is no solar controller that modifies the electricity generated by the photovoltaic modules, that is, no maximum power tracking. The photovoltaic modules therefore supply direct current I_F.
- The DC link has a constant voltage U_{DC}, which is not changed by the current flowing through the battery.

(Continued)

Task 15.1 (Continued)

- As an average over time, the power supplied by the battery is equal to the power drawn from the resistive load or the inverter.
- The inverter has no losses and does not absorb any reactive power. Its output voltage is 230 V AC at 50 Hz.
- Consider a period T of the mains frequency, that is, 20 ms.

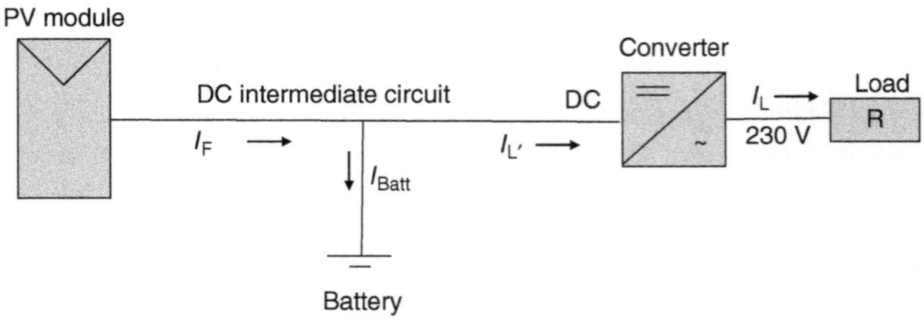

Figure 15.3 Block diagram of a photovoltaic system with battery and inverter to supply a 230 V resistive load.

Questions:

What is the ratio of the effective load current $I_{L,eff}$ and the photovoltaic current I_F at a battery voltage (= voltage of the DC link) of 24 V?

a) Calculate as a function of time the load current I_L and the current drawn by the inverter $I_{L'}$ as well as the ratio of $I_{L',eff}$ to I_F.
b) Calculate I_{Batt} and plot. I_{Batt} against time.
c) Assumption: The battery has no internal losses and is an ideal voltage source.
d) How large or small must the photovoltaic current I_F be so that I_{Batt} does not change direction during the time period T under consideration, that is, the battery is either charged or discharged?

Repeat the calculations for an inverter that supplies a capacitive or inductive load with reactive power (power factor 0.6) in addition to the resistive load and output of active power.

16

Starter Batteries for Vehicles (Starting, Lighting, Ignition – SLI)

The term "starting, lighting, ignition" (SLI) for starter batteries makes clear that the starting of combustion engines is not only one of the required functions but also the one whose failure is immediately noticed. For cost reasons and due to the high starting power required at low temperatures, the market continues to be dominated by lead–acid batteries. However, there are now also battery systems in which a lead–acid battery is connected in parallel with a lithium-ion battery or, in some exceptional cases, completely replaced by a lithium-ion battery.

16.1 Definition

This application area covers all starter batteries for vehicles (cars, trucks, rail vehicles, etc.), with a focus on the automotive sector. The term "starting, lighting, ignition" (SLI) indicates that, in addition to starting, the batteries are also responsible for lighting and all other loads in the electrical system of vehicles, including the supply to the spark plugs. If these vehicles have a rudimentary automatic start-stop system, they are so-called sub- or micro-hybrids that are assigned to this area of application. As soon as batteries also support the drivetrain (energy recuperation during braking and so-called boost processes, i.e. acceleration), they are assigned to the application area of batteries for electromobility. It is likely that the share of such vehicles will increase in the future. Starter batteries for emergency power systems differ significantly in terms of operating conditions and are dealt with in chapter 19 for stationary batteries.

This chapter includes separate on-board power supply batteries used in special vehicles (ambulances, motorhomes, special transportation vehicles, and construction machinery) in addition to the starter battery. Electric vehicles that do not require a starting process often contain such batteries, partly because many electrical loads in vehicles are optimized for a nominal voltage of 12 V and the voltage range of a 12 V lead–acid battery.

Battery Technology: Fundamentals of Battery Electrochemistry, Systems and Applications, First Edition. Alexander Börger and Heinz Wenzl.

Companion website: www.wiley.com/go/Börger/Battery_Technology

16.2 Requirements for the Battery

It should always be remembered that modern vehicles, except trucks or taxis, tend to be used for c. 1000 hours per year. The main operating conditions for batteries are therefore characterized by minimal discharge currents due to vehicle electronics.

16.2.1 Provision of Power for Starting

The most important task of a starter battery is to ensure the starting of a combustion engine, which depends on the state of health, both as regards capacity and internal resistance, temperature, and state of charge of the battery. The engine's ability to start is a complex function of many different factors: engine temperature, oil temperature, oil viscosity, age of the oil, time since the last start, position of the cylinders,[1] condition of the spark plugs, etc. Calculations of the battery's power profile during starting are therefore not used to size the battery. Tests in the laboratory using a typical power profile of a starting process are not practicable. Instead, for acceptance tests of new starter batteries, a high-current discharge with the so-called cold-test current, which is specified by the manufacturer, is carried out at a low temperature (−18 °C in Europe, −29 °C in the United Sates). The battery voltage must remain above a certain voltage level (e.g. 7.5 V for a battery with 12 V nominal voltage) for a given time (e.g. at least 90 seconds). After a short rest period without recharging, the battery is discharged again with a lower current, typically 60% of the cold-test current, until the same voltage limit is reached. The total capacity drawn during this test is called the cryogenic capacity. This test is a capacity test under special operating conditions and not an internal resistance test, and it has little to do with the actual power curve when starting a combustion engine.

The power profile during starting can be divided into several phases:

1) Initiation of the starting process
   In modern vehicles, the starter battery must supply the engine control unit (approx. 10 A) and, in the case of diesel engines, preheat the combustion chamber of the cylinders for several seconds and at higher currents.
2) Exceeding the breakaway torque of the motor
   The starter motor short-circuits the battery and a very high current flows for approx. 50–100 ms (several hundred amps, up to approx. 2500 A for large diesel engines). The resulting torque surpasses the static friction of the cylinders, and the crankshaft accelerates. In modern vehicles, the voltage dip of the battery during this high current pulse must not fall below the lower voltage limit of the engine control unit; otherwise, the starting process will be interrupted.
   In luxury vehicles, it is sometimes required that the voltage dip must not affect comfort functions (interior lighting, radio, etc.).

---

1 If the cylinder positions are known, fuel is injected into the cylinder that has the best position and the ignition is activated. The engine can be started faster and with less electrical energy ("direct start"). In a very narrow parameter range including temperature, it is even possible to start the engine without a starter motor.

3) Acceleration of the crankshaft up to a sufficiently high rotational speed
Depending on the motor control, the engine will continue to run automatically from a certain speed onward. The current is in the range of the cold-test current. This phase takes between c. 0.4 and c. 3 seconds for very large engines provided that the engine is in good condition (see Figure 16.1).

Obviously, such a power profile bears no similarity to the much more severe conditions during acceptance testing. It is easy to understand the difference: the starting

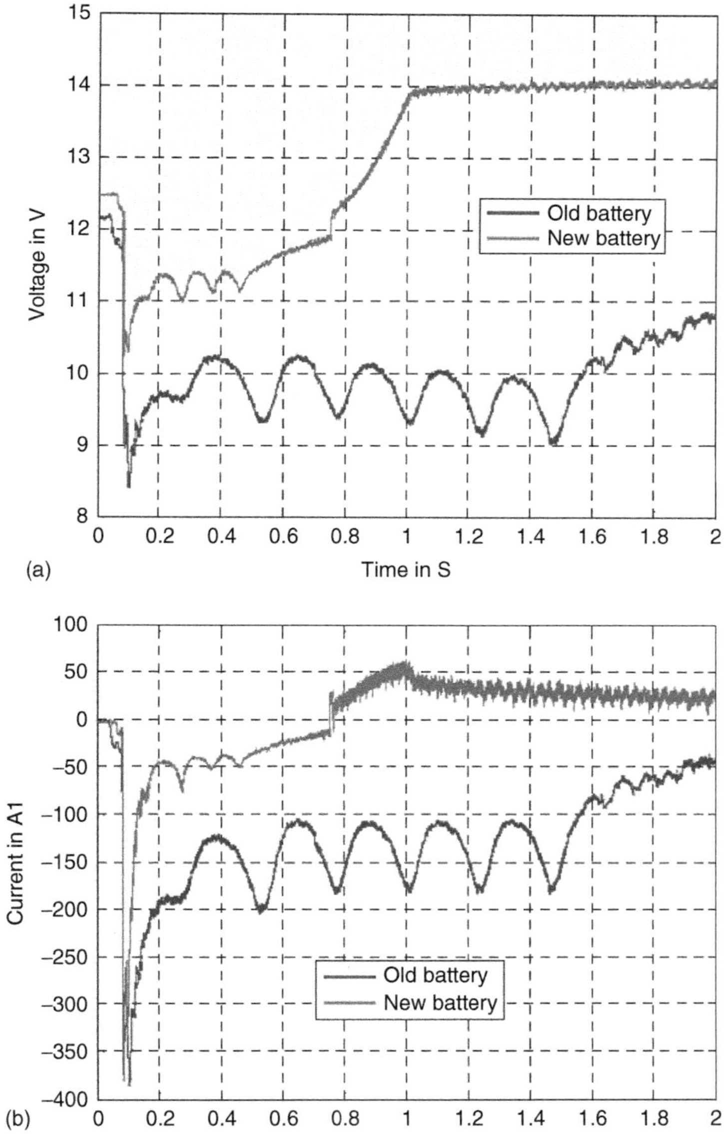

Figure 16.1 Current (a) and voltage (b) curve of a starting process with an aged battery and a new battery.

process must be successful even when the battery is aged, has a low temperature and state of charge, and the engine is in such poor condition that the starting process must be repeated a number of times. Empirically, it has been found that lead–acid batteries that meet the requirements of the acceptance test will provide a long service life (depending on average ambient temperature, more than c. six years).

The amount of charge and energy taken from the battery is approx. 0.5 Ah or approx. 6 Wh (including battery losses). It is therefore immediately clear that the main requirement for starter batteries is essentially a low internal resistance over the entire temperature and state of charge range expected during operation, but not primarily the capacity of the battery. Despite this low charge throughput for each start, recharging the battery in the subsequent driving phase is an issue if the vehicle is used only for short distances or operated in a start-stop mode. As a result, starter batteries tend not to be fully charged, see Ref. [1].

An increasingly important requirement for a battery system is the prediction of the starting capability, on the one hand, to always guarantee starting capability during start-stop, and on the other hand, to ensure that the combustion engine can be started in the following days even under unfavorable temperature conditions.

16.2.2 Supply of the Vehicle Electrical System

In addition to providing starting power, starter batteries are also required to stabilize the voltage and meet power demands during peak loads, as well as supply loads when the engine is not running. Originally, these were essentially only the vehicle lighting and safety functions (e.g. horn, warning signals), but now they also include comfort functions with minimal power consumption (e.g. remote control of doors, control units) or other comfort functions (such as radio, air conditioning, or ventilation). In some special types of vehicles, additional equipment (such as radios in police, fire, and rescue vehicles) must be supplied. In some of these vehicles, the total power consumption of important loads is so high that either batteries with very large capacities are used, or additional batteries must be installed, or the vehicles must be connected to a charger when stationary. The circuit topology of the supply and starter batteries varies. Sometimes such additional batteries are connected directly and permanently in parallel to the starter battery, sometimes they can be separated from each other by contactors, and sometimes they are completely separate. Figure 16.2 shows the voltage curve of a fire engine that must always be connected to a charger when not in use. These supply functions have led to the nominal capacity of lead–acid starter batteries being specified as a 20-hour capacity C_{20}. Nevertheless, the number of cycles required during acceptance testing is low (e.g. corresponding to 50 or 100 full cycles).

Vehicles with automatic start-stop systems, for example, used at traffic lights, must fulfill the requirements for supplying loads to a much greater extent when the engine is switched off. Start-stop operation leads to frequent cycles with low charge throughput due to the frequent short high-current discharges followed by short charging phases, the so-called partial-state-of-charge (pSoC) operation. This not only results in significantly higher charge throughput, but also increased robustness concerning

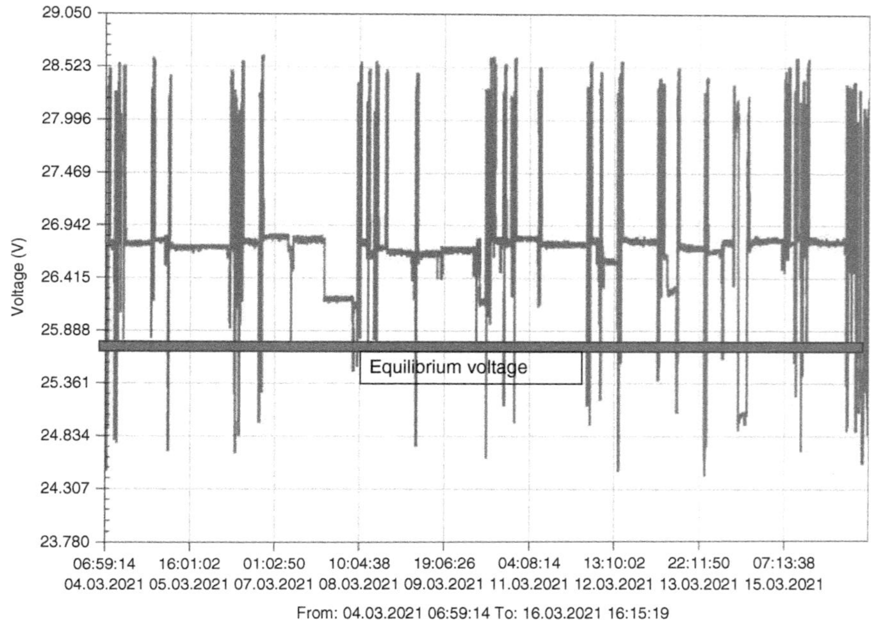

Figure 16.2 Voltage curve of a fire engine (24 V/140 Ah) over a period of almost two weeks. The data was recorded once a minute so the voltage dip during the starting process is not recorded in detail. Apart from starting processes, the battery voltage was always above the equilibrium voltage of approx. 25.6 V (Task 16.1). Source: Courtesy of H. Koch.

a reduced state of charge, preventing excessive sulfation and acid stratification in lead–acid batteries. Good charge acceptance is necessary so that the state of charge in *pSoC* mode can be increased as quickly as possible when the stop phase is over. A starter battery that is not fully charged is advantageous, as only such a battery allows storage of braking energy (recuperation) during braking phases. However, aging processes are accelerated by this mode of operation.

16.2.3 Ignition

For gasoline engines, the ignition of the air-fuel mixture in the cylinder requires a spark generated by the ignition coil. The battery is therefore constantly short-circuited at a frequency that depends on the engine speed. Figure 16.3 shows the current curve caused by this. The amount of current and energy drawn from the battery is so small that it can be drawn from the capacitor properties of the battery when fully charged and does not necessarily involve Faraday currents from electrochemical conversion processes.

16.2.4 Recharging

Immediately after starting the engine, the battery begins to be recharged via the vehicle's alternator. However, the small amount of charge that is required for

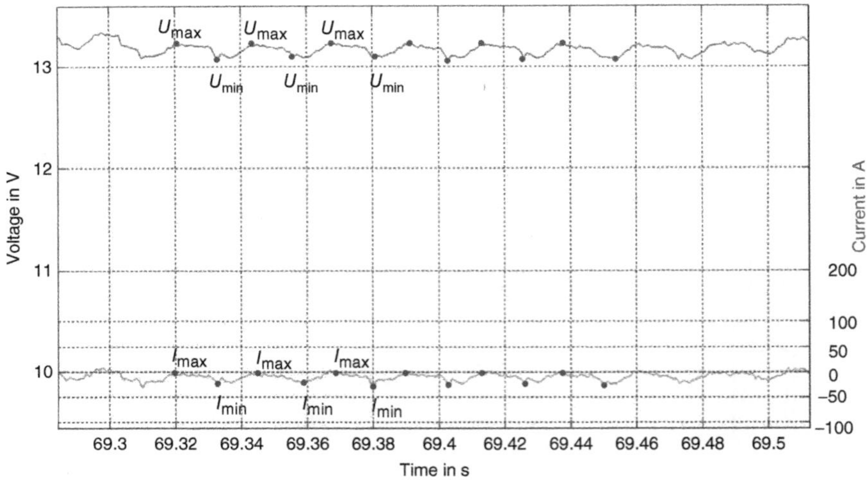

Figure 16.3 Current and voltage curve of a starter battery when supplying the ignition coil. The frequency of the current fluctuations is just under c. 30 Hz, corresponding to 450 rpm of a 4-cylinder engine idling.

starting can only be recharged slowly if the battery was previously fully charged because the charging current is regulated depending on the voltage. Particularly in the case of vehicles in constant short-distance operation, the engine operating time is often not sufficient to recharge the amount of charge removed by the starting process. In Ref. [1], data is given for modern vehicles in start–stop mode, which shows an average state of charge of approx. 90%. The more loads that are switched on, the more a negative energy balance can occur in the vehicle, for example, during slow journeys with heating, lights, other comfort functions, etc., so that the battery is slowly discharged despite engine operation. The sizing of the nominal power of the alternator to the battery and application is therefore important.

The charging strategy in modern vehicles is now very complex. Braking energy is recharged to the battery, the output of the alternator is reduced during acceleration to provide more driving power, the voltage is increased to achieve a full charge, the charging voltage is temperature-compensated, etc.

16.2.5 Weight, Installation Space, and Location

In the automotive sector, starter batteries have to meet, in particular, space and weight requirements. For example, they must be adequately protected against crashes, which, depending on the place of installation of lead–acid batteries, may require the battery acid to be immobilized (AGM technology), and in the case of lithium-ion batteries, mechanical protection of the battery housing. Test requirements include, for instance, that electrolytes must not leak from the battery container even if the battery is charged in an upside-down position.

Other aspects include weight, dimensions, protection of the battery from heat sources such as the engine and exhaust system, removal of gases from the battery into the environment and not into the vehicle interior, etc.

16.3 Choice of Battery Technology

In principle, the amount of charge and energy required to start an engine can be provided by ultracapacitors if the motor has good starting capability. When minor starting problems occur and energy needs to be supplied for initiating the starting process, the energy content of an ultracapacitor is no longer sufficient, so they cannot be used as the only energy storage device. However, there are some examples of the use of ultracapacitors in combination with lead–acid batteries – ultracapacitors serve as an intermediate buffer, for example, when starting, if very high starting currents are required. This is the case in trucks, for example, but also in streetcars and other rail vehicles.

Other battery technologies with very high low-temperature performance (e.g. nickel–zinc batteries) have not yet been used as standard. Nickel–cadmium batteries, which were used as standard in rail vehicles in, for example, arctic temperature zones, have been replaced by battery systems with integrated heating due to their high price and environmental concerns.

Lead–acid batteries dominate the starter battery sector, and unless legislative interventions occur, it is unlikely that this situation will change for some years to come. An obstacle to the replacement of lead–acid batteries by any other battery technology is the design of all electrical components of a vehicle for the voltage range of lead–acid batteries, 12 or 24 V nominal voltage. All attempts to change the on-board electrical system voltage, for example, to establish 42 or 48 V on-board electrical systems, have so far failed.

The advantages of lead–acid batteries are:

a) Relatively inexpensive
   The reason for this is that batteries are available worldwide in reasonably comparable quality because cars are sold worldwide, and car manufacturers are interested in or obliged to ensure a local supply, which is often provided by regional factories of well-known battery producers.
b) Raw materials
   Lead and all other alloys are abundant and cheap compared to a number of components of lithium-ion batteries. Recycling is well-established and exceeds 90% in many regions.
c) Production process
   The production of lead–acid batteries is technically comparatively simple.
   No clean room conditions and no supply of dehumidified air in the production process are necessary. The required technology is available from large mechanical engineering companies and can be used worldwide. Space and energy requirements, as well as environmental protection processes, are comparatively manageable.

d) Lead–acid batteries are robust
They are still efficient even at low temperatures and also tolerate comparatively high temperatures, as can occur when installed in the engine compartment.
e) Lead–acid batteries require no maintenance or monitoring
Modern flooded lead–acid starter batteries do not require topping up with water over their entire service life.
f) Simple system integration
In the past, lead–acid batteries were often used without any battery management system; nowadays, due to voltage prediction (i.e. the prediction of how far the battery voltage will drop during the next starting process), at least a rudimentary battery sensor is usually installed in vehicles, but its functionality is generally significantly lower compared to lithium-ion batteries. Battery voltage, temperature, and current are usually measured, but single-cell voltage measurements are neither necessary nor used.

Above, the performance requirements during acceptance testing for lead–acid batteries have been given. It seems unlikely that lithium-ion batteries will be able to have similar cold-test currents and cryogenic capacity at a reasonable size and cost.

Lithium-ion batteries are primarily used where lead–acid batteries are not an option for weight reasons (racing, sports cars, etc.) or when the engines of vehicles will never have to be started at temperatures below, for example, 0 °C. Five or six cells connected in series with a titanate anode, four cells with an iron phosphate cathode, or three or four cells with conventional cathode materials and graphite electrodes can be used to meet the voltage limits of the conventional vehicle electrical system, although only to a limited extent. The capacity of these batteries cannot be used fully because either the upper voltage limit or the lower voltage limit of the battery falls outside those of lead–acid batteries.[2] Due to the significantly higher price, such products are more likely to be found in more expensive cars and sometimes as accessories. It is to be expected that this niche will grow in the future, but the market share will nevertheless remain small for some years to come.

The desire to recover braking energy and the increase in powerful consumers has led to numerous proposals in the patent literature for the use of several different batteries with different nominal voltages, which can be separated from each other or used in parallel by means of DC/DC converters and switches. These proposals have not yet been used on a large scale.

It has also been proposed to operate lithium-ion batteries and lead–acid batteries in parallel. At high currents, the lithium-ion battery is then preferably charged or discharged, and the lead–acid battery provides the required total energy. Such combinations are used in some series of production vehicles.

2 Reengineering all electrical components of a car to optimize them to a slightly different voltage range is of course possible but extremely expensive.

16.4 Operation and Design

16.4.1 Operation

It has been emphasized both in chapters 1 and 15 how important it is to know the exact operating conditions for selecting the correct battery and for troubleshooting in the event of battery failures. There are clear differences, for example, between the following vehicle types and the use of batteries:

- Starter batteries for long-distance trucks are sometimes operated in such a way that the engine is constantly running for several days (i.e. because of cooling systems, etc.). There may then only be one or two starting processes per week. For the remaining operating time, which can reach 5000–6000 hours per year, the battery is charged in trickle charge mode. Similar to batteries for stationary applications, cycles are rare. The service life of such batteries therefore essentially depends on their ability to tolerate long phases of trickle charging.
  The high voltage applied most of the time leads to high corrosion attack, and the lifetime of starter batteries in this application is therefore only about two years – in contrast to the much longer lifetimes of lead–acid batteries for stationary applications.
- Batteries for cabs, on the other hand, are characterized by very many starts and cycles during standstill (radio, light, communication systems).
  The high energy throughput leads to a short lifetime of such batteries, in the range of one to two years.
- Cars for private use are often only used for a few hundred hours a year. The rest of the time, the vehicle is stationary, and the engine is switched off. For the starter battery, this means very long periods of self-discharge with very low parasitic loads (control units, clock, door opening) and very little time for a full charge. The above-mentioned average state of charge of approx. 90% appears to be too high for this type of use.
  Depending on the average summer temperature in the region of use, starter batteries in these applications have a lifetime of about five to seven years, sometimes longer. The authors are not aware of studies that have investigated the *SoH* of such batteries at the point of replacement, as the end of a lifetime is linked purely to an inability to start the engine.
- Starter batteries in many emergency service vehicles are permanently in trickle charge mode when at their base. The number of hours of use per year is extremely low. Although various advanced lead–acid batteries (e.g. the so-called ultra-battery with a carbon anode) were introduced, these have never been used in industrial series production.
  The lifetime of these batteries varies greatly but is usually not as long as those used in cars for private use.
  Lithium-ion batteries without a parallel lead–acid battery may be an option for this type of application if starting at very low temperatures can be excluded. Lithium-ion batteries may not require oversizing as a reserve for high energy throughput, as they are much more robust than lead–acid batteries in terms of

aging. As a result, they can be designed with smaller capacities than corresponding lead–acid batteries, which somewhat reduces the effect of their considerably higher cost. In terms of installation space, it is important to remember that extensive sensor technology and battery management systems, as well as any other safety measures, take up considerable space and add weight.

16.4.2 Design Options

The two most important design parameters for starter batteries are the battery capacity or, more precisely, the energy content during discharging, and the internal resistance for the whole temperature range of the vehicle, the cold-test current, and cryogenic capacity.

Of course, the two values are not completely independent of each other and thus a trade-off between capacity and internal resistance must be made. Low internal resistance is achieved by very large effective surface areas, for example, by using many and very thin plates and highly porous active materials. Increasing capacity using the same standard battery box requires fewer and thicker plates and less porous active materials, which also leads to a much longer service life when the battery is cycled. To minimize battery failures as a result of active mass shedding, a consequence of cycling, the positive electrode is placed into a pocket separator, which prevents active material from causing a short circuit with the negative electrode at the bottom of the battery box. When using pocket separators, it is no longer necessary to keep an electrolyte reservoir at the bottom, which could take up active material that falls down to the bottom (complete usage of battery volume). Sealed batteries even without pocket separators have places that go the whole depth of the battery box.

Modern starter batteries are often referred to as "maintenance-free,"[3] which means that topping up with water is not required for the whole service life, and often, it is either not possible to top up with water or it is not obvious how to do this. The use of sealed batteries solves this problem but at a higher purchasing price. Flooded batteries are categorized into three groups (W1, W2, and W3) based on alloys and additives, particularly to the negative electrode. The aim is to minimize water loss during float charging by minimizing the exchange current density of the hydrogen evolution reactions. Unfortunately, low gas evolution rates lead to difficulties in removing acid stratification, which is particularly important for batteries used in start-stop operation. So-called enhanced flooded batteries (EFB) use an insert that utilizes the movement of the electrolyte in the battery box for electrolyte circulation (mammoth pump principle). EFB batteries compete with sealed batteries as regards their suitability for start-stop operation but are less expensive.

Corrosion attack is an issue, as the battery voltage tends to be very high throughout its entire service life. The choice of materials for grids and group bars, as well as the welding processes to weld plates to the group bar and to connect cells in series

3 The term is misleading as it only refers to one aspect of servicing a battery as part of a technical system.

inside a battery box (intercell connector), are important. In some starter batteries, the intercell connectors are located as much as possible below the electrolyte level so that they will always be covered by electrolyte – even when the battery has aged.

For cost reasons, starter batteries are always produced as a 12 V (six cells in series) module. Starter batteries with other nominal voltages are very rare exceptions, and single cells, which are connected in series using external connectors, are not available.

Sealed batteries are technically superior to flooded batteries in virtually all cases. However, they are more complicated to manufacture and more sensitive to high temperatures. As a result, both flooded and sealed batteries will continue to exist side by side.

16.5 Monitoring of the Battery

The most important diagnostic tool for lead starter batteries is the starting voltage prediction, which is indispensable for starter batteries in *pSoC* and start-stop operation, so that the vehicle can always be restarted after a stop. This usually involves voltage and current diagnostics in combination with temperature measurement, which can be supplemented by further information, for example, data on the vehicle's history, data on the last starting voltage dips, (usually rudimentary) impedance data and much more; this depends on the manufacturer's expertise.

The prediction of the starting voltage dip is not only important because an excessive voltage dip prevents a restart, but also because the operating voltage of important, and in particular safety-relevant – control units in the vehicle must not fall below a certain voltage level. This then leads either to a restart or, in the worst case, even to a complete failure of the control unit, meaning that safety-critical functions would no longer be available.

Simple diagnostic procedures, such as measuring and comparing the voltage values of the first and second $n/2$ cells of the battery, which would be very easy to implement, especially for 24 V systems in commercial vehicles, are not used. Similarly, there is normally no special thermal protection against very low or very high temperatures. Obviously, lead–acid batteries are so robust that the additional expense is not economically worthwhile.

Lithium-ion starter batteries, on the other hand, require much more extensive monitoring with individual cell voltage diagnostics, temperature monitoring, etc., which differs little from the monitoring of larger electric traction batteries in terms of the state-of-the-art and the normative and legal requirements.

16.6 Other

A key aspect of the use of batteries is recycling and ensuring that lead cannot enter the environment. National legislation may differ greatly. As an example, the

German Battery Act based on the European Battery Directive, which has lately been revised, in combination with other regulations such as the End-of-Life Vehicle Ordinance, etc., has led to a very high recycling rate for lead starter batteries in Germany and throughout Europe, partly due to a battery deposit when buying a new battery. Unfortunately, strict regulations do not yet apply worldwide; in some developing countries, lead is still handled too laxly, which can lead to environmental pollution and health hazards.

Reference

1 Schaeck, S., Stoermer, A.O., and Hockgeiger, E. (2009). Micro-hybrid electric vehicle application of valve-regulated lead-acid batteries in absorbent glass mat technology: testing a partial-state-of-charge operation strategy. *Journal of Power Sources* 190: 173–183.

Tasks

Task 16.1

Describe the operation of the vehicle whose vehicle electrical system voltage is shown in Figure 16.2.

Task 16.2

Outline specific energy storage solutions for

a) An autonomously driving vehicle with a redundant supply of the 12 V on-board power supply, and
b) A rail vehicle with a diesel drive, which must supply the engine start as well as the train lighting and operationally necessary loads, such as control and communication equipment, and mark the specific features.

Task 16.3

What design changes result from *pSoC* operation for lead starter batteries in start-stop mode?

17

Batteries for Electromobility

The electrification of the drivetrain for vehicles takes place to varying degrees, so a distinction has to be made between sub-, micro-, mild-, full-hybrids (possibly plug-in hybrid electric vehicles [PHEVs], with external charging options), and pure electric vehicles (EVs), including fuel cell electric vehicles (FCEVs).

Market requirements, political will, and social pressure to introduce electric road vehicles dominate the technical development of these batteries. These factors can be expected to increase significantly in the coming years.

17.1 Definition

This area of application includes all vehicle batteries that are used to power electric vehicles (EVs), from hybrid vehicles to pure EVs.

Many vehicles now have a completely automatic start-stop system, including brake energy recuperation and a power assist mode to support acceleration. They are called micro-hybrids, and purely electric driving is often not possible. They are assigned to this area of application because the batteries need to meet different requirements compared to conventional starter batteries. Mild hybrids allow increased recovery of braking energy and support the drivetrain, for example, during acceleration processes and especially when starting up after stop phases. Full hybrids can drive (possibly only short distances) purely electrically. If full hybrids can be charged from a socket and/or charging station, they are referred to as plug-in hybrid electric vehicles (PHEV), which are usually already classified as purely EVs because, in these cases, the combustion engine is only used for longer journeys or when there are no recharging options from the power grid. Pure EVs no longer contain a combustion engine for propulsion. If such an engine is nevertheless present to extend the range of purely electric driving, this is referred to as a range extender, whose output only needs to be designed for the average output of the vehicle (approx. 15 kW for many vehicles). For an overview see Figure 17.1.

Battery Technology: Fundamentals of Battery Electrochemistry, Systems and Applications,
First Edition. Alexander Börger and Heinz Wenzl.

Companion website: www.wiley.com/go/Börger/Battery_Technology

	Internal combustion engine (ICE)			Hybrid vehicles			Pure electric vehicles		
Type	ICE			HEV		PHEV	Range Extender	EV	FCEV
Subtype	Classical ICE vehicle	Sub-hybrid	Micro-hybrid	Mild hybrid	Full hybrid	Plug-in full hybrid	Range extender	Pure electric vehicle	Fuel cell electric vehicle
Main Propulsion system	Combustion engine			Combustion engine		Combustion engine + battery	Battery + combustion engine	Battery	Fuel cell
Energy from	Hydrocarbons			Hydrocarbons		Hydrocarbons + power	Power	Power	H_2
Properties	Only combustion engine	+ Start/Stop	+ Recuperation	+ Boost	+ Pure electric drive possible	+ External charging	Range extension by combustion engine that assists electric powertrain	Pure battery vehicle	In reality: Fuel cell vehicle with traction battery

Figure 17.1 Hybridization concepts for vehicles; HC = hydrocarbons (i.e. petrol, diesel, and natural gas).

FCEVs, which represent a technological alternative to battery-powered vehicles, are usually equipped with an additional battery to recover braking energy and enable rapid acceleration. They are then referred to as fuel cell hybrid vehicles, and the requirements for the battery correspond to hybrid vehicles with combustion engines.

All hybrid vehicles, whether with a combustion engine or fuel cell, require algorithms that transmit the required drive power to the battery, combustion engine, or fuel cell to ensure that driving is energy-efficient and does not reduce the vehicle's service life. The concepts for the algorithms and how to couple the torques of the electric motor and combustion engine go beyond the scope of this textbook and are not covered.

Social and political demands to ban vehicles with combustion engines for environmental reasons will, when fully implemented, mean that there will only be purely EVs, that is, either battery-powered or fuel cell-powered vehicles. Whether such a scenario will actually prevail worldwide in the medium term is not yet foreseeable. In any case, hybrid concepts with partly electric, partly combustion engine driving remain an important concept on the way to further and possibly complete electrification of the powertrain.

17.2 Requirements for the Battery

The performance requirements for batteries vary considerably depending on the details of the application. Batteries for sub-, micro-, and mild-hybrid vehicles need to be able to start an internal combustion engine. For PHEVs and pure EVs, a high specific energy and energy density are required so that the driving range is maximized. Batteries for pure electric traction have an energy content of up to 100 kWh and more.

Concerning current amplitudes: batteries that are also used for starting combustion engines have to be able to supply very high currents (in the region of a few hundred amperes and even higher for supplying the breakaway torque) – albeit only for

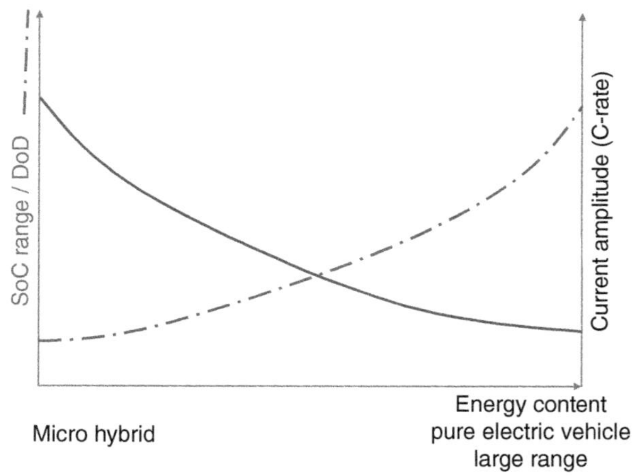

Figure 17.2 Schematic diagram of current amplitude and range of *SoC/DoD* as a function of energy content and driving range.

a very short time. The requirements and design options to meet these requirements were dealt with in chapter 16. The additional power requirements for recovering braking energy and boosting acceleration may reach similar values but much more frequently and for longer. In addition, batteries for micro- and mild-hybrid vehicles must also be able to absorb high charging currents. Of course, all of this tends to reduce the service life.

Electric traction batteries have to fulfill a variety of requirements, some of which derived from legal requirements, but mostly from performance and, above all, safety requirements.

- Target state of charge and range of state of charge range used,
- Energy current, energy throughput and current,
- Lifetime,
- Charging processes, and
- Safety.

Figure 17.2 schematically shows the relationship between requirements depending on usage.

17.2.1 Target State of Charge and Range of State of Charge

The greater the importance of recuperation of braking energy, particularly in micro- and mild-hybrid vehicles, the more important it is to keep the battery in a medium state of charge, for example, c. 60%. Braking energy can then always be absorbed by the battery, and drive power can always be drawn from it. The total range of the battery's state of charge used is therefore relatively small.

The longer the distance to be covered purely electrically, the more important it is to have a high target state of charge. Whether the battery is charged to 100% state of charge or, for reasons of service life, possibly only 95% is a question of the expected service life and driving range. The battery of plug-in hybrid vehicles and purely EVs will be fully charged (meaning 95–100%) whenever possible but remain at rest for most of the time. The range of state of charge during use is large. Full cycles (discharge of a fully charged battery to the end-of-discharge voltage) will be the exception. The range of state of charge differs greatly depending on use (see Figure 17.2).

17.2.2 Energy Content, Energy Throughput, and Current

Micro- and mild-hybrid vehicles only need to absorb and release small amounts of energy, so batteries with a low energy content (1–2 kWh) can be sufficient. However, the current in relation to the battery capacity is comparatively high.

Depending on the driving range, the energy content of the battery must be significantly higher for full hybrid vehicles and pure EVs. For PHEVs with an electric range of approx. 50 km approx. 5–10 kWh and for EVs with a range of several hundred kilometers, up to approx. 100 kWh are required. Obviously, the lower the total weight of a battery system, the lower the energy content can be for achieving the same range. The average current and the maximum peak currents in relation to the battery capacity are much smaller in comparison. A battery with 50 kWh energy content has approximately 100 kW output power at a 2 C current rate.

The energy throughput, respectively the total number of equivalent cycles per year, is relatively small for full EVs. A battery for c. 300 km, corresponding to just under 200 m, that is, a relatively short range, only undergoes only 100 full equivalent cycles for an annual driving distance of 30,000 km.

17.2.3 Lifetime

Guarantees for service life and total driving distance differ between manufacturers of EVs. Typical values are 180,000 km driving distance (i.e. 600 full equivalent cycles for a battery with a 300 km range) and eight years. Ambient temperature plays a significant role as many aging processes are temperature-dependent. A range extender can greatly extend the service life because the batteries do not have to be cycled so deeply and the target state of charge can be slightly reduced when fully charged.

17.2.4 Charging Processes

Major differences exist in the charging processes. For micro- and mild-hybrids, the battery is charged by regenerative braking or by diverting drive energy generated by the combustion engine while driving. In the case of pure hybrid vehicles, these are the only options for recharging. The charging currents during braking are highly dynamic, possibly interrupted by short discharge phases. Charge acceptance, that is, minimum losses during charging, is therefore very important.

For PHEVs and EVs, external charging processes at AC or DC charging stations (the latter are usually fast-charging stations) must be available and are the main energy source for recharging. External charging uses CC/CV charging characteristics in most cases with a highly smoothed charging current. It should be noted that fast charging is only necessary for purely EVs, because hybrid vehicles always have a second drive unit that can be charged with diesel or gasoline much faster.[1] The current during fast charging can reach up to 4 C and possibly more.

Pilot projects are now also testing inductive charging options and conductive charging methods during the journey (trucks with pantographs, trolleybus systems,[2] etc.). The challenge is to charge as quickly as possible while at the same time avoiding excessive heating, which would further accelerate the aging process caused by fast charging. This is pushing current battery technology to its limits.

17.2.5 Safety

Safety aspects for electric traction batteries are of paramount importance. These go far beyond the requirements that already apply to the transportation and usage safety of lithium-ion batteries, as additional sources of danger must be taken into account for vehicles:

- Accidents, especially collisions with other vehicles or objects, so special attention must be paid to the location of a battery and its mechanical protection.
- Protection against external fires from other sources (e.g. cable fires, fires caused by fuel, or even explosions in vehicles with an additional combustion engine).
- Loss of battery power while driving (e.g. due to short circuits, faulty cable connections, etc.).
- Weather conditions (lightning, storms, heavy rain, sandstorms, very low or very high ambient temperatures, etc.).

On the one hand, this requires extensive safety considerations according to common industry standards such as ISO 26262 (basic standard for functional safety), failure modes and effects analysis (FMEA), and fault tree analysis (FTA), and on the other hand, far greater testing than of starter batteries or many other components.

Of course, the battery management systems, which also take over communication with the vehicle, are an integral part of the safety and performance requirements. Extensive so-called residual bus simulations must therefore ensure that the vehicle behaves as desired in the event of error messages.

There are worldwide standards that must be adhered to. These do not demand that a battery must not undergo a thermal runaway event under any conditions, as this is considered not possible for the time being. Instead, it stipulates that the danger to passengers is minimized, for example, by providing warning signals at least five minutes before a thermal runaway and thermal propagation can endanger passengers (GTR 20, see chapter 13).

---

1 During filling up with gasoline or diesel, (chemical) power in the range of 10 MW flows into the tank.

2 Trolleybuses or electric streetcars only fall under the generic term electromobility if parts of the route do not have an overhead line for technical or urban planning reasons.

17.3 Choice of Battery Technology

In recent years, lithium-ion batteries have become almost the sole standard for electric traction batteries in the automotive industry, especially for pure electric and hybrid vehicles. This was not always the case: In the past, there were also experiments with lead–acid batteries, but these were not feasible due to the weight disadvantages. After that, there were various trials with high-temperature batteries, such as the sodium–sulfur battery, until the 1990s, when pure hybrid vehicles began to be used on a large scale, with nickel–metal hydride batteries being used for a long time (and in some cases to this day). It is unclear whether NiMH batteries, which are still being installed in new vehicles, will be used in the long term.

For micro-hybrids, which only have to absorb small amounts of braking energy and provide drive energy, lead–acid batteries continue to be used, mainly to provide starting power for the internal combustion engine. The need for a high state of charge and regular full charging for reasons of service life means that they often have to be operated with unfavorably high charging voltages during recuperation processes. Schindler [1] has carefully analyzed the mode of operation and suggested using lithium-ion batteries in parallel with a lead–acid battery to improve recuperation. It remains to be seen whether such dual-battery concepts will prevail, where cells with lithium ion phosphate (LFP) or lithium titanate (LTO) will be used and how the economic evaluation of alternatives to lead–acid batteries will develop.

The requirements described above cannot be met with the same cells and nominal battery voltages – on the contrary, the use of cells specially optimized for energy or power and the adaptation of nominal voltages are the rule in such applications.

Ultimately, lithium-ion batteries offer clear weight and energy density advantages for hybrid and pure EVs and despite the price disadvantages, the production of millions of electric cars has now become a reality. Today, nickel-manganese-cobalt (NMC) or nickel-cobalt-aluminium (NCA) graphite chemistry is predominantly used. Only with batteries of these technologies are acceptable ranges (>300 km, in some cases already significantly more) for pure electric cars technically possible and this technology also offers considerable advantages for plug-in or pure hybrid vehicles due to its comparatively high power density. With regard to range, it must be noted that despite the now standardized driving cycles (worldwide harmonized light vehicles test cycle [WLTC]), individual driving behavior and ambient conditions have a very strong influence on the real range – speeds of well over 100 km/h, low outside temperatures, the use of, for example, heating, infotainment, and reserves before recharging becomes possible, etc., have a very strong impact on the range that can be achieved.

LFP cells, which have a significantly lower specific energy due to the lower voltage per cell and require a larger volume than NCA or NMC graphite cells, ultimately only offer an advantage when adapted to a 12 V vehicle electrical system, despite the lower price, simpler production and certain safety advantages.[3] LTO cells

3 From a safety engineering point of view, the difference is small. Both LFP and NMC or NCA require single-cell voltage measurements, protection against external hazards, and can undergo a thermal runaway.

have an even lower voltage per cell, so they are not currently used for full hybrid and all-EVs but only for micro- or mild-hybrid vehicles, mostly parallel to lead–acid batteries.

17.4 Structure of the Battery System

While purely EVs from most manufacturers may have two or more strings connected in parallel because the energy content required for longer ranges cannot be achieved otherwise, high-performance hybrid vehicles with batteries of significantly lower capacities use an xs1p topology. The number of parallel strings and the capacity per cell affect not only the installation in a vehicle but also reliability and safety. If a thermal runaway occurs in a cell, the energy released per cell is comparatively low with many parallel strings, and the chances of preventing propagation may be better. On the other hand, the probability of a cell fault is greater when 6000 or more cells are installed, as is the case with some models, than with 200 cells installed. If, on the other hand, a cell in 60 parallel strings becomes highly resistive, the overall performance of the battery system hardly changes. However, the trend seems to be toward larger cells, partly because they are easier to monitor and have slightly higher specific energy (fewer connectors, lower total weight for the housing, etc.). The price advantage of small round cells such as the 18,650 cells often used by Tesla, which are also used in a similar way in the consumer sector and therefore offer major economies of scale, will diminish as the production figures for automotive cells for EVs increase.

Another important aspect of lithium-ion batteries is the question of the cell format, that is, prismatic, round, or pouch cells. There is still no clear trend. On the contrary, many battery and car manufacturers use different cell formats for different vehicle projects, which compete with each other. Pouch cells have the advantage of good space utilization, but may have problems with long-term stability and safety problems with gas formation due to aging processes. Prismatic cells also have good space utilization and, thanks to their stable housing, offer greater stability against certain aging phenomena. Round cells can only be packed together with an increase in volume. However, this disadvantage can be partly compensated for by well-adapted cooling concepts. In all cases, cells must have very similar temperatures during operation, as temperature differences would be detrimental to service life and also problematic in terms of safety.

A particularly important design issue for traction batteries is the choice of a modular design, that is, similar or identical sub-units ("modules"), or a one-piece battery that incorporates all cells (cell to pack). Modules have some cost advantages and are easier to repair or exchange but are likely to be heavier due to more complex housing for mechanical protection and cables. They may also provide better protection against thermal propagation, as thermal propagation from one module to the next may be easier to prevent. In addition, it may be easier to utilize space better. Development is ongoing, and the choice is likely to depend on many external factors.

When designing the battery, the mechanical structure is also important. This is not only about protection against mechanical destruction but also about the position of fuses and switches, as well as the removal of gases from cells so that harmful gases emitted, in particular CO, cannot enter the passenger compartment. Other aspects include insulation and contact protection, integration of the cooling system, and, if necessary, heating and thermal insulation.

17.5 Design and Operation

Due to the high electrical power requirements, it is necessary to use high-voltage batteries. Converter technologies and safety regulations require significant additional effort above 1000 V DC. The usual voltages are therefore 250–500 V with a trend to higher voltages. Only mild-hybrid vehicles, which allow very short distances to be driven purely electrically at low power, are designed for lower voltages, for example, 48 V. There are no standards concerning nominal voltages, similar to those for lead–acid batteries in industrial trucks (see chapter 18).

A nominal voltage of 12 V, corresponding to the current vehicle electrical system voltage and the input voltage of virtually all electrical loads in vehicles, leads to a power limitation. One hundred ampere or just over 1 kW of electrical power already requires cables with larger cross-sections with weight and cost disadvantages. The replacement of the 12 V on-board electrical system with a 48 V on-board electrical system, originally due to a lack of alternatives to the lead–acid battery, has been discussed for decades to supply loads with higher power requirements more easily. The use of a medium-voltage on-board electrical system has now found its way into some vehicles, but not as a replacement for the 12 V on-board electrical system, but as a supplement.

As traction batteries are used to power a vehicle, the first requirement is that they can be integrated into the vehicle architecture. This is by no means a completely trivial requirement, because in the past many traction batteries had to be adapted to existing installation spaces that had originally been developed for combustion engines. This often led to painful compromises between what was technically desirable and what was technically feasible. For some time now, however, the battery is planned first and then the car around it, so that technically optimized solutions can be used.

In addition to the available installation space, further boundary conditions must be taken into account, such as the permissible total weight (which is of course also relevant to range), the desired ranges for electrical driving and power even at lower temperatures, the activation of loads such as interior heating (and its mode of operation, e.g. heat pump), as well as charging.

It is also important to note that high peak loads on batteries and electric motors are required during acceleration processes, for example. The rated power of electric motors for EVs, for example, is in the region of 35 kW, but the maximum power during acceleration is closer to 100 kW. Because these phases are very short, this is not a problem for electric motors and should not be a problem for HV batteries either.

Aging prognosis is a particular challenge. Many manufacturers of EVs give guarantees for a certain mileage or a certain number of years up to which a defined percentage of the initial range will still exist (e.g. 75% after eight years or 120,000 km[4]). Since usage behavior and real driving profiles are very different, extensive aging studies are required both in experiments and in simulations to provide estimates of the necessary buffer required to prevent high warranty costs due to early battery failure. It must be taken into account that aging often occurs disproportionately strongly in the first months and years and then weakens (so-called root-t-laws often apply), that is, aging in terms of capacity or range availability decreases with $t^{0.5}$, see chapter 13 for details). However, this only applies if there are no surprises due to new aging mechanisms that were not realized in the design. It must be taken into account that not only the cells age, but also the cell connections, the housing (and its adhesive), bushings, sensors, and ultimately even the battery management system.

As a high number of cycles or a high energy throughput, high currents, and high temperatures accelerate battery aging (see chapter 13), electric cars should not undergo fast charging with high currents too frequently. Fast charging is to be expected mostly after an intensive phase of use, when the battery is already at a higher temperature, so coupling with the maximum cooling capacity of the battery system makes sense.

The average temperature naturally depends on the climate zone. Especially for very hot countries such as the Gulf States, the range available for EVs is therefore limited in some cases. Cooling the battery when stationary with an appropriately sized cooling unit is only possible with a mains supply. The same applies to cold climate zones.[5]

The effects of many cycles on service life are also one of the reasons why bidirectional charging or the use of electric traction batteries for the intermediate storage of electrical power for a private photovoltaic system or to stabilize the power grid has not been implemented. The authors consider this more of a warranty issue than a technical issue, as lithium-ion batteries in vehicles are not likely to fail as a result of exceeding their cycle life – 1500 equivalent full cycles at a range of 300 km would give 450,000 km over the whole service life, which most vehicles do not achieve. As a result, plans for implementation have so far largely remained at a concept level.

The standardized WLTC driving profile (WLTC) is used for vehicle type approval (i.e. vehicle homologations) and corresponding battery designs, for example, as a requirement that a battery must be able to achieve a certain energy throughput when used with such a profile.

Operating data is shown in the following, clearly showing that EVs do not always experience full cycles, and the batteries are therefore often in a partially charged state (Figures 17.3–17.5).

4 120,000 km correspond only to 400 full cycles for a battery with 300 km range. Currently more than 1000 cycles and more are standard in laboratory tests.

5 In arctic zones, it is also common for combustion engines to be connected to the electrical grid when the vehicle is stationary so that the starting process can take place.

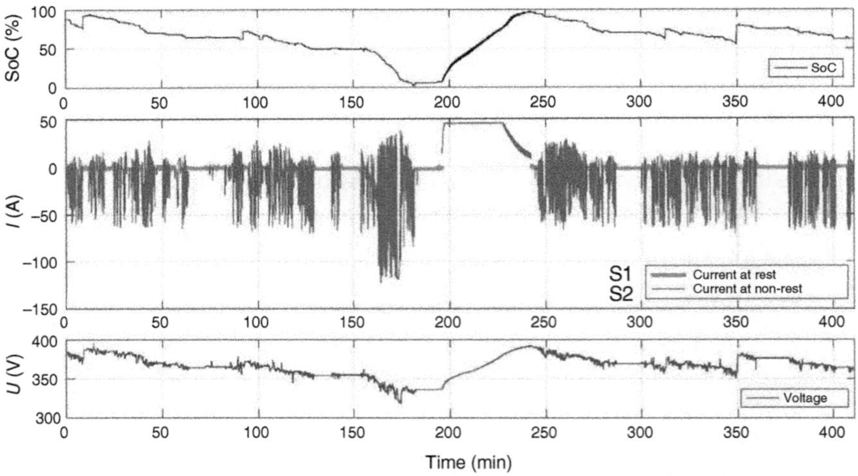

Figure 17.3 Current, voltage, and state of charge curve of an electric vehicle (13.8 kWh, 25 Ah, 400 V) over a period of approx. seven hours. The brighter area describes when the vehicle is stationary (the battery is not charged) and the darker area describes when the vehicle is in motion (the battery is charged, marked S2). Source: Adapted from Ref. [2].

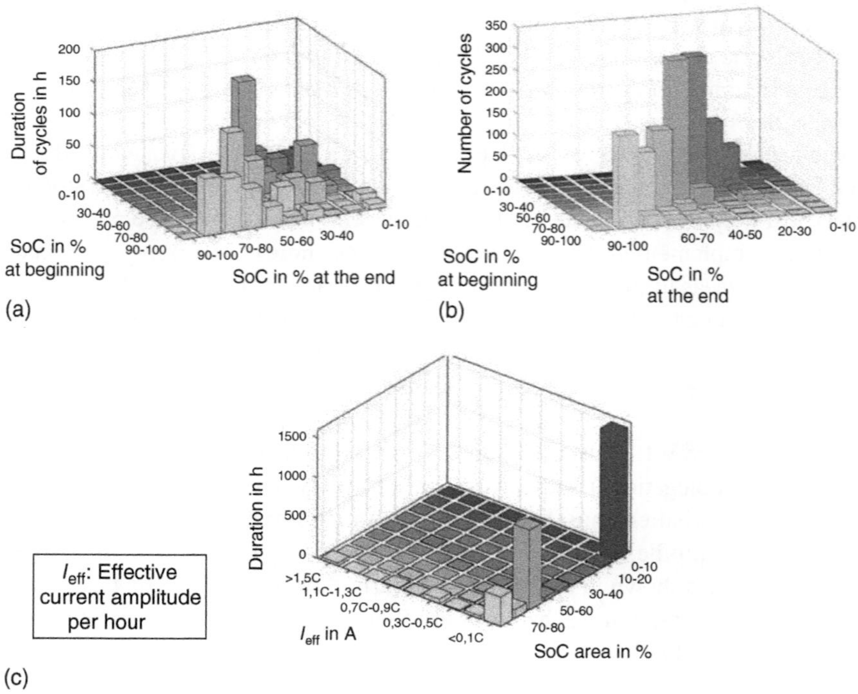

Figure 17.4 Diagrams showing 203 days of operating data from measurements of an electric vehicle (13.8 kWh, 25 Ah, 400 V), summarized by a "from-to" count and a two-parameter dwell time count [2].

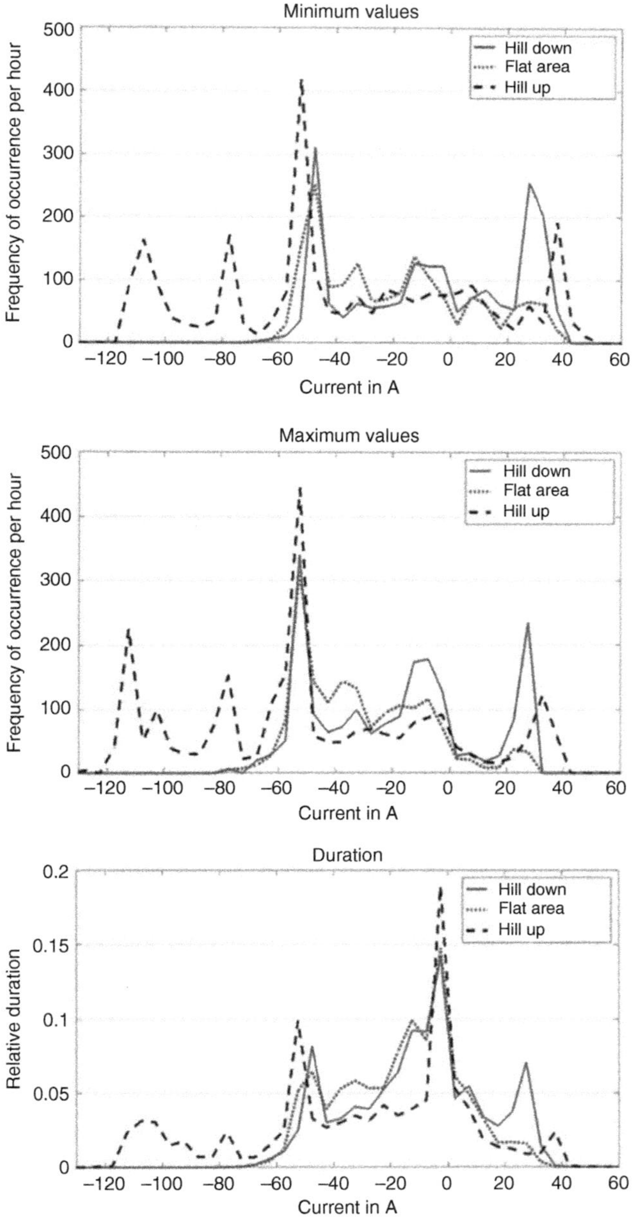

Figure 17.5 Histograms of three load profiles of an electric vehicle for the frequency of current loads for different driving events. Source: Adapted from Ref. [2].

Particularly worth mentioning in Figure 17.3 are the short phases without current, where the voltage assumes a constant value immediately after the current stops. This shows that an equivalent circuit diagram consisting of a voltage source and resistor, or a look-up table, provides a very good representation of the battery behavior in such applications, and that the determination of the state of charge is possible even during short rest periods.

Figure 17.4 shows histograms of usage over 203 days of operation. It can clearly be seen that most cycles during the use of the vehicle led to a relatively small change in the state of charge. The initial state of charge was mostly high, and there was not a single full cycle (100% discharge). Figure 17.5 shows further evaluations of usage. Obviously, these data are usage-dependent and cannot easily be transferred to other uses.

17.6 Monitoring the Battery

The monitoring of electric traction batteries takes place on several levels:

a) At the level of the individual cell, whose cell voltage and, in some cases, also temperature are recorded (possibly also further data).
   For high-capacity cells (from approx. 20 Ah, i.e. the so-called large-format cells), it is the state of the art that single-cell voltage measurements are carried out, that is, approx. 100 V measurements for a string with a 360 V nominal voltage. This is no longer practicable when using small round cells (up to approx. 10,000 cells in some vehicle models). The loss of safety when only measurement the voltage of several cells together is offset by the advantage of a lower energy content per cell, which may be advantageous if a thermal propagation event can be safely avoided. The extent to which the number of voltage measurements can be reduced in future through better evaluation algorithms without any loss of safety remains to be seen.
b) At module level (if available), where at least one temperature value and possibly other values (e.g. gas sensors, shock sensors) are recorded in addition to the module voltage.
c) At the level of the entire battery, where the information from all sensors is processed in the battery management system.
d) At the vehicle level, where the information provided by the battery management system is checked for plausibility and, if necessary, information from the vehicle (e.g. crash signals, cooling system) also triggers actions from the battery.

Level b) obviously only applies to batteries built from modules.

Battery monitoring must be carried out during driving and charging. In addition, even when the vehicle is at rest, at least rudimentary detection or forecasts of the existing state of charge, the voltage value, and, if necessary, other signals must be carried out so that safety-critical states (e.g. undervoltage or deep discharge) can be excluded as far as possible.[6]

In our view, safe and redundant battery management systems must check the function of the battery switches and the measured values of the sensors (voltage, temperature, and current) for consistency and plausibility, and thus communication lines, at the start of the journey.

---

6 When analyzing a fire in a large lithium-ion battery in a stationary application, it turned out that there was a considerable delay in detecting and thus potentially preventing the fire because the battery's monitoring system had been deactivated as the battery was not in use.

Successful battery monitoring also generates data that could be sent to servers. In some countries, such as China, this is already required; in other nations, the pressure is likely to increase to send at least safety-relevant information to servers under the control of the respective vehicle manufacturers to warn third parties if necessary.

17.7 Other Aspects

The field of electric traction batteries has been extremely dynamic in recent years, with a wide variety of technologies being used, further developed, discarded, and then reactivated. Due to the strong financial and legislative support for electromobility, further dynamic development can be expected.

However, it should not be overlooked that numerous basic issues have not yet been resolved. This applies in particular to the recycling of aged traction batteries, but also their possible reuse in stationary energy storage systems (the so-called second-life applications). In addition, the production of batteries has also come under scrutiny because the extraction of raw materials often takes place under problematic conditions (child labor, inhumane working conditions in war or crisis zones, etc.[7]) and the production is often not (yet) climate-neutral if large quantities of CO_2 are released in the process. Life cycle assessments are complex cross-scientific analyses and are beyond the scope of this textbook. For further information, see, for example, Ref. [3].

Electromobility currently enjoys considerable political support, for example, through tax benefits, premiums for purchasing, or legal bans on combustion vehicles, as electromobility can make a very significant contribution to making the world's economy CO_2-neutral and therefore climate-friendly. However, holistic considerations are crucial for this: Discussions must also take into account that:

- Electric cars, whose electricity comes mainly from coal-fired power plants and whose batteries were also produced using such electricity, do not contribute to climate protection.
- High-quality and powerful plug-in hybrids cause higher emissions than conventional vehicles because they consume more fuel when used over long distances due to their significantly higher weight than they reduce during pure electric drive.

Both are correct, but there is no reason not to push electromobility, because both represent transition phases from one major technology and infrastructure to another. Balancing emission reductions and cost benefits of a new technology against the existing technology in all phases of market introduction would only increase costs and take time. Nevertheless, holistic considerations in terms of energy and life cycle assessments, etc., are important. For example, existing alternative means of transportation, such as rail, must be included in overall transport concepts, and the development and expansion of the charging infrastructure must

---

7 It is necessary to point out that this also applies to the extraction of coal, gas, oil, uranium, and minerals used for other industrial uses and is not an aspect that only applies to lithium-ion or lead–acid batteries.

not be neglected. The discussion of new mobility concepts also belongs in this context. With regard to electric traction batteries, the battery exchange concept of the company Better Place is worth mentioning here, which wanted to establish large battery exchange stations. This concept has failed for the time being, but there are still vehicle manufacturers, such as the Chinese company Nio, who are promoting it. However, it seems more realistic to initially use battery exchange systems within a vehicle fleet (cabs, delivery vehicles, etc.). Battery exchanges across brands or even manufacturers are very problematic for technical, economic, and legal reasons.

References

1 Schindler, M. (2013). *Multi-battery system for micro-hybrid vehicles based on lead-acid and lithium-ion technology*. Dissertation. RWTH Aachen University.

2 Tchoupou Lando, E. (2015). Event-based model for the lifetime prediction of lithium-ion batteries. In: *Presentation at the 7th E-MOTIVE Expert Forum "Electric Vehicle Drives", Munich, June 09, 2015.*

3 Klöpffer, W. and Grahl, B. (2009). *Life cycle Assessment (LCA)*. Weinheim, New York: Wiley-VCH.

Tasks

Task 17.1

Which hybridization concept is the basis for the following vehicles?

a) Volkswagen ID.3,
b) Toyota Prius Hybrid,
c) Kia Sorento with hydrogen drive,
d) Audi A3 with diesel engine, automatic start-stop system, and brake energy recovery,
e) Chevrolet Volt, and
f) Renault Zoe.

Task 17.2

Which battery technology would you prefer for the following purposes and why?

a) Traction battery of a pure electric vehicle in the luxury class,
b) Hybrid battery for a mild-hybrid vehicle in a cost-sensitive market, and
c) Vehicle with a battery that is intended to perform start/stop and recuperation and some low power acceleration assistance (boosting).

18

Traction Batteries for Material Handling

Lead–acid batteries still dominate in industrial trucks, for example, forklift trucks for material handling, because their greater specific weight compared to lithium-ion batteries is an advantage, in most cases, as regards the stability of such vehicles. The significantly higher price of lithium-ion batteries can only be justified for special applications and vehicle types.

There is a large and well-established market for powering industrial trucks used for material handling, as well as for small special vehicles such as wheelchairs, cleaning machines, and other similar vehicles. These applications are treated separately from the electromobility application area because they differ significantly concerning technical and economic requirements for batteries. The market for industrial trucks is the fourth largest market for batteries after starter batteries and batteries for stationary and portable applications and is considerably larger than the market for batteries for electric road vehicles. In the following, a distinction is made between two areas of application:

- Batteries for industrial trucks for material handling, and
- Batteries for special vehicles with low drive power and range requirements (semi-traction batteries).

18.1 Industrial Trucks for Material Handling

Many different vehicles are used in industry, logistics centers, and retailing for the transport of goods. In the vast majority of cases, they are powered electrically because the vehicles are often used inside factories and other enclosed spaces. The use of combustion engine vehicles in enclosed spaces, even with highly effective filters, is not permitted in many countries for reasons of occupational safety and emissions. Only vehicles that are used almost exclusively outdoors may be powered by internal combustion engines (diesel, natural gas). The last tightening of emission values (Stage V for mobile machinery) in Europe came into force on January 1, 2020.

Battery Technology: Fundamentals of Battery Electrochemistry, Systems and Applications, First Edition. Alexander Börger and Heinz Wenzl.

Companion website: www.wiley.com/go/Börger/Battery_Technology

Examples of battery sizes are

- For the transport of shipping containers:
  720 V nominal voltage and 480 Ah capacity, corresponding to a nominal energy content of 345 kWh.
- Forklift trucks:
  nominal voltages of 80 (rarely also 96), 48, and 24 V with a battery capacity in the range of 1200 to 210 Ah. The energy content ranges from 100 to approx. 5 kWh.

The capacity of these batteries is so high that only single cells of lead–acid batteries, and not block batteries, can be used, and their handling during operation is only possible in battery trays. When using lithium-ion batteries, the use of a single string is only possible for applications with very small energy content if nominal voltage conventions are observed.

The economic importance of industrial trucks has led to a very large number of standards, which is in stark contrast to the lack of in-depth scientific studies and publications. If operating data is published at all, it can be found in industry-oriented trade journals.

18.1.1 Requirements

Industrial trucks of all sizes are part of production and logistics processes whose requirements in terms of efficiency, availability, and flexibility dominate. Important requirements derived from this are

1) Low life cycle costs (investment costs, maintenance costs, energy costs).
2) High cycle stability as part of the required low life cycle costs.
3) Production requirements dominate; in many cases, this means that the battery must be able to power the vehicle for at least one shift and can be recharged within one shift (eight hours). The capacity values for such batteries are therefore given for a five-hour discharge rate C_5, as there are always rest periods.
4) The installation of the battery must increase the stability of the industrial trucks and usually also serves as a counterweight for the loads to be lifted.

18.1.2 Choice of Battery Technology

Battery costs are approx. 30% of the costs of an industrial truck. For example, for a 3 ton[1] forklift truck and a lead–acid battery with 750 Ah capacity at 80 V nominal voltage, the weight approx. 1600 kg[2] and costs approx. 9000 €. In comparison, the use of a lithium-ion battery with comparable functional performance costs four to five times as much (Proppe, A. (2021), Forklift truck service, Proppe GmbH, Bevern, pers. communication).

---

1 Weight that can be lifted, without special gear usually c. 2 m high.

2 When such a truck is not used frequently, a battery with much lower energy content can be fitted.

The advantages of a lithium-ion battery – faster[3] and unproblematic rechargeability, lower maintenance requirements, longer expected service life, and smaller installation space with greater design freedom – must therefore be considerable to justify this additional price. Examples of possible advantages are

- For certain industrial trucks where small size and maneuverability are important, there are now models designed exclusively for operation with lithium-ion batteries.
- For vehicles used very intensively in multi-shift operation, the use of lead–acid batteries requires the availability of so-called exchange batteries and the infrastructure (hoisting equipment for changing the battery, charging space for the battery, equipment to refill with purified water). The additional price for lithium-ion batteries may be justified in such cases. However, the number of lithium-ion batteries currently used for such vehicles still appears to be very low, partly because there are technically mature complete systems for handling lead–acid batteries in battery changing stations.

The widespread dominance of lead–acid batteries is likely to continue as long as the cost of lithium-ion batteries remains so high compared to lead–acid batteries. A high cost reduction potential for lithium-ion batteries for industrial trucks can be expected if the existing costs and quantities in the electromobility sector are transferred to this market segment. It remains unclear when and to what extent lead–acid batteries will be replaced.

The use of fuel cells to power forklift trucks has also been discussed for many years. A few demonstration vehicles have been built. In addition to the high investment costs, it should be noted that fuel cell vehicles may require an additional battery to cover power peaks when lifting very heavy loads and absorb braking energy.

DIN EN 60254-1 [1] specifies the following requirements for the service life of lead–acid traction batteries with tubular plates[4] (PzS) as 1200 cycles (in this market segment, a discharge of 80% is referred to as a full cycle) and for traction batteries with tubular plates and immobilized electrolyte as 900 cycles. ZVEI data sheet no. 8, "Service life of traction batteries" [2] describes a calculation method for determining how long a traction battery is likely to last at which average temperature and depth of discharge per cycle. The calculation method corresponds to the weighted ampere-hour throughput model explained in chapter 6. Usual lifetimes are often in the range of approximately five to six years and are therefore not an economic problem, especially when compared with the current prices for lithium-ion batteries. Furthermore, a generous warranty is also common (the so-called pro-rata warranty), that is, the price of the new battery is reduced proportionately if it has to be replaced early, for example, in less than four years after purchase.

---

3 A charging phase with a high charging current during short breaks during use would also reduce the energy content of the battery, which must be made available for the intended use.

4 In this area of application, batteries with tubular plates are used almost exclusively in Europe because their significantly longer service life more than makes up for the additional cost. In other regions of the world, the use of tubular plate batteries is less pronounced.

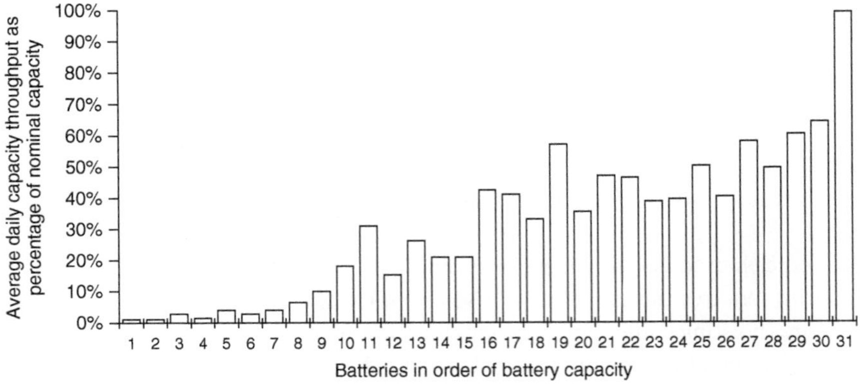

Figure 18.1 Average daily use of the battery capacity of a fleet of industrial trucks (battery numbers in order of battery capacity).

For these reasons, only lead–acid batteries are considered below, especially as the important advantage of lithium-ion batteries in many other applications, the significantly lower weight, not only plays no role in a large number of industrial trucks but is counterproductive. When using small lead–acid batteries, sometimes lead or concrete plates have to be installed in the battery trays to ensure sufficient stability.

18.1.3 Operation

A distinction must be made between the operation of vehicles with relatively little operating time per day and intensively used vehicles in multi-shift operation. Figure 18.1 shows measurements of the battery usage of an industrial company's vehicle fleet with a partial two-shift operation, which illustrates the wide range of usage.

It should be noted that even high-capacity vehicles with large batteries may be used little if their constant availability at a particular workplace allows the production process to be optimized. The following subdivision into vehicles with low and intensive use only represents the extremes of a broad spectrum of use. At this point, reference is again made to the explanations in chapter 15 and the need to analyze intensively the actual use of the battery during planning and troubleshooting.

The usual charging characteristic for flooded traction batteries is IUIa or, more rarely, WoWa. For an IUIa charge, the initial charging currents are usually 16–20 A/100 Ah. When 2.4 V/cell is reached (voltage at the terminals of the battery or output points of the charger, compensation of the voltage drop via the charging cable is not standard), the system switches to a constant voltage charging phase. When a charging current of 4 A/100 Ah is reached, charging continues at this current until an end-of-charging criterion is reached. The charging voltage increases to values of 2.65 to 2.75 V/cell for new batteries.

Figure 18.2 shows measurements of a 750 Ah battery (nominal current 150 A) that illustrates the highly dynamic operation. The maximum discharge current of the 750 Ah, 48 V battery used is just under 500 A, and the maximum current fed in by

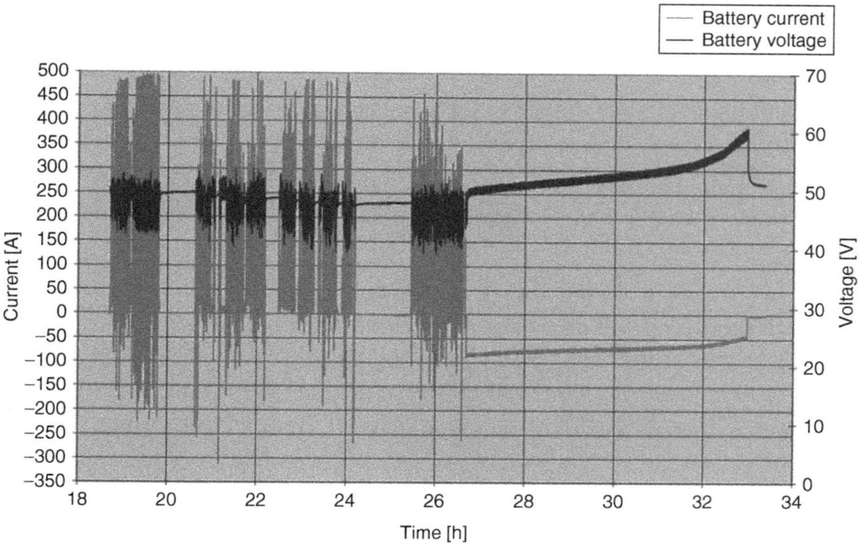

Figure 18.2 Current and voltage curve of a lead–acid battery (750 Ah/48 V) in an industrial truck. Note that the discharge current is shown as a positive value, in contrast to the usual conventions. Source: Courtesy of A. Proppe.

braking or lowering loads is over 200 A. The battery is charged after use at approx. 60 V, corresponding to approx. 2.5 V/cell. However, a full charge has not yet been achieved at this point, as the current and voltage have not yet reached constant values. The setting and/or the quality of the charger are therefore problematic.

18.1.3.1 Vehicles with Low Daily Operating Time

The operational requirements for batteries are easy to meet for relatively little-used vehicles with a battery of fairly low energy content but sufficient weight.

- Recharging takes place either at night during non-operating times at a convenient location for the charger (in the immediate vicinity of the place of use or in central charging stations) or several times a day after each use by reconnecting to the charger, known as opportunity charging. In the latter case, the battery is often massively overcharged with conventional chargers, because a full charge is carried out even if only a very small percentage of the capacity has been drawn. Our own measurements have shown that batteries of this type of charging strategy can have a charge factor, the ratio of the amount of charge charged into the battery to that discharged, of 5, and the energy efficiency can be in the region of approx. 10%. Despite the low usage, the service life of these batteries is rather short due to constant overcharging. A full charge with complete conversion of the discharged material and removal of the acid stratification could be guaranteed by using a suitable charging characteristic without such massive overcharging of the battery.
- Cost-effective concepts for topping up with water can be easily set up. There are automatic water topping-up systems in which, in addition to connecting the

battery plug to the charger, a hose coupling to the battery with a water supply system is all that is required during the charging process. Either a container with purified water or an ion exchanger connected to the drinking water supply is used. Where refilling with water cannot be reliably organized, sealed traction batteries, usually gel batteries, are often used. Their shortened service life is then accepted. Especially in companies with only a few hand pallet trucks and in large factories at individual, scattered locations, sealed batteries are predominantly used because refilling with water is difficult to organize reliably and cost-efficiently.

Lead–acid batteries in this market segment usually have no temperature problems and are rarely equipped with electrolyte circulation systems.

In principle, this area of application is suitable for replacing lead–acid batteries with significantly smaller lithium-ion batteries, possibly with additional weights, because deep discharging without the possibility of intermediate charging is an atypical application.

18.1.3.2 Intensively Used Vehicles

For intensively used vehicles, the vehicle battery is usually designed so that the vehicle can be operated for one shift without intermediate charging. Only in rare cases (e.g. loading docks in the beverage industry) does the battery not last the full shift.

Lead–acid batteries require frequent, complete recharging for reasons of service life and, in the case of flooded batteries, regular topping up with water. There are technically and economically mature solutions for these two aspects. To ensure economical operation, it is important to carefully plan the time and place of recharging and, in the case of multi-shift operation, the exchange of batteries. The high standardization of battery sizes makes it possible to use batteries from different manufacturers for trucks of different brands. So-called battery management systems for handling many batteries, in large car plants easily a few hundred, are commercially available, but should not be confused with battery management systems for individual lithium-ion batteries.

The operating conditions are characterized as follows:

- Within one shift (eight hours), a full charge with removal of the acid stratification is possible. Compliance with this charging time is considered more important than a gentler charge with a lower charging voltage and gassing rate.
- The temperature of the battery rises slowly during a week of continuous operation because the high charging voltages increase the temperature at the end of charging. Contrary to expectations, however, the temperature drops slightly during use. Design measures to avoid high temperatures, such as cooling or constructing the battery trays in such a way that natural convection enables even cooling of all cells, are not standard. The temperature difference between the cells in the battery tray is considerable.
- Intensively used traction batteries are often equipped with electrolyte circulation systems, which reduce water consumption and a rise in temperature. The charge factor is reduced from approx. 1.15 to 1.18 (for new batteries) to approx. 1.05 by

electrolyte circulation systems that are supplied with air via the charging plug. The lower gassing and temperature increase explains the observation of a longer service life for batteries when equipped with electrolyte circulation.

- Time and place of recharging for vehicles with drivers. Industrial trucks are generally equipped with a state-of-charge indicator. The simplest devices are only based on voltage monitoring (moving average values, measurement only during current flow, evaluated with expected average currents) but are sufficiently accurate. Failure to observe and falling below a state of charge of 20% considerably shortens the service life and is often technically prevented.
- In single-shift operation, recharging usually takes place at the end of the operating time. The battery then remains in the truck for changing, but precautions must be taken to remove hydrogen emissions.
- In multi-shift operation, the battery is charged outside the vehicle. To do this, the discharged battery is either removed from the vehicle manually using lifting gear and a charged battery is inserted, or this process takes place semiautomatically in a battery charging and changing station. The next battery to be inserted is automatically selected; involvement by staff in battery changing is minimal. Large battery charging stations with the exchange of batteries in large industrial companies have very large dimensions (up to 1000 m^2 area and more), taking up potential production space. There is sometimes intensive data recording with battery management systems that record the charging time, amount charged, time of reuse, etc. Figure 18.3 shows an example of this.
- Operation with opportunity charging when the truck is not used is rare, although there are examples of the advantages (time costs for the journey to the battery charging station, reduction in the number of batteries required[5]). Figure 18.4 shows usage data of a battery used intensively in multi-shift operation in opportunity charging mode. To achieve a regular full charge and to enable topping up with water, the battery or vehicle and battery must be taken out of operation briefly at regular intervals, for example, after a week or after a charge throughput of several times the nominal capacity. In most cases, the amount of energy removed between two charges is low, so a reduction in the capacity of batteries and their number per truck becomes possible and cost-effective, even if there might be a reduction in service life due to a higher energy throughput.
- Water topping up. The state of the art is the use of an automatic water refilling system which is sometimes even integrated into the charging plug. Toward the end of the charge, a valve is opened that feeds water into a water topping-up system installed on the battery. There is a valve on each cell that interrupts the water supply when the cell has the correct electrolyte level but does not cut off the supply to the neighboring cells. Manual refilling of each individual cell is rarely carried out for cost reasons. Sealed batteries, where it is neither possible nor necessary to top up with water, are more expensive than flooded batteries and more sensitive to temperature and have therefore not become as popular as might be expected.

5 In continuous three-shift operation, there may be more than two batteries per truck.

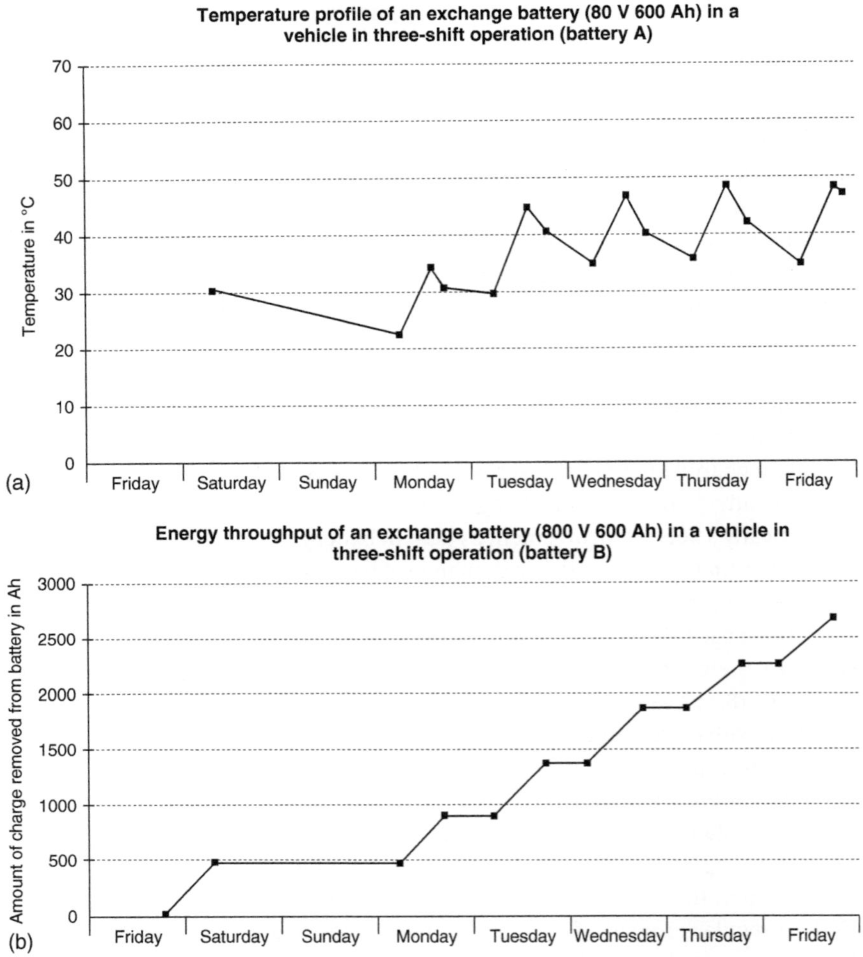

Figure 18.3 Utilization profile of a traction battery (80 V/600 Ah) where a battery exchange strategy is used in a company with three-shift operations. (a) The dots mark the temperature at the start of charging (lowest value in each case), the highest temperature during charging, and the temperature at the start of use. (b) Each point corresponds to the start or end of a discharge.

18.1.3.3 Operation of Driverless Transport Vehicles or Automated Guided Vehicles (AGV)

More and more factories and logistics centers are using driverless transport systems. These are partly conventional industrial trucks with minor technical adaptations, but also special vehicles. Charging takes place automatically either inductively or conductively at special charging points.

Operation with opportunity charging at the place of use and the organization of maintenance work (refilling with water) is easily possible.

The information above for intensively used vehicles is largely applicable.

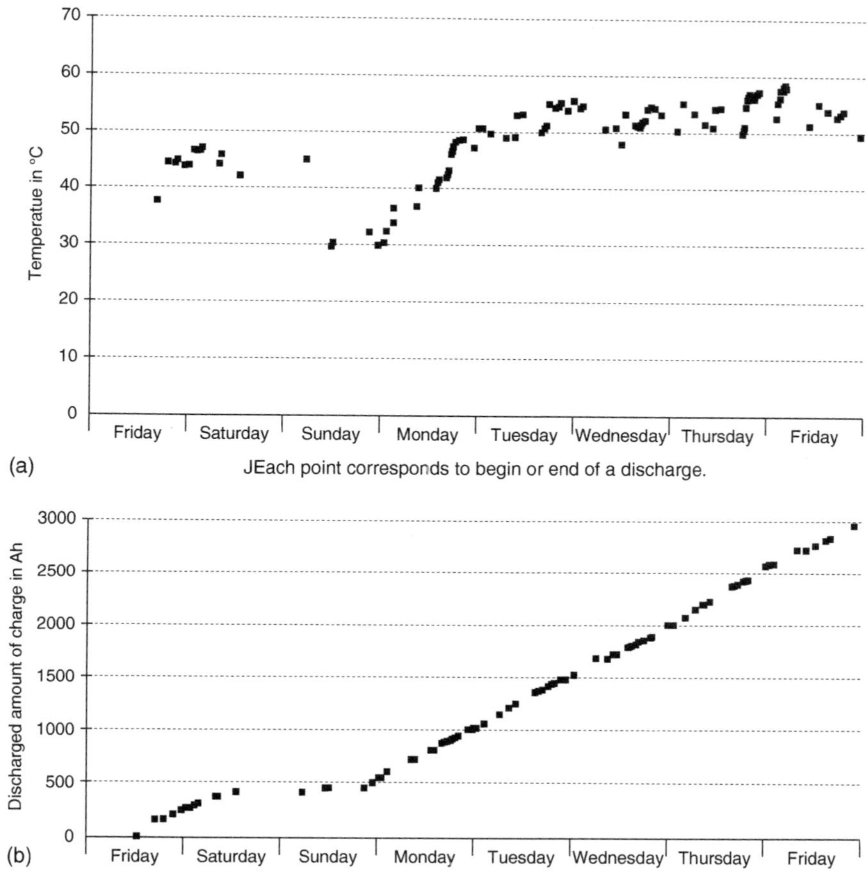

Figure 18.4 Usage data of a traction battery (80 V/450 Ah) in opportunity charging mode in a company with three-shift operation. Per day, 105% of the nominal capacity was drawn from the battery. Each point corresponds to either the start or the end of a discharge. (a) Temperature provide and (b) accumulated among of charge drawn from the battery.

18.1.4 Monitoring Batteries

For monitoring batteries, most chargers record usage data, which, when matching batteries to a charger, allows analyses of operating times, energy throughput, etc., similar to battery management systems of battery charging stations. In addition, they help with troubleshooting.

There are also monitoring systems for batteries available that are permanently installed on the battery and record the following data over a longer period of time:

- Start and end of a discharge,
- The amount of charge or energy removed,
- Temperature values reached,
- Voltage values, including voltage values at $n/2$ cells for fault detection, and
- Start and end of charging,
- Amount of charge during charging.

Furthermore, the state of charge is determined precisely even for opportunity charging operations. The *SoH* is calculated by comparing the measured discharge data with the expected discharge data.

There is no publicly available information on the use of batteries and vehicles obtained from such data and the resulting consequences for operation.

18.2 Small Traction Batteries

Batteries that are used for wheelchairs, cleaning machines, golf carts, mobile work platforms, etc. are much smaller traction batteries, sometimes referred to as semi-traction batteries. There is no clear distinction between on-board power supply batteries, which are used in addition to starter batteries, for example, for work platforms on trucks, lighting for warning beacons for road traffic, and also for camper vehicles.

In contrast to batteries for industrial trucks, these batteries are permanently installed in vehicles or appliances due to their relatively low energy throughput. Operating voltages are in the range of 12 or 24 V, and the capacity is up to approx. 180 Ah.

The difference to starter batteries concerns a significantly higher cycle stability, which is achieved by using much thicker plates. Standards require acceptance tests of c. 300 cycles instead of 50 for starter batteries.

18.2.1 Requirements

Apart from the need to provide a high counterweight, the requirements are comparable to the requirements for batteries for low-use industrial trucks but are lower overall.

18.2.2 Choice of Battery Technology

Lead–acid batteries still dominate in this area of application for cost reasons, although lithium-ion batteries would only have advantages, including the higher specific energy. Lead–acid batteries with pasted plates (flooded or sealed) are predominantly used instead of batteries with tubular plates. The achievable cycle stability of only 300–600 cycles is considered sufficient in this area of application.

Due to the comparatively high capacity, the batteries are sometimes offered as 6 V blocks (to minimize the weight per block) or 12 V blocks, often in a sealed design (mostly gel). Differences from comparable batteries for stationary applications and starter batteries with high cycle stability may be different materials and thicker electrodes (e.g. 2–6 mm). Manufacturers offer blocks with tubular plate electrodes of the appropriate size, but these are rarely used, although their service life would be significantly longer.

18.2.3 Operation

The batteries are almost always permanently installed in the vehicle or mobile device, and a charger is often integrated to enable charging at any electricity plug.

Due to the comparatively low costs, there is virtually no battery monitoring and maintenance, and the end of the service life is determined by the user on the basis of insufficient operating time. The battery is a “disposable product”. The authors are not aware of any analyses of the service life achieved in operation.

References

1 DIN EN 60254-1:2005 (2005). *Lead-Acid Motive Power Batteries – Part 1: General Requirements and Tests.*

2 ZVEI Information Sheet 8 (2013). *Service Life of Traction Batteries.*

19

Stationary Applications of Batteries

Batteries for securing the power supply in the event of voltage dips and power failures to protect people and property are one of the largest markets for batteries. In developed countries, batteries are rarely discharged due to the high quality of the public electricity supply system and are usually kept fully charged. In remote regions and in the electricity systems in some developing countries, however, more frequent, albeit short periods of discharging may occur. Numerous standards and national legal requirements exist and must be observed for batteries and secure power supply systems, for example, hospitals, elevators, and emergency lighting. For cost reasons, lead–acid batteries still dominate in some applications, but lithium-ion batteries are rapidly gaining market share due to their overall technical advantages.

The transition of the energy system from conventional electricity generation plants (nuclear, fossil fuels) to renewable energy systems (mostly photovoltaic [PV] and wind power) with their already high market penetration in some countries lead to more and more batteries in the electricity grid. The technical and economic use of these batteries is heavily dependent on legislation. Changes in the power generation mix and the further market penetration of PV and wind power will have an additional impact on the number and size of battery-based energy storage systems.

The description of this area of application is divided into four different groups, which lead to different requirements:

1) Standby parallel operation for mains backup and uninterruptible power supply (UPS) systems that can supply the connected loads with electrical energy for a longer period of time (approx. 10 minutes to several hours) in the event of a mains failure;
2) Batteries for starting diesel generators as emergency power systems;
3) Batteries for balancing electricity production and consumption over time (off-grid, autonomous energy systems, home energy storage systems [ESSs], and large stationary battery storage systems for the electricity industry); and

Battery Technology: Fundamentals of Battery Electrochemistry, Systems and Applications, First Edition. Alexander Börger and Heinz Wenzl.

Companion website: www.wiley.com/go/Börger/Battery_Technology

4) Batteries for system stability of the power supply system (frequency stability and optimization of the operation and dispatch of power plant and load management of the grid).

In most cases, batteries in these applications are connected to loads and to a power supply at the same time.

Various regulations apply to secure power supply systems for stationary battery applications in probably all countries worldwide, some of which are mandatory. In addition, sometimes very detailed regulations on the properties, installation conditions, and required inspections of the systems must be observed, particularly in buildings and hospitals where the safety of people may otherwise be in danger. Most of the technical rules and regulations do not primarily concern the design and properties of batteries, but rather the overall technical system in which they are embedded. The regulations listed in this chapter should not be regarded as a complete overview.

19.1 Standby Parallel Operation for Emergency Power Supply and UPS Systems

19.1.1 Definition of Terms

The purpose of the batteries is to provide energy to loads in the event of a power failure or voltage dips and to supply loads for some time with very high power that cannot be provided via the mains connection and the power supply units. DIN EN 50272-1 Annex A [1] makes the following distinctions:

1) Standby parallel operation A1.1 according to DIN EN 50272-1
   Standby parallel operation is characterized by continuous charging in trickle charge mode. The batteries are constantly kept in a fully charged state and the charging current covers the average value of both load current and charging current of the battery. In DIN EN 50272-1, it is permitted that the battery current is superimposed with an alternating current, which can also lead to short discharge currents. A schematic diagram of this is shown in the appendix to DIN EN 50272-1, and a measurement of the battery charging current of a UPS system for supplying AC loads is shown in Figure 19.1. When measuring the float charge current with high time resolution, in many cases such a current profile is observed: many and short discharge currents as a result of the control characteristics of the loads and the switching frequency of the semiconductor components used in the charger and the loads. The effective alternating current amplitude can be a multiple of the direct current (see also Figure 15.2). Discharging of the battery by this high-frequency current does not take place as long as the battery voltage remains above the equilibrium voltage and the battery current does not contain any Faraday current. The amount of charge during such discharges is usually below the mAh range.

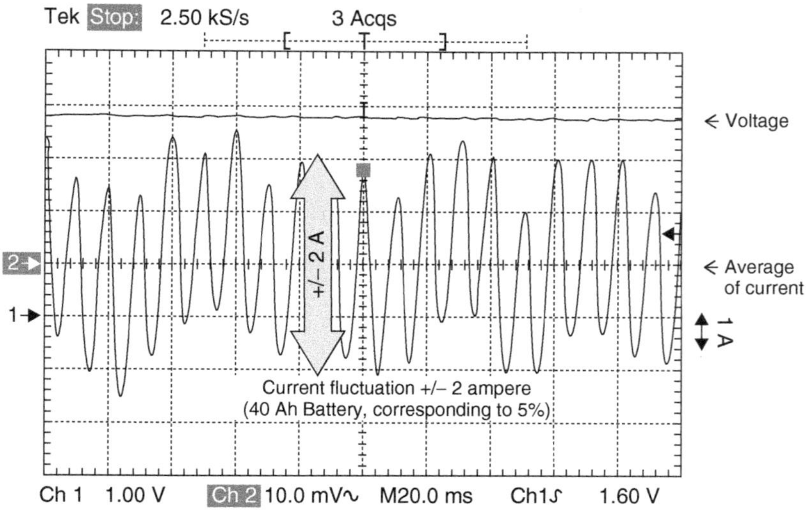

Figure 19.1 Measurement of the float charge current and voltage of a UPS system for supplying AC loads (216 V nominal voltage, 18 12 V blocks, 40 Ah capacity). +/+ 2 A corresponds to $5A_{eff}$/100 Ah capacity, 50 Hz alternating current causes a 100 Hz alternating current on the DC voltage (see Figure 15.1).

Energy is only drawn from the battery in the event of a mains power failure and during tests, so that in stable electricity grids the annual charge throughput, including test discharges, is likely to be less than the battery capacity.

2) Buffer operation A1-2 according to DIN EN 50272-1
   If the power supply units are tightly dimensioned, this leads to more frequent but short and shallow discharges of the battery. The battery is kept "at approximately full charge."
3) Operation with shallow cycles A1-3 DIN EN 50272-1
   Battery systems that are operated in parallel with the mains and are discharged more frequently and for longer periods when loads demand power due to insufficient power from the power supply unit are constantly subjected to more or less deep cycles. Specifications for a clear distinction to cyclic operation are not given.
4) Standby mode A2 according to DIN EN 50272-1
   The battery is constantly in float charge mode and is only coupled to the loads in the event of a power failure.

The conceptual distinction between these operating modes is easy, but the actual classification is not. For example, the question arises as to whether certain topologies of UPS systems are standby parallel operation or changeover operation because in some uninterrupted power supply systems the inverter is in operation to supply the loads but provides no power under normal operating conditions.

In modern battery-supported energy systems, care is taken to ensure that the charging system is dimensioned sufficiently big so that operation according to A1-2 and A1-3 should no longer occur. The categories are of little importance for the requirements of the battery and its design, as long as the cycle depth has no effect on the service life in the operating mode with weak cycles.[1] The operation of the battery under such conditions shows similarities with that of traction batteries for industrial trucks with low energy throughput in opportunity charging mode, which are constantly recharged during use without being fully charged each time.

For the following discussion, it is assumed that the target state of charge of the battery is 100%, that this target state of charge exists for the majority of the time, that the annual charge throughput is very low, and that calendar aging effects dominate.

19.1.2 Requirements

For the requirements that the battery has to meet, it is irrelevant whether or not there is a brief interruption between the mains failure and the battery taking over the supply.[2]

With the exception of operation with weak cycles (A1–3), the batteries are permanently kept in a fully charged state and only discharged in the event of mains failures, voltage dips, or tests. For emergency lighting systems and some other special systems, a test is required every year for the required bridging time,[3] in which the full bridging time must be demonstrated. In developed countries, a charge throughput in the range of the nominal capacity per year is likely. In the case of very long-life batteries (design life over 12 years), a maximum of approx. 15 times the nominal capacity over the entire service life can therefore be expected. The batteries must be designed for a long service life in float charge mode, not for high cyclical loads. The charge turnover may be somewhat higher when operating with weak cycles, if this type of operation still occurs at all in modern battery systems, and with very frequent short test discharges.[4] Here, too, the requirement is for a long service life in float charge mode but not for a high number of cycles.

Continuous trickle charging without any discharges makes it difficult to detect faults, so robust systems and easy diagnosis of the battery condition are of great importance.

Depending on the loads that need to be protected, nominal voltages between 24 and approx. 800 V and cell capacities of approx. 7–3000 Ah are used. The capacity of the battery system can be provided by several parallel strings. The total energy

1 One of the authors was told of a case in which a lead–acid battery was designed for standby parallel operation but was partially discharged during the day due to an inadequate charging system, which was not capable of recharging the battery during the day between frequent but shallow discharge cycles. In the evening the battery was regularly partly discharged and could only be recharged at night. A full charge was not always achieved at night. The battery was in fact in cyclic operation and, worse, even cycled in a partially charged state – the worst operating conditions possible for lead–acid batteries.

2 The differences concern switching devices and the algorithms for detecting a power failure, not the battery, its charger, and its monitoring.

3 Depending on the design, the amount of charge removed at the beginning of the service life only corresponds to approx. 60–80% of the capacity.

4 Some manufacturers test the battery function daily by means of the voltage response to a short discharge pulse. Usually, the charging voltage is switched off or reduced for a few seconds so that the battery continues to supply the loads. A weekly test discharge is prescribed for emergency lighting and alarm systems (fire and burglary).

stored per battery system can be up to 1000 kWh and may be achieved by connecting complete systems (batteries and converters) in parallel.

Stationary battery systems are usually located in buildings at room temperature. However, they can also be located in unheated buildings (outdoor high-voltage transformer and switchgear sites, etc.), so the battery size must be designed for the minimum expected temperature. Times at high temperatures must be taken into account when estimating the service life.

19.1.3 Choice of Battery Technology

At this point in time (2024), lead–acid batteries still dominate the market for stationary applications. Depending on the region, the market share may still be close to 90%. The most important aspects for the dominance of lead–acid batteries are as follows:

- Weight and volume restrictions are rarely relevant.
  With lead–acid batteries, the total weight of large systems can amount to many tons. Particularly in the case of expansion measures, a further increase in the weight of the battery can lead to structural problems that may favor replacement by lithium-ion batteries.
- The service life of high-quality lead–acid batteries for stationary applications is well over 10 years under good operating conditions and good maintenance and is comparable to the expected service life of many lithium-ion batteries, which have to be kept at a very high state of charge.
- Cost advantage
  Lithium-ion batteries are currently (2024) around 3–3.5 times more expensive than lead–acid batteries in terms of stored energy.
- For systems with a high energy content (e.g. 250 kWh, a standard industrial size), several parallel strings of lithium-ion batteries must be used because cells with sufficient capacity are not available. This increases complexity and costs.
- The life cycle costs are still significantly lower than those of lithium-ion batteries, despite the maintenance requirements, especially for flooded lead–acid battery designs.
- The reduced performance of lithium-ion batteries at very low temperatures does not play a role in most installation conditions.
- There are more safety concerns for lithium-ion batteries due to thermal runaway and thermal propagation risks. Regulations differ between different countries, see, for example, UL 9540 ESS Requirements – Evolving to Meet Industry and Regulatory Needs and a report of the US National Fire Protection Association, Hazard Assessment of Lithium-Ion Battery ESSs, which raise concerns about the installation of large ESSs in high-rise buildings.
- With very short bridging times, e.g. 5–15 minutes, only a small proportion of the total installed capacity of lead–acid batteries (Peukert coefficient approx. 1.15–1.2) can be used. For lithium-ion batteries, the removable capacity decreases much less, so that the cost disadvantage is smaller when very short bridging times are needed.

The following therefore always assumes the use of lead–acid batteries.

Both sealed and flooded batteries are used. Flooded batteries require toppling-up with water once a year but empirically are found to be more robust. "Pure lead" batteries that are slightly more expensive but have a higher service life are gaining market share.

There are very detailed test procedures for batteries to evaluate their suitability for use in emergency power supply systems. For sealed batteries, they are specified in IEC 60896-21:2004 and IEC 60896-22:2004 and include

- Gas emissions
- High current tolerance
- Short-circuit current and direct current (DC) internal resistance
- Content and durability of required markings
- Flammability rating of materials
- Discharge capacity
- Charge retention during storage
- Recharge behavior
- Impact of a stress temperature of 55 °C
- Abusive overdischarge
- Dimensional stability at elevated internal pressures and temperatures

This level of standardization is a sign of a very mature industry.

19.1.4 Sizing of Batteries

The maximum discharge current, the minimum voltage for supplying the loads, and the maximum voltage for charging the battery as well as the required bridging time in the event of a power failure are the key design parameters for the battery. Bridging times range between 5 and 15 minutes for telecommunications and computer systems, which can be shut down sufficiently quickly or continue to be supplied by starting a diesel generator, and several hours for certain types of buildings and special applications.

In most cases, the maximum and minimum voltage of the battery is specified by the application and the voltage range of the connected loads and chargers.

The battery is usually only designed on the basis of the required bridging time in the event of a power failure and the maximum load to be expected. To compensate for the loss of capacity during battery aging, the capacity is often oversized by 25%. For installation conditions with low temperatures, the capacity available at the operating temperature must be used as the basis. Highly fluctuating loads are not to be expected in emergency lighting systems and UPS systems for telecommunications equipment and computers. A procedure for designing complex load profiles and nonstandardized system concepts is explained in section 19.3.

Special features only need to be taken into account when protecting loads that can feed back high power during operation. High-power recuperation of braking energy during a mains failure can increase the DC link voltage to such an extent that it can cause control systems and chargers to fail.

A frequent discussion concerns the use of several parallel battery strings instead of just one battery string. Important aspects are as follows:

- Availability and price of cells and blocks
  For lead–acid batteries, cells up to 3000 Ah are commercially available as single cells for stationary applications. For lithium-ion batteries, the largest available cells are in the range of approx. 250 Ah. For safety power supplies using lead–acid batteries, there is therefore no technical need to use parallel battery strings.[5] However, the price of cells depends to a large extent on the number of units produced, so that two cells or blocks may cost less than one cell or block of twice the capacity.
- Functional safety
  The use of parallel strings is said to be more reliable because a high-impedance fault in one string should not reduce the function of the overall system as much as with a battery string. However, the maximum power may no longer be available if one string fails and the bridging time is then significantly reduced. Strings connected in parallel require more installation, maintenance, and monitoring.[6]

In parallel-connected strings, the individual cells or blocks are often also connected in parallel. In normal operation, no currents flow through these connections, but if there are differences in the internal resistance of parallel cells or cell groups, the low-resistance part of the battery string takes over a higher percentage of the power than the higher-resistance part (see Figure 19.2). The higher load on the lower-resistance cell can therefore increase aging processes. The higher the number of parallel strings, the lower the effect and the greater the functional reliability and the lower the risk of failure due to a defective cell. However, the risk increases due to the significantly larger number of cell connectors and additional components during assembly and fault detection using additional voltage measurements after $n/2$ cells becomes worse. The authors are not aware of any reports on the higher reliability of

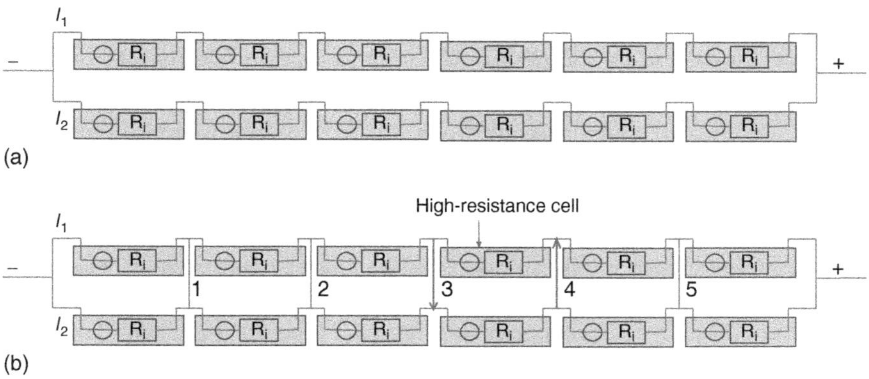

Figure 19.2 Schematic diagram of the options for connecting parallel battery strings: (a) pure parallel connection and (b) with additional connectors between cells or modules.

5 At 800 V nominal voltage and 3000 Ah cell capacity, the battery has an energy content of 2400 kWh.

6 A general engineering principle is using as few parts as possible.

parallel strings. It seems though that large, redundant UPS systems with very high safety requirements are now often equipped with two parallel battery strings per UPS system. This is also suggested in TR 62060 to improve system behavior.

19.1.5 Operation

Except during tests and mains faults, batteries are only operated in trickle charge mode to keep the battery in a fully charged state or, in the case of lithium-ion batteries, in a high target state of charge.

Due to the availability of side reactions in the charging and discharging direction, which cause a high current carrying capacity and comparatively low damage,[7] it is possible to charge series-connected cells or blocks of a lead–acid battery with a constant charging voltage without a battery management system (BMS), even if several hundred cells are used for very high DC link voltages. The charging voltage is specified by the battery manufacturer so that a sufficiently high charging current flows through all cells to compensate for self-discharge and minimize corrosion. In addition, a temperature-dependent charging voltage is often required for batteries that have a fluctuating temperature environment due to their installation conditions. A very similar temperature of all cells or blocks is always necessary. TR 62060 [2] specifies the charging currents that can be expected with the selected charging voltage.

To ensure the functional safety of lead–acid batteries, a short phase of higher charging voltage is regularly recommended. The aim is to fully charge all electrodes and eliminate sulfatation. These phases are referred to as equalization charging if individual cell voltages become too low or deviate too far from the average value.

Depending on the battery type, the float charge current of a new battery is in the range of 10–60 mA per 100 Ah battery capacity. At the same time, with a short bridging time (e.g. 10 minutes for UPS applications), discharge currents in the range of 300 A per 100 Ah can be expected. It is technically extremely difficult to measure the trickle charge current precisely. Measuring shunts for very low currents are readily available, but then cause an unacceptably high-voltage drop at very high discharge currents and generate heat. Current clamps and the like, on the other hand, are inaccurate at such low currents and can rarely be overloaded with high discharge currents.

An important aspect is the current ripple of the float charge current generated by the charger and the connected loads. DIN EN 60896-11 [3] requires that the effective value of the alternating current component does not exceed 5 A per 100 Ah battery capacity, while IEE 1188 [4] requires that the heat generated by the alternating current should be less than 3 K. Alternating current with very high amplitude during trickle charging mode quickly damage the battery.

---

7 High overcharging currents or charging voltages do not immediately lead to a loss of capacity and an increase in internal resistance. A significant and rapid loss of performance is only to be expected with frequent, permanently occurring, and very high voltages.

After discharging, the aim is to recharge the battery quickly. To achieve this, an IUoU characteristic curve is usually used. The charging voltage in the first constant voltage charging phase is in the range of 2.35–2.4 V, sometimes with a current specified to be in the range of 20 A/100 Ah, but sometimes without the specification of a current limit. At the end of this phase, the battery is largely fully charged. Full charge is then achieved at a charging voltage that is optimal for the service life in the range from 2.23 V (flooded batteries) to approx. 2.3 V (sealed batteries).

Expected and recommended float charge currents and voltages are specified in DIN 50272 and TR 62060.

19.1.6 Monitoring the Battery

The determination of the expected remaining runtime in the event of a power failure is standard in most stationary battery systems. The remaining runtime is determined by measuring the discharge current and comparing discharge curves of the new battery at different current amplitudes with the actual discharge curve, assuming a constant load until the final discharge voltage is reached. An analysis of the data also allows one to assess the battery capacity and its state of health. Figure 19.3 illustrates the basic procedure used in commercial UPS systems, whereby fluctuating discharge currents can often also be evaluated. The further

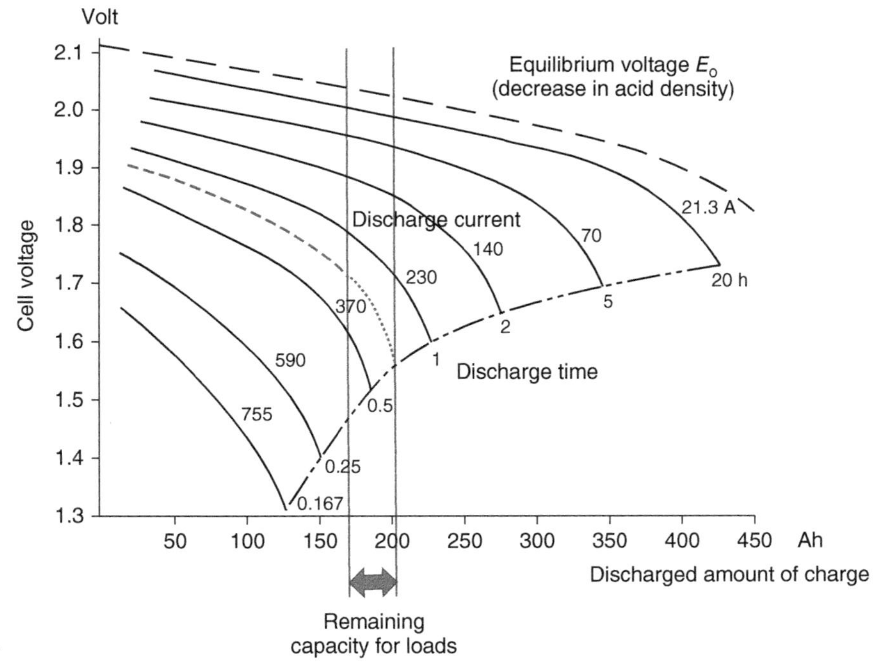

Figure 19.3 Schematic diagram for determining the remaining runtime of a battery. The characteristic discharge curves of the battery must be known from measurements of new batteries. When discharging with a constant current, the voltage curve during further discharging and the remaining runtime can be estimated.

course of the measured voltage is extrapolated up to the end of discharge voltage and then the deviation from the expected voltage value for the measured current amplitude is calculated. This deviation is then proportional to the deviation of the capacity from the capacity that existed for the new battery during the measurements for the discharge characteristics.

It must also be possible to monitor the battery in trickle charge mode without discharging the battery. The results of monitoring by service personnel and monitoring systems that continuously or automatically record battery parameters at short intervals (daily, weekly) must be sufficiently meaningful to ensure that the battery can continue to function until new measurement data is available or the next discharge.

General guidelines are contained in DIN EN 50272, ZVEI data sheets, and manufacturers' instructions for use, but usually without further details and instructions for carrying out and evaluating the measurement data. TR 62060, on the other hand, describes in detail how a battery can be monitored so that it can deliver the expected power for the required bridging time in the event of a mains failure. The concept is based on electrically detectable measured values of individual cells or blocks, and the expected changes of which are assessed during aging until the end of the service life or in the event of faults. The measurement of acid density is not mentioned as a measurement method.[8] The proposed procedure contains precise specifications, flow charts, and information on the expected measured values and their interpretation. The steps are as follows:

1) Temperature measurement of the battery room (to evaluate the difference between cell temperature and ambient temperature) and measurement of the cell temperatures,
2) Measurement of the float charge current,
3) Measurement of the individual voltages of all cells or blocks,
4) Measurement of the internal resistance of all cells or blocks, and
5) Measurement of discharge characteristics.

19.1.6.1 Float Charge Current

The float charge current is in the range of approx. 10–60 mA/100 Ah when new, depending on the battery type. With normal aging, the value increases by a factor of five to six. Measurement is difficult with these low currents when very high discharge currents will also exist. The authors are not aware of any battery system in which the float charge current is measured with sufficient accuracy and resolution and evaluated when the battery is serviced. This is very regrettable because a decrease in the float charge current in sealed batteries is a clear warning signal of future failure, see Figure 19.4.

8 The reason for not measuring acid density is the focus on measuring methods using electrical parameters that can easily be automated. Measuring acid density is a time-consuming task, even with high-quality measuring devices, and comparatively low additional diagnostic power for batteries that are fully charged most of the time.

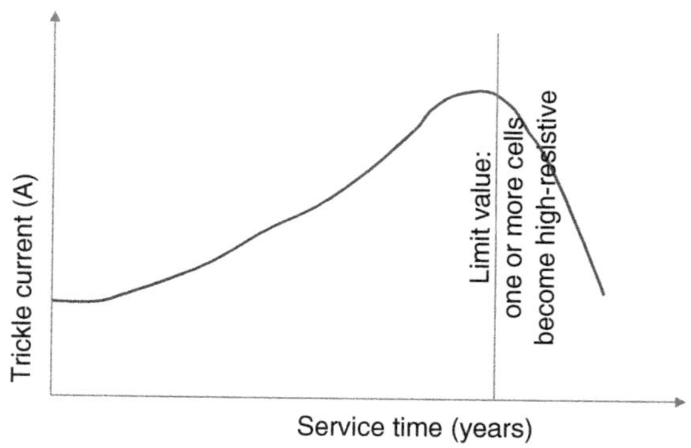

Figure 19.4 Schematic diagram of the float charge current over the operating time in accordance with TR 62060. A decrease shows that at least one cell has an increased resistance.

19.1.6.2 Internal Resistance Measurement

Frequently, the measurement of individual cell voltages is combined with the measurement of the individual internal resistances in one step. However, internal resistance measuring devices draw the excitation energy from the measured cell, which is discharged as a result. During the measurement of all cells in a string, the float current therefore increases and the voltage measurement of the first and last cell takes place under different conditions. In practice, the difference may be small and be neglected.

IEEE 1188 [4] proposes that only the increase in internal resistance should be evaluated for sealed batteries, not the voltages in float charge mode.

Figure 19.5 in accordance with TR 62060 schematically shows the relationship between the internal resistance and capacitance of cells. The relationship has been confirmed in many publications.

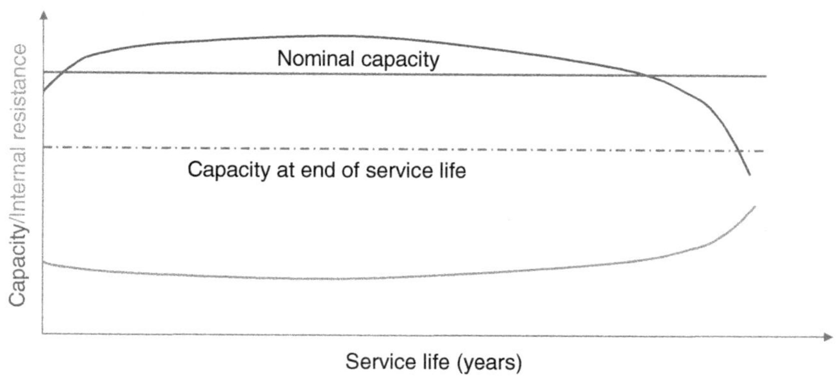

Figure 19.5 Diagram showing schematically the internal resistance and capacity of cells during their service life according to TR 62060, according to figures 10 and 11 in the technical report TR 62060 of the Ínternation Electrochecnical Commission (IEC).

19.1.6.3 Discharge Characteristics

The term discharge characteristics in a technical report by the IEC (TR 62060) refers to a discharge that is ideally carried out under constant conditions (current, resistance, or power) or using the connected loads with, possibly, fluctuating current. Discharging up to the final discharge voltage is not necessary. The test according to the proposed procedure is therefore often not a regular standard capacity test of the battery, which must be carried out with constant current up to the final discharge voltage.

The determination of the discharge characteristics of the battery is intended as the final test, which provides the most meaningful information about the condition of the battery. However, this requires the measurement of the individual voltages during the test, the measurement of the current or, in the case of parallel strings, the measurement of the respective string currents, compensation of the temperature, and comparison of the data with previous tests. To avoid risks from a discharged battery in the event of a power failure immediately after the test, the test must be carried out at times with the lowest possible risk to people, for example, in shopping centers outside opening hours.

In addition to the requirements of TR 62060, it is advisable to install a voltage measurement point after $n/2$ cells per string, to read out and archive the fault log of the system, and to scan the battery for hot spots (terminals, connectors, and housing sides) with an infrared camera at the end of discharge tests, if they are part of the maintenance procedure. Using the maximum design currents has the advantage of shortening the duration of a discharge test and facilitating fault detection.

For emergency lighting, a discharge test is mandatory once a year. For cost reasons, the test is usually carried out with the connected loads and only over the required bridging time. This is not a capacity test but rather the determination of a discharge characteristic value in accordance with TR 62060 when measuring current and individual voltages. For small battery systems for burglary, fire, and emergency lighting systems, the only requirement is that a short load test is carried out automatically at regular intervals [5] and that the battery must be replaced after four years [6]. Maintenance and detailed inspections do not have to be carried out in the preceding years. In many applications, however, no tests are prescribed and the operators are therefore not obliged to discharge the battery or carry out any other checks on the battery systems or converters.

Figure 19.6 shows the voltage curve of a UPS system with a nominal voltage of 216 V, whose voltage fell below the lower voltage limit of the loads after approx. 65 ms. By dividing the string into three subgroups, the two defective blocks could be easily identified. The battery system did not have a voltage measurement point after $n/2$ cells, so that a slow increase in internal resistance could not be detected in time before the system failed. Even very short mains power failures would have led to voltage differences between parts of the string.

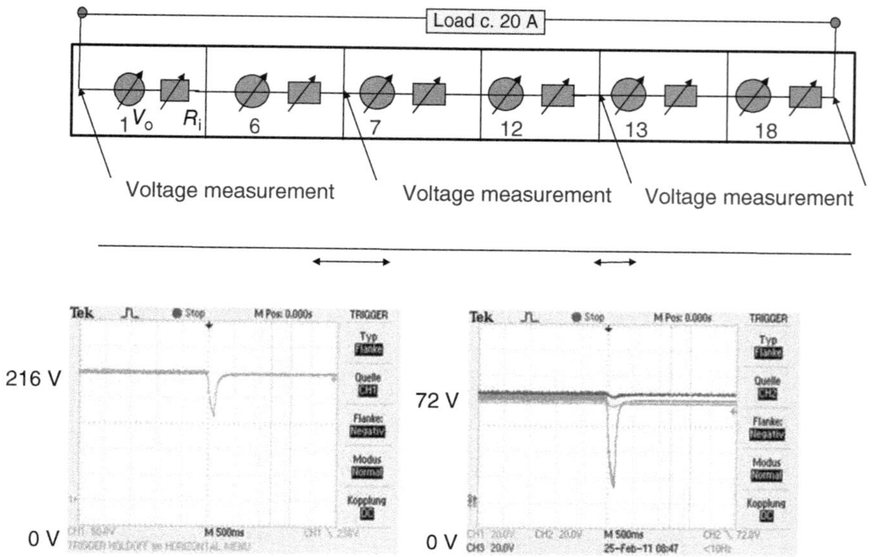

Figure 19.6 Voltage curve of a battery with a nominal voltage of 216 V (18 blocks with 12 V) and identification of the defective blocks by dividing the string into three parts, each with a nominal voltage of 72 V.

Figure 19.7 also shows the discharge curves of all other 16 blocks. It can be seen that a short load test of the string could not have detected the failure of two further blocks. However, if the procedure described in TR 62060 had been applied in full, these two blocks would have been identified by internal resistance measurement.

Batteries for particularly sensitive and valuable loads are increasingly monitored using online systems. The voltage of each cell or block is measured continuously and deviations are reported. Sometimes the internal resistance is measured additionally and recorded at regular intervals, for example, once a week.

Depending on the conditions of use and installation, adjustments to the proposed procedure often have to be made and restrictions observed, for example,

- Due to installation restrictions, it is not possible or not permitted for safety reasons to measure the individual voltages of cells or blocks of the battery in trickle charging mode. Sometimes it is necessary to disconnect the battery from the charger before measurement.
- For reasons of time or work organization, it is not possible to carry out measurements on individual cells or blocks before a battery discharge. Measurements of individual cell voltages after a discharge, irrespective of the amount of energy drawn, followed by a short pause to prepare for the measurement, are very difficult to interpret.

Careful planning and detailed work instructions are always necessary to ensure that measurements over the whole service life are comparable and the results can be easily interpreted.

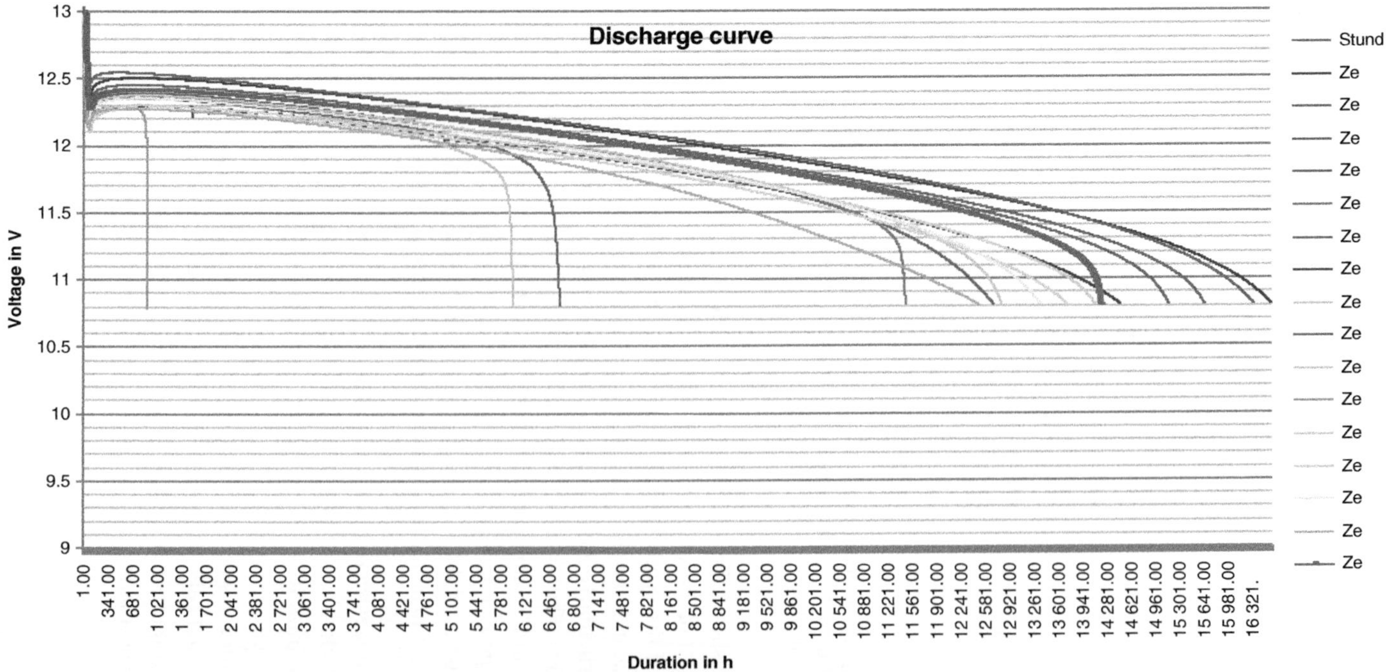

Figure 19.7 Discharge curves of sixteen 12 V blocks from a UPS system that were discharged individually. Two further blocks of the 18 block string had such a high impedance that no capacity test was possible. The voltage curves show that the behavior at the beginning of the discharge has little significance as regards the actually available capacity.

19.1.7 Other Information

The concept of "energy efficiency" is problematic for these applications because, depending on the details of the application, no energy is drawn from the battery for very long periods of time. The ratio of energy withdrawn to energy charged is therefore very small. It is better to speak of standby losses. The standby losses of the batteries (for lead–acid batteries dominated by the trickle charge mode and for lithium-ion batteries by the power consumption of the BMS) are, however, very small in relation to the standby losses of the inverters and, in the case of larger systems, their cooling and ventilation requirements.

19.2 Diesel Start for Emergency Power Supply Systems

If, in the event of a power failure

- very high power is required for a longer period of time (range of hours) and/or
- particularly high safety requirements exist,

emergency power supply systems are equipped with a diesel generator,[9] in many cases in addition to stationary battery systems[10] for UPS until the diesel generator can take over.

In the event of a power failure, the diesel generators are usually started with batteries[11] and, after the end of the starting process, supply the connected loads continuously until the mains power is restored or the fuel is completely used up. The batteries for the starting process are kept fully charged at all times and are only discharged for a few seconds at high power during diesel startup. As soon as the diesel generator has taken over the secured loads, the batteries are recharged via the chargers connected to the secured grid. The nominal voltage of the batteries is usually 12, 24, or 48 V, depending on the power and design of the starter motor.

19.2.1 Requirements

The only requirement for the battery is a very reliable supply of high starting currents for starting combustion engines (see chapter 16) at moderate lifetime costs. In contrast to starter batteries for vehicles, however, the batteries used for emergency power systems do not have to supply any other loads – there is no "on-board power supply system." The starting process of a diesel engine that starts easily (good maintenance condition and preheating if necessary) only takes a few seconds and, depending

9 Diesel engines are cheaper and more robust than gasoline and gas engines, and fuel cells are not yet commercially available.

10 For UPS systems in the range of 1 MW or more, flywheels compete with batteries. Their energy content is much small; however, the cost of providing high power for a few seconds only during the start of the diesel generator is much lower.

11 Alternatives for special requirements are, for example, compressed air or hard mechanical coupling to rotating units, such as flywheels.

on the size of the diesel engine, a charge of 0.5–2 Ah is required per starting process. DIN 6280-13 1994-12 stipulates that the battery must be able to carry out three diesel starts with 10 seconds duration and 5 seconds rest in between. In principle, a battery with a capacity of approx. 10 Ah including reserve for aging would therefore be sufficient. However, to provide the high currents of well over 1000 A in the first phase of the starting process for large diesel engines, the internal resistance must be in the range of 1–0.5 mΩ/V or 2–1 mΩ per cell. Such low internal resistances are only achieved by batteries with a significantly higher capacity. It should be noted that excessively high starter currents due to the use of an oversized battery with too low an internal resistance can damage the magnets used in the starter motor. The internal resistance of the battery and the starter motor must therefore be matched.

Depending on legal regulations or internal company specifications, the diesel engine is started regularly. In some cases, a test is carried out every month. Nevertheless, the charge throughput of a battery per year remains low and will rarely exceed the nominal capacity of the battery over its entire service life. The batteries must therefore be designed for a long service life in float charge mode and a permanently low internal resistance, not for cycle stability.

In most cases, the batteries are at room temperature. Very low and very high temperatures are the exception and require a temperature-compensated charging voltage. For special operating conditions and locations, as with stationary battery systems, further requirements such as earthquake resistance may exist.

19.2.2 Choice of Battery Technology

Double-layer capacitors are suitable for starting the engine in principle. However, their energy is not sufficient for several starting processes and possibly additional energy consumption of the starting system (engine electronics and preheating). The technical advantages of lithium-ion batteries, such as cycle stability, low weight, and volume, do not play a role in starting diesel engines for power generation; the disadvantage of lower power output at low temperatures is irrelevant due to the usual temperatures at the place of use.

The costs of both lithium-ion batteries and double-layer capacitors are very high compared to lead–acid batteries. For this reason, but also because they meet all requirements well, lead–acid batteries have been used almost exclusively up to now.

Technically, conventional starter batteries for trucks can be used to provide power for starting. However, these are not designed for several years of trickle charging and therefore have a comparatively short service life in continuous charging mode. Corrosion attacks on the positive grids, intercell connectors, and group bars pose the risk of interrupting current flow, so that the engine can no longer be started, as well as the risk of explosions if the current is interrupted at high discharge currents during diesel start. Although the number of explosions may be low overall, it is an application problem when using conventional starter batteries. The damage is considerable, as splashing sulfuric acid may require the replacement of entire system components. Due to the low cost[12] of starter batteries for trucks, they continue to be used where permissible. Replacement of truck starter batteries after two years at the latest is recommended.

More suitable than starter batteries for trucks are stationary batteries that have thicker grids, group bars, and terminals and are designed for continuous charging operation. The use of starter batteries for trucks is not permitted for systems in accordance with DIN 6280. Because the internal resistance of stationary batteries is higher than that of starter batteries for trucks with the same capacity, it is necessary to choose stationary batteries with a higher capacity. The battery costs can more than double compared to conventional starter batteries due to the combination of significantly higher capacity coupled with the more complex design. However, the significantly longer service life of stationary batteries suggests that the life cycle costs of stationary batteries are lower despite the higher investment costs.

19.2.3 Maintenance and Fault Diagnosis

Flooded batteries must be topped up with water as part of regular maintenance. To avoid refilling with water, sealed batteries with tubular plates or pasted plates are now increasingly being used as standard.

When using lead–acid batteries, ventilation regulations must be observed.

The specifications for monitoring stationary batteries in accordance with TR 62060 are suitable for monitoring batteries for diesel starting. Products that evaluate the starting capability of the battery from the voltage and current curve of the starting process during tests are not commercially available. The reasons for this may lie in the dependence of the current and voltage curve on various properties of engines.

19.3 Batteries for Balancing Electricity Demand and Supply Over Time

19.3.1 Application Groups

Three areas of application, namely

(a) off-grid, autonomous energy supply systems,
(b) domestic energy storage, and
(c) large stationary energy storage facilities of the electricity industry or industries with very high electricity consumption

are discussed together because the operating conditions of the batteries are similar. In all three types of applications, the batteries are used to balance electricity supply and demand over timecaused by the fluctuating supply of electricity from renewable energy sources or strongly fluctuating electricity prices and/or the cost of using the grid. In all three types of applications, the state of charge of the battery cannot be

12 The high-cost sensitivity that leads to the use of conventional starter batteries also illustrates the difficulty for lithium-ion batteries to enter the market for this area of application.

Table 19.1 Overview of similarities and differences between batteries for off-grid autonomous energy systems, home ESSs, and large stationary ESSs in the electricity industry.

	Off-grid autonomous energy systems	**Home energy storage**	**Large stationary energy storage systems in the electricity industry**
Requirements	Long bridging time, depending on the operating and investment costs of a generator; depending on the required degree of autonomy, long cycle times, possibly lasting several days, can be expected	Daily cycles, the mains connection protects the battery from deep discharges	Predominantly short and only short-term predictable cycles of low discharge depth but high power
Criteria for charging and discharging	Electricity supply from renewable energy systems, load management options, costs of electricity generation through additional diesel generator if necessary	Storage of electricity when there is a local surplus and discharging when no electricity from renewable energy systems is available	Fluctuations in the electricity price over time; supply despite grid restrictions
Reasons for the investment	Cost-effectiveness of the power supply due to high costs for a grid connection or high costs for an exclusive supply with diesel generators	Maximizing the degree of self-sufficiency of the power supply and minimizing electricity procurement costs	Minimization of energy procurement and grid connection costs
Legal framework	Possible local building regulations	Profitability depends on subsidies and legislation	Profitability depends on subsidies and legislation, in particular, the allocation of grid costs

planned or predicted over a longer period of time. Neither electricity demand nor the availability of electricity from renewable energy sources nor electricity prices are easy to forecast. Table 19.1 provides an overview.

In Ref. [7], the possible uses of batteries in grid-connected energy systems were carefully examined and the conclusion was drawn that strengthening the grid and the use of electricity for other energy applications, especially for the heating sector (power to heat), is a more economical and ecologically sensible use of local electricity surplus than storage in batteries.

The only market segment that is not heavily dependent on subsidies and government regulations is the market for autonomous, remote-area power supply systems.

19.3.2 Requirements

The development of the energy throughput over time cannot be predicted for these applications. In Ref. [8], the operating data of batteries in autonomous energy systems have been classified into four categories with significantly different operating conditions.

- Cat. 1: The state of charge is very high from spring to fall, with a full charge during the day and a shallow discharge cycle at night. In winter, a very long and deep discharge takes place. Examples of this are parking ticket machines, measuring stations, etc.
- Cat. 2 and 3 are characterized by continuous cycling in a partially discharged state. A defined full charge is not possible (cat. 2) or only possible if a diesel generator is used (cat. 3).
- Cat. 4: The battery is cycled daily and is also frequently fully charged.

Figure 19.8 shows the typical annual state of charge curve. At the same time, it can be assumed that batteries for this application are often exposed to the outside temperature and can therefore become very cold in winter and very warm in summer.

There is no comparable published measurement data on the use of batteries for domestic energy supply or large stationary ESSs in the electricity industry. Depending on the load profile, whereby average load profiles may not be used because the differences between different days and seasons cannot be neglected, the batteries are

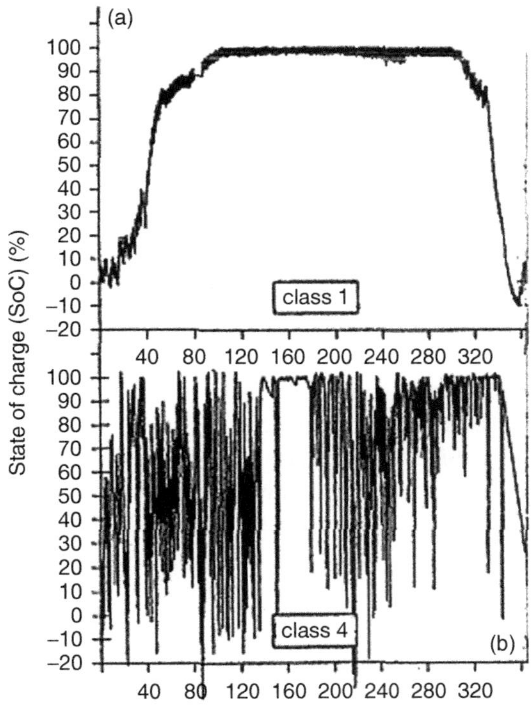

Figure 19.8 State of charge curves over 1 year for batteries in off-grid autonomous energy supply systems for the four categories defined by [8]. (a) Cat. 1 parking ticket machines; (b) Cat. 4 energy system with photovoltaic system, wind power, and diesel generator with largely daily full cycle. Source: Reproduced with permission from Ref. [8]/Elsevier.

cycled more or less deeply. A full charge of the battery can be achieved due to the availability of a grid connection or an auxiliary power unit such as a diesel generator. Detailed analyses and simulations are necessary so that the battery can be chosen and sized according to the expected load profile. There are high cyclic loads, frequent short partial cycles, rarely deep cycles, and rarely the possibility of fully charging the battery. The choice of the most suitable battery depends to a large extent on the type of operating conditions.

19.3.3 Choice of Battery Technology

The use of batteries for off-grid, autonomous energy supply systems has been established for several decades, long before lithium-ion batteries with sufficiently high energy content became commercially available. Lead–acid batteries were therefore the only technically and economically possible choice. Home energy storage and large stationary battery storage systems in the electricity industry, on the other hand, are relatively new concepts that emerged parallel to the availability of lithium-ion batteries.

Lithium-ion batteries have significant advantages and, apart from the higher investment costs, few disadvantages. Relevant points are as follows:

- Cycling in a partially discharged state
  In contrast to lithium-ion batteries, prolonged operation without a full charge is a critical operating condition for lead–acid batteries. When using lead–acid batteries, a full charge must be carried out regularly, for example, every four weeks at the latest, for reasons of service life (sulphatation and formation of acid stratification in flooded batteries). For off-grid systems, this is only possible for systems with diesel generators or renewable energy systems capable of generating a lot of energy compared to the energy demand and energy content of the battery, possibly coupled with an energy management system that allows load shedding. At times, when a lead–acid battery is charged fully, it is only available to supply the loads to a limited extent.
- High cycle stability and high charge throughput
  The charge throughput depends very much on the details of the application. Lead–acid batteries of the right technology (not starter batteries, but batteries designed for cyclic applications, possibly sealed batteries to reduce acid stratification) also allow a high energy throughput but normally lower than that which may be achieved when using lithium-ion batteries.
- High charging efficiency
  The higher internal resistance of lead–acid batteries leads to a greater voltage difference between the discharging and charging voltage and side reactions at high charging voltage reduce the charging efficiency.
- In contrast, operation at very low voltages and states of charge, sometimes even with a negative state of charge,[13] is critical for lithium-ion batteries and must

13 A negative state of charge occurs when the amount of charge removed exceeds the nominal capacity. This can occur at very low discharge currents without causing safety problems in lead–acid batteries.

be prevented by switching off loads. Although such measures make sense for lead–acid batteries, they are not relevant to safety. A very deep discharge, if it occurs only rarely, can therefore be tolerated more easily, see the state of charge curve of Cat. 1 in Figure 19.8.

- Investment costs

  The investment costs for lithium-ion batteries are significantly higher than for lead–acid batteries, so the authors expect that lead–acid batteries will continue to be used for some applications, in particular for autonomous off-grid installations. Home ESSs on the other hand now seem to be dominated by lithium-ion batteries. As weight and volume considerations are of little relevance in these applications, lithium iron phosphate (LFP) cells have an advantage over cells with nickel manganese cobalt (NMC) or nickel cobalt aluminium (NCA) chemistries.

Sauer [8] proposed an allocation of various lead–acid battery technologies to the four categories shown in Figure 19.8.

19.3.4 Sizing of Batteries

The sizing of a battery (see Task 19.1) for the three applications discussed here is complicated. Detailed analyses of the power flow through the battery are necessary for determining their required energy content. Software programs for calculations are available as freeware, for example, Homer. The battery size ultimately depends on the desired time of self-sufficiency or bridging time, in which there is no or too little power supply from renewable energy systems or, in the case of large stationary ESSs, the power supply from the grid is restricted for cost reasons. The ratio of the battery's energy content to the maximum current through the battery and the amount of energy available during a relevant time period are the most important criteria. The specification of the bridging time and the load profile using load management systems has a major impact on costs. The cost of charging by a diesel generator for autonomous systems without a grid connection or the availability and the cost of the external power connection for grid-connected systems are the second major design parameters. It is not advisable to use standard load profiles or time-averaged load profiles and standard feed-in profiles except for home ESSs, which are virtually always grid connected. Sizing of the battery is carried out in the following steps:

1) Determination of the maximum and minimum permissible DC link voltage of the loads or inverters.
2) From this and the maximum required charging voltage (for lead–acid batteries including phases of boost charging) and the minimum permissible cell voltage, it is possible to determine the number of cells or the nominal voltage. It is advisable to use the maximum possible number of cells, as this opens up reserves at the end of the discharge in the event of cell faults. These first two steps are not necessary when using standard voltages.

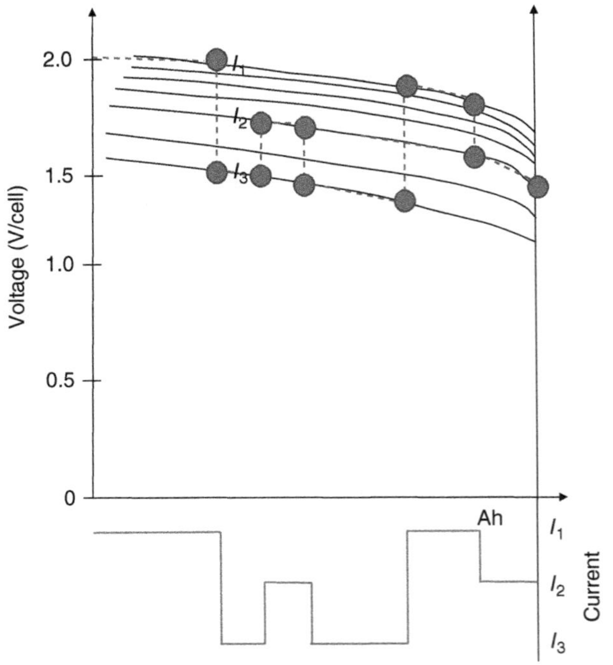

Figure 19.9 Schematic diagram showing the voltage curve of batteries with strongly fluctuating load currents.

3) Conversion of the load profile that the battery is to cover into a state of charge and voltage curve of the battery.
   The required battery capacity is particularly difficult to determine when very high power requirements have to be met at the end of discharging while at the same time complying with voltage limits. Some battery manufacturers provide tools that can be used for this purpose.
   A manual sizing procedure for fluctuating loads (Figure 19.9) is based on the discharge curves of the battery, which are based on the percentage of capacity removed and not time. The power curve can be plotted on these curves and thus simultaneously provides an estimate of the voltage curve. The battery capacity and the number of cells must be selected in such a way that a sufficiently large distance to the required minimum voltage is maintained at all times.
4) Adjusting the capacity to nominal capacity data of the cells to the expected discharge currents and ambient temperature, particularly when low temperatures may exist. Normally, the calculated value is increased by 25% so that the full bridging time is also available at the end of the service life (usual definition: capacity has fallen to 80% of the nominal value).
   If the battery must be able to supply sufficiently high currents to trigger fuses in the event of a fault (in some cases a mandatory requirement for battery-supported central power supply systems for hospitals), then further consideration must be given to the short-circuit power over the entire range

of permissible temperature and state of charge. The short-circuit currents of the battery can be obtained from good data sheets. The voltage drops during a short circuit in the energy system suddenly for a very short time and possibly significantly below the permissible end of discharge voltage of the cells. It must be checked whether these short-circuit currents can be controlled by an inverter that may be installed between the battery and the loads, and if so, how high the line resistance between the battery and the fuse is.

5) Calculation of ventilation requirements (lead–acid batteries).
6) Calculation of the climate control requirements (maximum and minimum temperature in the battery room or in the area surrounding the battery, particularly important for lithium-ion batteries).
7) Ensuring structural requirements (fire protection, floor conditions, space required for installation, maintenance, fastening of power cables due to high forces as a result of high inrush currents, etc.)

There are mandatory technical guidelines for some of the design points shown here.

19.3.5 Operating Strategy

Despite the difficulties of predicting generation and load curve, the operating strategy has to

- ensure that the battery remains within its operating limits, which is important for a long service life,
- minimize the need to switch off loads in off-grid, autonomous energy systems or minimize the costs of operating a diesel generator,
- minimize the amount of electricity drawn from the grid for home storage systems, or
- enable the electricity industry to comply with grid restrictions for large stationary batteries and minimize the costs of external electricity procurement.

To optimize the operating strategy and determine the target state of charge depending on the external conditions and load forecasts, sometimes very complex programs have to be used that include the entire energy system into which the battery storage system is integrated.

19.3.6 Monitoring

Determining the state of charge is particularly important for operation. Algorithms with which this can be done with sufficient accuracy are commercially available.

At the same time, the state of health and internal resistance must be estimated. As the battery systems in most cases have frequent and sufficiently deep discharge cycles, this is easily possible using a model-based procedure and comparison of expected and measured values.

19.4 Batteries for Stabilizing the Energy Supply System

19.4.1 Examples of Large Battery Storage Systems in the World and Evaluation

The stability of the public power supply system is based on compliance with narrow voltage and frequency bands. Deviations between the generation and consumption of active and reactive power lead to voltage and frequency changes and, if the deviations from the target values are too large, to grid failure, more or less widespread blackouts. In large interconnected systems, system stability was and is ensured by rapidly increasing and decreasing the output power of power plants and using the rotational energy stored in the rotating masses of the generators and turbines.

Several large battery storage systems have been built around the world in the past to support changes in the output of power plants in small energy systems and improve the efficiency and stability of the electricity supply. The most important of these are as follows:

1) Bewag energy storage system 17 MW/17 MWh@1 hour:
   Used for stabilizing frequency and optimization of power plant scheduling in West Berlin until 1989. West Berlin's connection to the European grid after 1989 eliminated the need for local energy optimization.
   Flooded lead–acid batteries with copper-expanded metal grids, cooling, and electrolyte circulation were used.
2) Fairbanks Alaska 45 MW/15 MWh@20 minutes:
   The ESS installed there stabilizes the local power supply system. The long overland line from Anchorage is not sufficient to stabilize the power grid in conjunction with small local power plants. The number of power outages fell dramatically after the battery storage system was installed.
   The plant was commissioned in 2003.
   Alkaline nickel batteries are used.
3) BigBattery Lausitz 50 MW/53 MWh@1 hour:
   The BigBattery Lausitz was put into operation at the Schwarze Pumpe power plant site at the turn of 2020/2021. Its usable capacity is 53 MWh and it can supply up to 50 MW of primary control power. The size is 110×62 m, where 13 battery containers with a total of 8840 battery lithium-ion modules are used [9].

It is important to emphasize that the aim of these systems was not to take advantage of price differences when purchasing electricity (charging when prices are low and discharging when prices are high) or to limit the maximum power drawn from the grid but to increase the stability and reliability of the electricity supply by providing additional power instantaneously. This is in marked contrast to the battery storage systems described in section 19.3 for balancing the supply and demand of electricity over time.

There is a final report for the Bewag plant in which the performance curve, energy throughput, and economic effects were analyzed. From the report, it is particularly important to note that the investment was classified as cost-effective and that the lead–acid batteries used there had reached a capacity throughput of approx. 7000 times their nominal capacity by the time the plant was shut down. The battery had still not reached the end of its service life.

Public subsidies were used for the design and construction of the BEWAG and Fairbanks systems. However, the operators of the battery storage systems in Berlin and Fairbanks were or are of the opinion that the systems have demonstrated their economic viability.

By contrast, the battery storage systems described below are a consequence of the liberalization of the electricity market and new markets for providing primary and secondary control poweroutside the operation of conventional power plants. In the market for primary control power, power of at least 100 kW must be either supplied or absorbed within 10 seconds for a maximum of 15 minutes if the frequency in the grid deviations more than 20 mHz from its nominal value. In the market for secondary control power, the supply or consumption of at least 1 MW of power must happen within 15 minutes for 1 hour. In the future, a market for instantaneous reserve that immediately supplies or absorbs power in the event of frequency deviations, similar to rotating masses of turbines and generators in large power plants when the rotational speed changes, may be created.

The provision of balancing power by battery storage competes with the operational management of power plants, electrically operated heat storage systems for rapid consumption of large amounts of energy and power plants with highly dynamic generation and high efficiency at partial load (gas motors).

19.4.2 Requirements

Batteries for frequency stability (primary reserve power) and secondary reserve power for restoring energy balance must be able to both take-up and release energy.[14] The marketing of batteries for these purposes follows a complicated tendering process based on prequalification, in which the supplier's ability to provide the required, precisely specified power must be demonstrated. The requirements are different in different countries and in any case require large electrical outputs, for instance, at least 1 MW. Pooling of power from several locations and technologies is permitted in some countries.

Requirements that battery systems have to fulfill are specified by the transmission system operators, who are responsible for the stability of the electricity grid, and approved by regulatory authority. For example, when prequalifying systems for the provision of primary reserve power, it must be demonstrated that the maximum power can be provided twice in succession for 15 minutes each time. This means that an energy content of 1 MWh must be provided per megawatt of nominal output at a target state of charge of 50%.

14 In combination with other energy technology systems such as heat storage systems, which can only absorb energy, it may be possible in the future to modify these requirements so that batteries only have to release energy.

Depending on the tendering conditions, the target state of charge of these batteries is either approx. 50%, so that energy can be both absorbed and released, or approx. 100% if only energy has to be released, or approx. 0% if energy only has to be absorbed. The market rules allow or entail phases in which the batteries cannot be used for ancillary services. These phases can be used for systems with lead–acid batteries in order to be able to carry out a full charge.

The batteries are usually located in a container that is insulated against the outside temperature and may also be cooled or heated so that extreme temperature conditions are not to be expected.

The capacity throughput of the batteries is low. Estimates show that the capacity for primary reserve power will only be used approximately 10 times a year and even less when used for secondary reserve power. According to the current technical guidelines, the maximum currents in both the charging and discharging directions correspond to the 1-hour current.

19.4.3 Choice of Battery Technology

Lithium-ion batteries can have a considerable technical advantage over lead–acid batteries due to the 50% target state of charge, which is required under some market rules. However, in conjunction with other energy technology systems (heat generators and heat storage systems, gas engines, combined heat and power plants, etc.), the possibility of regularly not participating in tenders, and the significantly lower investment costs, lead–acid batteries can still be an option.

19.4.4 Other Aspects

For management and monitoring, essentially the same points apply as for battery systems for applications where the battery balances electricity supply and demand (section 19.3).

References

1 DIN EN 50272-2:2001-12 (2001). *Safety Requirements for Batteries and Battery Systems – Part 2: Stationary Batteries.*

2 TR 62060:2001-09 (2001). *Secondary Cells and Batteries – Monitoring of Lead Acid Stationary Batteries.* User Guide.

3 DIN EN 60896-11:2003-07 (2003). *Stationary Lead-Acid Batteries – Part 11: Sealed Batteries; General Requirements and Test Methods (Incl. Corrigendum of 2006).*

4 IEEE 1188-2005 (2005). *Recommended Practice for Maintenance, Testing and Replacement of Valve-Regulated Lead-Acid (VRLA) Batteries for Stationary Applications.*

5 EN 54-2:2016-03 (2016). *Fire Detection and Fire Alarm Systems – Part 2: Fire Alarm Control Panels.*

6 VdS (German Association of Property Insurers) 2192 (1998). *Surge Protection.*

7 ESPEN (2016). *Energieforschungszentrum Niedersachsen: Potentials of Electrochemical Storage in Electrical Grids in Competition with Alternative Technologies and System Solutions (Original German title: Potenziale elektrochemischer Speicher in elektrischen Netzen in Konkurrenz zu alternativen Technologien und Systtemlösungen) (ESPEN, Final Report).* Goslar: EFZN.

8 Sauer, D.-U., Bächler, M., Bopp, G. et al. (1995). Which battery for which application. *Journal of Power Sources* 64: 197–201.

9 https://www.tagesspiegel.de/wissen/stromspeicherung-fuer-die-energiewende-die-riesenbatterie-von-schwarze-pumpe/26061888.html (last accessed 05.06.2022).

10 DIN 6280-13:1994-12 (1994). *Generating Sets – Reciprocating Internal Combustion Engine Driven Generating Sets – Part 13: For Emergency Power Supply in Hospitals and in Structures Designed to Accommodate Large Numbers of People.*

11 VDE 0510-485-5:2021-12 (2021). *Safety Requirements for Secondary Batteries and Battery Systems – Part 4: Safe Operation of Stationary Lithium-Ion Batteries.*

Task

Task 19.1

Design of a battery for a decentralized energy supply system

The Clausthal Energy Park, a research center at Clausthal University of Technology in Germany, has a high-quality (voltage and frequency stability, harmonics) stable power grid that is alternatively mains connected or fed from renewable and decentralized energy sources in off-grid mode. The battery of the overall system serves to stabilize the grid, as the control speed of dispatchable generation units (biogas engine, Stirling engine, rapeseed oil engine, etc.) is not sufficient to compensate for the load fluctuations of the institute and changes in output of the installed wind power and photovoltaic systems.

Set points for the electricity production of the dispatchable generation units are adjusted to the load requirement and the load forecast at intervals of a few minutes, taking into account minimum runtimes and minimum downtimes for each unit. The procedure is largely identical to that of a grid control center with power plant scheduling.

When fully charged, the battery should be able to guarantee the power supply for a period of 30 minutes in the event of a failure of all generation units and a failure of the public grid. When all generation units in the Clausthal Energy Park are in operation, an "unlimited" supply to the institute is possible, independent of the public energy supply.

(Continued)

Task 19.1 (Continued)

Data for the design:

Maximum average energy requirement per hour	**180 kWh, design value 200 kWh**
Minimum energy requirement per hour	15 kWh
Average energy requirement per hour	50 kWh
Peak power in the grid (start-up of motors, etc.) and design of the inverter	700 kVA
Short-term (a few minutes) maximum active power	250 kW
Probable deviation between the forecast value for load demand and the forecast value for renewable own generation	Max. 20% of the active power per forecast interval (10 minutes), on average 10%
Minimum voltage for inverters	350 V
Maximum voltage for inverters in operation	650 V
Maximum voltage for inverters without supplying the loads	700 V

Take battery data from manufacturers. The example solution is created with data sheet information from BAE for lead–acid batteries and Toshiba for lithium-ion batteries. Please assume a maximum charging voltage of 2.5 V/cell for lead–acid batteries for equalization charging and 4.2 V/cell for lithium-ion cells.

Questions:

1) What is the permissible range for the nominal battery voltage (number of cells in series) and what nominal voltage should be aimed for?
2) What battery size (Ah) is required depending on the nominal voltage?
3) How much energy must the battery be able to absorb or release at any given time?
4) What is a possible and what is a sensible target state of charge for the battery?
5) How big would the battery have to be if 100 kWh had to be available for 30 minutes at any time?
6) What energy throughput must be expected for the battery?
7) Which battery technology do you choose? How do you justify the higher price of the lithium-ion battery?
8) What is the average energy throughput and what thermal load can be expected during operation and, if lead–acid batteries are used, during equalization charging?

9) What bridging time can be safely expected from the battery if all generation units fail at the maximum average energy demand of 180 kWh?
10) What is the maximum current load of the battery for times at maximum power?
11) Is the short-circuit capacity of the battery sufficient to trigger fuses?
12) What thermal load on the battery (temperature increase of the cells) and heat development in the installation room can be expected?
13) What are the ventilation volume requirements for lead–acid batteries?

20

Batteries for Portable Applications

In addition to secondary batteries, there are many primary batteries used in portable applications, but these will not be discussed in this chapter. Today, lithium-ion batteries dominate this area, but there are numerous examples of other technologies (excluding lead–acid batteries due to their high specific weight).

20.1 Definition

This chapter deals with all rechargeable batteries used in portable applications in commercial and consumer markets: cell phones, tablets, laptops, radios, wearables, digital cameras, video cameras, toys, drones, power tools (e.g. cordless screwdrivers), electric toothbrushes, hearing aids,[1] electric razors, e-cigarettes, robotic vacuum cleaners, robotic lawnmowers, traffic and weather monitoring stations, etc. Special medical applications such as implants are not included because of very special regulatory requirements.

The need to develop suitable battery solutions for such portable applications has greatly stimulated battery development since the 1980s; the development of lithium-ion batteries is hardly conceivable without these applications.

Not considered here in detail are non-rechargeable so-called primary cells that are mostly based on manganese oxide/zinc but in some cases also on lithium technology predominantly with a lithium metal anode. It should be mentioned that there has been some progress in recent decades in achieving certain reversibility in alkaline manganese oxide/zinc systems. The corresponding cell reaction equation is as follows:

$$Zn + 2MnO_2 + 2H_2O \rightleftharpoons 2MnOOH + Zn(OH)_2 \; (U_N \text{ c.}1.55 - 1.65\,\text{V}). \tag{20.1}$$

1 Hearing aids are currently undergoing a substitution process from zinc–air primary cells to lithium-ion secondary cells, which is described as cost-neutral in terms of lifetime costs and reduces handling problems.

Battery Technology: Fundamentals of Battery Electrochemistry, Systems and Applications, First Edition. Alexander Börger and Heinz Wenzl.

Companion website: www.wiley.com/go/Börger/Battery_Technology

Table 20.1 Overview of mono (D), baby (C), mignon (AA), micro (AAA), and other battery formats.

Name	Designation	Length [mm]	Diameter [mm]
AAA	Micro	44.5	10.5
AA	Mignon	50.5	14.6
C	Baby	50.0	26.2
D	Mono	61.5	342

Source: Adapted from Ref. [3]/IEC 61951-2, 2017.

So far, however, the number of cycles that have been achieved is in the low double-digit range, which is why this technology will not be discussed further in this book.

Consumer batteries shape the common image of batteries, for example, through the dimensions defined in various standards, see Table 20.1 for typical formats for nickel–metal hydride batteries that are interchangeable with primary batteries. The standard sizes listed in Table 20.1 can only be used for cells that can safely be interchanged as they have similar cell voltages and electrical behavior. Cells of the same size but completely different properties which impact safety, for example, when being deep discharged or charged, are a danger to consumers and professional users alike. For this reason, lithium and lithium-ion batteries must have different dimensions. The "18,650" format, which is often used in consumer products and the automotive sector, is not interchangeable with any of the sizes given in Table 20.1. The 18 refers to 18 mm in diameter, and 65.0 mm in its length. Standardization for lithium-ion battery sizes and cell chemistries is still in progress. In addition to round cells, there are many different sizes and shapes that can only be used for certain products, that is, smartphones or notebooks with sufficiently large production volumes.

Overall, primary batteries account for around 20% of the total battery market, but are increasingly competing with rechargeable batteries in various fields of application. Nevertheless, there are large areas in which primary batteries are not only the cheapest, but also tend to be the most ecological solution due to good storage options and long-term use with environmentally friendly production (e.g. remote controls, smoke detectors, etc.). A huge market increase for batteries for portable applications however is not straightforward since, for example, smartphones make many other portable devices superfluous (e.g. pocket calculators, portable CD players, …).

20.2 Requirements for the Battery

The main requirements in these applications are high specific energy, high energy density, cost, and also cell safety in an environment where the use and handling of cells are outside the control of manufacturers. Compared to the applications described so far, batteries in portable applications have very low energy contents (only a few – if any – cells in series, low capacity per cell). The safety objective is then fulfilled by precautions in each individual cell (CID element, fuse, etc.) in conjunction with a BMS that is usually very simple compared to the applications mentioned so far. It would be risky and difficult to justify if no monitoring of any kind were provided for series connections of at least two cells.

Compared to electric traction and many other applications, the general requirements for batteries are significantly lower in terms of calendar life and number of cycles. Other requirements however depend greatly on the application and the user. Examples are:

- Batteries for radio communication in toys (walky talky) do not require high power levels at low temperatures, and their operation in cold climates is simply impossible. Batteries for the same application with the same voltage and energy content for emergency services and some types of commercial use must be able to function at rather low temperatures, for example, −20 °C.
- The ability to withstand very high temperatures without the risk of catastrophic failure is important. In contrast to all the applications discussed above, controlling the maximum temperature that the device and battery are subjected to is difficult or impossible, particularly in consumer products.
- Safe storage conditions as the path between manufacturer and end-user is difficult to control and may be lengthy.
- Users must be prevented from abusing cells. In contrast to all other applications where it can be assumed that the correct charger is used (in terms of voltage and power), unauthorized use of chargers is conceivable in the area of portable batteries and suitable precautions must be taken against this. For example, when rechargeable NiMH batteries were introduced for children's toys, there was a discussion about zinc–brownstone primary cells being inserted into the chargers and how measures could be taken to prevent this. The range of requirements that have to be met is therefore not only application-specific but depends also on the type of user. A detailed discussion is not possible in the context of this book. We would like to remind the reader of the importance of understanding the application and operating conditions of a battery.

Service life expectations are significantly reduced: Hardly any cell phones have been used for more than 10 years, and even if they are, replacing the battery after a few years should by no means be considered a disaster. Manufacturer warranties beyond the legal requirements are usually limited to a maximum of two years, which often corresponds to the useful life of the device.

An interesting aspect concerning portable batteries is included in the new battery regulation by the European Union (EU Battery 2023/1542 Regulation). From February 2027 onwards, it must be possible for end users to easily remove and replace batteries in portable devices. Obviously, this makes it necessary to design safeguards so that end-users can only replace a used battery with a battery of the correct type.

20.3 Choice of Battery Technology

The basic decision to use a rechargeable battery system essentially depends on the required power and energy content, which makes secondary cells indispensable for all devices that consume the energy of the battery within a short time, while primary batteries dominate in radio remote controls, conventional wristwatches and many toys, as the required currents are often only in the range of a few μA (a capacity of 1 Ah will provide 10 years of operation at 1 μA) or the device is only used intermittently with long rest periods after each use. Once this basic decision has been made, the next step is to select the appropriate secondary battery technology.

In the early days of consumer electronics, nickel–cadmium batteries, then nickel–metal hydride batteries dominated until lithium-ion batteries with their weight advantage began to compete. Lead–acid batteries never had any serious significance in this area. For a transitional period, nickel–cadmium batteries were able to have a significant market share in the power tool segment because no suitably powerful alternative battery technology was available. However, technological advances in lithium-ion batteries have made them obsolete there too.

Even though lithium-ion batteries now dominate this segment, there are always certain (not so small) niche markets where alkaline nickel batteries or other technologies can be a cost-effective alternative (e.g. due to their low-temperature properties, lower safety risks, etc.). This is partly due to the considerable cost disadvantages of lithium-ion batteries, which can be up to 50% or higher for the same energy content of the cell, especially if cells with high-priced and production-intensive nickel-manganese-cobalt (NMC) or nickel-cobalt aluminium (NCA) chemistry are chosen.

Due to lower raw material prices and simpler production, lithium iron phosphate (LFP) cathodes are of significant importance in this area. Electric traction batteries need comparatively expensive raw materials such as cobalt so competition between LFP and NMC or NCA is likely to intensify in the future. In the consumer electronics sector, it is comparatively easier to switch from NMC (or NCA) to the alternative LFP. Titanate is hardly ever used as an anode material due to its significantly lower specific energy; its use is only conceivable for special requirements with very high demands on low-temperature properties.

While prismatic cells are frequently found in the field of electromobility, either round cells due to their ease of production or pouch cells, due to their optimum use of the available installation space dominate in portable applications.

20.4 Design and Operation

When selecting and designing batteries for portable applications, space requirements and weight play a major role in addition to the desired device runtime, power output, service life, and cycling requirements (including charging times) that are considered acceptable. Depending on the application there may be further requirements: Splash water protection, shock resistance, and reliability even at low air pressure, as exists in mountainous regions. Most of these requirements are also tested as part of the UN 38.3 tests for the transportation of dangerous goods (see Ref. [1]). Safety issues, such as preventing the use of inappropriate chargers, mechanical protection, etc. are not a battery design issue but an issue of the design of the device. In the future, the ability to remove and replace batteries easily and safely by end-users will also become a design issue.

The nominal voltage of a battery technology plays an important role in the selection process. While alkaline nickel batteries have a voltage range of around 1.2 V, lithium-ion technology provides voltage levels around ca. 3.5 V (nominal: 3.6 or 3.2 V). This voltage level rules lithium-ion cells out for many applications in which one or two series connected aqueous-based primary batteries compete with alkaline nickel batteries, as the higher voltage level is unlikely to be compatible with existing devices and may even cause hazards. To avoid the risk of accidentally exchanging NiMH cells with lithium-ion cells, their format is different – they simply do not fit, single lithium-ion cells are not easily available to consumers compared to the readily accessible range of primary and alkaline nickel secondary cells in many shops, and lithium-ion cells can often not be accessed as easily. In addition to the voltage level, however, the voltage curve during discharging is also important: There are applications in which a constant output voltage is required, which can only be achieved with batteries with a flat discharge curve.

But the higher voltage per cell also has advantages in some cases. Instead of requiring three cells in series, only one lithium-ion cell is required.

The self-discharge rate is an important issue, as it can often take a relatively long time between manufacturing a cell and commissioning a consumer electronics device. Here, modern lithium-ion, nickel–metal hydride, and many primary battery systems based on zinc/manganese dioxide have reached a very high level,[2] which represents a major hurdle for the market entry of completely different technologies in the future.

As in all other applications, the real-life operation of portable devices varies greatly. While smartphones and laptops are often in use around the clock, other devices such as cordless screwdrivers or e-book readers may only be used sporadically. For most applications with lithium-ion batteries, however, a type of trickle charging that keeps the battery almost always fully charged is not conducive to battery life. When recharging is only carried out occasionally, full-cycle operation is usually to be expected, whereby experience has shown that a certain deep discharge tolerance (or at least a sufficient reserve after reaching the final discharge voltage) must be provided.

2 Even in the past, primary batteries had no difficulty ensuring storage times in the region of years, which is also achievable today for modern nickel–metal hydride batteries. The self-discharge of lithium-ion batteries is greater, which is why the so-called "shelf life" (storage time before use without recharging) should be in the range of several months and not years.

20.5 Monitoring the Batteries

In general, batteries for portable applications require significantly less monitoring effort than electric traction batteries. This is primarily due to the fact that the energy content of the smaller batteries is lower. Nevertheless, the potential damage effects are considerable, as shown by the case of the Samsung Galaxy S7 batteries, which repeatedly caused fires and were then banned from being carried on airplanes before the device was completely withdrawn [2].

Monitoring is particularly important with regard to charging, which should be as fast as possible but also as gentle on the battery as possible. A large number of special and proprietary charging processes have therefore become established in the field of portable batteries, in which more or less complex pulse sequences are used, some of which are also adapted using impedance measurements.

Many portable devices indicate the state of charge respectively remaining run time. Mostly very simple voltage-based algorithms are used and the load currents do not vary much in many devices.

20.6 Other Aspects

The field of consumer electronics will continue to evolve, with virtual reality headsets, smartwatches, and other wearable devices already available. This certainly brings further challenges for battery development, as highly flexible materials that still function even when twisted around 180° are now becoming of interest. While such things may initially be regarded as gimmicks, the medium-term use of such materials elsewhere is certainly within the realm of the imaginable. It is also not unlikely that the market penetration of lithium solid-state batteries, and perhaps also lithium–air batteries, will first take place in the consumer electronics sector before the quality requirements of the automotive industry and other sectors can be met.

Recycling must be improved, but this must also include the development of efficient collection systems. The much larger quantities that can be expected from the automotive industry in the near future will be advantageous here to achieve economic efficiency more quickly. While battery recycling has now achieved very high return rates in Europe for starter batteries, for example, this has not yet been achieved in the area of batteries for portable applications despite similar legal requirements.[3]

Batteries for portable applications tend to be lighter and easier to manufacture using simpler technologies than batteries for other sectors where the energy content of battery systems is magnitudes greater and the potential risks in case of failure much higher. In this respect, an extension of cell production to other countries, especially emerging countries, is to be expected. However, environmental standards

---

3 The ban on NiCd batteries in these areas of application has significantly reduced the health problems associated with unregulated disposal.

in production must also be transferred and global sustainability aspects, such as those required by the Supply Chain Sustainability Act, must be taken into account.

References

1 UN (2021). https://unece.org/rev-21-2019.

2 Wikipedia (2021). https://de.wikipedia.org/wiki/Samsung_Galaxy_Note_7.

3 IEC 61951-2:2017 (2017). *Secondary Cells and Batteries Containing Alkaline or Other Non Acid Electrolytes – Secondary Sealed Cells and Batteries for Portable Applications – Part 2: Nickel-Metal Hydride.*

Tasks

Task 20.1

Explain what advantages the development of lithium-ion batteries for portable applications has had for other areas of application, such as the automotive industry. Are any disadvantages conceivable?

Task 20.2

Which developments in the field of electric vehicles could influence the development of batteries for portable applications in the future?

Task 20.3

Pick out some examples of portable battery applications from your household. Determine which battery technology was used and try to identify reasons for the battery technology (primary vs. secondary cells, Li-ion vs. other technologies, etc.)!

Appendix A

Overview of Terms

This appendix explains relevant and frequently used terms in battery technology. The following categories are used:

1) Galvanic elements,
2) Cells, blocks, modules and batteries,
3) Reactions that take place to convert energy,
4) Terms used to describe electrochemical reactions,
5) Components of galvanic elements,
6) Characteristics of cells and batteries,
7) Operating modes (according to DIN EN 50272-2), and
8) State variables.

A.1 Galvanic Elements

- Fuel cells (conversion of fuel gases into electricity and water as a reaction product) and electrolyzers (use of electricity to convert substances, in energy technology, conversion of water into hydrogen and oxygen) are energy conversion systems that convert electrical energy but cannot store chemical energy as electrical energy and release it again as electrical energy. Despite close scientific similarities, fuel cells and electrolyzers therefore do not belong to the generic term "accumulators."
- Galvanic primary element, primary cell
  Non-rechargeable electrochemical energy storage device.
- Galvanic secondary element, secondary cell, accumulator
  Rechargeable electrochemical energy storage device.

A.2 Cells, Blocks, Modules, and Batteries

In line with common usage, the term "batteries" is often used in this textbook. If it is not clear from the context what is meant, the following subdivisions are made:

- Battery
  Group of cells, blocks, or modules

Battery Technology: Fundamentals of Battery Electrochemistry, Systems and Applications,
First Edition. Alexander Börger and Heinz Wenzl.

Companion website: www.wiley.com/go/Börger/Battery_Technology

In many applications, cells, blocks, or modules are connected in series (string) to achieve sufficiently high voltages. In some applications, cells are also connected in parallel (parallel strings). With lithium-ion batteries in particular, modules or battery packs are often made up of several strings, each consisting of several cells connected in series. The term "4*s*2*p*" refers to a battery with two strings (parallel) with four cells connected in series.

- Battery system
  Complete unit with all additional components required for safe and reliable operation (housing, heating and/or cooling system, battery management system, fuses, etc.)
  The charger for recharging the battery is often not considered a component of a battery system. In the case of mobile or portable battery systems where the charger is an integral part of the battery system, comparisons with non-electrochemical systems may result in incorrect conclusions if the charger is not included in terms of cost, weight, and volume.
- Block
  Several interconnected cells in a common housing as the smallest production unit, respectively product. The term is mainly used for lead–acid batteries and some alkaline battery systems but rarely for lithium-ion batteries.
- Cell
  The cell is the smallest electrochemical unit that can be used technically and contains all the components required for safe and reliable operation, for example, fuses.
- Module
  Several interconnected cells are mounted in a common housing for easier handling. Modules may already contain additional units for monitoring and controlling the module, which communicate with a higher-level battery management system.

A.3 Reactions for Energy Conversion

Incomplete reactions take place at the interface between two materials (in the case of galvanic elements, materials that are conductive for electrons and materials that are conductive for ions):

- Anodic reactions – the electrode material is oxidized
  After the reaction, ions without the counterions required for electrical neutrality remain in the electrolyte, and electrons without the nuclear charges required for electroneutrality remain in the electron-conducting electrode material. Educts and products of the reaction are in equilibrium.

$$aA + bB \leftrightarrow cC + dD + ne^- + m\,\mathrm{Ion}^{+l}. \tag{A.1}$$

A, B, C, and D are atoms, molecules, ions, or electrons; a, b, c, and d are moles or the number of particles. The number of electrons n and ions (Ion^{+l}) that are formed during the reaction is the same if A, B, C, and D are uncharged atoms or molecules, and the number of charges l of the cation is 1.

- Cathodic reactions – the electrode material is reduced
  Ions from the electrolyte and electrons from the electron-conducting electrode material are consumed:

$$eE + fF + pe^- + q\,\mathrm{Ion}^{+l} \leftrightarrow gG + hH. \tag{A.2}$$

  The meaning of the symbols corresponds to the symbols for the anodic reaction.

- Voltage level of the reaction
  The charge carrier separation (excess ions in the electrolyte and excess electrons in the electron-conducting electrode material) creates a voltage between the electrode and the electrolyte, the electrode voltage (E_0), which can be measured externally using special measuring devices (reference electrode). The further reaction of the excess ions in the electrolyte and excess electrons in the electrode does not occur, or only very slowly, because electrons and ions are in different places: the ions in the electrolyte and the electrons in the active material of the electrodes.

A.4 Terms Used to Describe Electrochemical Reactions

- Activity
  The voltage level of the reaction and the speed at which the reaction can take place depends on the availability of the reactants. Because these are often surrounded by a solvation shell, the concentration of the reactants must be related to their activity, that is, an effective concentration. Activities are measured and cannot be calculated, except in a few special cases.
- Butler-Volmer (BV) equation
  The BV equation makes it possible to calculate the voltage difference between the equilibrium potential and the actual electrode voltage when a current is flowing through an electrode.
  The charging current is usually defined as a positive value, and the discharging current is a negative value. For the BV equation, this means that the exponential term with a negative sign represents the discharge current and, in the usual notation, the transfer coefficient α stands for the charging direction and $(1 - \alpha)$ for the discharging direction.
- BV resistance
  Conversion of the charge transfer voltage or charge transfer overvoltage into an electrical resistance using the BV equation.
- Charge transfer voltage
  Difference between the electrode voltage and the equilibrium voltage of an electrode is caused by the passage through the electrode/electrolyte interface.

- Equilibrium voltage
  The equilibrium voltage is the voltage level of an electrochemical reaction that results from the thermodynamic data.
  By definition, the equilibrium voltage always refers to conditions existing after the end of all equilibration and diffusion processes and only to one reaction. It is rarely possible to measure the equilibrium voltage directly on cells because there are often other reactions, and the time until the end of all equalization processes can be very long.
- The equilibrium voltage given in tables usually refers to one-molar aqueous solutions, normal pressure, and room temperature. For electrochemical systems without aqueous solutions, it is difficult to determine the equilibrium voltage from thermodynamic data.
- Exchange current and exchange current density
  In equilibrium, an equal number of reactions take place in both directions. Because charge carriers are generated or consumed during the reaction, the number of reactions taking place can be understood as a current, which is the same in both directions at equilibrium. This current is referred to as the exchange current and is a characteristic parameter of all electrochemical reactions. The exchange current is usually related to the effective area of the electrode/electrolyte interface and is referred to as the exchange current density.
- Faraday current
  The current through a battery that is linked to electrochemical reactions – that is, main and side reaction currents – is called Faraday current. This term is important for differentiation when considering effects resulting from the capacitance of a battery.
- Faraday number F:
  F is the ratio of the number of molecules in 1 mol of a substance (6×10^{23} molecules, Avogadro's number) to the number of electrons per coulomb (6.25×10^{18} electrons). A Coulomb (= As) was measured electrochemically by Faraday as the amount of current that separates 1 mg of silver from a silver nitrate solution.
  The value of F is 96,500 As/mol or 26.8 Ah/mol. F is the key value of conversion between chemical and electrotechnical quantities.
  For each reaction, n electrons are released during discharge (one electron for lithium, two for lead, and three for aluminum). The theoretical specific capacity per mole is therefore nF, either 26.8, 53.6, or 80.4 Ah/mol for the various electrochemical systems.
- Gibbs [free] energy
  One of the thermodynamic values used to describe the energy of reactions is Gibbs [free] enthalpy. It can be used to directly calculate the equilibrium voltage of the reaction.
- Half cells
  A laboratory cell for a more detailed investigation of the reaction mechanisms.
  A reference electrode is used to measure the voltage (the voltage of the reference electrode relative to the electrolyte is the same for all test conditions), and

a counter electrode is used to measure the current. The counter electrode needs to have a significantly greater capacity than the electrode that is the focus of the investigation, so that its voltage relative to the electrolyte does not change during the experiment. It is not possible to examine the behavior of an electrode alone without a counter electrode.

- Main and side reactions
  Several different reactions can take place at the interface between electrodes and electrolyte, including simultaneous anodic and cathodic reactions.
  The reaction that is used to release and absorb energy is referred to as the main reaction. All other reactions that can also take place are referred to as side reactions and cannot be used to release and store energy in the cell. Main reactions differ from side reactions in that the exchange current density is an order of magnitude higher.
  The electrode at which the main reaction during discharge is anodic, that is, the electrode material is oxidized, is referred to as the anode or negative electrode.
  The two main reactions at the two electrodes are independent of each other. During charging and discharging, however, they are coupled with each other by the same number of ions that are formed or consumed and the same type of ion.
- Nernst voltage
  Concentration-dependent equilibrium voltage.
- Open-circuit voltage (OCV)
- The requirement for determining the OCV is that no external current flows through the cell. The voltage measured after the end of all equalization and diffusion processes, or after a precisely defined time, depending on the pre-treatment, is called the OCV. In electrochemical systems with side reactions, the OCV differs from the equilibrium voltage, albeit often only very slightly.
- Overvoltage
  Differences between the equilibrium voltage and the voltage without current flow are referred to as overvoltage. Analogous to voltage sources in the field of electrical engineering, this voltage deviation can be understood as the internal resistance of the voltage source formed by the reaction.
  In this book, the term overvoltage is usually specified by further information on the causes for them. The following differentiations can be made:
  - Charge transfer overvoltage η_T
    If the above equilibrium reactions only take place in one direction (charging or discharging), then the electrode voltage differs from the equilibrium voltage. The difference depends on the current and can be calculated using the BV equation.
  - Diffusion overvoltage η_D
    Voltage differences are caused by concentration gradients on the electrode surface that slowly decay due to diffusion and other equalization processes.
  - Ohmic voltage drop η_Ω
    The voltage drop caused by ohmic resistors, which can be calculated using specific resistance, length, and area, is also referred to as overvoltage in some books, but in this book, it is normally referred to as ohmic voltage drop.

- Polarization
  Similar to the term "overvoltage," "polarization" refers to the difference between the equilibrium or OCV and the terminal voltage (with and without current flow). In this book, the term polarization is usually specified by further information on its causes.
- Reference electrode
  A special electrode for measuring the potential of an electrode against a fixed potential that is independent of all other conditions. The reference electrode must be located in the electrolyte of the cell.
- Thermodynamic values
  The equilibrium voltage and heat of reaction can be calculated from the thermodynamic values of reactions.
  Convention for the heat of reaction for the discharge direction (negative value: heat is removed from the environment – the cell cools down despite current flow).

A.5 Components of Galvanic Elements

- Active mass or active material
  Material that is converted from a charged to a discharged state (discharge) or from a discharged to a charged state (charge) during the electrochemical reaction, or where intercalation and deintercalation processes take place.
  PAM: active mass of the positive electrode.
  NAM: active mass of the negative electrode.
- Electrode
  The electrode is the place where the actual electrochemical processes take place. In the case of galvanic elements (without redox flow systems), it is also the component that contains the active material required for the reaction.
- Electrolyte
  Ion-conducting but not electron-conducting material (liquid, solid, liquid bound in gel, or fleece/fiber mat) between the electrodes.
  In commercially available batteries, the electrolyte is an acid, alkali, or a salt dissolved in an (organic or inorganic) solvent. Other alternatives such as ionic liquids or solid electrolytes are not yet commercially available on a large scale.
- Positive and negative electrode
- Components of the cell where the chemically bound energy is released or electrical energy stored, and the actual reactions take place
- Separator
  Highly porous material that is impregnated with the electrolyte to be conductive for ions. The separator prevents the electrodes from touching each other. Contact between the positive and negative electrodes within a cell can lead to violent reactions (risk of fire, etc.).
  A separator can only be dispensed with if the electrolyte has an appropriate strength or structure. A separator can usually never be dispensed with for galvanic elements.

A.6 Characteristics of Cells and Batteries

- Bridging time
  The ratio of energy content (kWh) and power (kW), with the unit of time in hours (h), describes the bridging time of the battery. The bridging time is the time that a battery can provide energy at the assumed constant power.
- Capacity
  The amount of charge in ampere-hours (Ah) that can be drawn from a battery until the end of discharge is reached (minimum permissible voltage value).
  The capacity of batteries remains the same when connected in series. Depending on the definition, a battery has different capacities: nominal capacity, practical capacity, actual capacity, etc. – for details, see chapter 3.
- Capacity calculation
  The measured capacity of the battery is calculated as discharge current × discharge time, based on a fully charged battery according to the manufacturer's specifications.
- Capacity data
  For lithium-ion batteries, only a capacity C based on a discharge time of one hour is specified.
  With lead–acid batteries, the capacity is greatly dependent on the discharge current, and the capacity specification is always linked to the discharge duration. C_n indicates the amount of charge (Ah) that can be drawn from lead–acid batteries when the battery is discharged for n hours at a constant current to the end-of-discharge voltage. The specifications regarding the amount of charge and end-of-discharge voltage are always given by the battery manufacturer. Common specifications are $C_{0.25}$, $C_{0.5}$, C_1, C_3, C_5, C_{10}, C_{20}, C_{100}, or C_{240} (depending on the application of the battery).
- Charging characteristics
  Charging characteristics are a specified current and voltage curve for charging the battery.
  There are many different charging characteristics. The most common are charging curves with a constant current up to a specified terminal voltage of the battery, followed by a constant voltage that is either maintained permanently or switched off when the current falls below a certain value. This charging characteristic is referred to as IU (lead–acid batteries) or CC/CV (constant current/constant voltage, the usual term for lithium-ion batteries).
- Charging current
  See discharge current.
- Charge factor
  The ratio of the amount of charge removed to the amount charged until the same state of charge, for example, full charge, is reached.
- Coulombic efficiency
  The ratio of the amount of charge removed to the amount recharged to the same state of charge.

- Deep discharge
  Discharging close to or beyond the limit defined by the manufacturer (voltage, possibly depending on the discharge current; sometimes also depending on the calculated state of charge).
- Discharge current
  The discharge current is given a negative sign.
  The discharge current is often related to the capacity of the cell. The specification differs between lead–acid and lithium-ion cells.
  Lead–acid batteries: I_n is the discharge current that can be constantly drawn from the battery for n hours until the final discharge voltage is reached. As this current is not known without discharging the battery several times, the current is always calculated from the manufacturer's capacity specification by: $I_n = C_n/n\text{h}$.
  Lithium-ion batteries. The discharge current is always related to the capacity value C, for example: $2C = C/0.5\,\text{h}$.
- Discharge limit/end of discharge
  To prevent premature damage to a battery, batteries may only be discharged until an end-of-discharge voltage set by the manufacturer. For lithium-ion batteries, the end-of-discharge voltage is usually fixed; for lead–acid batteries, it is current-dependent.
- End-of-charge voltage
  Voltage value at which the charging voltage is kept at a constant value and the charging current decreases.
- Energy density
- Amount of energy related to the volume (Wh/l)
- Energy efficiency
  The ratio of energy withdrawn to energy recharged to the same state of charge.
  The efficiency rating can refer to cells, batteries, and complete systems including chargers, etc.
- Internal resistance of the battery
  There are very different effects that lead to a difference between the OCV and the terminal voltage, and sometimes these have a pronounced dependence on current amplitude and direction. It is not possible to calculate the internal resistance of the battery by dividing the cell voltage through the current, but only by dividing the voltage difference between equilibrium voltage and terminal voltage through the current.
- Nominal temperature
  Temperature specified by the manufacturer, in addition to the nominal capacity, at which the capacity is to be determined under standard conditions.
- Nominal voltage
  Voltage value assigned to a cell or battery system. The actual voltage during charging and discharging changes approx. $\pm 30\%$ of the nominal voltage, depending on the electrochemical system.
- The nominal voltage often corresponds to the average discharge voltage when discharging with standard discharge currents

- Overcharging
  The reaction that takes place during charging, which leads to the storage of energy (main reaction), is supplemented or replaced by other reactions (side reactions) with different properties, and these are the only remaining reactions after full charge has been achieved.
  Overcharging can be necessary and desirable (aqueous electrolytes), impossible (NaS or NaNiCl), or destructive (lithium-ion batteries).
  In batteries with an aqueous electrolyte, the side reaction current is also referred to as the gassing current because it leads to the decomposition of water and the formation of hydrogen and oxygen.
- Overdischarge
  The normal reaction that takes place during discharge is supplemented or replaced by another reaction (with different properties) in the case of very deep discharges.
- Peukert coefficient
  The Peukert coefficient describes the extent to which the capacity depends on the discharge current and is always greater than one. The closer the value is to one, the less the capacitance depends on the discharge current.
- Ragone diagram
  A representation of the specific energy on one axis and the specific power on the other (sometimes also energy and power density) is called a Ragone diagram and has a characteristic shape for all batteries: the slower a battery is discharged (the lower the power drawn), the higher the amount of energy that can be drawn from the battery.
- Rated current
  The discharge current is based on the nominal temperature T_n and nominal capacity C_n (nominal conditions): I_n is the rated current with which a cell with the nominal capacity C_n can be discharged to the discharge limit ($I_n = C_n/nh$)
  When specifying test conditions, it is important to always specify the discharge current as a multiple of the nominal current, for example, $4 \times I_n$ for a battery whose nominal capacity is specified as C_n. This is particularly important for lead–acid batteries because the capacity depends significantly on the discharge current.
  For lithium-ion batteries, the nominal current is often specified as the *C-rate* ($1C = C/1\,\text{h}$, $2C = C/(1/2\,\text{h})$) with C as the value of the nominal capacity.
- Self-discharge
  Internal conversion from charged to discharged material is either reversible or irreversible with a permanent reduction of the capacity.
- Service life
  Service life until the cells or the entire battery is replaced; a distinction is often made between the calendar service life (aging without current load) and the cyclical service life (aging due to discharging and charging processes).
- Specific capacity (Ah/kg)
- The specific capacity describes the degree of utilization of the active mass, that is, what proportion of the active mass can actually be used during discharging. The same comments as those for the term specific energy apply. Specific energy (Wh/kg) and specific power (W/kg)

Careful consideration must be given to what the weight refers to: a cell, a module, or the complete battery system with all components used for operation, including the charger if applicable.

It should also be noted how the energy content (the amount of energy delivered to an external load during discharge) is defined. In general, the energy content increases when the battery is discharged more slowly. In addition, in many applications, the energy content cannot be fully utilized because of service life restrictions. It may be useful to include this aspect in the term specific energy or power. The major differences in published values for the same electrochemical system are often due to such differences.

For the specific power, details of the power measurement must be taken into account (duration, condition of the battery, charging or discharging direction, etc.).

- Theoretical specific energy
  Maximum total energy that can be released by the chemical reaction (conversion of the entire material, without any losses and without taking into account any additional components such as current collectors, etc.). For the calculation, the definition of the voltage at which the system is discharged is important and difficult. Equilibrium conditions, that is, an infinitely slow discharge, are often assumed.
- Theoretical specific power
  The term theoretical specific power is not commonly used.

A.7 Operating Modes (According to DIN EN 50272-2)

- Buffer operation
  Like standby parallel operation, but with short discharges during operation in the event of short-term high-power consumption of the loads or in the event of a power failure.
- Charge retention operation, float charge, or trickle charge operation
  The battery is permanently kept in a fully charged state. The battery is not discharged during trickle charging. However, short discharge currents (lasting a few seconds at most) may occur if the current ripple generated by the charger and/or the loads is too high.
- Cyclical operation
  The battery is constantly being charged and discharged.
- Standby parallel operation
  The battery is connected directly or via a DC/DC converter to a DC link, which is supplied by a rectifier and to which loads are connected. The battery is only discharged if the grid or rectifier fails, or if (rarely and for a short time) the rectifier is unable to provide the power drawn from the loads (power limitation of the rectifier, control speed).
- Standby operation
  The battery is either connected to the charger or the load, for example, in many emergency lighting systems.

A.8 State Variables

The following terms require extensive discussion. For further details, see chapter 7 and the literature mentioned therein.

- Depth of discharge (*DoD*):
  1-SoC (in percent)
  However, the term *DoD* is often also used to denote a difference in the state of charge, and additional information on the average state of charge must then be provided.
- Fully charged state:
  For lead–acid batteries: Complete conversion of the discharged material into charged material and elimination of all effects resulting from discharging and charging (inhomogeneities, passivation layers, etc.).
  Full charging in this sense is only possible with battery technologies where there are no destructive side reactions. Measuring the state of full charge is not easy.
  For lithium-ion batteries: The state of full charge is based on the degree of lithiation of the anode defined by the manufacturer and is not a clearly defined physical quantity or property of a cell.
- State of charge (*SoC*):
  The ratio of the amount of charge that can still be drawn from a battery under nominal conditions to the amount of charge that can be drawn from a fully charged battery under nominal conditions (expressed as a percentage).
- State of energy (*SoE*):
  This term refers to the amount of energy that can still be drawn from a battery compared to the amount of energy from a fully charged battery. The term corrects the SoC value by the voltage drop that occurs during discharge.
- State of function (*SoF*):
  The ability of a battery to fulfill the required function in its current state, for example, to start an internal combustion engine under existing conditions (e.g. low temperature and low *SoC*).
- State of health (*SoH*):
  Reduction of the capacity of an aged battery compared to a new battery (*SoH* value related to the capacity) or (much less frequently used) change in the internal resistance compared to a new battery (*SoH* value related to the internal resistance).
- State of safety (*SoS*):
  So far only inadequately defined and quantifiable value that is intended to indicate how much the current operating conditions differ from critical battery states (only for lithium-ion batteries).

Appendix B

Safe and Environmentally Friendly Handling of Batteries

B.1 General Information

Various chapters have highlighted the risks associated with batteries and their toxic and/or chemically aggressive ingredients. In this appendix, measures to minimize the risks and to handle batteries in a safe and environmentally compatible manner are dealt with in as much detail as possible in a textbook, regardless of whether the risks have electrochemical, electrical, or chemical causes. Only those risks that exist under normal operating conditions or as a result of events that are considered unavoidable irrespective of their cause are dealt with but not risks as a result of poor engineering, inadequately designed battery management systems, etc. The definition of safety and what might be meant by the term "state of safety" (*SoS*) are explained in section 13.19 and are not discussed further in this appendix [1, 2].

Measures for battery safety and environmental compatibility during use affect all areas of application to varying degrees and must often already be taken into account when designing cells, but at the latest when integrating them into overall systems. The ISO 26262 series of standards (2018 edition) emphasizes the organizational aspect of safety [3], which also includes the company's safety culture and processes for identifying and assessing hazards (see also chapter 13, State of Safety). ISO 26262 relates to road vehicles and is therefore applicable to traction batteries for electric vehicles, among other things. The general approach and methodology for ensuring safety apply to all other safety issues irrespective of technology and application.

Safety issues can be divided into five groups. They apply to all battery technologies and applications, although to varying degrees.

1) Electrical safety,
2) Fire protection,
3) Explosion protection,
4) Requirements for site of installation and transportation, and
5) Environmental impact and disposal.

Safety and environmental aspects must be taken into account for your own protection, the protection of others, high property values, and for the protection

Battery Technology: Fundamentals of Battery Electrochemistry, Systems and Applications,
First Edition. Alexander Börger and Heinz Wenzl.

Companion website: www.wiley.com/go/Börger/Battery_Technology

of the environment. In some cases, technical and legal requirements must be complied with when installing, operating, transporting and disposing of battery systems. The appendix contains an overview of standards that also serve as the basis for many laws, accident prevention regulations and occupational safety guidelines. Risk assessments and work instructions must always be drawn up when working with batteries. In some countries, for instance, lithium-ion batteries are considered hazardous materials and companies are required to follow guidelines for storing them. The transport of cells and batteries is regulated worldwide by the United Nations Recommendations for the Transport of Dangerous Goods UN 38.3, and it is mandatory to follow their rules.

Hazard reduction measures mentioned in the following are sometimes achieved by additional technical components. The possibility of maintaining and, if necessary, repairing or replacing these must therefore also be taken into account when designing and installing battery systems. The possibility to carry out maintenance and repairs is very product-specific and therefore, despite being a significant part of achieving high safety levels, is not dealt here.

B.2 Electrical Safety

The lower the internal resistance of cells and the more cells are connected in series or the higher the nominal voltage of the battery system, the higher the currents that can flow in the event of a fault or unintentional contact and the greater the resulting hazards. The usual precautionary measures in electrical engineering are required, for example, protective masks, insulating and nonflammable protective clothing, and safety shoes. Batteries differ from all other electrical systems because they cannot be switched off – the terminals are always live and dangerous to touch – and their low internal resistance is much lower than that of many electrical installations.

All electrical systems are divided into voltage and internal resistance ranges. In Ref. [4], for protection against direct contact, it is required that batteries with a rated voltage up to or equal to DC 60 V do not require protection against direct contact, provided that the entire system complies with the conditions for safety extra-low voltage (SELV) and protective extra-low voltage (PELV). Protection against short circuits may be required. Above this voltage, special measures are required, which are further tightened from 120 V DC to 1000 V DC. Batteries almost always fall into the protection ranges with the lowest internal resistance and usually have a significantly lower internal resistance than most other electrical components. Exceeding 1000 V DC requires considerably more complex and costly precautions and as a result, there are only very few battery systems worldwide with a higher voltage level.

Most work on batteries involves touching electrically conducting parts of the battery (replacing cells or modules, connecting or disconnecting cables to terminals, etc.). In most countries, this will require special training, supervision by certified engineers, etc., to minimize risks to service personnel and to avoid fire hazards.

1) Protection against short circuits
The short-circuit currents of a new battery in a fully charged state are specified in many data sheets. The short-circuit current flows for many seconds and can release very high thermal energy at the short-circuit point and may cause a fire. Fuses suitable for DC current and the installation of short-circuit-proof cables up to a fuse are therefore necessary and required in many technical regulations.
2) Protection against sparks
Sparks occur when high currents can flow during connection and disconnection. All connected loads and the charger must therefore be switched off before they can be connected to the battery. However, as capacitors are often used for radio interference suppression in vehicle electrical systems or the vehicle electrical system has capacitive properties, there is a residual risk even when loads are switched off. Sparks are possible even with low-voltage batteries. Sparks must be avoided at all costs in potentially explosive atmospheres, but they are also dangerous for people in the absence of an explosion due to sudden movement caused by a spark. Switches should therefore be provided near the battery terminals.
3) Protection against arcing faults
Depending on the atmosphere, length, material, and surface of the contact points, the voltage dropped across an arc is at least approx. 20 V. A spark can therefore only lead to an arcing fault, that is, a technically unintended arc, if the nominal voltage of the battery is sufficiently high and the internal resistance is sufficiently low – which can always be assumed for industrial applications of lead–acid and lithium-ion batteries, even with low capacity. In addition to the risk of fire from an arc, serious burn injuries can also be caused to people. An arc can only be extinguished by switches in the circuit if there are no fuses. Correctly selected and positioned fuses can extinguish an arc in fractions of a second.
4) Fuses
Fuses in the immediate vicinity of the battery terminals must be installed as standard. To protect against insulation faults within modules, fuses are sometimes installed in the middle of battery strings after $n/2$ cells within a module. DC fuses must be used. Short-circuit-proof cables that can carry very high currents without risk of damage to objects in the vicinity must be used between the battery terminals and the fuses. If the battery has to be operated close to the short-circuit current, for example, during starting engines, all cables to this load must be short-circuit-proof. The design of fuses is problematic because they must also safely and quickly switch off the current flowing in an aged battery and at the lowest operationally permissible temperatures and states of charge. It is therefore advisable to design the fuse for currents that cannot be drawn by the connected loads during normal operation as the lower limit and for the maximum possible short-circuit current of the battery as the upper limit. In addition, a switch that can be activated by the battery management system that switches off the battery when the battery voltage reaches a non-permissible voltage range should be installed.[1]

---

1 During the investigation of a battery fire by one of the authors, it was found that the battery management system measured a battery voltage of approx. 50% of the nominal voltage for more

5) Tools and measuring devices
   When working on batteries, tools and measuring tips that are insulated as completely as possible should be used. All measuring devices and sensors used must have sufficiently large fuses so that the short-circuit current of the battery in the measuring device can be safely switched off in the event of a fault in the measuring device or incorrect operation. The specification of category III measuring devices[2] is not sufficient; fuses with a 30,000 A cut-off current and a rated voltage of 1000 V DC are recommended.
6) Protection against current flow through the body
   Dangerous body currents can occur above the regulated limit of 60 V DC (voltage limit of SELV devices). The protection principle is that potential equalization without high currents takes place between the body when touching a live battery part. Simultaneous contact with a second battery part with a different voltage must be safely avoided so that no current flows through the body. The following must always be observed to avoid dangerous body currents:
   - Resistance of the floor between 50 kΩ ≤ R ≤ 10 MΩ at voltages ≤500 V, otherwise 100 kΩ ≤ R ≤ 10 MΩ, so that electrostatic charging is prevented on the one hand and body currents are reduced to a safe minimum on the other. The protective effect of the floor is linked to wearing shoes that prevent electrostatic charging and are electrically conducting.
   - The battery should not be connected to parts of the electrical system which have a fixed potential so that potential equalization can take place with low currents in the event of touching battery parts.
   - If the battery is galvanically isolated from the charger and the loads, the battery should then not be connected to ground potential. With step-up or step-down converters[3] there is often no galvanic isolation, so the potential of the battery is fixed. In these cases, greater precautions must be taken, which may include disconnecting both poles of the battery from loads and chargers before working on the battery.
   - There must be no grounded components in the working area that can be touched, for example, radiators.
7) Protection concepts for battery cabinets
   Due to overriding safety considerations, metallic battery cabinets must be earthed. When working on batteries in earthed racks and battery cabinets, the battery must therefore always be disconnected from all loads and the

---

than one minute and classified it as a battery fault, without activating the battery's main switch. The battery voltage then rose again to a value within the normal range as the short circuit was cleared by the high current. Iin the meantime a fire had started.

2 Measuring devices are classified according to their voltage limit and the internal resistance of the test object. There are four categories. Categories III and IV both have a voltage limit of over 1000 V, but both are only designed for an internal resistance of the equipment to be measured of 2 mΩ. The standard fuses of such measuring devices are not able to safely disconnect possible short-circuit currents of a battery due to a fault in the measuring device.

3 DC/DC converters without galvanic isolation by a transformer either increase or decrease the output voltage compared to the input voltage.

charger at both terminals. If only very little space is available for installation, this precaution should also be observed for batteries with floating potential (no ground connection). It may then be necessary to wear insulating protective gloves in addition to normal insulating protective clothing. Special protective clothing is available for such work (e.g. arc-resistant jacket, insulating head and foot protection, insulating rubber gloves) and insulated tools, which are mandatory for high nominal voltage levels, low internal resistance, or when a limit value for the releasable energy is exceeded. It is likely that different rules will apply in different countries, but unlikely that there are no standards or safety rules governing this aspect.

8) Insulation monitoring
   Reduced battery insulation due to damage, age, or environmental influences leads to leakage currents. These pose a risk due to the potential development of smoldering fires and the possibility of electric shock on contact. In addition, different charging currents can occur in the cells connected in series, leading to different states of charge.

It is immediately clear that the requirements mentioned here must already be taken into account when designing cells, but above all when designing and assembling complete battery systems. Furthermore, it is important that work on batteries in which cells or modules are replaced or cables are disconnected or connected is only ever carried out using detailed work instructions and in compliance with the relevant standards.

B.3 Fire Protection

Batteries must not cause fires, and they must also not increase the risk of fire in the event of external fires. The amount of flammable material in lithium-ion cells (organic solvents as part of the electrolyte, possibly graphite as anode material) is significantly higher than in lead–acid batteries, whose polypropylene housings can even be made flame-retardant for hazardous areas by adding additives. There are three causes of fire:

1) External fires in the surrounding area
   The usual measures to minimize the fire load in modules by using noncombustible or flame-retardant materials and, in the case of stationary battery systems and storage, fire protection walls and reducing the number of cells in storage.
2) External short circuits and operation outside voltage and current limits
   With the exception of insulation faults within a module, these causes of fire can be completely avoided by safe and redundantly designed battery management systems. Overcharging of a cell in a module or a string can and must be prevented by a BMS.
3) Internal short circuits and thermal runaway
   At present, it does not seem possible to rule out internal short circuits which then may lead to a thermal runaway as a potential cause of fire with certainty.

It does, however, seem possible to detect cells with internal short circuits early and take precautionary actions. When a thermal runaway occurs, the temperature of the gas generally exceeds the self-ignition temperature of the gases, just under 600 °C for the main combustible components of the gas stream. The gas starts to burn without an external source of ignition as soon as the concentration of the gases in the surrounding atmosphere exceeds a threshold. Maximum flame temperatures are in the range of 1900 °C. The following propagation to other cells and spreading of the fire to other components is then difficult to prevent.

Measures are therefore aimed at managing the flow of hot gases after cell venting, reducing the propagation from one cell to the next by taking appropriate steps, for example, by an appropriate location of the rupture valve, rapid dissipation of the heat generated to avoid high temperatures, etc.

In principle, internal short circuits that cause a thermal runaway as the cause of a fire are also conceivable in lead–acid batteries. It has been observed, for example, that cell or block housings of lead–acid batteries for UPS systems stick together, which indicates temperatures in the range of at least approx. 115 °C. However, the development of fires has not been reported.

When lithium-ion batteries catch fire, hydrofluoric acid (HF) and its reaction products as well as organic phosphorus compounds, which are similar to those used for insecticides and pesticides, and carbon monoxide (CO), carbon dioxide (CO_2) can be produced. All of this is extremely harmful to human health. The objects contaminated by the fire in the vicinity of the fire must therefore be carefully cleaned or constitute hazardous waste. Due to contamination with lead, burnt lead–acid batteries are also hazardous waste.

As a general rule, the potential fire load (also in the vicinity of a battery) must always be as low as possible (highest possible fire protection classes of the materials used), any heat generated must be dissipated as quickly as possible and there must be no sources of ignition.

Lithium-ion cells and batteries are considered by many insurance companies as hazardous products as a result of possible fire hazards. Storage facilities of cells in industry therefore have to fulfill various requirements – number of cells, fire alarm systems, etc. Consumer advice bodies, too, recommend that batteries and battery-powered products are kept safe.

Standard fire protection protocols should be followed.

B.4 Explosion Protection

Battery explosions are very rare events whose frequency and effects can be reduced, but cannot be completely avoided according to the current state of the art.

B.4.1 Explosion Protection for Lead–Acid Batteries

Due to the importance of explosion protection for lead–acid batteries in normal operation, the design considerations are described in detail in the following. In addition, the description is intended to serve as an example of the level of detail and the practical feasibility of the protective measures to be taken in accordance with the applicable standard.

Hydrogen can no longer explode in air at a concentration of more than 75% or less than 4%. To avoid a risk of explosion, dilution to at least 4% with a safety factor of 5 is therefore prescribed.

The standard [5] summarizes the requirements.

B.4.1.1 Calculation of the Air Exchange Volume for Ventilation

The gas produced must be safely discharged to the outside air. The required air exchange volume is calculated in Ref. [4] using the following formula:

$$Q = v \cdot q \cdot s \cdot n \cdot I_{\text{gas}} \cdot C_n \cdot 10^{-3} \ (\text{m}^3/\text{h}), \tag{B.1}$$

with

- Q Air volume flow in m^3/h,
- v Required dilution factor for hydrogen $(100-4)/4 = 24$,
- q Volume of hydrogen released per 1 Ah is 0.42×10^{-3} (m^3/Ah),
- s General safety factor of 5,
- n Number of cells,
- I_{gas} Current that causes the gas development, in mA/Ah, and
- C_n Capacity (C_{10} up to the final discharge voltage of 1.8 V/cell at 20 °C) in Ah.

The combination of v, q, and s results in

$$Q = 0.05 \cdot n \cdot I_{\text{gas}} \cdot C_n \ (\text{m}^3/\text{h}), \tag{B.2}$$

$$I_{\text{gas}} = I_{\text{float/charge}} \cdot f_s \cdot f_g, \tag{B.3}$$

with

- $f_s = 5$ Safety factor to take into account increased temperatures and defective or aged cells in series connection.
- f_g Proportion of the current that causes gassing (1 for closed batteries and 0.2 for sealed batteries).

For I_{gas}, typical currents are given for lead–acid batteries at various voltages (2.23 and 2.27 V/cell and 2.4 V/cell for equalization charging) and battery types, which must be adapted for other cell voltages and temperatures. It should also be noted that for other characteristic curves, for example, IUIa for traction batteries, the maximum possible charger current must be used.

B.4.1.2 Discharging of Gas

The gas/air mixture produced must be discharged into the environment outside the building. Natural ventilation is preferred. The opening for the supply air should be near the floor and the opening for the exhaust air near the ceiling on the opposite side. The standard contains further information for calculating the size of the openings and distances for openings on the same side of the room.

The following requirements must be met to discharge the gases produced:

- Warning message in the event of ventilation failure, for example, message to the fire department or building services for rapid intervention (e.g. $<1\ h$) of the fault or
- Coupling of the charger to the function of the ventilation, for example, by means of a flap in the ventilation opening, which falls down if the ventilation fails and triggers a switch.

The air that is extracted is already diluted to such an extent that the fan does not need to be explosion-proof.

B.4.1.3 Safety Distance to Cell Openings

The calculation is based on the assumption that the oxyhydrogen gas spreads evenly into a hemisphere around the cell opening. There is a risk of explosion up to a distance from the gas opening at which sufficient dilution of the gases has occurred. Annex B4 of DIN EN IEC 62485-2:2018 contains a formula for calculating the required distance:

$$d\,(\text{mm}) = 2.8 \cdot \sqrt[3]{I_{gas}}, \tag{B.4}$$

where I_{gas} is the maximum current during overcharging for the battery (multiplied by the number of cells with a common gas vent).

In the case of a six-cell starter battery with a high capacity, this safety distance can be approx. 1 m. In addition, no measuring devices or other electrically operated devices and tools[4] may be placed in the immediate vicinity of cell openings, for example, directly above battery plugs.

The air in the battery room is not explosive at a sufficient distance from cell openings. Therefore, technical systems in the battery room (chargers, switch cabinets, lighting, etc.) do not have to be explosion-proof and lamps may also be installed below the height of the exhaust air opening.

B.4.1.4 Ignition Sources

In addition to the absolute ban on naked flames and smoking in the vicinity of batteries, the following safety measures must be observed:

- No naked flames or sparks, for example, from welding or metalworking in the vicinity
- Avoidance of electrostatic charge
  Before touching the battery housing, tools and hands or items of clothing must have the same potential as the battery. Work shoes and floors with the correct discharge resistance are sufficient for battery systems. In many other cases (vehicle batteries, transportation of batteries) this rule is not easy to comply with. Covers at battery potential etc. are used for this purpose as removing them leads to equalization of potentials.

4 With the exception of intrinsically safe devices, whose energy content is not sufficient to ignite a mixture of oxygen and hydrogen gas in the event of a fault.

To avoid electrostatic charging, batteries may only be cleaned with damp cloths and never wiped with dry cloths.

- Sparks when connecting and disconnecting
  The battery must always be connected before the circuit is closed by the battery switch or the battery must be disconnected afterward. In vehicles in particular, capacitors are installed in the vehicle electrical system, so that even if all loads are switched off, considerable currents can flow during contacting and cause sparks. The safety distance between the cell openings and the contacting point must be observed.
- Corrosion attack in the cell
  Corrosion attacks on grids, terminal lugs, terminals, and intercell connectors can cause ignition sparks in the cell, especially when loads with high power and therefore high current requirements are connected or the circuit is suddenly interrupted under high current loads. Table B.1 shows the battery current, the voltage dropping across a corroded intercell connector and the thermal power at the corrosion point in a very simplified form. If the current path breaks during a high current pulse, which causes mechanical stress via electromagnetic field effects, a spark can occur in the potentially explosive area of the cell between the cell cover and the electrolyte level. In batteries with a high nominal voltage, very high voltage drops can occur between a terminal and the current collectors of an electrode (group bars) within the cell container due to corrosion, even in trickle charge mode. The corrosion point reduces the trickle charge current, which leads to a reduction in the charge transfer voltage for all intact cells and electrodes. Despite only very small voltage differences between the cell terminal and the electrodes before disconnection, an explosion can only be ruled out if the inductively or capacitively stored energy released during disconnection is less than the ignition energy.
  It is very difficult to reliably detect a corrosion attack in the cell in the current path in good time so this cause of explosion appears unavoidable even if all protective measures are observed.

To avoid or reduce the risk of explosion, design measures are taken as already described in chapter 12.

Table B.1 Schematic calculation of the voltage across a resistor (here intercell connector) within a cell under high current load (starter motor for engine start).

Starter motor resistance (mΩ)	Intercell connector resistance (mΩ)	Electricity (A)	Voltage drop across starter motor (V)	Voltage drop via intercell connector (V)	Heat output at the intercell connector
10	0.001	1200	12	0	0 W
10	2	1000	10	2	2 kW
10	10	600	6	6	3.6 kW

B.4.2 Explosion Protection for Lithium-Ion Batteries

There are no processes in lithium-ion batteries that can lead to an explosion under normal operating conditions. The maximum pressure in the cell as a result of gas generation by aging processes, volume expansion of electrodes, and high ambient temperatures is in the range of c. 3 bar. Explosions of lithium-ion batteries are therefore the result of a thermal runaway which produces large quantities of very hot, explosive gases. The gases may explode after mixing with the ambient air if the gases are emitted into a closed or poorly vented space. If the gases are emitted into an open space, it is more likely that a fire will be started.

Chapter 13 describes design measures to avoid explosion risks. Knowledge comparable to that of lead–acid batteries, with which the risk of explosions can be reduced or the effects minimized through the design of the system environment, is currently undergoing rapid development. There are no detailed standards comparable to those for lead–acid batteries which are a much more mature technology.

B.5 Requirements for Site of Installation and Transportation

Rooms for battery charging stations and stationary battery systems are often understood as electrical operating rooms that are subject to the corresponding general electrical engineering regulations, for example, access restrictions and fire protection. Other aspects include:

- Quantity limits, distance regulations, and fire protection concepts for the storage of lithium-ion batteries.
- Floor coating for lead–acid batteries to protect against leaking aqueous sulfuric acid.

Industry associations of insurance companies have issued guidelines that should be followed. A discussion with the respective insurance companies is therefore advisable to maintain insurance coverage.

Batteries are subject to transport regulations as hazardous goods both when new and during transportation for disposal. These regulations apply worldwide (UN 38.3).

B.6 Environmental Impact and Disposal

Materials used in batteries are in most cases toxic and/or chemically aggressive. The environmental standards and occupational health and safety regulations that must be followed during production are not the subject of this textbook.

In normal operation, there are no problems with lithium-ion batteries, only with lead–acid batteries. These are easy to solve. When cleaning corroded terminals or during assembly work, small pieces of lead can become detached from the battery terminals. Lead parts and the cleaning cloths etc. used for this purpose constitute hazardous waste. Cleaning rags and absorbent materials used to absorb aqueous sulfuric acid that has leaked for some reason are normal industrial waste.

At the end of their service life, batteries must be disposed of properly. The term "circular economy" is used worldwide to describe the concept of existing and future legislation. A multi-stage procedure is proposed: reuse, repair, recycling of materials, and recover energy before disposal. Special requirements are contained for instance in the Battery Regulations of the European Union. These requirements have led to a deposit for lead–acid batteries in Germany, for example, which is only refunded when another battery is returned for recycling. The "second life" concept for lithium-ion batteries, that is, the use of batteries in a second application with lower or different requirements, also meets the requirement of a circular economy and is not just economically advantageous.

Batteries at the end of their service life have value because of the materials in the battery, which is an economic incentive to pass them on to recycling companies. The recycling rate for lead–acid batteries is therefore extremely high and the processes are well controlled. For lithium-ion batteries, on the other hand, there have only been a few pilot projects to date and recycling has been far less successful. The valuable ingredients, especially copper, nickel, cobalt, graphite, and lithium itself, cannot yet be extracted at sufficiently low cost and there is still no selling price for used lithium-ion batteries. However, industry and recycling technology are changing fast not only because of regulatory pressure.

References

1 ISO Guide 64:2008-08. Guide for Addressing Environmental Issues in Product Standards (2008).

2 ISO/IEC Guide 51:2014-04. Safety Aspects – Guidelines for their Inclusion in Standards (2014).

3 ISO 26262:2018. Road Vehicles – Functional Safety. [12-volume series of standards] (2018).

4 DIN EN 50272-2:2001-12. Safety Requirements for Batteries and Battery Systems – Part 2: Stationary Batteries (2001) [Note: this standard has now been withdrawn].

5 DIN EN IEC 62485-2:2019-04. Safety Requirements for Secondary Batteries and Battery Installations – Part 2: Stationary Batteries (2019).

References

Appendix C

Overview of Standards

C.1 Importance and Role of Standards and Technical Regulations

Standards[1] are documents issued by standardization institutions and are intended to help describe the state of the art in terms of minimum requirements, not technical solutions. At the same time, they facilitate economic growth, because many companies can provide products within a precisely described framework of requirements and work according to similar principles and processes on similar products. Standards are generally not binding unless they are explicitly referred to in legal texts. However, in many product liability cases proof of compliance with a standard can be helpful or even decisive in proving that a product has been developed in accordance with the state of the art. In contrast to standards, laws and official regulations derived from laws are instructions issued by authorized bodies (parliament, government, administrative authorities, etc.) that must be followed, otherwise sanctions, that is, penalties, may be imposed.

Standards are created in a complex coordination process in which the national institutes act as platforms that organize the process of updating and writing new standards.

Standards are developed and adopted by consensus[2] of working groups with the participation of all interested parties (e.g. occupational health and safety organizations, professional associations, accident and property insurance institutions and companies, trade unions, supervisory authorities of governments, environmental

1 The term standard is used here as a generic term for all norms and comparable standards. Standardized procedures, processes, descriptions, etc. are not only published by officially commissioned standardization institutions, but also by industry associations, companies, academic institutions, independent institutes and many more. With regard to batteries, the standards of the Underwriters Laboratories (UL) and the SANDIA National Laboratories from the United States should be mentioned in particular. Such documents also describe the state of the art in science and technology and may serve as a reference when designing products and making decisions concerning operating conditions, maintenance, etc.

2 While important laws of nations may be adopted with very close majorities of one single vote, the acceptance criteria for standards are usually higher and require, e.g. at least two thirds or even a unanimous decision.

Battery Technology: Fundamentals of Battery Electrochemistry, Systems and Applications,
First Edition. Alexander Börger and Heinz Wenzl.

Companion website: www.wiley.com/go/Börger/Battery_Technology

protection and consumer protection organizations, manufacturers, operators, and scientific institutions). Participation in a working group for the development or revision of a standard is open to any individual or company with a declared interest and is usually subject to a fee for companies. Details governing the formation of working groups differ between countries and organizations.

New topics for standards can be proposed by anyone. At first, an attempt is made to integrate the topic into an existing committee (so-called Technical Committee, TC) and work on it there. The decision as to whether a new working group is set up to develop a standard is made by a steering committee.

There is close cooperation between national, supranational (e.g. the European Committee for Standardisation (CEN) and the Comité Européen de Normalisation Electrotechnique (CENELEC) which are associated with the European Union), and international standardization organizations (e.g. the International Electrotechnical Commission [IEC] and International Organisation for Standardization [ISO]). In some areas, particularly for the transport of lithium-ion batteries and electric road vehicles, there are also standards under the leadership of the United Nations (e.g. UN38 for the transportation of lithium-ion batteries).

Standards always have a certain scope of applicability, both in technical terms and in relation to the standards organizations that issue them. The regional applicability of standards is marked. DIN norms only apply to Germany, DIN EN norms are valid and applicable within Europe, while the DIN EN IEC marking means that the standard's applicability is worldwide. National versions should not contain any differences in content. However, national additions are possible. This facilitates the free movement of goods and competition. Norms that carry only the abbreviation for one country are standards that have no equivalent at the transnational level and either contain additional requirements or deal with aspects for which there is no equivalent in other countries or national and international organizations have not been able to agree. It is permissible for companies to use such norms to describe that the product fulfills certain requirements and companies from different parts of the world can agree on using such norms to specify their contractual obligations.

There may be conflicting requirements in different standards, in particular between ISO standards and SAE standards[3] (which are more widespread in North America). This means that there are parallel documents that apply in different regions and, as long as the standards have not been explicitly named in laws, companies can agree on which standard they want to use. Currently (2024), for example, there is the DIN EN 50342 group of standards for lead–acid starter

3 Strictly speaking, the Society of Automotive Engineers (SAE) was founded as an industrial and/or professional organization. However, it has also become important for standardization. Since SAE does work on the level of individual's expertise and not – like ISO – of national delegations, the outcome of the standards might vary significantly. The reason for less acceptance of ISO standards in North America is the feeling that the single US vote in national voting on ISO level might lead to an underrepresentation of US interests.

batteries, as well as the IEC 60095 standard, which contains small but possibly decisive difference for a given application.

For this reason, it is advisable to observe the relevant applicable standards and their date of publication when designing, selecting, operating, and maintaining products.

In addition to the minimum requirements for products and processes agreed in standards, there are also technical requirements based on international agreements at the level of the United Nations or its sub-commissions. Relevant to lithium-ion batteries are, for example, the transport regulations according to UN 38, which deals with the transportation of batteries, and the GTR 20 (Global Technical Regulation, GTR) for the safety of electric vehicles. Other technical vehicle regulations such as ECE R100 on the safety of electric vehicles and their traction batteries apply internationally, but not worldwide. There may also be national regulations, for example, the conversion of the European Battery Directive into national law. Standards based on such international agreements are, in contrast to other standards, legally binding.

As a preliminary stage to standards, there are documents referred to as "Technical Specifications" (e.g. DIN TS), which have to be reviewed at shorter intervals than standards, and so-called "Technical Reports" (TR) at the international level. An important report for lead–acid batteries is the "Guideline on the reduction of explosion hazards for rechargeable batteries, Part 1: Lead–acid batteries (TR 61430)" and the "Guideline for the use of monitoring systems for stationary lead–acid batteries (TR 62060)." The existence of such a TR may indicate that either no consensus could be reached in standardization or that the state of the art is not yet mature enough for standardization to be worthwhile (because one disadvantage of standards is that they can also hinder technical progress). For detailed questions, these TR and possibly also preliminary stages of finished standards, so-called draft standards, may be of interest for certain technical aspects.

In addition to standards, legally binding or not, it is important to point out the wealth of information and solutions to problems contained in documents by professional and industrial organizations such as:

- Underwriter Laboratory (UL)
- National Fire Protection Association (NFPA)
- The European Battery Association (Eurobat)
- Battery Council International
- Zentralverband der Elektrotechnischen Industrie (ZVEI)

to name just a few

C.2 Overview of Standards and Other Relevant Documents

The following overviews are a list of a number of standards that are relevant to the design, operation and use of batteries.

C.2.1 Terms Used in Battery Technology

DIN 40729:1985-05: Accumulators; Galvanic secondary elements; Basic concepts
DIN EN 61429:1997-11: Marking of accumulators and batteries with the international recycling symbol ISO 7000-1135 (IEC 61429:1995)
DIN EN IEC 62902:2020-06: Secondary batteries – Symbols for marking to identify their chemistry (IEC 62902:2019)

C.2.2 Recycling and Life Cycle Assessment

prEN IEC 63330:2023-11; Requirements for reuse of secondary batteries (standard under development)
EN IEC 63338:2024; General guidance for reuse of secondary cells and batteries (standard under development)
IEC 63369 ED1 Methodology for the Carbon Footprint calculation applicable to Lithium-ion batteries (project to develop a standard by 2024)

C.2.3 General Safety Aspects

a) Overarching:
   DIN EN IEC 62485-1:2019:01 Safety requirements for secondary batteries and battery systems – Part 1: General safety information
b) Lithium-ion batteries:
   DIN EN 62485-5:2017-06 Safety requirements for batteries and battery systems – Part 5: Lithium-ion batteries for stationary applications

C.2.4 Starter Batteries

a) Lead–acid batteries:
   DIN EN 50342-1:2019-05 Lead–acid accumulator starter batteries – Part 1: General requirements and tests

Note: Parts 2–7 of the series of standards relate to dimensions, dimensions, components and special areas of application (e.g. 36 V systems and motorcycle batteries)
IEC 60095-1:2018 Lead–acid starter batteries – Part 1: General requirements and methods of test

Note: Parts 2–7 relate to dimensions, dimensions, components, and special areas of application. This standard differs from DIN EN 50342 in a few important points and must be observed, particularly outside Europe.

a) Lithium-ion batteries:
   DIN EN 63118:2018 Secondary cells and batteries with alkaline or other non-acid electrolytes – Performance of secondary lithium batteries for use in road vehicles other than traction (draft)
   DIN EN IEC 63057:2020-12 Secondary cells and batteries with alkaline or other non-acid electrolytes – Safety requirements for secondary lithium batteries for

use in road vehicles other than propulsion (IEC 63057:2020); German version EN IEC 63057:2020

C.2.5 Electromobility

a) General:
DIN EN 62485-3:2015-09 Safety requirements for batteries and battery systems – Part 3: Traction batteries for electric vehicles

b) Lead–acid batteries:
DIN EN 61982:2013-04 Secondary batteries (excluding lithium batteries) for the propulsion of electric road vehicles – Capacity and cycle life tests
DIN EN IEC 63193:2021-11 Lead–acid batteries for propulsion of light vehicles – General requirements and test methods (IEC 63193:2020)

c) Lithium-ion batteries:
DIN EN IEC 62660-1 VDE 0510-33:2020-07 Secondary lithium-ion cells for the propulsion of electric road vehicles. Part 1: Performance testing
DIN EN IEC 62660-2 VDE 0510-34:2020-07 Secondary lithium-ion cells for the propulsion of electric road vehicles. Part 2: Reliability and abuse testing
DIN EN 62660-3 VDE 0510-49:2017-05 Secondary lithium-ion cells for the propulsion of electric road vehicles. Part 3: Safety requirements
ISO 6469-1:2019-04 Electrically propelled road vehicles – Safety specifications – Part 1: Rechargeable energy storage system (RESS)
ISO 6469-2:2018-02 Electrically propelled road vehicles – Safety specifications – Part 2: Vehicle operational safety
ISO 6469-3:2021-10 Electrically propelled road vehicles – Safety specifications – Part 3: Electrical safety
DIN EN IEC 62485-6:2021-12 Safety requirements for secondary batteries and battery systems – Part 6: Safe operation of lithium-ion batteries in traction applications (IEC 62485-6:2021)
DIN EN 50604-1:2017-05 Secondary lithium batteries for light electric vehicle applications – Part 1: General safety requirements and test methods

C.2.6 Traction Batteries (Not or Not Specifically Electric Vehicles; Here Only Lead–Acid Batteries)

DIN 40540 2007-09: Lead–acid accumulators – Traction batteries 24 V, 48 V, 80 V with cells according to DIN EN 60254-2 for industrial trucks for use in potentially explosive atmospheres (Group II, Category 2, and Category 3) – Dimensions, weights, design

DIN EN 61044 1994-05: Intermediate charging of lead traction batteries
DIN EN 60254-1 2006-01: Lead–acid traction batteries – Part 1: General requirements and tests
PD IEC/TR 61431 2020-09-29: Guidelines for the use of monitor systems for lead–acid traction batteries

IEC TR 62540:2009: Radio frequency identification (RFID) for stationary lead–acid cells and monoblocs – Tentative requirements

C.2.7 Stationary Applications

a) General:
   DIN EN IEC 62485-2:2019-04 Safety requirements for secondary batteries and battery systems – Part 2: Stationary batteries
   DIN EN 61427-1:2014-02 Rechargeable cells and batteries for renewable energy storage – General requirements and test methods – Part 1: Photovoltaic off-grid application
   DIN EN 61427-2:2016-09 Rechargeable cells and batteries for renewable energy storage – General requirements and test methods – Part 2: Grid integrated applications
b) Lead–acid batteries:
   DIN EN 60896-11:2003-07 Stationary lead–acid accumulators – Part 11: Sealed batteries; General requirements and test methods (incl. corrigendum of 2006)
   DIN EN 60896-21:2004-12 Stationary lead–acid accumulators – Part 21: Sealed types – Test methods (incl. corrigendum of 2007)
   DIN EN 60896-22:2004-12 Stationary lead–acid storage batteries – Part 22: Sealed types – Requirements
   DIN 40740:1991-04 Lead–acid accumulators; stationary batteries, accessories; ceramic funnel plugs, ceramic plugs, electrolyte level indicators
   DIN 40742:2015-09 Lead–acid accumulators – Stationary sealed cells with positive armored plates and fixed electrolyte (gel)
   DIN 40744:2015-09 Lead–acid accumulators – Stationary sealed batteries with positive armored plates and fixed electrolyte (gel)
c) Lithium-ion batteries:
   DIN EN IEC 63056:2021-04 Secondary cells and batteries with alkaline or other non-acid electrolytes – Safety requirements for lithium secondary cells and batteries for use in electrical energy storage systems
   IEC 62897 Stationary Energy Storage Systems with Lithium Batteries – Safety Requirements [under development]
   DIN EN IEC 62485-5:2021-12 Safety requirements for secondary batteries and battery systems – Part 5: Safe operation of stationary lithium-ion batteries

C.2.8 Portable Applications

a) Lead–acid batteries:
   DIN EN 61056-1:2013-06 Lead–acid batteries for general use (sealed) – Part 1: General requirements, characteristics – Test methods
   DIN EN IEC 62485-4:2019-04 Safety requirements for secondary batteries and battery systems – Part 4: Sealed lead–acid batteries for use in portable applications

b) Lithium-ion batteries:
UNE-EN IEC 63218:2021-11-01 Secondary cells and batteries containing alkaline or other non-acid electrolytes – Secondary lithium, nickel cadmium and nickel-metal hydride cells and batteries for portable applications – Guidance on environmental aspects
DIN EN 61960-3:2017-12 Secondary cells and batteries with alkaline or other non-acid electrolytes – Lithium secondary cells and batteries for portable applications – Part 3: Prismatic and cylindrical lithium secondary cells and batteries made from them

C.2.9 Other Standards

a) Lead–acid batteries:
DIN EN 62877-1:2016-12 Electrolytes and water for sealed lead–acid batteries – Part 1: Requirements for electrolytes (incl. corrigendum of 2021)
DIN EN 62877-2:2016-12 Electrolytes and water for sealed lead–acid batteries – Part 2: Requirements for water
DIN 43534:1983-11 Lead–acid accumulators; maintenance-free sealed accumulators with grid plates and fixed electrolyte; capacities, voltages, main dimensions, design features, requirements
b) Lithium-ion batteries:
DIN EN 62619:2017-11 Accumulators and batteries with alkaline or other non-acid electrolytes – Safety requirements for lithium accumulators and batteries for use in industrial applications
DIN EN 62620:2015-09 Accumulators and batteries with alkaline or other non-acid electrolytes – Lithium accumulators and batteries for industrial applications

Appendix D

Electrochemical Impedance Spectroscopy (EIS)

To fully determine the electrical properties of any system, the frequency-dependent resistance, or impedance, must always be measured, except in the case of ideal resistors. The measurement of the frequency-dependent impedance Z (alternating current resistance), impedance spectroscopy, is used in many different areas of engineering and the natural sciences.

D.1 Overview of Terms

When a sinusoidal voltage signal or sinusoidal current signal is applied to any system, the result is a current or voltage response as long as the excitation takes place within the linear range, that is, at low excitation amplitudes.[1]

$$I(t) = U(t)\sin(\omega t + \vartheta), \tag{D.1}$$

with ϑ as the frequency-dependent phase shift and ω as frequency. A diagram of an excitation and response signal has already been shown in Figure 9.6. The excitation can take place either by applying a sinusoidal current signal (galvanostatic) or by applying a sinusoidal voltage signal (potentiostatic). For a given frequency, the AC resistance $U(t)/I(t)$ is characterized completely by the maximum of the current and the voltage, their respective peak values, and the phase shift. Both parameters change with frequency.

Usually, complex values (Gaussian number plane) are used for current and voltage because this simplifies the visualization and calculation methods. The frequency-dependent impedance is $Z(\omega)$:

$$Z(\omega) = \frac{U(\omega)}{I(\omega)}, \tag{D.2}$$

$$Z = Z' + \mathrm{i} \cdot Z'' = |Z| \cdot \mathrm{e}^{\mathrm{i}(\vartheta)}. \tag{D.3}$$

1 Capacitors and coils do not exhibit linear behavior, so the assumption of linear behavior is an approximation whose quality depends on the amplitude of the excitation and frequency and is only possible at low amplitudes.

Battery Technology: Fundamentals of Battery Electrochemistry, Systems and Applications, First Edition. Alexander Börger and Heinz Wenzl.

Companion website: www.wiley.com/go/Börger/Battery_Technology

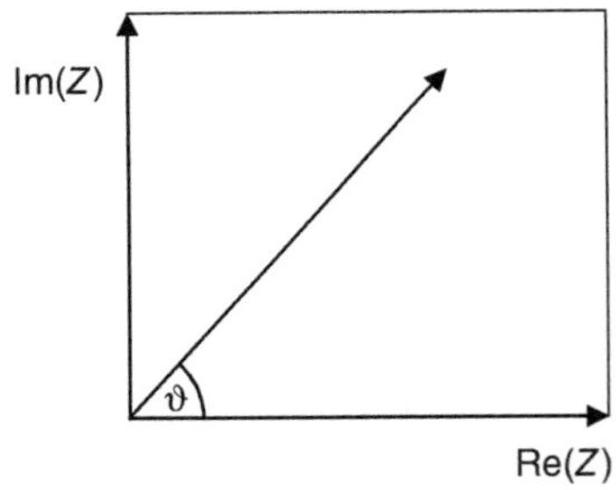

Figure D.1 Diagram showing the complex impedance in the complex number plane.

$|Z|$, the ratio of the maximum value of the current (peak value) to the peak value of the voltage, is referred to as amplitude.

$$\text{The real part of impedance: } \mathrm{Re}(Z) = Z' = |Z| \cdot cos(\vartheta). \tag{D.4}$$

$$\text{The imaginary part of impedance: } \mathrm{Im}(Z) = Z'' = |Z| \cdot sin(\vartheta). \tag{D.5}$$

$$\text{The absolute value of the impedance (apparent resistance): } |Z| = \sqrt{Z'^2 + Z''^2}. \tag{D.6}$$

$$\text{Phase angle: } \vartheta = arctan(Z''/Z'). \tag{D.7}$$

Figure D.1 shows a diagram of the complex number plane to illustrate the imaginary part, real part, and phase angle.

It should be mentioned at this point that, although the use of a complex AC resistance as impedance is most widely used in battery technology, there are also alternative descriptions which, for example, use conductivity as a complex quantity (as so-called admittance) or quantities proportional to impedance or admittance. The generic term for all these functions is immittance. Definitions of immittance values are as follows:

$$\text{Admittance: } Y = 1/Z = Y' + iY''. \tag{D.8}$$

$$\text{Modulus function: } M = i\omega CZ = M' + iM''. \tag{D.9}$$

$$\text{dielectric permittivity: } \varepsilon = 1/M = \varepsilon' + i\varepsilon''. \tag{D.10}$$

Series or parallel connections of an ohmic resistor and a capacitor result in characteristic impedance or admittance curves. Impedance and admittance are also subdivided into resistance (or effective resistance) R and reactance X or in conductance (or active conductance) G or susceptance B respectively:

$$Z(\omega) = R(\omega) + iX(\omega). \tag{D.11}$$

$$Y(\omega) = G(\omega) + iB(\omega). \tag{D.12}$$

For the characterization of electrical systems and thus also electrochemical systems, it is necessary to measure the impedance values over a large frequency range – this is called impedance spectroscopy. In electrochemical impedance spectroscopy as an examination method for detailed investigations, the frequency range often extends from 0.01 mHz to 100 kHz.

D.2 Representation of Results

Common representations of results of impedance measurements are:

a) Nyquist plot (see Figure D.2), in which the real and imaginary parts of the impedance are plotted against each other, and
b) Bode plot, in which the amplitude of the impedance calculated from the real and imaginary parts and the phase angle are plotted against the frequency (see Figure D.3).

The Nyquist plot is a very compact representation but has the disadvantage that it does not contain any direct information on the frequency. Often, therefore, a Nyquist plot provides information on the frequency by marking points on the curve with the corresponding frequency values. In contrast, Bode plots provide information on the frequency directly but always require two curves. For this reason, both representations are often used in practice and there are different preferences. The information content of the two representations is identical.

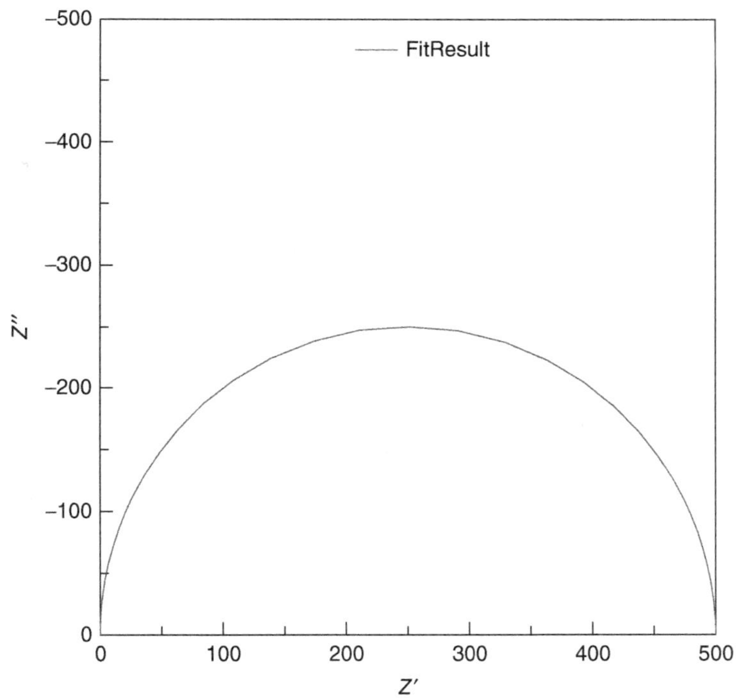

Figure D.2 Nyquist plot of an equivalent circuit diagram with a resistor and a capacitor connected in series.

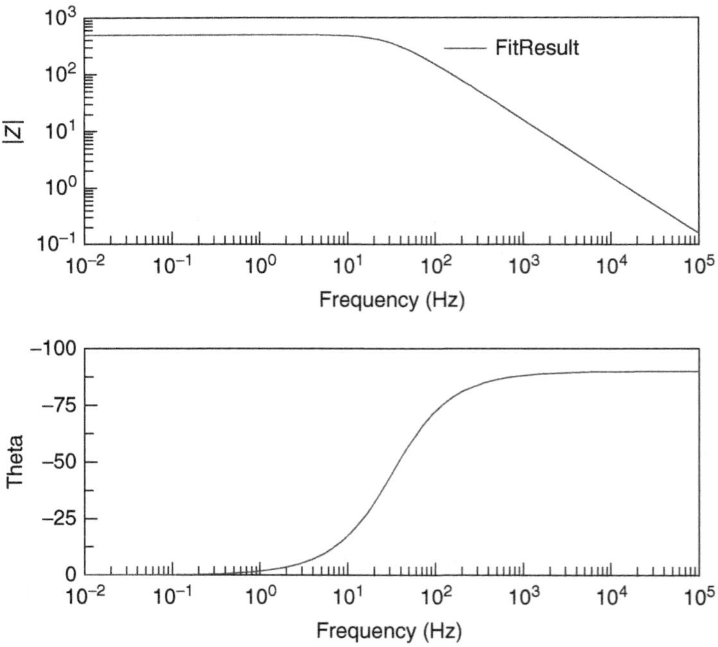

Figure D.3 Bode plot of an equivalent circuit diagram with a resistor and a capacitor connected in series.

D.3 Determination of Cell Parameters Using Impedance Spectroscopy

The measured values of impedance spectroscopy alone do not provide a physicochemical explanation of the processes in the battery, but can only be interpreted in combination with an electrical equivalent circuit diagram (EECD). Only in this way is it possible to assign real physical processes the model parameters, which can be calculated from the measured spectrum. This is shown schematically in Figure D.4.

To make it easier to understand the general form of impedance spectra, the spectra resulting from individual elements of an equivalent circuit diagram and resistance/capacitance (RC)circuits are shown below. The impedance at a certain frequency results from the superposition of the three basic impedance types, namely the classic ohmic resistance, the capacitive resistance, and the inductive resistance (see Table D.1).

An overview of the elements used and some combinations of them can be found in Table D.2.

Ultimately, modeling of impedance spectra is about determining the parameters of the model used and interpreting them physicochemically.

To determine the model parameters, the first step is to create an algorithm that can be used to calculate the impedance spectrum for the model used. Per definition,

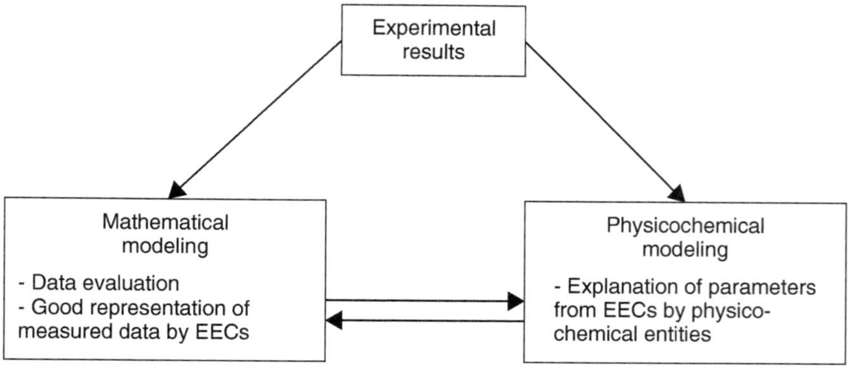

Figure D.4 Relationship between experimental results, mathematical model using an electrochemical equivalent circuit diagram (EECD) and physicochemical modeling of electrochemical impedance spectra.

Table D.1 Fundamental types of impedance.

Impedance element	Z'	Z''	Phase shift
Ohmic	R	0	0°
Inductive	0	ωL	+90°
Capacitive	0	$-1/\omega C$	−90°

none of the elements have a frequency-dependent value. Skin effects and the changing inductance of the electrochemical system due to changes in the current inhomogeneity are therefore not taken into account. The impedance spectrum is then calculated for the same frequencies as the measured spectrum and the parameters are varied until there is maximum agreement for all measured frequencies. Typical parameters used to evaluate the quality of the fitting process are so-called R values (R for reliability). It should be noted that the results depend on the software used, initial values for the variables, and the number of iteration steps and convergence criteria of the numerical calculation method. Standard software is available from equipment manufacturers.

Unfortunately, finding a suitable and physicochemically meaningful equivalent circuit diagram that reflects the measurements well is a complex and by no means a trivial task. Often the Randles model consisting of a resistor and two RC circuits (see chapter 8) with an additional inductance is used. In general, however, further elements may often be required in the equivalent circuit diagram to obtain sufficient agreement between measured impedance values and calculated model parameters of the chosen physicochemical model. In addition to further RC circuits, the two most important and most commonly used elements are the so-called Warburg impedance and the constant phase element (CPE).

Table D.2 Equivalent circuit diagrams, their impedance spectrum, and explanations.

Description	Equivalent circuit diagram/Nyquist plot
The eigenvector of an ideal resistance corresponds to a point in the Gaussian number plane independent of the frequency.	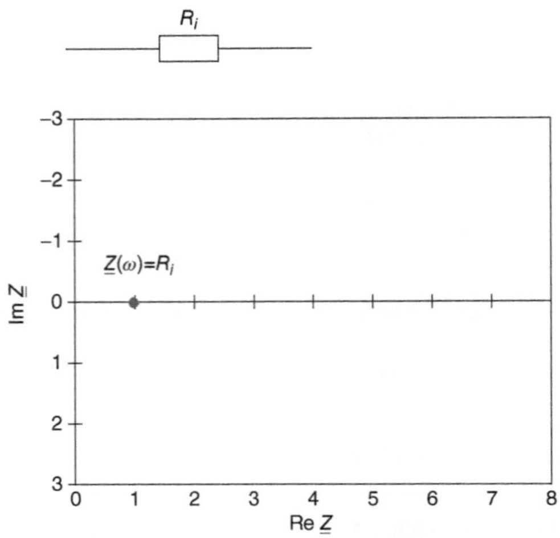
The impedance of an ideal inductance is a vertical straight line that coincides with the y-axis (the value on the x-axis is 0) but has only a positive imaginary part. Its value is the inductive reactance $\omega \cdot L$ and is purely imaginary and dependent on the frequency.	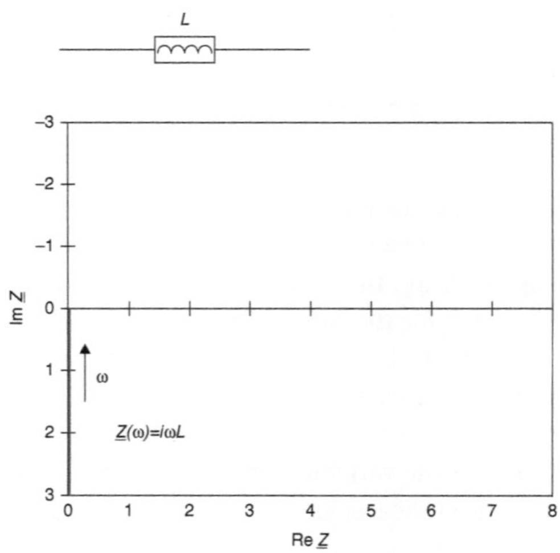

Table D.2 (Continued)

Description	Equivalent circuit diagram/Nyquist plot
If an ideal resistance and an ideal inductance are connected in series, the eigenvector in the Gaussian number plane is a vertical straight line from the "resistance point" with a positive imaginary part.	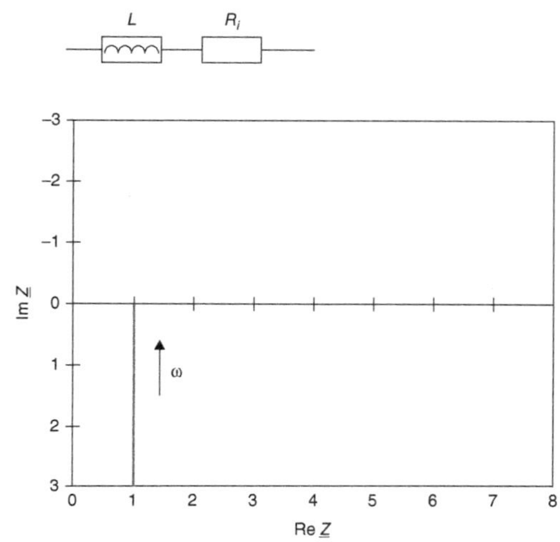
The impedance of an ideal capacitor is a vertical straight line (the value on the x-axis is 0) with a negative imaginary part in the Gaussian number plane; the value is the capacitive reactance $\frac{1}{\omega \cdot C}$ and is purely imaginary and dependent on the frequency.	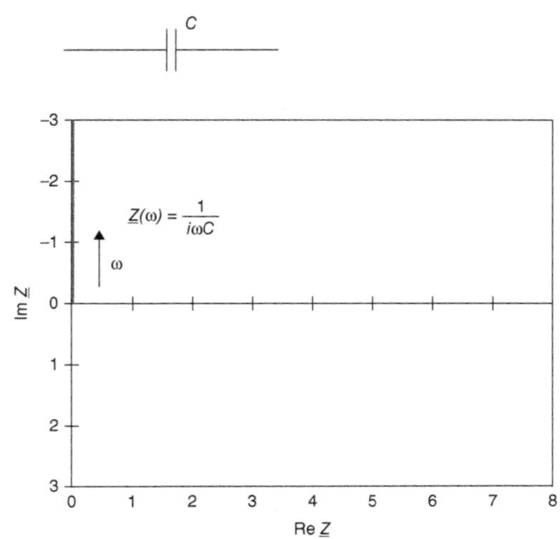

(Continued)

Table D.2 (Continued)

Description	**Equivalent circuit diagram/Nyquist plot**
If an ideal resistor and ideal capacitor are connected in parallel and the angular frequency is $\omega = \frac{1}{R\cdot C}$, you can see from the formula $z = \frac{R}{1+j\cdot\omega\cdot R\cdot C}$ that the eigenvector in the Gaussian number plane is a semicircle from the origin to "resistor-R" with a negative imaginary part.	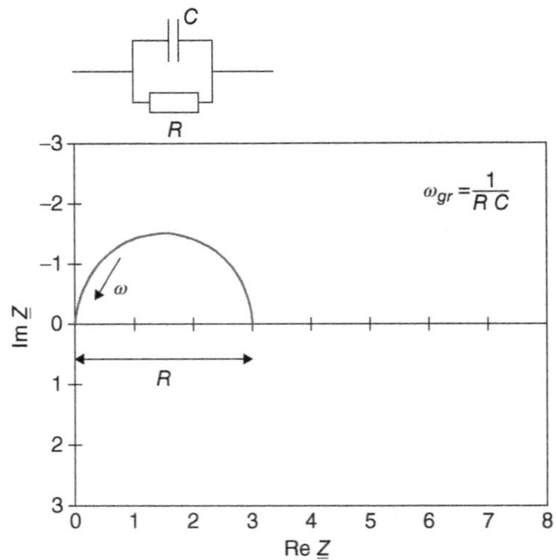
Two RC elements in series form a curve that represents a superposition of two semicircles; depending on the difference in resistance, the two semicircles are more or less separated and can be identified as separate semicircles and can describe the curve of the Warburg impedances in the Gaussian number plane.	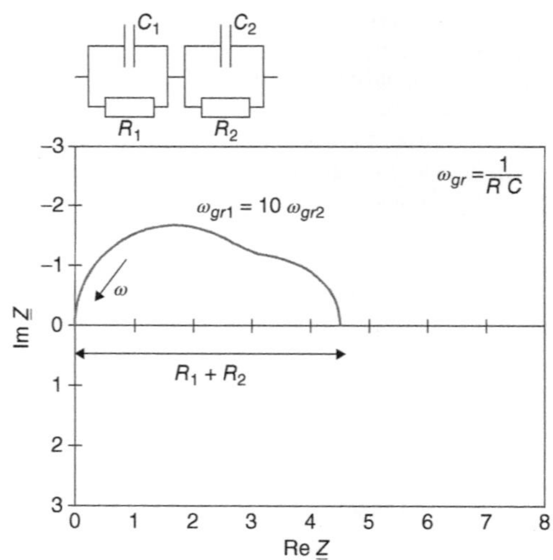

Table D.2 (Continued)

Description	Equivalent circuit diagram/Nyquist plot
An ideal inductance, an ideal resistor, and two RC elements connected in series, the curve often reflects experimental results on batteries well.	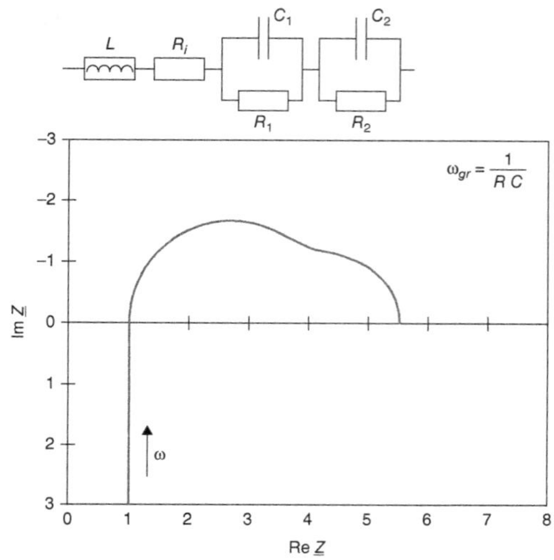

The mathematical formula for the Warburg impedance is as follows:

$$Z_W = \frac{A_W}{\sqrt{\omega}} + \frac{A_W}{i\sqrt{\omega}}. \tag{D.13}$$

As can be seen, the Warburg impedance is a one-parameter element with the parameter A_W to be fitted to measurements. The Warburg impedance is used in particular for modeling diffusion processes; under certain conditions, diffusion coefficients can even be extracted directly from the parameter A_W, see, for example, Ref. [1].

The element with constant phase, on the other hand, is a two-parameter element with pure ohmic resistance, pure capacitance, pure inductance, and Warburg impedance as limiting cases:

$$Z_{CPE} = \frac{1}{T(i\omega)^{\varphi}}, \tag{D.14}$$

where T is the capacitance parameter and φ is referred to as the exponent of the constant phase. The element with constant phase is primarily an empirical element that is used, for example, as a generalized capacity due to its flexibility in fit processes. In particular, it is used for modeling local distributions (dispersions) of impedances, such as those that occur with non-uniform (fractal) surfaces. Since batteries often have large surfaces or phase boundaries, CPEs are often used to model the electrochemical impedance spectra of batteries.

The physical meaning of Warburg impedances and elements with constant phase and differences in their values with changing cell states (temperature, state of charge) are difficult to interpret. The sole use of these elements and more than two RC circuits for a better agreement between measurement and model parameters calculated using numerical fit processes appears to be of little help.

D.4 Quality of Parameter Determination

The quality of the modeling is calculated according to the following equation

$$\Delta Z = Z_{i,\text{obs.}} - Z_{i,\text{calc.}} = \varepsilon_{i,\text{fit}} + \varepsilon_{i,\text{stoch}} + \varepsilon_{i,\text{bias}}, \tag{D.15}$$

where the symbols mean the following:

$Z_{i,\text{obs.}}$	the observed impedance at the i-th discretization point,
$Z_{i,\text{calc.}}$	the calculated impedance at the i-th discretization point,
$\varepsilon_{i,\text{fit}}$	systematic error due to an inadequate fit model,
$\varepsilon_{i,\text{stoch}}$	the stochastic error of the experiment,
$\varepsilon_{i,\text{bias}}$	systematic error of the experiment due to drift, device artifacts, etc.

Different fit strategies can be used (least square error, simplex methods, and genetic algorithms). When modeling, it must be strictly ensured that additional free parameters will not only lead merely to a numerical improvement of the fitting results, but that the improvement is significant and ideally (according to Figure D.4) a physicochemical interpretation of these parameters is possible.

A meaningful evaluation of impedance spectra is only possible if the measurement system fulfills the following requirements during a measurement:

a) linearity of the system response, that is, the system response L to a composite signal:

$$L(af_1 + bf_2) = aL(f_1) + bL(f_2). \tag{D.16}$$

This also means that the excitation must be chosen dependent on the cell capacity and internal resistance. With very large cells, a current signal that is too small will result in a voltage response that is barely measurable and a voltage signal may result in a current response that is too high.

a) Causality, that is, the response of a system is caused exclusively by the excitation,
b) Stability, that is, the test system cannot deliver more energy than was introduced by the disturbance and, in particular, there are no changes in the system over time due to drift or irreversible reactions,
c) Finiteness, that is, the real and imaginary parts of the impedance are finite in the entire frequency range $0 \leq \omega \leq \infty$ and tend toward a constant real value for $\omega \rightarrow 0$ and $\omega \rightarrow \infty$.

A typical impedance spectrum of a lithium-ion battery is shown in Figure D.5 which must be compared with the spectra calculated above to determine model parameters. One characteristic feature is the linear increase in the Nyquist plot which is measured at low frequencies (mHz) and corresponds to diffusion processes

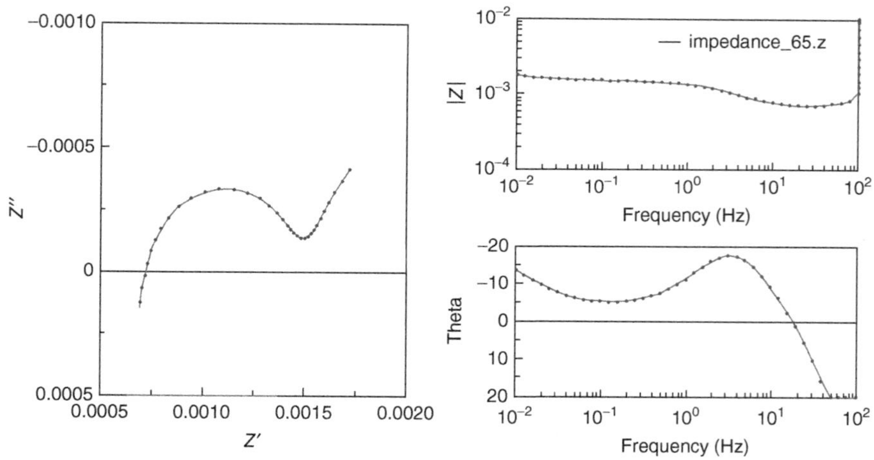

Figure D.5 Impedance spectra of a commercial lithium-ion cell.

(this part of the diagram is often modeled by a Warburg impedance). The semicircle in the hertz to kilohertz range is described in the simplest case by one RC circuit, but usually by a superposition of several RC circuits. A lot of electrochemical information is available in an appropriately chosen model, for example, in relation to the electrolyte resistance, the contact resistance of the electrodes, the double-layer capacitance of the electrode surface, etc.

At high frequencies (left part of the Nyquist plot), inductances dominate, which are mainly due to electrical cables and the size of the electrodes as well as the fact that measurements are only possible under very homogeneous conditions. The larger the electrodes of a cell and the number of cells in series or parallel, the more the measured impedance spectrum represents the superposition of individual spectra with slightly different material properties and it becomes increasingly difficult to interpret results.

It is important to understand what impedance spectroscopy can do in the hands of experts. At the same time, however, these explanations should also help to ensure that this method is only used after careful preparation and intensive further study.[2]

In general, the potential of impedance spectroscopy in the field of battery analysis and diagnosis has not yet been fully exploited. This is partly due to measurement difficulties: In the megahertz range, where measurements are very fast, it is mainly wiring inductances that are measured. In the range of a few mHz and lower frequencies, where diffusion processes often dominate, measurements take several minutes to hours, which takes too long for practical purposes.

2 Once an experimental setup for impedance spectroscopy measurement has been selected, it should not be changed.

References

1 Barsoukov, E.R. and Macdonald, J.R. (2005). *Impedance Spectroscopy. Theory, Experiment and Applications*, 2e. Wiley.

2 Christophersen, B.J. et al. (2012). Rapid impedance spectrum measurements for state-of-health assessment of energy storage devices. *SAE International Journal of Passenger Cars – Electronic and Electrical Systems* 5: 1.

3 Howey, D.A. et al. (2014). On-line measurement of battery impedance using motor controller excitation. *IEEE Transactions on Vehicular Technology* 63: 2557.

4 Lasia, A. (2014). *Electrochemical Impedance Spectroscopy and Its Applications*, 1e. Springer.

5 Lvovich, V.F. (2012). *Impedance Spectroscopy: Applications to Electrochemical and Dielectric Phenomena*, 1e. Wiley.

6 Orazem, M.E. and Tribollet, F. (2008). *Electrochemical Impedance Spectroscopy*, 1e. (ECS).

Appendix E

Acid Stratification

The development of acid stratification proceeds in the following steps, starting from a cell with fully charged electrodes and homogeneous acid density. The steps are described for flooded batteries with liquid electrolytes but are also applicable to sealed batteries in which the sulfuric acid is immobilized in gel or AGM. Acid stratification develops much more slowly in gel or AGM batteries and is usually of no particular significance for their performance and service life [1–3].

1) When discharging the cell with homogeneous acid concentration and homogeneous state of charge (Figure E.1a), sulfuric acid is consumed as an active element in the electrochemical reactions of both electrodes and the concentration of sulfuric acid decreases in the area between the electrodes. It can be assumed that the concentration of aqueous sulfuric acid above and below the electrode stack is not directly influenced by the discharge current.
   Due to the drop in resistance along the electrodes, the current density distribution is inhomogeneous in the vertical direction (Figure E.1b). An analytical calculation of the current density distribution is possible with homogeneous initial conditions [4]. In the vicinity of the current collectors at the top of the electrodes near the terminals, the current density is greater than in the lower area, so the local state of charge and the local sulfuric acid concentration in the upper area is lower than in the lower area between the electrodes (Figure E.1c).
   The higher the resistance of the current collectors, the more pronounced the current inhomogeneity.
2) The local differences in acid density and state of charge in different electrode areas change the local Nernst voltages and the local electrical resistance of the electrolyte so that the current inhomogeneity changes further. An analytical calculation is no longer possible.
3) The acid above the electrodes is not directly affected by the discharge of the electrodes. However, due to the effect of gravity, it sinks downwards and displaces the acid of lower density below (Figure E.2a). This process is all the more pronounced the higher the acid level in a cell above the electrodes. It is important to note that the inhomogeneous current distribution in the electrode contributes to the formation of the acid stratification, but even with a homogeneous current distribution, the acid concentration between the electrodes would decrease

Battery Technology: Fundamentals of Battery Electrochemistry, Systems and Applications,
First Edition. Alexander Börger and Heinz Wenzl.

Companion website: www.wiley.com/go/Börger/Battery_Technology

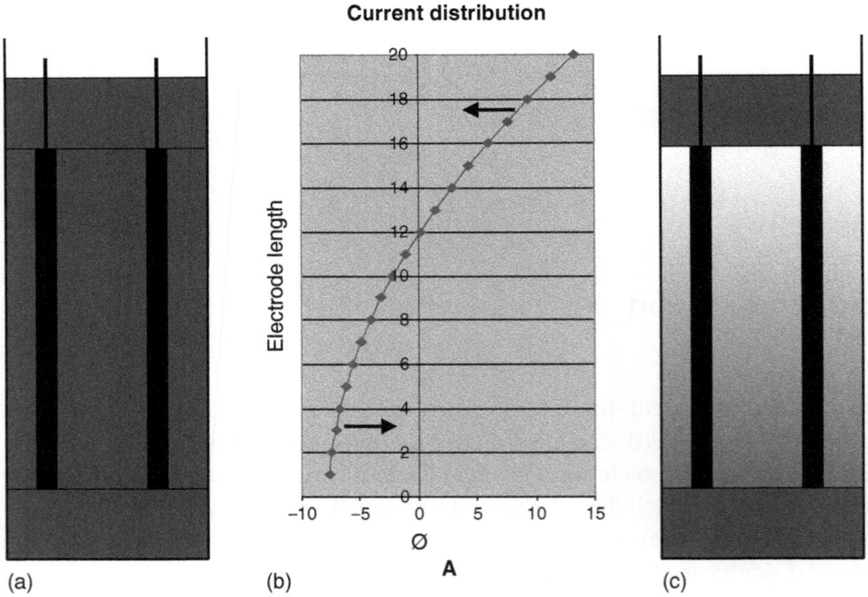

Figure E.1 Schematic diagram showing the development of acid stratification and local differences of state of charge during discharging. (a) Initial state with homogeneous acid concentration and state of charge; (b) diagram showing schematically the current distribution between the electrodes at the beginning of discharge (can be calculated analytically) and schematic diagram of the change in current distribution during discharging (arrows); (c) development of the acid gradient during further discharging.

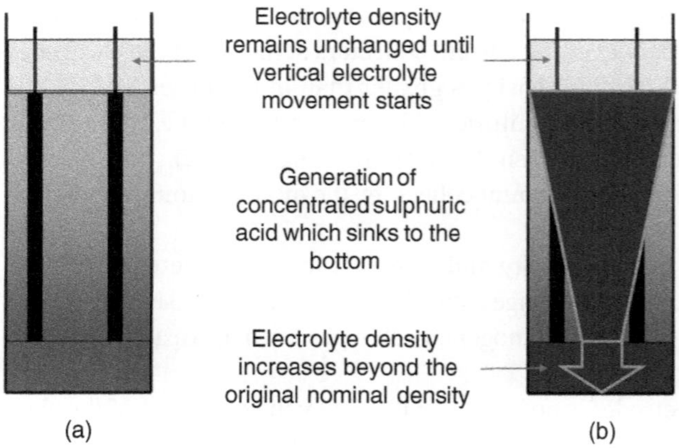

Figure E.2 Schematic diagram showing the development of acid stratification and local differences of state of charge during charging in flooded batteries. (a) After the end of discharging, the acid concentration at the top above the electrodes is at its lowest; (b) Generation of highly concentrated sulphuric acid and its downwards movement due to the effect of gravity.

because the electrolyte is an active component of the reactions taking place in the lead–acid batteries.

The local acid concentration and the local state of charge no longer match, the current density distribution changes again due to the change in the electrolyte resistance and the local Nernst voltage.

With sealed batteries, the process is very slow because there is hardly any electrolyte above the electrodes and the vertical mobility of the electrolyte is very low. The state of charge and acid density curve then correspond more closely to Figure E.1c.

4) After a certain waiting time after the end of discharging, the following condition exists:
   - The acid below the electrode stack still has the initial concentration.
   - The acid density between the electrodes decreases toward the top.
   - The lowest acid density is in the electrolyte above the electrodes. Measuring the acid density in the area above the electrodes no longer provides any meaningful information about the condition of the battery.

   Figure E.3 shows measurements in which a battery was discharged very deeply to demonstrate the effect [5].
5) Due to the low diffusion coefficient of sulfuric acid in water, this acid stratification is only reduced very slowly (months, not hours) by diffusion. Driving movements can only slightly reduce the acid stratification in starter batteries unless special measures are taken (see section 12.10).

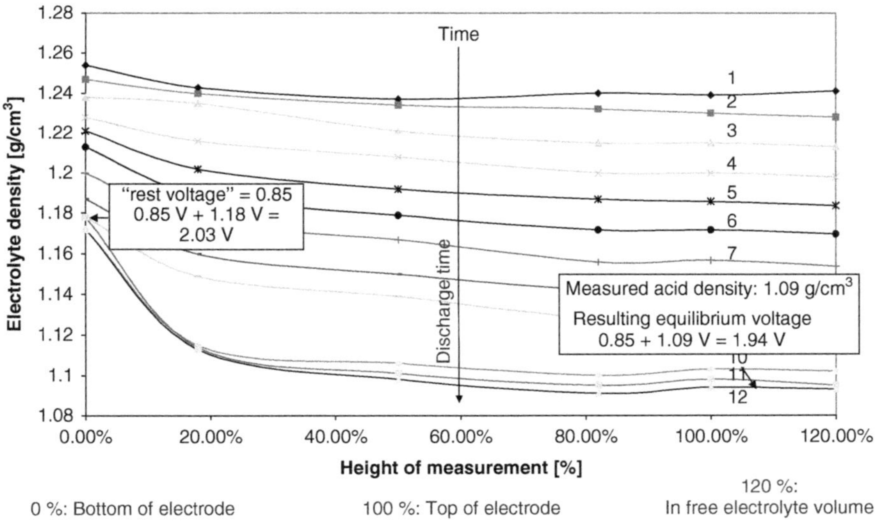

Figure E.3 Changes in the acid concentration along the length of the electrodes during discharging [5]. The measurement shows a slightly increased acid concentration in the lower part of the cell container at the beginning of discharging – an indication of how difficult it is to completely remove an acid stratification once it has formed. Source: Adapted from Ref. [5]/Elsevier.

6) The local Nernst voltages in the various electrode areas differ because the acid concentrations are different. This results in equalizing currents within the electrode, even without an external current flow (see the Nernst voltages marked in Figure E.3). The electrode areas at the bottom with the higher acid density discharge and charge the electrode areas at the top have a lower acid density. However, this process does not reduce the differences in acid density because the concentrated sulfuric acid produced in the upper area by the charging process sinks downwards due to gravity.
   It should be noted that the normally existing relationship between the state of charge and acid density is thereby further decoupled. It should also be noted that these equalizing currents mean that there is a constant current flow between electrode areas so the concept of open-circuit voltage must be viewed critically. The voltage measured at the cell terminals corresponds to the charging voltage of the upper electrode area. Methods for determining the state of charge that use the open-circuit voltage as an input variable are problematic once acid stratification has developed.
7) When charging the cell, the current density distribution is characterized by the state of charge dependent resistance of the active masses and the concentration-dependent resistance of the electrolyte, the resistance of the current collectors, and, above all, the concentration-dependent Nernst voltage of the different electrode areas. The higher Nernst voltage in the lower electrode area has the greatest influence, as the difference between charging voltage and Nernst voltage there is smallest and, as a result, the charging current there is smallest. The upper part of the electrode is preferentially charged because of a low Nernst voltage there (see Figure E2b).
   The charging reaction produces concentrated sulfuric acid at both electrodes and consumes water molecules. The highly concentrated local sulfuric acid sinks downwards due to the influence of gravity, causing the concentration in the lower electrode area to become higher and higher and increases the electrolyte concentration below the electrodes (see Figure E2b). The sulfuric acid in the area above the electrolyte remains unaffected by this process and does not change its concentration. Toward the end of the charge, before gassing reactions become relevant, the following state is reached (Figure E.4):
   - The acid density below the electrodes is above the initial concentration at the start of the discharge.
   - The concentration of sulfuric acid between the electrodes slowly decreases toward the top.
   - Toward the end of the charge, but before the end of a full charge, the concentration of acid above the electrodes corresponds to the state at the start of the charge and has the lowest local value in the cell. The acid concentration and the local state of charge no longer show any correlation. It is no longer possible to deduce an acid density value from the open-circuit voltage measurement and vice versa.
8) A charging phase at or above the "gassing voltage" leads to gas development at both electrodes, which mixes the electrolyte. Further measures may have to be

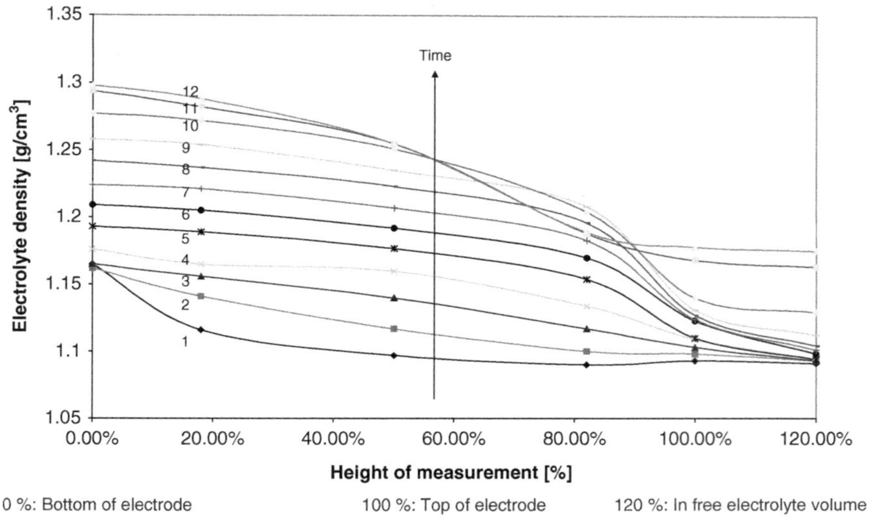

Figure E.4 Development of the acid concentration along the length of the electrode during charging. Toward the end of charging, the measurement shows an acid concentration of almost 1.30 g/cm^3 compared to the initial concentration of just under 1.26 g/cm^3 in Figure E.3. The greatest difference in concentration is reached shortly after the "gassing voltage" of 2.4 V is reached.

employed to achieve a completely homogeneous acid density throughout the cell, e.g. long phases of overcharging or electrolyte circulation. However, it is not easy to reduce the high acid concentration below the electrode, particularly if the distance between the bottom of the cell container and the electrodes is great.

At the end of a suitable charging process, the relationship between acid density and state of charge is restored. Only then is the term "fully charged" again meaningful.

If acid stratification is not removed, the service life will be considerably shortened due to increased corrosion and the formation of large lead sulfate crystals which are difficult to recharge.

References

1 Döring, H. (1995). PV operation and failure mode of different storage batteries and recommendations for operation management (Original German title: PV-Betrieb und Fehlermodus verschlossener Speicherbatterien und Empfehlungen für die Betriebsführung). In: *10th Symposium on Photovoltaic Solar Energy, Staffelstein.*

2 Ebner, E., Börger, A., Gelbke, M. et al. (2013). Temperature-dependent formation of vertical concentration gradients in lead-acid batteries under PSoC operation – Part 1: acid stratification. *Electrochimica Acta* 90: 219–225.

3 Ebner, E., Gelbke, M., Zena, E. et al. (2018). Temperature-dependent formation of vertical concentration gradients in lead-acid batteries under PSoC operation – Part 2: sulfate analysis. *Electrochimica Acta* 262: 144–152.

4 Euler, K.J. (1965). Current distribution in and on porous electrodes (Original German title: Stromverteilung in und auf porösen Elektroden). *Chemie Ingenieur Technik* 37: 626–631.

5 Schulte, D., Sauer, D.-U., Ebner, E. et al. (2014). Stratificability Index – a quantitative assessment of acid stratification in flooded lead acid batteries. *Journal of Power Sources* 269: 704–715.

Index

Battery Technology: Fundamentals of Battery Electrochemistry, Systems and Applications, First Edition. Alexander Börger and Heinz Wenzl.

Companion website: www.wiley.com/go/Börger/Battery_Technology

f

g